Modern Livestock and Animal Health

Modern Livestock and Animal Health

Editor: Johann Casini

R CALLISTO
REFERENCE

www.callistoreference.com

Callisto Reference,
118-35 Queens Blvd., Suite 400,
Forest Hills, NY 11375, USA

Visit us on the World Wide Web at:
www.callistoreference.com

ISBN: 978-1-64116-290-6 (Hardback)

Cataloging-in-Publication Data

Modern livestock and animal health / edited by Johann Casini.
     p. cm.
Includes bibliographical references and index.
ISBN 978-1-64116-290-6
1. Livestock. 2. Veterinary medicine. 3. Animal health. 4. Animals--Diseases. I. Casini, Johann.
SF41 .M63 2020
636.009--dc23

Table of Contents

Preface

This book has been a concerted effort by a group of academicians, researchers and scientists, who have contributed their research works for the realization of the book. This book has materialized in the wake of emerging advancements and innovations in this field. Therefore, the need of the hour was to compile all the required researches and disseminate the knowledge to a broad spectrum of people comprising of students, researchers and specialists of the field.

Livestock refers to the domesticated animals in an agricultural setting which are raised by humans to produce commodities such as leather, fur, dairy products, wool, etc. Some common examples of livestock are cows, goats, sheep, buffaloes and pigs. The field which involves breeding, maintenance and slaughtering of livestock is known as animal husbandry. It is one of the major components of modern agriculture. There are many diseases and health issues that affect the livestock such as classical swine fever, scrapie, foot mouth disease, etc. Many measures are taken to avoid such diseases. Vaccines and antibiotics are widely used when required. Good husbandry practices, proper feeding, and hygiene are incorporated which contribute to good livestock health. This book unravels the recent studies in the field of modern livestock and animal health. Different approaches, evaluations, methodologies and advanced studies on modern livestock and animal health have been included in this book. It will serve as a reference to a broad spectrum of readers.

At the end of the preface, I would like to thank the authors for their brilliant chapters and the publisher for guiding us all-through the making of the book till its final stage. Also, I would like to thank my family for providing the support and encouragement throughout my academic career and research projects.

Editor

Slaughter of pregnant cattle in German abattoirs – current situation and prevalence

Patric Maurer[1*], Ernst Lücker[1] and Katharina Riehn[2]

Abstract

Background: The slaughter of pregnant cattle and the fate of the foetuses are relatively new subjects in the field of animal welfare. The Scientific Committee on Veterinary Measures relating to Public Health (SCVPH), however, does not believe this topic to be a critical issue because of the hitherto supposed rare occurrence of this practice. Some previous studies though, contradict this assessment, emphasising its relevance to animal welfare. With regard to the heterogeneous study design of previous investigations, the objective of this study is to evaluate the current situation concerning the slaughter of pregnant cattle in different German abattoirs. Additionally, the prevalence was assessed semi-quantitatively on the basis of a cross-sectional, voluntary and anonymous survey that was conducted amongst senior veterinary students of the University of Leipzig from 2010 until 2013.

Results: Of 255 evaluable questionnaires, 157 (63.6 %) mention the slaughter of pregnant cattle, corresponding to 76.9 % of all visited abattoirs. Slaughter of pregnant cattle is reported often (>10 % of females) in 6 (3.8 %), frequently (1–10 % of females) in 56 (35.7 %), and rarely (<1 % of females) in 95 (60.5 %) of all cases (n = 157) respectively. About 50 % of these animals were reported to be in the second or third stage of gestation. 15 (10.6 %) of 142 questionnaires providing information about the foetus, state that the foetus showed visible vital signs after the death of the mother, but in one case the foetus was euthanized subsequently.

Conclusions: The results show that the slaughter of pregnant cattle is a common and widespread practice in German abattoirs. The SCVPH's assumption that pregnant cattle are only slaughtered in rare exceptional cases can no longer be maintained. The high proportion of foetuses in the second and third gestational stage must also be considered. In this context the implementation of suitable studies and detailed analysis of the current situation is indispensable to ensure the high standards in animal welfare in Germany and Europe.

Keywords: Veterinary public health, Animal welfare, Pregnancy, Gravid cattle, Foetus, Abattoir, Slaughterhouse, Consumer protection

Background

Current developments in public opinion and politics

Germany has one of the strictest animal welfare laws worldwide. There is no other country in the European Union (EU), which integrated animal welfare into its constitution – in Germany animal welfare was made a national objective already in 2002 [1].

In recent years the slaughter of pregnant cattle has gained importance in the public debate on animal welfare. Due to scientific work by, amongst others, Peisker et al., Riehn et al., Di Nicolo, Lücker et al. [2–6], a political and scientific debate has emerged about whether the slaughter of pregnant animals can be reconciled with the requirements for a humane killing of animals. One of the major ethical concerns in this context is the perceptual awareness and viability of foetuses during the slaughter of the mother. Marahrens and Schwarzlose of Germany's Federal Research Institute for Animal Health promote the opinion that foetuses in the 3rd trimester will have a relevant reduced

* Correspondence: patric.maurer@vmf.uni-leipzig.de
[1]Institute of Food Hygiene, Centre of Veterinary Public Health, Faculty of Veterinary Medicine, University of Leipzig, An den Tierkliniken 1, 04103 Leipzig, Germany
Full list of author information is available at the end of the article

welfare during the slaughter of their mother [7]. The authors state, that more research is needed to assess the observed changes in physiological, electrophysiological, and endocrinological parameters of the foetuses that are exposed to such a treatment.

In March 2014 public interest was attracted by a television report that addressed the suffering of foetuses during slaughter [8]. The increased media interest on animal welfare issues related to slaughter in Germany and abroad has caused German politicians to focus on these topics. The on-going dialogue led to a request in the course of the Standing Committee on the Food Chain and Animal Health meeting in Brussels, April 8th 2014 [9]. In response to this, a mandate was given to the European Food Safety Administration (EFSA) by the European Commission in order to investigate the scope of this problem and develop possible solutions. Additionally, if needed, the council regulation (EC) No 1099/2009 on the protection of animals at the time of slaughter should be reviewed with respect to protection of the unborn life [9].

Appearance of the problem

The slaughter of pregnant cattle was considered negligible by the Scientific Committee on Veterinary Measures relating to Public Health in 1999 (SCVPH, [10, 11]). According to the SCVPH "meat consumption from pregnant heifers is exceptional as usually these animals are not slaughtered" [11]. Some studies, however, have since contradicted these

assumptions and demonstrated that the actual numbers are much higher. Table 1 gives an overview of study results for the prevalence of slaughter of gravid cattle in Germany and some other countries [3–7, 9, 12–16]. Prevalence in German abattoirs ranges from 0.2 % ([7] referring to results of the National Association of Meat Hygiene, Animal Welfare and Consumer Protection) up to 15 % [3, 4]. In Luxembourg, Belgium and Italy the proportion of pregnant animals is 5.3 %, 10.1 %, and 4.5 % respectively [5]. In two British abattoirs, Singleton and Dobson observed that 23.5 % of the cows were pregnant [15]. As depicted in Table 1, most of the cows were in the 2nd and 3rd trimester at the time of slaughter (e.g. [13]). The fact, that hardly cows in the 1st trimester were observed may be attributed to the circumstance that early pregnancies may be overlooked on both farm and abattoir level. Generally, the data from the different studies should be interpreted with care, as different study designs may impede their comparability.

However, the problem cannot be underestimated in its significance, since not only cattle but also other species are effected as demonstrated by Fayemi and Muchenje [17].

Legal and ethical aspects

There are no regulations, neither in National nor in Community law, prohibiting the slaughter of pregnant animals and governing the fate of the foetuses. Due to a supposed underestimation of the total prevalence by the SCVPH and the lack of reliable data legislators at

Table 1 Prevalence of slaughtering gravid cows in Germany and other countries – results of different authors

Number	Prevalence of slaughter gravid cows	Stage of gestation	Reference
Germany			
1	Up to 10.8 %, mean 4.3 % of cows and heifers	NS	Lücker et al. [6]
2	4.9 %	Mostly in 5th month; 38 % and 62 % in 2nd and 3rd trimester, respectively	Di Nicolo [5]
3	Up to 15 %, mean 9.6 %, median 7.1 % of cows and heifers	90 % in 2nd or 3rd trimester	Riehn et al. [3, 4]
4	0.2 % and 1.2 %	NS	Marahrens and Schwarzlose [7] referring to results of the National Association of Meat Hygiene, Animal Welfare and Consumer Protection
5	3.5 %	56 % in 2nd or 3rd trimester, 0.8 % in the 3rd trimester	German Government [9] referring the German Association of the Meat Industry
Other countries			
6	approximately 5 % (USA)	NS	Kushinsky [12]
7	NS	13.1 % in 1st, 62.6 % in 2nd and 24.3 % in 3rd trimester (Canada)	Herenda [13]
8	8.6 % (Pakistan)	NS	Khan and Khan [14]
9	23.5 % (Great Britain)	26.9 % in 3rd trimester (Great Britain)	Singleton and Dobson [15]
10	5.3 % (Luxembourg), 10.1 % (Belgium), 4.5 % (Italy)	In 3rd trimester: 36 % (Luxembourg), 15 % (Italy)	Di Nicolo [5]
11	1.5–2.1 % (Nigeria)	NS	Ademola [16]

Different authors described the prevalence of slaughtering gravid cows and their stage of gravidity in Germany and some other countries. (NS: not specified)

European level have not issued any mandatory regulations on how to handle the problem [3, 4]. In addition, no options are granted to the Member States to adopt any regulations with regard to this issue at national level. Currently the only statute protecting females around the time of calving is the EU Transport regulation – Regulation (EC) 1/2005 Annex I, Chapter I Nr. 2 c [18]. It prohibits the transport of "pregnant females for whom 90 % or more of the expected gestation period has already passed, or females who have given birth in the previous week". Riehn et al. [3, 4] indicate that determining the correct percentage of gestation is almost impossible on abattoir level, where only limited examination opportunities are available during *pre-mortem* inspection. The authors also state, that the Regulation (EC) 854/2004 [19] obliges the official veterinarian nevertheless to do a *pre-mortem* inspection (Annex I, Section I, Chapter II, B, Nr. 2 a) "to verify compliance with relevant Community and national rules on animal welfare, such as rules concerning the protection of animals at the time of slaughter and during transport" (Annex I, Section I, Chapter II, C).

The fate of foetuses is also not specifically mentioned in any regulation. However, Chapter II, Article 3 of the Regulation (EC) 1099/2009 [20] states that "animals shall be spared any avoidable pain, distress or suffering during their killing and related operations"(1) and that "business operators shall, in particular, take the necessary measures to ensure that animals (a) are provided with physical comfort and protection" and (d) "do not show signs of avoidable pain or fear or exhibit abnormal behaviour"(2a, d).

To date there is a dispute if and from which developmental stage on foetuses are conscious and sensitive to stress. Former studies, especially by Mellor et al. (e.g. [21, 22]) assume that foetuses lack such abilities. This opinion may be reviewed in the light of new scientific evidence. Bellieni and Buonocore [23] report that foetuses are able to feel stress and pain from the second half of the gestation on. The Experimental Animals Directive 2010/63/EU revising Directive 86/609/EEC on the protection of animals used for scientific purposes [24] already considers these new research results and states in recital 9 that "(...) there is scientific evidence showing that such [foetal forms of mammals] in the last third of the period of their development are at an increased risk of experiencing pain, suffering and distress, (...)." Altogether, it cannot be ruled out that foetuses feel pain, distress and other forms of suffering and therefore the slaughter of pregnant cattle in advanced gestational stages should be considered as an animal welfare problem.

The reasons on farm level for slaughtering gravid cows can only be assumed in the absence of valid data. The Federal Association of Veterinary Officers Germany presumes the following reasons (in decreasing order of importance): Slaughter by mistake with unknown pregnancy; injured animals, which cannot be used any more; economic reasons [25]. The relevance of economic aspects in this context is stressed by many authors e.g. Riehn et al., Cordes, Münch and Richter, Tierärztliche Vereinigung für Tierschutz e.V. [3, 8, 26, 27].

Objective

Considering the heterogeneous data and study designs of previous German studies while respecting their relevance for further political development, our objectives were to i) determine the prevalence of the slaughter of pregnant cattle in Germany semi-quantitatively and ii) to investigate the occurring gestational stages and fate of the foetuses.

Methods
Survey design and questionnaire

For this investigation, an observational, cross-sectional study design was used. A voluntary and anonymous survey with specific questions was conducted amongst all final-year-students of the Faculty of Veterinary Medicine of the University of Leipzig from January 2010 until September 2013.

In Germany the study of veterinary medicine is regulated by the "Verordnung zur Approbation von Tierärztinnen und Tierärzten" (TAppV, [28]). The regulation stipulates that every student has to absolve a slaughterhouse-internship of at least 100 hours in three weeks under the supervision of the local competent authority. In the course of the internship, students are required to learn and practice the *ante-* and *post-mortem* inspection of cattle and pigs, taking particular care of animal welfare. Our study is therefore aimed at the observations of the veterinary trainees during their meat hygiene internship.

General information and preparation for the internship was conducted during a short obligatory briefing as part of the meat hygiene lecture. In addition to the general organizational information, students were asked to pay attention to specific aspects such as the slaughter of pregnant cattle and write a voluntary, anonymous internship report.

From January 2010 until September 2013 following their final exam in meat hygiene, all students of the faculty of veterinary medicine of the University of Leipzig were asked to participate in a survey. Participation was voluntary, anonymous, and had no influence on exam results.

The survey consisted of two questionnaires– one for cattle and one for pigs – each containing five groups of questions about *ante-* and *post-mortem* inspection and animal welfare. For each species and abattoir the students were asked to indicate the date and abattoir of their internships. One set of questions addressed the prevalence of the slaughter of female cattle and the frequency of pregnancy amongst the cows and heifers including their stage of gestation as recognized by the students. Additionally, questions to vital signs and the fate of the foetuses were asked (see Fig. 1). Answers could be given on a nominally dichotomised or ordinal scale and for some questions free-text responses were required. The ordinal scales were based on data of previous oral reports.

Ethical approval

Research activities involving human participants will require approval prior to their commencement. Therefore the opinion of the data protection officer of the University of Leipzig has been obtained. He confirmed an exemption from these formal requirements with regard to the following criteria: (i) The current study involves no foreseeable risk of harm or discomfort to participants and any further foreseeable risk would involve no more than inconvenience to participants, and (ii) the study involves the use of existing collections of data or records that contain only non-identifiable data about persons, entities and bodies. No names will be published.

Data analyses

Analyses were carried out using Microsoft® Excel® 2013 (Windows)[1] and IBM® SPSS® Statistics 22.0.[2] The specific data of each abattoir were encrypted and abattoirs were grouped by region in order to ensure their anonymity. Only completed questionnaires were included in the initial analysis. Due to the step-by-step design of the questionnaire, the number (n) of answers for each question differs as non-responders matched the exclusion criteria and hence were dropped out from further analyses. The Somers-D-test was used to determine the association between ordinal variables. A P-value of < 0.05 was considered significant for all comparisons.

Results

Participants and submission (persons and the involved abattoirs)

The survey generated 286 responses in the period from January 2010 to September 2013. All participants were students at the Faculty of Veterinary Medicine, University of Leipzig, Germany. Students were interviewed after finishing their final examination in the subject "meat hygiene". Participation was voluntary and anonymous. Of 286 cattle-questionnaires, 263 (92 %) described the situation at one abattoir. 255 (97 %) of those 263 confirmed the slaughter of female cattle and thus fulfilled the inclusion criteria to be examined further as shown in Fig. 2. The related internships took place between 2007 and 2013. The majority however, in 2010 ($n = 75$; 29 %), 2011 ($n = 65$; 25 %) and 2012 ($n = 49$; 19 %), as depicted in Fig. 3. The students did their internships in 67 different slaughterhouses in Germany ($n = 66$) and abroad ($n = 1$). The 66 German abattoirs were located in 12 of the 16 federal states. The majority of the abattoirs ($n = 39$; 58 %) was located in 3 federal states; Bavaria ($n = 14$; 21 %), North Rhine-Westphalia ($n = 14$; 21 %) and Baden-Wuerttemberg ($n = 11$; 16 %).

Numbers of slaughtered cattle and prevalence of gravid cows and heifers

The reported numbers of slaughtered female cattle are described by using a breakdown of ordinal categories. The numbers differ between the 67 abattoirs. Of the 255 eligible questionnaires, 86 (33.7 %) cases slaughtered between 1–10, 60 (23.5 %) between 11–50, 37 (14.5 %)

CATTLE

Abattoir: Date (approx.):

1.1 Approximately how many cows and heifers were slaughtered in a day?

☐ 1-10 ☐ 11-50 ☐ 51-100 ☐ 101-200 ☐ 201-300 ☐ 301-................

1.2 How frequent was slaughter of gravid cows?

☐ never ☐ rarely (<1%) ☐ frequently (1-10%) ☐ often (>10%)

1.3 Were advanced gestational stages (second/ third trimester) present? ☐ yes ☐ no

1.4 Did the foetuses/calves show life signs? ☐ yes ☐ no

1.5 Were special measures taken to kill the foetuses/ calves? ☐ yes ☐ no
If so, which:

Fig. 1 Part of the questionnaire given to the veterinary students. This translated extract of the survey shows the relevant questions for this study. It consists of 5 questions with one follow-up question

Criteria / Question	Eligible (n)		Not eligible (n)	
	Questionnaires	Abattoirs	Questionnaires	Abattoirs
Numbers of completed questionnaires	286	71		
Numbers of questionnaires referring to one abattoir	263	68	23	3
Slaughter of female cattle	255	67	8	1
Slaughter of gravid cattle	247	65	8	2
Reported slaughter of gravid cattle	157	50	90	15
Occurrence of advanced gestational stages	152	50	5	0
Occurrence of vital foetal signs	142	49	15	1
Special measures of killing the foetuses	132	44	25	6

Fig. 2 Scheme of (n) evaluable and non-evaluable questionnaires and encoded abattoirs of the veterinary students' survey. Only correctly completed questionnaires are included in the analysis. Due to the design of the questionnaire the number (n) of answers for each question can differ as non-responders were dropped from analyses

the question about the occurrence and frequency of slaughtered gravid cattle. In 157 cases (63.6 %), the slaughter of pregnant cattle was reported, although in 90 cases (36.4 %), this was never actually seen by the trainees themselves.

An analysis of the data from the related 65 slaughter-houses shows, that there are 30 abattoirs (46.2 %) that are always reported to slaughter pregnant cattle, 20 (30.8 %) where reports vary and 15 (23.1 %) that are never reported to slaughter gravid cattle. In summary, 50 out of 65 abattoirs (76.9 %) that slaughter females also slaughter pregnant ones (see Fig. 4). However 95 of the 157 positive cases (60.5 %) reported, that the slaughter of pregnant cattle occurs rarely (<1 % of females), 56 cases (35.7 %) described it as frequently (1–10 %), and only in 6 cases (3.8 %) pregnant cows were slaughtered often (>10 %). The proportion of pregnant cattle slaughtered rises with the total number of slaughtered females (see Table 2 and Fig. 5). The correlation is significant (Somers-D 0.470, $p < 0.01$).

Enquiry was also made about the prevalence of slaughter cows in advanced gestational stages (second or third trimester). Of 152 eligible questionnaires, 78 (51.3 %) reported advanced gravidities amongst the slaughter cows. In Fig. 6, the correlation between the total number of gravid slaughter cows and the number of animals in higher gestational stages is depicted. The rarer the slaughter of pregnant cattle was, the less often advanced gravidities were reported. The correlation between those two aspects is significant (Somers-D 0.499, $p < 0.01$).

Foetuses – vital signs and fate
Another aspect of the survey focused on the fate of the foetuses. 15 (10.6 %) of 142 respondents reported vital signs of the foetuses, such as independent movement and pulsation of the umbilical vessels. Irrespective of vital signs, one (0.8 %) of 132 respondents – as shown in Fig. 2 – reported about special measures that were taken

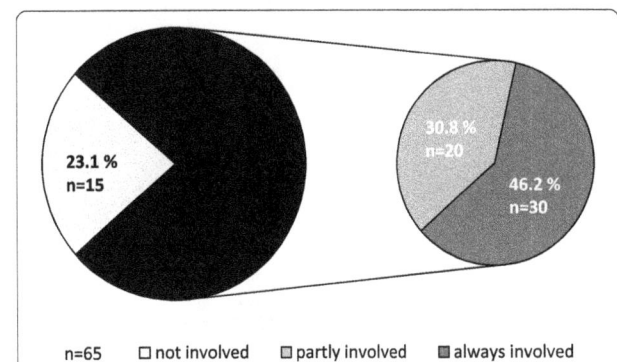

between 51–100, 43 (16.9 %) between 101–200, 17 (6.7 %) between 201–300 and 12 (4.7 %) more than 300 female cattle per day. The majority of abattoirs are therefore small and medium-sized enterprises with a slaughter capacity of ≤50 cows/day.

In 247 (96.9 %) of the 255 questionnaires that refer to the slaughter of female cattle, the participants answered

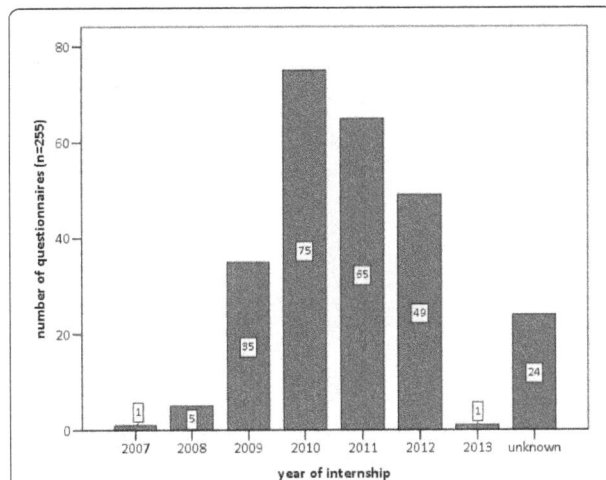

Fig. 3 Number of valid internship questionnaires (n = 263) fielded in year of internship. 263 valid questionnaires have been examined. Most of the associated internships took place in 2010 (n = 76), 2011 (n = 66) and 2012 (n = 52). 25 dates are not specified

n=65   □ not involved   ▣ partly involved   ■ always involved

Fig. 4 Percentage of abattoirs (n = 65) that are associated with slaughtering pregnant cattle. As described in the student survey, these abattoirs (n = 65) are never, sometimes or always associated with slaughtering pregnant cattle

Table 2 Number of slaughtered female cattle in relation to the frequency of slaughtered pregnant cattle (n = 247)

		Approximately how many cows and heifers were slaughtered a day?						Sum
		1–10	11–50	51–100	101–200	201–300	>300	
How frequent was slaughter of gravid cows?	never (0 %)	60	16	5	6	3	0	90
	rarely (<1 %)	22	30	16	14	7	6	95
	frequently (1–10 %)	4	9	14	19	7	3	56
	often (>10 %)	0	1	0	2	0	3	6
Sum		86	56	35	41	17	12	247

The reported proportion (never, rarely, frequently, often) of pregnant cattle slaughtered in relation to the total number of slaughtered females per day is shown in this table

to kill the foetus (euthanasia). The methods of euthanasia were not questioned, indeed. One student indicated that upon request, the official veterinarian stated that the foetus would die due to the cessation of uterine blood supply and that no special measures are needed. Moreover, the slaughter of pregnant farm animals also occurs to other species, as described for sheep by another student.

Discussion
Limitations
This survey is an observational, cross-sectional study based on students' reports on their compulsory slaughterhouse internship, which took place 1 to 18 month before their final exam in "meat hygiene". These time delays as well as the post exam situation are possible confounding factors that may affect the results of the study. However, all students attended a special briefing that focused on both general and specific aspects of their slaughterhouse internship. In addition, they were asked to write a voluntary, anonymous report. Therefore students were well prepared for the internship and their

report served them as a helpful reminder for both, the exam preparation and the survey. In their lectures of embryology and dairy reproduction as well as in their clinical internships, the students learned to classify the stage of gravidity on the basis of the crown-rump length, eruptions of teeth, development of fur and teeth and position and size of the foetus. Thus, knowledge of these fundamentals was assumed to be compulsory.

Although the survey was conducted without knowledge of the abattoirs and the local competent authorities, this approach is acceptable. Student internship reports are common practice in many professions and the survey of students and official veterinarians on their experiences during the slaughterhouse internship is carried out since 2007. Special announcement of certain key issues would lead to bias and distortion of the data. Privacy is respected by investigation of a large number of different and anonymously evaluated abattoirs. Additionally, one should note that due to the structure of the internship (students should visit different areas of the abattoir in order to see all different aspects of meat production and inspection) the reports represent only a

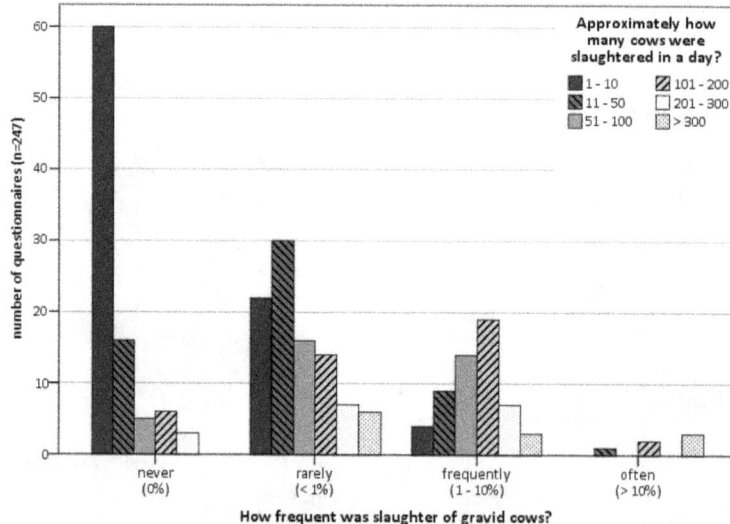

Fig. 5 Frequency of slaughtering gravid cattle in relation to total number of female cattle slaughtered daily. This chart shows a positive tendency to slaughter gravid cows by rising number of daily slaughtered cows

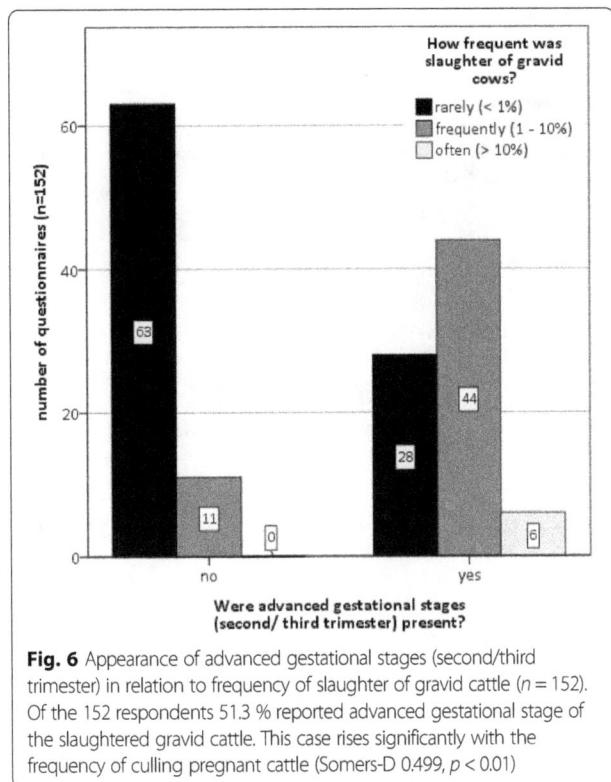

Fig. 6 Appearance of advanced gestational stages (second/third trimester) in relation to frequency of slaughter of gravid cattle (n = 152). Of the 152 respondents 51.3 % reported advanced gestational stage of the slaughtered gravid cattle. This case rises significantly with the frequency of culling pregnant cattle (Somers-D 0.499, p < 0.01)

snapshot rather than a complete overview of the actual situation in cattle slaughtering.

Key results and interpretation

Based on these "snapshots", however, results show that in nearly every abattoir for cattle, females are slaughtered. Only in 8 of the 263 questionnaires (3 %), none of the given ordinal options have been selected, which could either mean that no female cattle were slaughtered or that the student did not recall or refused to answer the question. The total ratio of females amongst all cattle slaughtered is unknown and irrelevant for this survey.

With 157 of 247 eligible cases (63.6 %), the share of pregnant cattle in this study is far higher than those reported by Riehn et al., Di Nicolo, Lücker et al., Singleton and Dobson, Herenda and others [3–6, 13, 15]. Due to study design though, the data should be interpreted with care because students during their internship have limited experience regarding the assessment and interpretation of findings in comparison to an official veterinarian. In addition, these 157 questionnaires were related to only 50 abattoirs. Besides, 20 of this 50 slaughterhouses were also reported in other questionnaires, where the participant never actually saw it directly. So the information can be repeated in the same and also in other abattoirs. Moreover, a high level of attentiveness and

empathy might have influenced their perception particularly with regard to health or welfare problems.

On the other hand, the aforementioned points of criticism may also be interpreted as an advantage for the present study: it offers the rare opportunity to collect information from external but yet competent observers.

Even though the semi-quantitative approach of the study does not permit an exact conclusion on the actual prevalence of the slaughter of pregnant cattle, the high number of positive reports (62 from 157; 39.5 %) shows unequivocally that the slaughter of pregnant cattle is not an exceptional event but occurs regularly even though at a different frequency of occurrence. This might also be due to the fact that large shares of the abattoirs (n = 146; 57.2 %) are small and medium-sized holdings with slaughter volumes of less than 50 female cattle per day. These small holdings may operate on a more local level are therefore not affected in the same way by fluctuations of suppliers on farm and transportation level as bigger, supra-regionally operating enterprises. This may result in above- or beyond-average shares of pregnant cattle in smaller plants depending of the practises in the supplying farms and the subsequent stages of the food chain. However, the present study includes many more abattoirs of different sizes and locations than former investigations on this issue.

The high proportion of questionnaires (n = 78; 51.3 %) reporting the slaughter of cattle during advanced gestational stages supports the findings of previous studies reporting the same phenomenon [4, 5]. However the data should be interpreted with care because differences in methodology may influence the results. An important point in this context is without doubt, that during the *ante-* and *post-mortem* inspection of slaughter cattle, later stages of gestation (late second and third trimester) are easier to diagnose then earlier stages. Furthermore, the study provides only qualitative data and is therefore not directly comparable to other studies.

A completely new aspect of this study is the focus on foetal life signs. Respondents report only in 15 of 142 cases (10.6 %) that foetal vital signs were recognizable. In only one of these cases the foetus was euthanized. This is not compatible with the animal welfare guidelines regarding the culling of pregnant animals in case of epizooty [29]. These guidelines recommend a separate euthanasia of foetuses by injection of Pentobarbital after electrical stunning of the cow. The drug diffuses through the placenta and causes a loss of consciousness and - in a high dose – paralyses the respiratory and circulatory centre leading to a rapid death [29]. Without such measures foetuses die by hypoxia. Although a number of studies [e.g. 21, 22] assume that due to low levels in foetal circulation, and the actions of other suppressors, it is unlikely that awareness occurs in the foetus during the death caused by hypoxia, a suffering of foetuses can not be completely ruled out [3, 4, 7]. It is for example questionable in this context whether only the cortex is actually involved in conscious perception or also other structures like the brainstem. In addition, is still controversial whether the prenatal measurements of EEG, which form the basis for the studies of Mellor et al. [21, 22], are comparable with the EEG reactions of an adult animal [7]. Because of these reasonable doubts the German government has tried to initiate a change of the existing European law to prevent a possible suffering of foetuses in later gestational stages [9].

Conclusions

The results of the present study show that the slaughter of pregnant cattle is a common and widespread practice in German abattoirs. Of 255 evaluable questionnaires, 157 (63.6 %) mention the slaughter of pregnant cattle, corresponding to 76.9 % of all visited abattoirs. Slaughter of pregnant cattle is reported often (>10 %) in 6 (3.8 %), frequently (1–10 %) in 56 (35.7 %), and rarely (<1 %) in 95 (60.5 %) of all cases ($n = 157$) respectively. Hence, the SCVPH's assumption that pregnant cattle are only slaughtered in rare exceptional cases [10, 11] can no longer be maintained. About 50 % of these animals were reported to be in the second or third stage of gestation. The high proportion of foetuses in the second and third gestational stage emphasises the relevance for animal welfare. In addition, 15 (10.6 %) of 142 questionnaires, providing information about the foetus, state that the foetus shows visible vital signs after the death of the mother. Where at present pain, suffering, and injury of the foetus during the slaughter of the mother cannot be excluded with absolute certainly, slaughter of pregnant farm animals must be considered as a pressing issue related to animal welfare. In this context suitable studies regarding (i) the prevalence of the slaughter of pregnant cattle in German and European abattoirs, (ii) the state of gestation of these animals, and (iii) the condition of the mother and the foetus with regard to welfare related parameters have to be initiated promptly. In addition, studies, regarding the sensitiveness and perceptiveness of foetuses during late pregnancy and at the time of birth have to be performed in order to elucidate the lack of scientific data in this context.

Endnotes

[1]Microsoft Corporation, One Microsoft Way, Redmond, WA 98052–6399, USA <https://products.office.com/de-de/home>

[2]IBM Deutschland GmbH, IBM-Allee 1, 71139 Ehningen, Germany <http://www.ibm.com/de/de/>

Abbreviations
SCVPH, Scientific Committee on Veterinary Measures relating to Public Health

Acknowledgements
The authors thank all participating students for their help – please keep your eyes open. We also gratefully thank Catharine van Maanen for proof reading and language support.

Funding
This study has been financed out of budgetary resources.

Authors' contributions
PM – analysed and interpreted the data and wrote the main part of the manuscript; EL – designed the study, participated in collecting the data and helped with writing the manuscript; KR – designed the study, participated in collecting the data and helped with writing the manuscript. All authors have read and approved the final manuscript.

Competing interests
The authors declare that they have no competing interests.

Consent for publication
Not applicable.

Author details
[1]Institute of Food Hygiene, Centre of Veterinary Public Health, Faculty of Veterinary Medicine, University of Leipzig, An den Tierkliniken 1, 04103 Leipzig, Germany. [2]Faculty of Life Sciences, Hamburg University of Applied Sciences, Ulmenliet 20, 21033 Hamburg, Germany.

References
1. Anon.: Grundgesetz für die Bundesrepublik Deutschland in der im Bundesgesetzblatt Teil III, Gliederungsnummer 100–1, veröffentlichten bereinigten Fassung, das zuletzt durch Artikel 1 des Gesetzes vom 11. Juli 2012 (BGBl. I S. 1478) geändert worden ist: GG; 2012.
2. Peisker N, Preissel A, Henke J. Kritische Aspekte bei der Tötung gravider Nutztiere. Tierarztl Umsch. 2012;67:214–8.
3. Riehn K, Domel G, Einspanier A, Gottschalk J, Lochmann G, Hildebrandt G, Luy J, Lücker E. Slaughter of pregnant cattle - aspects of ethics and consumer protection. Tierarztl Umsch. 2011;66:391–405.
4. Riehn K, Domel G, Einspanier A, Gottschalk J, Hildebrandt G, Luy J, Lücker E. Schlachtung gravider Rinder-ethische und rechtliche Aspekte. Fleischwirtschaft 2010;8:100–106.
5. Di Nicolo K. Studie zum zusätzlichen Eintrag von Hormonen in die menschliche Nahrungskette durch das Schlachten von trächtigen Rindern in der Europäischen Union am Beispiel von Luxemburg und Italien. Dissertation med. vet. Universität Leipzig, Veterinärmedizinische Fakultät; Institut für Lebensmittelhygiene; 2006.
6. Lücker E, Bittner A, Einspanier A. Zur toxikologisch-hygienischen Bewertung der Exposition mit hormonell wirksamen Stoffen bei Schlachtungen trächtiger Rinder unter verschiedenen Produktionsbedingungen. In: Proceedings 44. Arbeitstagung DVG Lebensmittelhygiene 2003. Edited by Deutsche Veterinärmedizinische Gesellschaft (DVG). Gießen. 2003. p.628–33.
7. Marahrens M, Schwarzlose I. Stellungnahme zu einem möglichen Empfindungsvermögen und der Lebensfähigkeit entwickelter Feten. Tierschutz/Schlachten gravider Tiere - Az. 331 - 34600/016. Celle; 2013.
8. Cordes B. Leidvoll: Das Schlachten trächtiger Kühe 2014 [http://www.ndr.de/nachrichten/Leidvoll-Das-Schlachten-traechtiger-Kuehe,kaelber113.html]. Accessed 31 Jul 2014.
9. German Government: Antwort der Bundesregierung auf die Kleine Anfrage der Abgeordneten Bärbel Höhn, Friedrich Ostensdorff, Nicole Maisch, weiterer Abgeordneter und der Fraktion Bündnis 90/Die Grünen - Drucksache 18/1391 -. Schlachtung tragender Kühe. Drucksache 18/1535 [http://dipbt.bundestag.de/doc/btd/18/015/1801535.pdf]. Accessed 25 Jul 2014.

10. European Commission, Scientific Committee on Veterinary Measures relating to Public Health: Review of previous SCVPH opinions of 30 April 1999 and 3 May 2000 on the potential risks to human health from hormone residues in bovine meat and meat products [http://cordis.europa.eu/docs/publications/5549/55496421-6_en.pdf]. Accessed 3 Aug 2014.

11. European Commission, Scientific Committee on Veterinary Measures relating to Public Health: Assessment of potential risks to human health from hormone residues in bovine meat and meat products [http://ec.europa.eu/food/safety/docs/cs_meat_hormone-out21_en.pdf]. 23 Jul 2014.

12. Kushinsky S. Safety aspects of the use of cattle implants containing natural steroids: Syntex Research; 1983.

13. Herenda D. An Abattoir Survey of Reproductive Organ Abnormalities in Beef Heifers. Can Vet J. 1987;28:33–7.

14. Khan MZ, Khan A. Frequency pf pregnant animals slaughtered at Faisalabad abattoir. J Islamic Acad Sci. 1989;2:82.

15. Singleton GH, Dobson H. A survey of the reasons for culling pregnant cows. Vet Rec. 1995;136:162–5.

16. Ademola AI. Incidence of Fetal Wastage in Cattle Slaughtered at the Oko-Oba Abattoir and Lairage, Agege, Lagos, Nigeria. Vet Res 2010;3:54–57.

17. Fayemi PO, Muchenje V. Maternal slaughter at abattoirs: history, causes, cases and the meat industry. SpringerPlus 2013;2:125. doi:10.1186/2193-1801-2-125.

18. Council Regulation (EC) No 1/2005 of 22 December 2004 on the protection of animals during transport and related operations and amending Directives 64/432/EEC and 93/119/EC and Regulation (EC) No 1255/97; 2005.

19. Regulation (EC) No 854/2004 of the European Parliament and of the Council of 29 April 2004 laying down specific rules for the organisation of official controls on products of animal origin intended for human consumption; 2004.

20. Council Regulation (EC) No 1099/2009 of 24 September 2009 on the protection of animals at the time of killing; 2009.

21. Mellor DJ, Diesch TJ, Gunn AJ, Bennet L. The importance of 'awareness' for understanding fetal pain. Brain Res Rev. 2005;49:455–71.

22. Mellor DJ, Gregory NG. Responsiveness, behavioural arousal and awareness in fetal and newborn lambs: experimental, practical and therapeutic implications. N Z Vet. 2003;51:2–13.

23. Bellieni CV, Buonocore G. Is fetal pain a real evidence? J Matern Fetal Neonatal Med. 2012;25:1203–8.

24. Directive 2010/63/EU of the European Parliament and of the Council of 22 September 2010 on the protection of animals used for scientific purposes; 2010.

25. Kulow W. Erkenntnisse zur Schlachtung gravider Rinder; 2013

26. Münch T, Richter T. Abgänge und Abgangsursachen bei Milchkühen in Baden-Württemberg unter dem Blickwinkel des Tierschutzes und der Ökonomie. Tierarztl Umsch. 2012;67:68–74.

27. Tierärztliche Vereinigung für Tierschutz e.V. (TVT): Codex Veterinarius: Ethische Leitsätze für tierärztliches Handeln zum Wohl und Schutz der Tiere. 2. überarbeitete Fassung Juli 2009. Bramsche; 2009.

28. Verordnung zur Approbation von Tierärztinnen und Tierärzten vom 27. Juli 2006 (BGBl. I S. 1827): TAppV; 2006.

29. Tierärztliche Vereinigung für Tierschutz e.V. (TVT): Töten größerer Tiergruppen im Seuchenfall: (Schwein, Rind, Schaf, Geflügel). Merkblatt Nr. 84. Bramsche; 2011 [Merkblatt].

2

Effect of ATP and Bax on the apoptosis of *Eimeria tenella* host cells

Zhiyong Xu[1,2], Mingxue Zheng[1*], Li Zhang[1], Xuesong Zhang[1], Yan Zhang[1], Xiaozhen Cui[1], Xin Gong[1], Rou Xi[1] and Rui Bai[1]

Abstract

Background: *Eimeria tenella* (*E. tenella*) is a species of *Eimeria* that causes haemorrhagic caecal coccidiosis, resulting in major economic losses in the global poultry industry. After *E. tenella* infection, the amount of ATP and Bax in host cells showed highly significant changes. Therefore, it is necessary to investigate the effects of ATP and Bax on the apoptosis of *E. tenella* host cells.

Results: The ATP-treated group and the V5-treated group had higher *E. tenella* infection rates than the untreated group at 24, 48, 72, 96, and 120 h after infection with *E. tenella*. The results of flow cytometry showed that compared with the control group, the mitochondrial permeability transition pore (MPTP) opening in the untreated group was highly significantly increased ($P < 0.01$) at 4, 24, 48, 72, 96, and 120 h. Moreover, results from Hoechst-Annexin V-PI staining and flow cytometry showed that the rates of early apoptosis, late apoptosis, and necrosis in the untreated group were significantly lower ($P < 0.05$) or highly significantly lower ($P < 0.01$) than those of the control group at 4 h, while the rates of early apoptosis, late apoptosis, and necrosis in the untreated group were higher at varying degrees than those in the control group at 24–120 h ($P < 0.05$ or $P < 0.01$). After treatment with ATP and Bax inhibitors, the rates of early apoptosis, late apoptosis, and necrosis, in addition to the MPTP opening in both the ATP-treated and V5-treated groups, were significantly lower ($P < 0.05$) or highly significantly lower ($P < 0.01$) than those in the untreated group.

Conclusions: ATP and Bax play important roles in regulating the apoptosis of *E. tenella* host cells.

Keywords: *Eimeria tenella*, Host cell, Apoptosis, ATP, Bax

Background

Eimeria tenella (*E. tenella*) is the most common species of *Eimeria*, and it parasitises chicken intestinal mucosa epithelial cells. After infection with *E. tenella*, the mortality rates in chickens increase up to 80%, whereas weight and egg production significantly decrease [1, 2]. These conditions result in major economic losses in the global poultry industry [3, 4].

E. tenella primarily damages the chicken caecum. The apoptosis rate of duodenal mucosal cells infected with *Eimeria acervulina* reaches the highest at 0.5 and 5 days [5]. *E. tenella*-infected embryo caecal epithelial cells have shown decreased apoptosis at early developmental stages (24 h or less) but have conversely shown increased

apoptosis at the middle and late developmental stages (24–120 h) [6]. The degree to which the mitochondrial membrane KATP channel opens has been shown to decrease the apoptosis rate of cells infected with *E. tenella* and promote the development of *E. tenella* in chicken caecum epithelial cells [7]. The mitochondrial apoptotic pathway is regulated by concentration changes in Ca^{2+} outside of *E. tenella* host cells and endoplasmic reticulum Ca^{2+} channels [8]. The amount of apoptosis in intestinal epithelial cells infected with *E. tenella* is consistent with the severity of injury to the mitochondrial structure. These observations indicate a positive correlation of apoptosis in cells infected with *E. tenella* with changes in mitochondrial structure [9].

The mitochondrial permeability transition pore (MPTP), a compound channel composed of multiple proteins, is located between inner and outer mitochondrial membranes. A previous study showed that MPTP is a key node that

* Correspondence: zhengmingxue288@163.com
[1]College of Animal Science and Technology, Shanxi Agricultural University, Taigu 030801, China
Full list of author information is available at the end of the article

plays a predominant role in the mitochondrial apoptosis pathway in host cells induced by *E. tenella* [10]. Cyclophilin D has previously been identified as an essential component of the MPTP structure [11]. Other studies have also suggested that MPTP potentially comprises a voltage-dependent anion channel (VDAC) and an adenine nucleotide transporter (ANT) [12]. Adenosine triphosphate (ATP) is the sole supplier of energy in living organisms. To maintain cell metabolic activity, ATP is transported into the cytoplasm via ANT, whereas cytoplasmic ADP is transported to mitochondria via ANT, which provides the raw material for oxidative phosphorylation [13]. During ischaemia and hypoxia, decreased levels of ATP result in increased concentrations of cytoplasmic phosphorus and calcium ions and the production of a large number of superoxides, thus further promoting MPTP opening and eventually leading to cell death [14]. Bcl-2 family proteins can be divided into two categories as follows: pro-apoptotic proteins, such as Bax, Bak, Bad and Bid, and anti-apoptotic proteins, such as Bcl-2 and Bcl-xl. Bax primarily resides in the cytoplasm. Apoptosis stimuli increase BH3 expression, which enhances the effects of Bax and Bak by combining with Bcl-2 and Bcl-xl, further promoting cell apoptosis [15]. A previous study demonstrated that Bax could promote cell apoptosis by combining with VDAC [16].

The caspase-9 inhibitor Z-LEHD-FMK can significantly increase the infection rate of *E. tenella* by inhibiting host cell apoptosis [6]. In animal models, the inhibition of MPTP by either cyclosporin A (CsA) or the genetic ablation of CyP-D provides strong protection from both reperfusion injury and congestive heart failure [17]. Other evidence also suggests that apoptosis can be reversed by anti-apoptotic drugs, which can rescue cells and provide new directions for the protective treatment of an organism by avoiding or controlling harmful processes [18]. The control of host-cell apoptosis had been demonstrated as complementary in the treatment of parasitic diseases [19].

In a recent study, we showed that the Bax amount in *E. tenella* host cells visibly decreased during the early developmental stages of *E. tenella* and, conversely, remarkably increased during the middle and later developmental stages [20]. The ATP content decreased at all developmental stages of *E. tenella* [20]. In the present study, we further investigated the effects of ATP and Bax on the apoptosis of *E. tenella* host cells in vitro by flow cytometry (FC), Hoechst-fluorescein isothiocyanate-conjugated Annexin V-propidium iodide (Hoechst-Annexin V-FITC-PI) staining and primary chick embryo caecum epithelial cell culture techniques. These results can provide a theoretical foundation for studying the mechanism of *E. tenella*-induced apoptosis in host

cells and exploring anti-apoptotic adjuvant treatment of *E. tenella* infection in chickens.

Methods

Experimental animals

A total of twenty 1-day-old chicks and one hundred 15-day-old specific pathogen-free (**SPF**) chicken embryos were used in the present study and were provided by Beijing Meri Avigon Laboratory Animal Technology Co., Ltd. (Beijing, China). The 1-day-old chicks were raised under strict pathogen-free conditions (Isolator. Temperature and pressure: 1–3 d, 35–36 °C, 25 Pa; 4–7 d, 32–35 °C, 25–35 Pa; 8–14 d: 29–32 °C, 35–45 Pa; 15–21 d, 21–25 °C, 55–75 Pa; 22–30 d, 21–25 °C, and 55–75 Pa. Humidity: 1–10 d, 65–70%; 11–30 d, 60–65%).

Parasites

The *E. tenella* Shanxi virulent strain (**EtSX01**) used in the present study was obtained from the Laboratory of Veterinary Pathology in the College of Animal Science and Technology (Laboratory of Veterinary Pathology, Shanxi Agricultural University; Taigu, China).

Preparation of *E. tenella* Sporozoites

E. tenella was amplified by passage through twenty 20-day-old SPF chicks previously infected orally with 6000 sporulated *E. tenella* oocysts. The resulting oocysts were obtained from the faeces of chickens at seven to eight days post-infection. After the oocysts were isolated and sporulated, the sporozoites were excysted as previously described [6]. The chicks were euthanized with cervical dislocation under deep Nembutal anaesthesia [45 µg/g of body weight (BW), intraperitoneal injection; Shanghai Chemical Factory, Shanghai, China].

Primary culture of Chick embryo Caecal cells and parasite infection

One hundred 15-day-old SPF chick embryos (Merial Vital Corp, Beijing, China) were euthanized with cervical dislocation under deep Nembutal anaesthesia (45 µg/g of BW) for sample preparation. Chick embryo caecal epithelial cells were collected and cultured as previously described [7]. The caeca were briefly removed and placed in PBS, minced to 1 mm³, digested by 50 mg/l thermolysin (Sigma, California, America) at 37 °C for 2 h, rinsed with PBS, and centrifuged at 220 g for 5 min to remove single cells. Prior to plating, based on the adherence speed of each cell, the precipitated dissociated cells were cultivated for 70 min at 41 °C in a humidified, 8% CO_2-air incubator to remove all other cells, except the caecal epithelial cells in culture medium (DMEM; Sigma, California, America) supplemented with 10% FBS [6]. Non-adherent caecal epithelial cell aggregates were maintained at 41 °C in a humidified incubator with 8%

CO_2. The cells were subsequently plated at a concentration of 2×10^5 live cell aggregates per well onto a 6-well tissue culture plate in culture medium supplemented with 2.5% FBS [6]. Confluent chick embryo caecal epithelial cell monolayers were infected with 4×10^5 freshly excysted *E. tenella* sporozoites per well.

Experimental protocol

When the adherence rate reached 90%, chick embryo caecal epithelial cells in 6-well tissue culture plates were randomly divided into four experimental groups, as follows: (1) the control group (group C); (2) the untreated group (cells infected with *E. tenella* sporozoites, group T0); (3) the treated group I [cells infected with *E. tenella* sporozoites and treated with 30 μmol of ATP (Sigma, California, America), group T1]; and (4) the treated group II [cells infected with *E. tenella* sporozoites and treated with 150 μmol of V5 (H-VPMLK-OH, Bax-Inhibiting Peptide; Merck, New Jersey, America), group T2]. The cells were collected at 4, 24, 48, 72, 96, and 120 h after infection. In addition, the culture liquid was changed every 48 h. Cells in group T1 that were sampled at 4–24 h were treated with ATP at the time of infection with *E. tenella* sporozoites, whereas the cells sampled at 48–120 h were treated with ATP 24 h prior to sampling. Cells in group T2 that were sampled at 4 h were treated with V5 at the time of infection with *E. tenella*

sporozoites, while the cells sampled at 24–120 h were treated with V5 4 h prior to sampling.

Haematoxylin and eosin stain (HE staining)

At 4, 24, 48, 72, 96, and 120 h after infection, the chamber slides of groups C and T0 were collected and subsequently stained with Lillie-Mayer's haematoxylin (Solarbio) and 1% eosin (Solarbio) as previously described [6]. Sample *E. tenella* infections were observed in 200 randomly selected cells by light microscopy.

The infection rates at each time point (%) = the number of infected cells at each time point/200 × 100.

Dynamic detection of MPTP opening in *E. tenella* host cells

Fluorescent Calcein AM (Life Technology, New York, America; Cat.C3099 Lot: 1,311,548) and $CoCl_2$ (Sigma, Lot: 232,696 No: BCBG0246V) markers were employed to detect the dynamic changes of MPTP opening using FC. At 4, 24, 48, 72, 96, and 120 h after infection, chick embryo caecal epithelial cells were harvested using 0.25% trypsin, rinsed in PBS, centrifuged at 600 g for 5 min, suspended in 200 μl of binding buffer (Sigma, California, America), and incubated with Calcein AM and $CoCl_2$ (15 min, 37 °C, in the dark). Subsequently, 250 μl of binding buffer was added to the cells, and the mixture was subjected to FC analysis. Flow cytometry

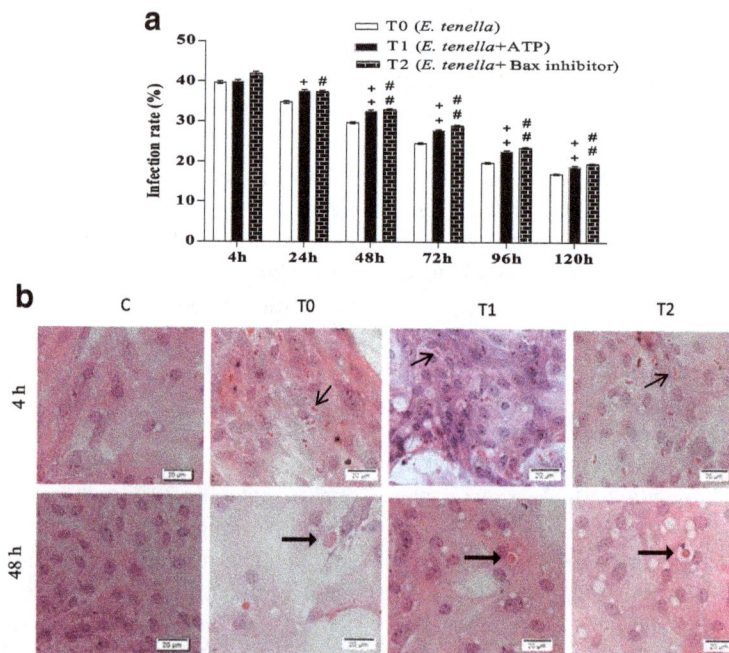

Fig. 1 *E. tenella* infection rates. **a** Quantitative determination of *E. tenella* infection ($n = 5$). **b** Caecal epithelial cells in groups C, T0, T1, and T2 at 4, and 48 h, respectively. $^+P < 0.05$ vs. T0, $^{++}P < 0.01$ vs. T0; $^\#P < 0.05$ vs. T0, $^{\#\#}P < 0.01$ vs. T0, as indicated below the figures. "⟶" represents sporozoites, "➡" represents trophozoites. Magnification 400×

(American BD, FACSCalibur) was performed as previously described [21]. The results were analysed using CellQuest software. Fluorescent intensity reflects a change in MPTP opening [22].

Dynamic detection of apoptosis in *E. tenella* host cells

The Annexin V-FITC/PI (Invitrogen, New York, America; Lot: 1,223,786, model: V13241) was used in the present study. The methods used to harvest and incubate the cells were consistent with those used to detect MPTP opening. Subsequently, chick embryo caecal epithelial cells were re-suspended and incubated with 5 μl of Annexin V-FITC and 1 μl of PI for 30 min at room temperature in the dark. Next, the cells were added to 200 μl of binding buffer and subjected to FC analysis. Annexin V-/PI- quadrant: viable cells; Annexin V+/PI- quadrant: early apoptotic cells; Annexin V+/PI+ quadrant: late apoptotic and necrotic cells.

Annexin V-FITC/PI and Hoechst 33,342 (Beyotime, Shanghai, China; cat: C1025) markers were employed to detect the apoptosis rate under a fluorescence microscope (OLYMPUS, Japan). The procedures were performed according to Venkatanarayan [23]. Briefly, at 4, 24, 48, 72, 96, and 120 h after infection, the culture media from groups C and T0 were aspirated into suitable centrifuge tubes. Subsequently, chick embryo caecal epithelial cells were harvested using 0.25% trypsin, centrifuged at 1000 g for 5 min, and suspended with 1 ml of ice-cold PBS. The cells were centrifuged at 1000 g for 5 min and gently suspended in 400 μl of 1× binding buffer (1 × 10^5 cell density), to which 5 μl of 2 μg/ml Hoechst 33,342 was added at 37 °C. The mixture was allowed to stand for 15 min, centrifuged at 1000 g for 5 min, and gently suspended in 400 μl of 1× binding buffer. Subsequently, 5 μl of Annexin V-FITC was added in the dark at room temperature. After 15 min, 10 μl of 100 μg/ml PI was added in the dark on ice; after 5 min, the resulting mixture was again centrifuged at 1000 g for 5 min. The cells were suspended in 50 μl of 1× binding buffer. Images were captured using a fluorescence microscope and CellSens software. Apoptotic cells were observed in 200 randomly selected sample cells and 5 parallel samples at each time point. The cells were distinguished as normal cells (Hoechst 33,342+/Annexin V+), early apoptosis cells (Hoechst33342+/Annexin V++), and late apoptosis and necrosis cells (Annexin V++/PI++).

Statistical analysis

All quantitative data were analysed by ANOVA in SPSS 19.0 (SPSS Inc., Chicago, Illinois, USA) and expressed as the means ± SE. A *p*-value of < 0.05 was considered significant.

Results

ATP and Bax-inhibitor increased *E. tenella* infection rates

The *E. tenella* infection rate was detected by HE staining (Fig. 1a-b; Additional file 1). The results indicated that although no significant difference in *E. tenella* infection rates was observed among T0, T1 and T2 groups at 4 h after infection by 4×10^5 *E. tenella* sporozoites per chamber slide ($P > 0.05$), and groups T1 and T2 exhibited higher *E. tenella* infection rates than group T0 ($P < 0.05$ or $P < 0.01$) at 24, 48, 72, 96 and 120 h after infection.

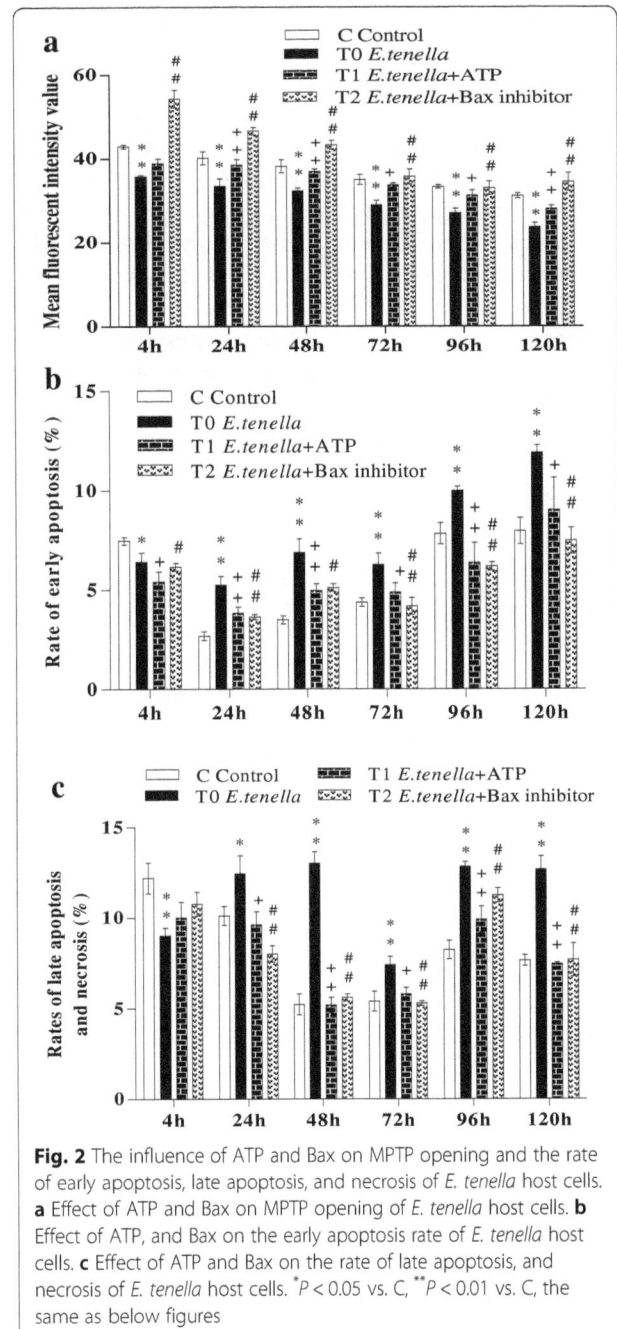

Fig. 2 The influence of ATP and Bax on MPTP opening and the rate of early apoptosis, late apoptosis, and necrosis of *E. tenella* host cells. **a** Effect of ATP and Bax on MPTP opening of *E. tenella* host cells. **b** Effect of ATP, and Bax on the early apoptosis rate of *E. tenella* host cells. **c** Effect of ATP and Bax on the rate of late apoptosis, and necrosis of *E. tenella* host cells. *$P < 0.05$ vs. C, **$P < 0.01$ vs. C, the same as below figures

ATP and Bax-inhibitor decreased MPTP opening in *E. tenella* host cells

A higher fluorescence value indicates a lower degree of MPTP opening. MPTP opening of the cells in groups C, T0, T1 and T2 were detected by FC. The results showed that MPTP opening of group T0 was higher ($P < 0.01$) than that of group C at 4, 24, 48, 72, 96, and 120 h (Figs. 2a and 3; Additional file 2). The results also indicated that MPTP opening of group T1 was visibly lower ($P < 0.05$ or $P < 0.01$) than that of group T0 (Figs. 2a and 3). MPTP opening of group T2 was also lower ($P < 0.01$) than that of group T0 (Figs. 2a and 3; Additional file 2).

ATP and Bax-inhibitor decreased apoptosis rate of *E. tenella* host cells

Early apoptosis, late apoptosis and necrosis rates in groups C, T0, T1 and T2 were detected by FC and Hoechst-Annexin V-FITC-PI staining. The results of FC showed that 4 h after infection with *E. tenella*, the early apoptosis, late apoptosis and necrosis rates in group T0 were lower than those in group C, whereas the early apoptosis, late apoptosis and necrosis rates in group T0 were higher than those in group C to varying degrees at 24, 48, 72, 96 and 120 h (Figs. 2b-c, and 4; Additional files 3 and 4). The

results of Hoechst-Annexin V-FITC-PI staining showed that the early apoptosis rate in group T0 significantly decreased at 4 h ($P < 0.01$) compared with that in group C. However, the early apoptosis rate in group T0 significantly increased at 24, 48, 72, 96 and 120 h ($P < 0.05$ or $P < 0.01$) compared with that in group C (Fig. 5a and c; Additional file 5). Compared with rates of late apoptosis and necrosis in group C, those in group T0 significantly decreased at 4 h ($P < 0.05$) and did not significantly differ at 24 h ($P > 0.05$) but highly significantly increased at 48, 72, 96 and 120 h ($P < 0.05$ or $P < 0.01$) (Fig. 5b and c; Additional file 6). Group T1 showed significantly lower or highly significant rates of early apoptosis, late apoptosis and necrosis than group T0 (Figs. 2b-c, 4, and 5a-c; Additional files 5 and 6). The rates of early apoptosis, late apoptosis and necrosis in group T2 were significantly lower or highly significantly lower compared to those in group T0 (Figs. 2b-c, and 4, and 5a-c; Additional files 5 and 6).

Discussion

As an intracellular parasite, the infection rate of *E. tenella* was the basic condition of this experiment. The results showed that group T0 host cells featured high rates of *E. tenella* infection (Fig. 1a-b; Additional file 1),

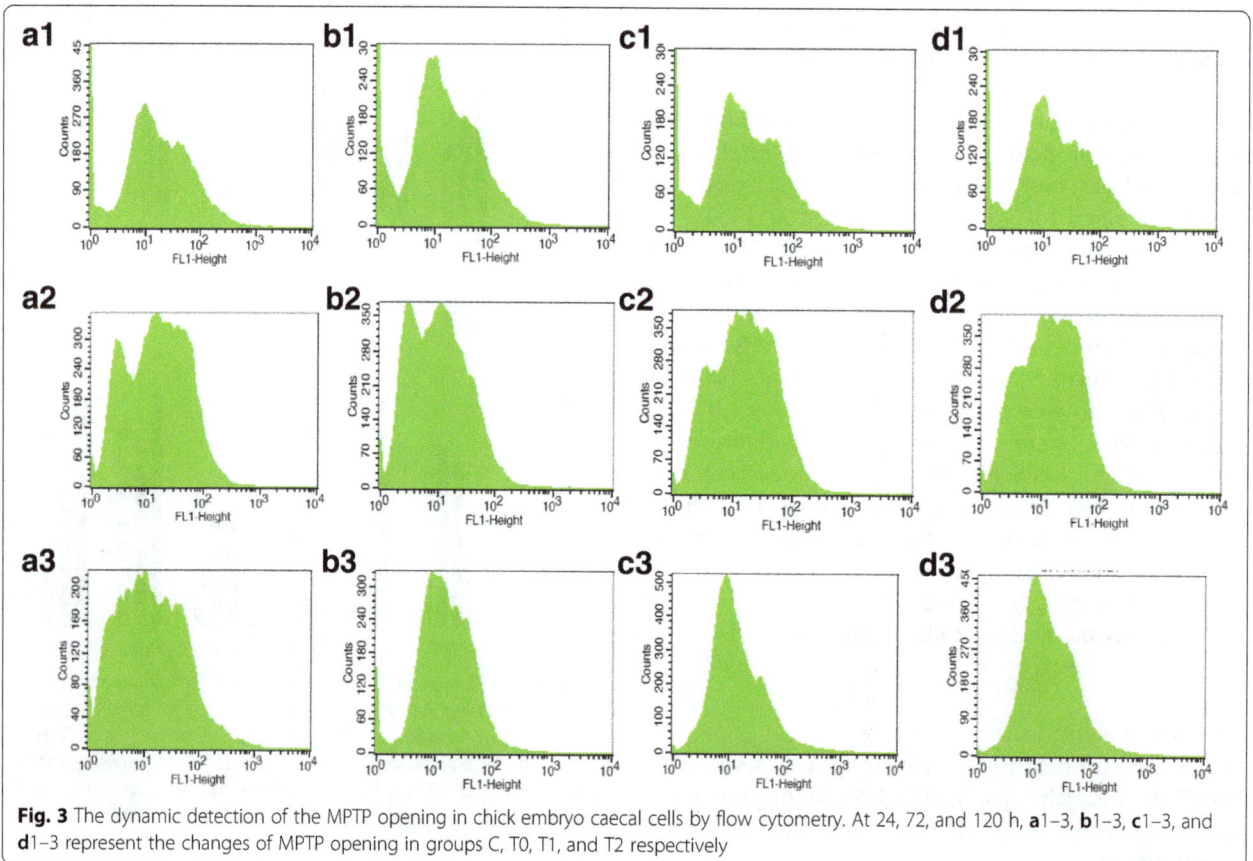

Fig. 3 The dynamic detection of the MPTP opening in chick embryo caecal cells by flow cytometry. At 24, 72, and 120 h, **a**1–3, **b**1–3, **c**1–3, and **d**1–3 represent the changes of MPTP opening in groups C, T0, T1, and T2 respectively

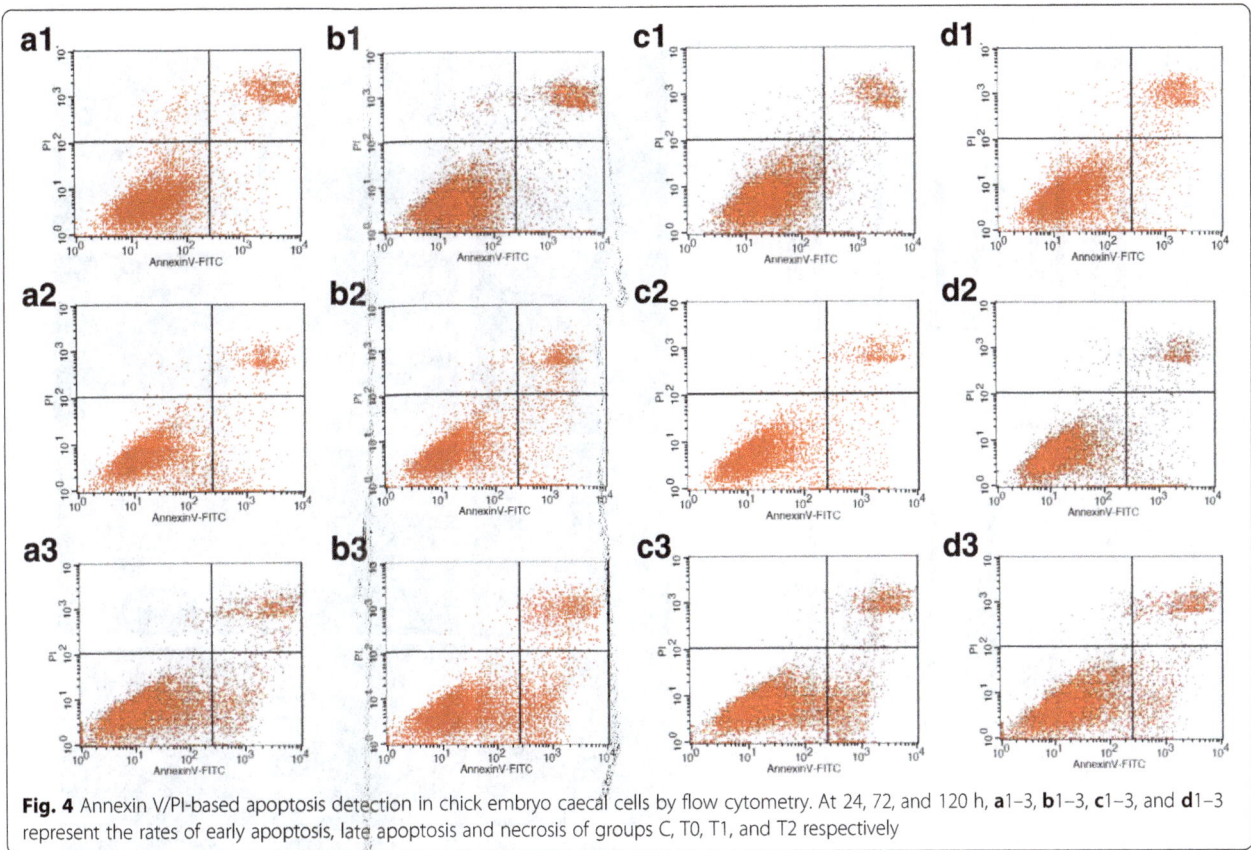

Fig. 4 Annexin V/PI-based apoptosis detection in chick embryo caecal cells by flow cytometry. At 24, 72, and 120 h, **a**1–3, **b**1–3, **c**1–3, and **d**1–3 represent the rates of early apoptosis, late apoptosis and necrosis of groups C, T0, T1, and T2 respectively

and the infection rates decreased with prolonged infection time, consistent with the results obtained in previous reports [6]. After treatment with ATP and Bax inhibitor, *E. tenella* infection rates remarkably increased, indicating the successful construction of experimental models.

After infection with *E. tenella*, the results of the two methods were consistent and showed that *E. tenella* inhibited host cell apoptosis during the early stages of *E. tenella* development and promoted host cell apoptosis during the middle and late developmental stages, supporting previous results [6, 24]. *E. tenella* also increased host-cell MPTP opening.

ATP plays a key role in maintaining normal cellular functions. The results by FC showed that ATP inhibited MPTP opening. The results of the two other methods used were consistent in this experiment, showing that ATP decreased the host cell apoptosis rate. ATP synthesis mainly depends on the mitochondrial membrane potential, and antioxidants can inhibit cell apoptosis by preventing decreases in ATP [25]. The exchange between ATP in mitochondria and ADP in the cytoplasm occurs via ANT. High concentrations of ATP can stabilise ANT in M constellation to inhibit MPTP opening, thus further preventing cell apoptosis [26]. Upon serious

stimulation, MPTP of the cell irreversibly opens, uncoupling oxidative phosphorylation, inhibiting ATP synthesis, and promoting ATP decomposition, eventually leading to necrocytosis [27]. A decrease in ATP results from the exhaustion of the membrane potential gradient caused by MPTP opening, and the release of apoptosis factors occurs due to swelling and the fracture of mitochondria caused by MPTP opening [28, 29]. In the present study, after treatment with ATP, MPTP opening in host cells was significantly inhibited, and the apoptosis rate of host cells significantly decreased, showing that ATP is a key factor in regulating the apoptosis of *E. tenella* host cells. These results are consistent with those of previous studies [25, 26]. At 4 h after infection with *E. tenella*, MPTP opening was increased in group T0 cells, whereas the apoptosis rate in group T0 decreased (Fig. 2a-c; Additional file 2). This result was due to a notable increase in the amounts of nuclear factor-κB and Bcl-xl, which can inhibit host cell apoptosis during the early stages of *E. tenella* development [30]. The involvement of other factors in this process must be further studied.

Bax is an important pro-apoptotic protein in the Bcl-2 family. V5 is a Bax-inhibitor and can protect cells from Bax-mediated apoptosis. The results of FC showed that

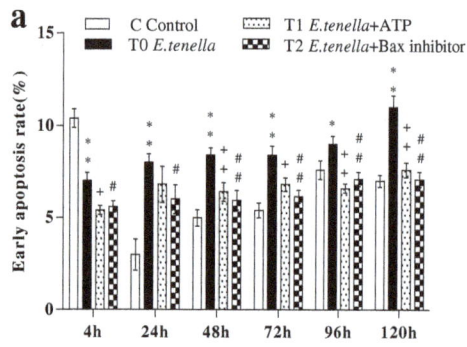

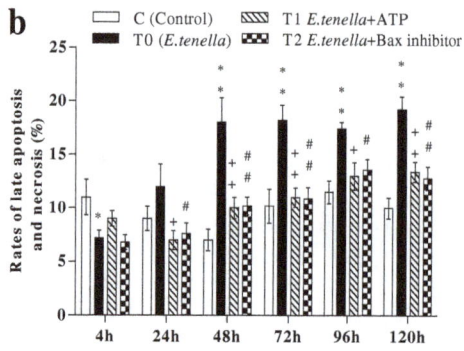

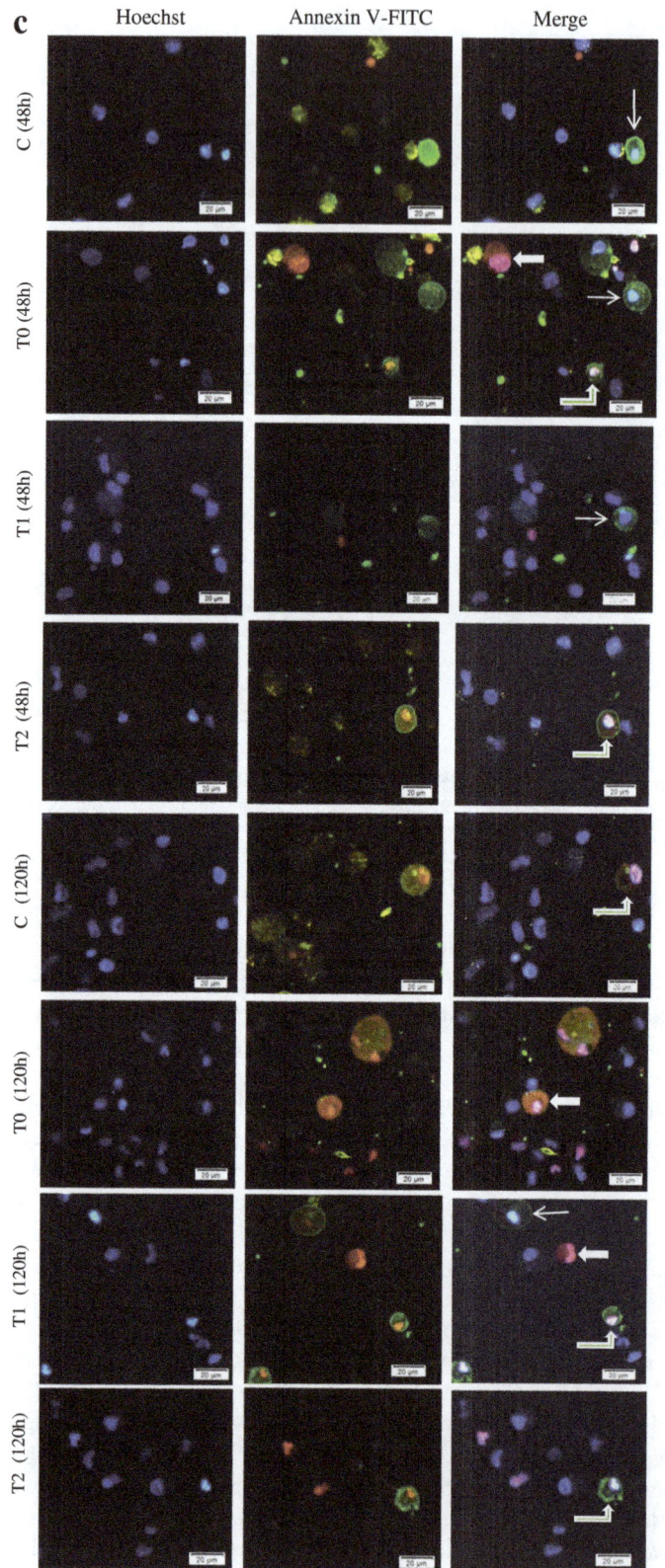

Fig. 5 (See legend on next page.)

(See figure on previous page.)
Fig. 5 Hoechst-Annexin V/PI-based apoptosis detection in chick embryo caecal cells. **a** Quantitative determination of early apoptosis (n = 5). **b** Quantitative determination of late apoptosis and necrosis (n = 5). **c** Hoechst staining (blue) and Annexin V/PI staining (green/red) to detect apoptosis cells in group C, T0, T1 and T2 at 48 and 120 h, respectively; merge is Hoechst staining/Annexin V/PI staining overlay. "⟶" represents early apoptosis cell, "◢" represents late apoptosis cell, "➡" represents necrosis cell. Magnification 400

the Bax inhibitor inhibited MPTP opening. The results of the other two methods used were consistent in this experiment, showing that the Bax inhibitor decreased the host cell apoptosis rate. Bax may regulate the mechanism of cell apoptosis in the following manner: first, Bax mediates changes in mitochondria permeability, further causing the synthetic obstruction of ATP; subsequently, Bax changes cell oxidation-reduction; third, Bax releases relevant factors that activate the signalling pathways of the caspase family [31]. Bax is assumed to regulate mitochondrial membrane permeability in two ways: first, Bax directly regulates MPTP opening [32]; second, Bax forms pores in the mitochondrial membrane and regulates mitochondrial membrane permeability. Brenner reported that Bax can combine with ANT and form pores in an artificial lipid bilayer in vitro, although the combination of Bcl-2 and ANT can inhibit the formation of Bax channels, thus further inhibiting changes in mitochondrial membrane permeability [33]. Studies have demonstrated that the oligomerisation of Bax subfamily proteins is promoted by the interactions of BH-3 and Bax subfamily proteins, and these complexes enter mitochondria to release cytochrome C, leading to apoptosis [34]. During the early stages of infection, *Cryptosporidium* inhibits host cell apoptosis by promoting the expression of the anti-apoptotic protein Bcl-2 but decreasing the expression of the pro-apoptotic protein Bax and reducing the release of apoptosis-related mitochondrial proteins. Conversely, during the late stages of infection, the development of polypide can inhibit Bcl-2 expression and increase Bax expression, which promotes MPTP opening and further induces cell apoptosis [35]. The results of the present study demonstrated that *E. tenella* promotes MPTP opening and further increases the cell apoptosis rate. However, V5 significantly decreases the cell apoptosis rate and inhibits MPTP opening, showing that Bax is a key regulator of apoptosis in *E. tenella* host cells. These results are consistent with those reported for other pathogens [35].

Conclusions

In conclusion, ATP and Bax inhibitors inhibited the MPTP opening and decreased the rate of early apoptosis, late apoptosis, and necrosis of *E. tenella* host cells. This finding shows that ATP and Bax play important roles in regulating the apoptosis of *E. tenella* host cells.

Additional files

Additional file 1: *E. tenella* infection rates. $^+P < 0.05$ vs. T0, $^{++}P < 0.01$ vs. T0; $^#P < 0.05$ vs. T0, $^{##}P < 0.01$ vs. T0, as indicated below the figures.

Additional file 2: The influence of ATP and Bax on MPTP opening of *E. tenella* host cells by flow cytometry. $^*P < 0.05$ vs. C, $^{**}P < 0.01$ vs. C, the same as below figures.

Additional file 3: The influence of ATP and Bax on the rate of early apoptosis of *E. tenella* host cells by flow cytometry.

Additional file 4: The influence of ATP and Bax on the late apoptosis, and necrosis of *E. tenella* host cells by flow cytometry.

Additional file 5: The influence of ATP and Bax on the rate of early apoptosis of *E. tenella* host cells by Hoechst-Annexin V/PI-based apoptosis detection.

Additional file 6: The influence of ATP and Bax on the late apoptosis, and necrosis of *E. tenella* host cells by Hoechst-Annexin V/PI-based apoptosis detection.

Abbreviations
Annexin V-FITC: Fluorescein isothiocyanate-conjugated Annexin V; ANT: Adenine nucleotide transporter; ATP: Adenosine triphosphate; BW: Body weight; CsA: Cyclosporin A; *E. tenella*: *Eimeria tenella*; FC: Flow cytometry; HE: Haematoxylin and eosin Stain; MPTP: Mitochondrial permeability transition pore; PI: Propidium iodide; SPF: Specific pathogen-free; V5: H-VPMLK-OH, Bax-Inhibiting Peptide; VDAC: Voltage-dependent anion channel

Acknowledgements
Not applicable

Funding
This study has been funded by a grant from the National Natural Science Foundation of China (Grant No. 31272536) and the Graduate Student Innovation Fund of Shanxi Province (Grant No. 2016BY069).

Authors' contributions
XZY performed most of the experiments and drafted the manuscript, and should be considered as first author. ZMX critically revised the manuscript and conceived the experimental design. ZL, ZXS, ZY, CXZ, GX, XR, and BR assisted with the experiments. All authors have read and approved the final version of the manuscript.

Consent for publication
Not applicable.

Competing interests
The authors declare that they have no competing interests.

Author details
[1]College of Animal Science and Technology, Shanxi Agricultural University, Taigu 030801, China. [2]College of Animal Science and Technology, Henan Institute of Science and Technology, Xinxiang 453003, China.

References

1. Bachaya HA, Raza MA, Khan MN, Iqbal Z, Abbas RZ, Murtaza S, Badar N. Predominance and detection of different *Eimeria* species causing coccidiosis in layer chickens. J Anim Plant Sci. 2012;22(3):597–600.
2. Dantán-González E, Quiroz-Castañeda RE, Cobaxin-Cárdenas M, Valle-Hernández J, Gama-Martínez Y, Tinoco-Valencia JR, Serrano-Carreón L, Ortiz-Hernández L. Impact of Meyerozyma guilliermondii isolated from chickens against *Eimeria sp.* protozoan, an in vitro analysis. BMC Vet Res. 2015; doi:10.1186/s12917-015-0589-0.
3. Abdelrahman W, Mohnl M, Teichmann K, Doupovec B, Schatzmayr G, Lumpkins B, Mathis G. Comparative evaluation of probiotic and salinomycin effects on performance and coccidiosis control in broiler chickens. Poult Sci. 2014;93(12):3002–8.
4. Jatau ID, Lawal IA, Kwaga JK, Tomley FM, Blake DP, Nok AJ. Three operational taxonomic units of Eimeria are common in Nigerian chickens and may undermine effective molecular diagnosis of coccidiosis. BMC Vet Res. 2016; doi:10.1186/s12917-016-0713-9.
5. Major P, Tóth Š, Goldová M, Revajová V, Kožárová I, Levkut M, Mojžišová J, Hisira V, Mihok T. Dynamic of apoptosis of cells in duodenal villi infected with *Eimeria acervulina* in broiler chickens. Biologia. 2011;66(4):696–700.
6. Yan Z, Zheng MX, Xu ZY, Xu HC, Cui XZ, Yang SS, Zhao WL, Li S, Lv QH, Bai R. Relationship between *Eimeria tenella* development and host cell apoptosis in chickens. Poult Sci. 2015;94(12):2970–9.
7. Yang SS, Zheng MX, Xu HC, Cui XZ, Zhang Y, Zhao WL, Bai R. The effect of mitochondrial ATP-sensitive potassium channels on apoptosis of chick embryo cecal cells by *Eimeria tenella*. Res Vet Sci. 2015;99:188–95.
8. Cui XZ, Zheng MX, Zhang Y, Liu RL, Yang SS, Li S, Xu ZY, Bai R, Lv QH, Zhao WL. Calcium homeostasis in mitochondrion-mediated apoptosis of chick embryo cecal epithelial cells induced by *Eimeria tenella* infection. Res Vet Sci. 2016;104:166–73.
9. Gu SP, Zheng MX, Li BJ. Analysis on apoptosis of chicken Cecal epithelial cells infected by *E. tenella*. Acta Veterinaria et Zootechnica Sinica. 2010;41(11):1322–7.
10. Xu ZY, Zheng MX, Zhang Y, Cui XZ, Yang SS, Liu RL, Li S, Lv QH, Zhao WL, Bai R. The effect of the mitochondrial permeability transition pore on apoptosis in *Eimeria tenella* host cells. Poult Sci. 2016;95(10):2405–13.
11. Devalaraja-Narashimha K, Diener AM, Padanilam BJ. Cyclophilin D gene ablation protects mice from ischemic renal injury. Am J Physiol. 2009;297(3):749–59.
12. Kubli DA, Gustafsson ÅB. Mitochondria and mitophagy: the yin and yang of cell death control. Circ Res. 2012;111(9):1208–21.
13. Li CY, Liu JZ. Structure and function of ANT and its relation to disease. J Int Pathol Clin Med. 2006;26(2):173–6.
14. Cui YM, Cheng HX, Zeng XM, Zeng QT, Gao W, Duan ML, Xu JG. Effects of hydrogen-rich saline on hippocampus mitochondrial permeability transition pore and apoptosis of rats with global cerebral ischemia-reperfusion injury. Chin Pharmacol Bull. 2012;28:853–8.
15. Adams JM, Cory S. Bcl-2-regulated apoptosis: mechanism and therapeutic potential. Curr Opin Immunol. 2007;19(5):488–96.
16. Pastorino JG, Shulga N, Hoek JB. Mitochondrial binding of hexokinase II inhibits Bax-induced cytochrome c release and apoptosis. J Biol Chem. 2002;277(9):7610–8.
17. Halestrap AP, Pasdois P. The role of the mitochondrial permeability transition pore in heart disease. Biochim Biophys Acta. 2009;1787:1402–15.
18. Ehrenreich H, Timner W, Sirén AL. A novel role for an established player: anemia drug erythropoietin for the treatment of cerebral hypoxia/ischemia. Transfus Apher Sci. 2004;31(1):39–44.
19. Bienvenu AL, Gonzalez-Rey E, Picot S. Apoptosis induced by parasitic diseases. Parasit Vectors. 2010;3 doi:10.1186/1756-3305-3-106.
20. Xu ZY, Zheng MX, Zhang Y, Cui XZ, Yang SS, Liu RL, Li S, Xi R, Gong X, Bai R. Dynamic changes in the main regulatory genes of mitochondrial permeability transition pore in *Eimeria tenella* host cells. Exp Parasitol. 2016; 171(10):42–8.
21. Raveche E, Abbasi F, Yuan Y, Salerno E, Kasar S, Marti GE. Introduction to flow cytometry. In: Litwin V, Marder P, editors. Flow Cytometry in drug discovery and development. New Jersey: Wiley; 2010. p. 93–9.
22. Mironov SL, Ivannikov MV, Johansson M. [Ca^{2+}] signaling between mitochondria and endoplasmic reticulum in neurons is regulated by microtubules. From mitochondrial permeability transition pore to Ca^{2+}-induced Ca^{2+} release. J Biol Chem. 2005;280(1):715–21.
23. Venkatanarayan A, Raulji P, Norton W, Chakravarti D, Coarfa C, Su X, Sandur SK, Ramirez MS, Lee J, Kingsley CV, Sananikone EF, Rajapakshe K, Naff K, Parker-Thornburg J, Bankson JA, Tsai KY, Gunaratne PH, Flores ER. IAPP driven metabolic reprogramming induces regression of p53-deficient tumours in vivo. Nature. 2015;517:626–30.
24. Deng JJ, Wang LX, An J. *Eimeria Tenella* sporozoite inhibits apoptosis in infected host cells. J Beijing Agricult Coll. 2011;26(2):24–7.
25. Jin J, Zhang F, Yang LL. Study progress of mitochondrial Uncoupler. Chin Bull Life Sci. 2013;25:707–15.
26. Doerner A, Pauschinger M, Badorff A, Noutsias M, Giessen S, Schulze K, Bilger J, Rauch U, Schultheiss HP. Tissue-specific transcription pattern of the adenine nucleotide translocase isoforms in humans. FEBS Lett. 1997;414(2):258–62.
27. Guo JY, Li Y, Pan JQ, Tang ZX. The relation of mitochondrial permeability transition pore and cell apoptosis. Prog Vet Med. 2009;30:101–5.
28. Kowaltowski AJ, Smaili SS, Russell JT. Elevation of resting mitochondrial membrane potential of neural cells by cyclosporin a, BAPTA-AM, and Bcl-2. Am J Physiol Cell Physiol. 2000;279(3):C852–9.
29. Michael RD. Mitochondria in health and disease: perspectives on a new mitochondrial biology. Mol Asp Med. 2004;25(4):365–451.
30. del Cacho E, Gallego M, López-Bernad F, Quílez J, Sánchez-Acedo C. Expression of anti-apoptotic factors in cells parasitized by second-generation schizonts of *Eimeria tenella* and *Eimeria necatrix*. Vet Parasitol. 2004;125(3–4):287–300.
31. Borkan SC. The role of BCL-2 family members in acute kidney injury. Semin Nephrol. 2016;36(3):237–50.
32. Cook SA, Sugden PH, Clerk A. Regulation of Bcl-2 family proteins during development and in response to oxidative stress in cardiac myocytes: association with changes in mitochondrial membrane potential. Circ Res. 1999;85:940–9.
33. Brenner C, Cadiou H, Vieira HL, Zamzami N, Marzo I, Xie Z, Leber B, Andrews D, Duclohier H, Reed JC, Kroemer G. Bcl-2 and Bax regulate the channel activity of the mitochondrial adenine nucleotide translocator. Oncogene. 2000;19(3):329–36.
34. Green DR. Apoptotic pathways: paper wraps stone blunts scissors. Cell. 2000;102(1):1–4.
35. Liu J, Deng M, Lancto CA. Biphasic modulation of apoptotic pathways in *Cryptosporidium parvum*-infected human intestinal epithelial cells. Infect Immun. 2009;77(2):837–49.

3

Revelation of mRNAs and proteins in porcine milk exosomes by transcriptomic and proteomic analysis

Ting Chen[1†], Qian-Yun Xi[1†], Jia-Jie Sun[1], Rui-Song Ye[1], Xiao Cheng[1], Rui-Ping Sun[1,2], Song-Bo Wang[1], Gang Shu[1], Li-Na Wang[1], Xiao-Tong Zhu[1], Qing-Yan Jiang[1] and Yong-Liang Zhang[1*]

Abstract

Background: Milk is a complex liquid that provides nutrition to newborns. Recent reports demonstrated that milk is enriched in maternal-derived exosomes that are involved in fetal physiological and pathological conditions by transmission of exosomal mRNAs, miRNAs and proteins. Until now, there is no such research relevant to exosomal mRNAs and proteins in porcine milk, therefore, we have attempted to investigate porcine milk exosomal mRNAs and proteins using RNA-sequencing and proteomic analysis.

Results: A total of 16,304 (13,895 known and 2,409 novel mRNAs) mRNAs and 639 (571 known, 66 candidate and 2 putative proteins) proteins were identified. GO and KEGG annotation indicated that most proteins were located in the cytoplasm and participated in many immunity and disease-related pathways, and some mRNAs were closely related to metabolisms, degradation and signaling pathways. Interestingly, 19 categories of proteins were tissue-specific and detected in placenta, liver, milk, plasma and mammary. COG analysis divided the identified mRNAs and proteins into 6 and 23 categories, respectively, 18 mRNAs and 10 proteins appeared to be involved in cell cycle control, cell division and chromosome partitioning. Additionally, 14 selected mRNAs were identified by qPCR, meanwhile, 10 proteins related to immunity and cell proliferation were detected by Western blot.

Conclusions: These results provide the first insight into porcine milk exosomal mRNA and proteins, and will facilitate further research into the physiological significance of milk exosomes for infants.

Keywords: Porcine milk exosomes, RNA-seq, Proteomic analysis

Background

Milk is the primary source of nutrition for newborns, and breastfeeding is known to make a valuable contribution to infant health [1]. Breast milk contains a potent mixture of diverse components including milk fat globules (MFG), immune competent cells, antibodies, soluble proteins, cytokines, and antimicrobial peptides [2] that together protect young infants against infections [3]. In addition, the milk contains growth factors which could promote intestinal development [4] and may protect infants against developing allergies [5]. Meanwhile, milk also contain many microvesicles, such as milk-derieved exosomes, who was reported to transfer contained RNAs to living cells and influenced the development of calf's gastrointestinal and immune systems [6].

Exosomes are small membrane vesicles (30–100 nm) which released from producing cells into the extracellular environment [7]. Many different cell types have the capacity to produce and release exosomes [8–13]. Additionally, milk-derived exosomes have been reported in humans, cows and pigs [14–17] and which involved in many biological processes. Exosomes contain proteins, mRNAs, miRNAs and lipids. Recent studies revealed that human [18], bovine [19], pig [20], and rat [21] milk contain miRNAs, and mRNAs have also been identified in

---

* Correspondence: zhangyl@scau.edu.cn
†Equal contributors
[1]National Engineering Research Center For Breeding Swine Industry, Guandong Provincial Key Laboratory of Agro-Animal Genomics and Molecular Breeding, Guandong Province Research Center of Woody Forage Engineering and Technology, South China Agricultural University, 483 Wushan Road, Guangzhou 510642, China

whey [6, 21–23]. 10,948 mRNA transcripts were detected in rat milk, and some immune and development-related mRNAs showed time-dependent expression [21]. 19,320 mRNAs were detected by microarray analyses in bovine milk exosomes, and they had possible effects of human cells [24]. Additionally, Cecilia Lässer et al. demonstrated that mRNAs in breast milk exosomes could be taken up by human macrophages [25].Until now, the components of mRNAs in porcine milk exosmes are still unclear.

Proteins in exosome were dependented on the specific cell-type [26], the dendritic cell-derived exosomes contain several cytosolic proteins [8]. Body fluid derived exosomes CD24, CD9, Annexin-1 and Hsp70 were as positive marker proteins [27]. Anti-MHC-class II- and anti-CD63 beads were used to isolate human breast milk exosmes [28]. In bovine milk exosomes 2,107 proteins were identified, and all major exosome protein markers were abundant [29], as were milk fat globule membrane (MFGM) proteins. Another report showed 2,350 proteins in bovine milk exosome via iTRAQ, and 90 exosomal proteins were found to be differentially regulated by infections [30].

In our previous study, miRNAs in porcine milk exosomes have been revealed by deep sequencing [17], but up to now, porcine milk exosomal mRNAs and proteins remains unknown. Therefore, we further performed RNA-sequencing and proteomic analysis of porcine milk exosomes in order to understand new physiological functions, especially immunity and proliferation related regulation of porcine milk.

Methods

Milk sample preparation
Fresh porcine milk samples were collected from 10 healthy Landrace female pigs that had been lactating for 1 to 5 days (after parturition) at the pig farm of the South China Agriculture University (Guangzhou, China). Milk samples were frozen immediately and kept at −80 °C until used.

Isolation of milk exosomes
Porcine milk exosomes were separated as previously described [17]. Briefly, about 80–100 mL fresh raw porcine milk samples were centrifuged at 2,000 g for 30 min at 4 °C to remove milk fat globules (MFGs) and mammary gland-derived cells [18]. Defatted samples were then subjected to centrifugation at 12,000 g for 30 min at 4 °C to remove residual MFGs, casein, and other debris [6]. From the supernatant, the membrane fraction was prepared by ultracentrifugation at 110,000 g for 2 h using an SW41T rotor (Beckman Coulter Instruments, Fullerton, CA). Then, the exosome purification steps were as previously described [29, 30].

RNA isolation
Total RNA was isolated from porcine milk exosome samples by Trizol reagent (Invitrogen, Carlsbad, CA) according to the manufacturer's protocol. The quality of RNA was examined by 2% agarose gel electrophoresis and with a BiophotometerNanoDrop 2000 (Thermo, USA), as well as further confirmed using a Bioanalyzer (Agilent Technologies, Santa Clara, CA).

RNA-sequencing
The collected RNA samples were analyzed by Illumina-HiSeq™ 2000 analyzer at Beijing Genomics Institute(BGI, Shenzhen, China) as previously described [31]. Firstly, poly (A) mRNA was isolated from total RNA sample with Oligo(dT) magnetic beads. Secondly, the purified mRNA was fragmented by the RNA fragmentation kit (Ambion), the first-strand cDNA synthesis was performed using random hexamer primers and reverse transcriptase, and the second-strand cDNA was synthesized using RNase H and DNA polymerase I. Then the cDNA libraries were prepared using the Illumina Genomic DNA Sample Prep kit (Illumina) following the manufacturer's protocol. Finally, the library was sequencing using Illumina HiSeq™ 2000.

Sequencing analysis
The porcine reference genome sequence and annotated transcript set were downloaded from the ensemble database (Sscrofa10.2, http://asia.ensembl.org/Sus_scrofa/Info/Index). After quality control (QC) step of raw reads, then removing low quality reads, reads containing Ns > 5 and reads containing adapters, clean reads were aligned to the reference pig genomic database (Sscrofa 10.2,) with SOAPaligner/SOAP2 [32] and allowing up to 5 mismatches in 90-bp reads. The alignment data were utilized to calculate distribution of reads on pig gene database (http://www.ncbi.nlm.nih.gov/), and the numbers of reads per kilobase of dexon region in a gene per million mapped reads were used as the value of normalized gene expression levels [33]. The unalignment data carried out novel transcript prediction, reads are at least 200 bp away from annotated gene, the transcript is of length over180 bp and the sequencing depth is no less than 2 for novel transcript unit analysis.

qPCR identification of known mRNAs in porcine milk exosome
Total RNA (identical with the RNA-sequencing sample) was first digested with DNase I (Promega, American), and 2 μg of total RNA was reverse transcribed by oligo (dT).The cDNA was diluted by 2-fold with ddH$_2$O, and PCR was performed on a Bio-Rad system (BIO-RAD, USA) in a final 20 μL volume reaction, containing 2 μL PCR cDNA, 10 μL of 2× PCR

Mix (Roche, Switzerland) and 1 mM of each primer. The real-time PCR thermal profile was as follows: 5 min at 95 °C, 40 cycles of 30 s at 94 °C, 30 s at the corresponding annealing temperature (Tm) and 72 °C for 30 s, followed by 72 °C at 10 min, and 5S ribosomal RNA was used as an internal control for the PCR [17, 34]. The mRNAs primers were designed with Primer 5.0 (Table 1).

Total protein extraction

RIPA lysis buffer was used to extract porcine milk exosomal proteins according to the assay kit protocol (Bioteke, Beijing). Briefly, 1 mM PMSF was added to the RIPA lysis buffer and 100–200 μL was added to porcine milk exosomes. Following complete exosome lysis, the sample was centrifuged at 10,000–14,000 g for 3–5 min

and the supernatant was subjected to further analysis. Proteins were stored at −80 °C until used.

Protein separation by 1D SDS-PAGE and in-gel digestion

Porcine milk exosome proteins were resolved by 12% polyacrylamide gel. The gel was stained with Coomassie blue R-250. 20 bands were excised and destained using 50 mM ammonium bicarbonate in 50% ACN. And then the gel pieces were performed incubating with 10 mM DTT in 25 mM ammonium bicarbonate for 1 h at 60 °C to reduce disulfide bonds and incubating the samples with 55 mM iodoacetamide in 25 mM ammonium bicarbonate for 45 min at room temperature in dark for Alkylation of cysteines. Then, using the Trypsin Gold (Promega, Madison, WI, USA) for digested (37 °C, 16 h) the gel bands. After the peptides sequentially extracted

Table 1 Primers for qPCR

ID	Primer	Sequences (5'to3')	Products length (bp)
ENSSSCG00000000207	LOC100739053-F	CAAAGGAAGCCTACAAGAA	198
	LOC100739053-R	CACGGTAGTCCAGCAGA	
ENSSSCG00000003930	RPS8-F	GAGAAAGCCCTACCACA	191
	RPS8-R	CGTCAATAATCCTCGTCT	
ENSSSCG00000004489	EF1ALPH-F	GATTGTTGCTGCTGGTGT	226
	EF1ALPH-R	TGCTACTGTATCAGGGTTGT	
ENSSSCG00000004177	RPS12-F	TCTACCCGTAACCCACC	219
	RPS12-R	CCTCCACCAACTTGACATA	
ENSSSCG00000015103	RPS25-F	GCCCAAGGACGACAA	109
	RPS25-R	GCCTTTGGACCACTTC	
ENSSSCG00000006249	RPS20-F	ACCGCTGTTCGCTCTTC	211
	RPS20-R	GTCCCTTCACTTTGAGGTTCT	
ENSSSCG00000029830	RPL8-F	CGAGCGACACGGCTACAT	255
	RPL8-R	GGCTTCTCCTCCAGACAACAC	
ENSSSCG00000009267	CSN3-F	CACCTGAGACCACCACT	140
	CSN3-R	TGACTGAAGGCAGATAA	
ENSSSCG00000013907	UBA52-F	ACGGGCAAGACCATCAC	196
	UBA52-R	GCAGACGAAGCACCAAGT	
ENSSSCG00000001502	RPS18-F	AGGGTGTAGGACGGAGAT	134
	RPS18-R	CTTGTATTGGCGAGGATT	
ENSSSCG00000024825	RPL6-F	CAGAGGCAAGAGGGTCA	128
	RPL6-R	TGGTGGAGGTGGCAATA	
ENSSSCG00000004970	RPLP1-F	GCACGACGATGAGGTTAC	131
	RPLP1-R	TGAGGCTCCCGATGTT	
ENSSSCG00000010328	RPS24-F	TTGATGTCCTTCACCCTG	269
	RPS24-R	CATTCTGTTCTTGCGTTCT	
ENSSSCG00000025527	FABP3-F	GCTGGGATTGAAGTTTGA	163
	FABP3-R	GTGGGTGAGTGTCAGGATG	
5S	5S-F	TCTACGGCCATACCACCCTGAA	83
	5S-R	GGCCCGACCCTGCTTAG	

from gel bands by 0.1% formic acid in 50% ACN twice, using 100% ACN twice, the extracted peptides were dried and stored at –80 °C until LC-MS/MS analysis.

Protein sequencing

Protein samples were analyzed using a Q-EXACTIVE mass spectrometer at the Beijing Genomics Institute (BGI, Shenzhen, China). Briefly, samples were separated by 1D SDS-PAGE and in-gel digestion was performed to generate peptides for LC-MS/MS analysis. Peptide fractions were initially separated on a LC-20 AD nanoHPLC (Shimadzu, Kyoto, Japan), then subjected to nanoelectrospray ionization followed by tandem mass spectrometry (MS/MS) using a Q EXACTIVE (ThermoFisher Scientific, San Jose, CA) coupled online to the HPLC.

LC-ESI-MS/MS analysis based on Q EXACTIVE

After a series of processing, we regulated each fraction at the average final concentration of peptide at 0.5 μg/uL and loading 10 uL on a LC-20 AD nanoHPLC (Shimadzu, Kyoto, Japan) by the autosampler onto a 2 cm C18 trap column. Then 10 cm analytical C18 column (inner diameter 75 μm) was used for eluted the peptides. After the sample was loading to the trap column, then bring into the analytical column, and finally the separated peptides were subjected to nanoelectrospray ionization followed by tandem mass spectrometry (MS/MS) in a Q EXACTIVE (ThermoFisher Scientific, San Jose, CA) coupled online to the HPLC. Resolution of 7,000 on Orbitrap was used to detect the intact peptides. Peptides were selected for MS/MS using high-energy collision dissociation (HCD) operating mode with a normalized collision energy setting of 27.0; ion fragments were setting of a resolution of 17,500. A data-dependent procedure that alternated between one MS scan followed by 15 MS/MS scans was applied for the 15 most abundant precursor ions above a threshold ion count of 20,000 in the MS survey scan with a following Dynamic Exclusion duration of 15 s. The electrospray voltage applied was 1.6 kV. The Automatic gain control (AGC) which used to optimize the spectra generated by the orbitrap was target for full MS was 3e6 and 1e5 for MS2. For MS scans, the m/z scan range was 350 to 2,000 Da. For MS2 scans, the m/z scan range was 100–1,800. All those works were carried out in Beijing Genomics Institute (BGI, Shenzhen, China).

Protein data analysis

All raw data were acquired using an Orbitrap, converted to MGF files using Proteome Discoverer 1.2 (PD1.2, Thermo), and the Mascot search engine (Matrix Science, London, UK; version 2.3.02) was used to search against a database containing 25,152 sequences(ftp://ftp.ensembl.org/pub/release-73/fasta/sus_scrofa/pep/).Non-intact (>20 ppm)

peptides and fragmented ions (0.6 Da) were removed, with allowance for one missed cleavage in trypsin digests. Next, the fixed carbamidomethyl (C) modification, and potential variable modifications Gln- > pyro-Glu (N-term Q), oxidation (M), deamidation (NQ), and +2 and +3 charge states were considered. Mascot was used to search the automatic decoy database by choosing the decoy checkbox, with the decoy checkbox set to generate a random sequence of database and test for raw spectra, as well as the actual database. Finally, only peptides with significance scores ≥20 at the 99% confidence interval in the Mascot probability analysis were counted as identified proteins [29]. All identified proteins included at least one unique peptide.

Western blot identification

Protein samples (20–30 μg) were measured by BCA assay [35], and separated using 10–15% SDS-PAGE, transferred to a 0.22 μm or 0.45 μm polyvinylidenedifluoride membrane (Millipore), and incubated with specific and HRP-conjugated secondary antibodies, and detected with an enhanced chemiluminescence kit (Roche, Switzerland) using FluorChem M (Proteinsimple) [36]. Anti-EGF (AB20578b), anti-TGFB-3 (AB20578b), anti-MSTN (AB60418a), connective tissue growth factor (CTGF) (AB60212a), anti-PDGFA (AB61078b), anti-CD63 (D260973), anti-IGFBP-7 (AB60509b), anti-CD9 (AB54118), anti-HTRA3 (AB61337a), and anti-THBS1 (AB61391a) were purchased from BBI Antibody (SangonBiotch, Shanghai, China). Lactoferrin (C-15) and β-actin were purchased from Santa Cruz (Santa Cruz, American). Protein' concentrations were determined using the Pierce BCA Protein Assay Kit (Thermo Fisher, American) using a BSA standard.

Bioinformatics analysis

We performed functional annotation using Blast2GO to search the non-redundant protein database (NR; NCBI) and the COG database (http://www.ncbi.nlm.nih.gov/COG/), which was used to classify and group the identified proteins. All the known mRNAs and proteins were performed Gene Ontology, KEGG pathway and Tissue-specific using DAVID6.7 bioinformatics resources (http://david.abcc.ncifcrf.gov/).

Results

Identification of exosomes by western blotting and extraction of RNA and protein from porcine milk exosome

We previously isolated exosomes from porcine milk and analyzed them using transmission electron microscopy [17]. In the present study, we observed exosomal marker proteins CD63 and CD9 by Western blotting (Fig. 1a). We extracted total RNA from the pellets after ultracentrifugation and examined the RNA by Agilent 2100, and the results showed that the porcine milk exosome contained RNAs and small rRNAs (Fig. 1c), which is consistent with

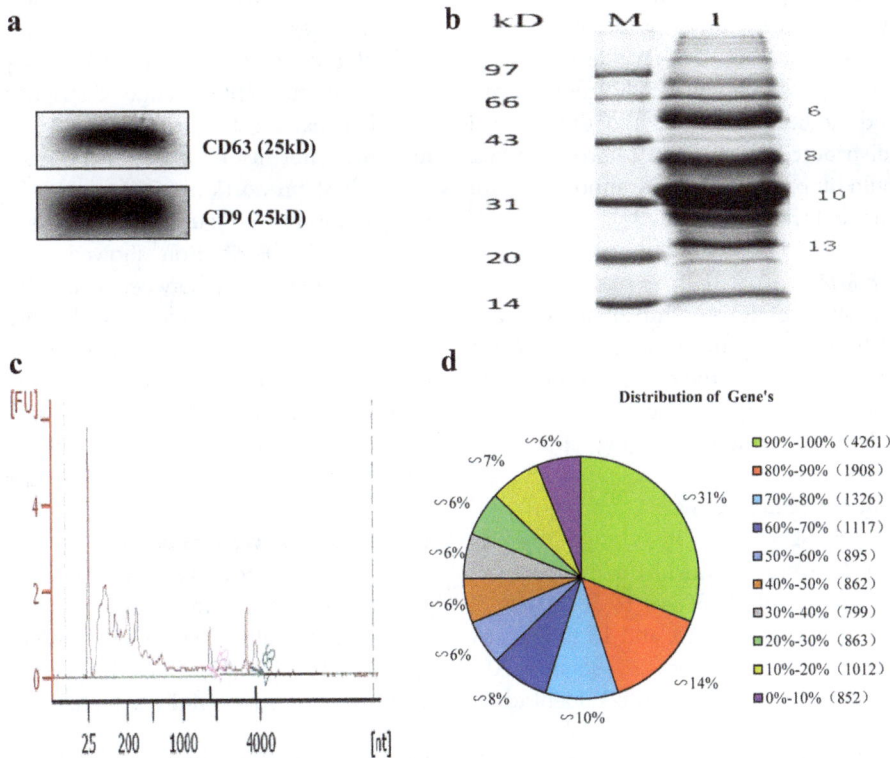

Fig. 1 Identification of proteins and mRNAs in porcine milk exosomes. **a** detection of the exosomal marker proteins CD63 and CD9 by Western blotting. **b** SDS-PAGE. **c** RNA sample analyzed by the Agilent Bioanalyzer 2100. **d** distribution of genen's coverage

previous studies [4, 6, 17, 20]. Porcine milk proteins were extracted using RIPA lysis buffer and resolved using SDS-PAGE (Fig. 1b), which proteins covered a large molecular weight range, but most of them were fell into the 20–25, 28–35, 35–40 and 43–55 kDa ranges, and these ranges were considered separately.

Transcript sequencing and analysis
Transcript sequencing
We totally obtained 77,106,888 raw reads, which mapped to porcine genome (**sscrofa10.2,** www.ensembl.org/Sus_-scrofa/). The mapped proportion was 63.76% accounting for 49,161,814 reads, and the perfect match reads were 33,863,808 (43.92%) and the unique match reads were 45,080,932 (58.47%). By blast searching the 77,106,888 reads against pig coding gene database (http://www.ncbi.nlm.nih.gov/), 57,413,016 total match reads (represented 74.46%) and 53,836,128 (69.82%) unique matched reads were identified (Table 2). All the reads represent 13,895 genes (Additional file 1), the subsequent distribution of genes' coverage analysis showed the number of genes' coverage >50% contained 9,507 genes and represented 69% of 13,895 genes, the 4,261 (representing 31%) genes coverage are 90%–100% (Fig. 1d).

Table 2 Alignment statistics of RNA-seq data map to reference genome and gene database

	Map to Genome		Map to Gene	
	Reads number	Percentage	Reads number(control)	Percentage
Total Reads	77,106,888	100.00%	77,106,888	100.00%
Total Base Pairs	6,939,619,920	100.00%	6,939,619,920	100.00%
Total Mapped Reads	49,161,814	63.76%	57,413,016	74.46%
perfect match	33,863,808	43.92%	45,400,239	58.88%
<=5 bp mismatch	15,298,006	19.84%	12,012,777	15.58%
unique match	45,080,932	58.47%	53,836,128	69.82%
multi-position match	4,080,882	5.29%	3,576,888	4.64%
Total Unmapped Reads	27,945,074	36.24%	19,693,872	25.54%

Novel mRNAs predicted in pig exosome milk

Then we performed a novel transcript prediction and annotation according to the criteria described in Method. Results showed we obtained 2,409 novel transcripts (Fig. 2a and Additional file 2), and those novel transcripts were distributed in all the 19 chromosomes. These results would improve the gene annotations of the porcine genome and transcriptome [31].

qPCR identified for mRNAs

After a series of analysis of RNA sequencing, we randomly selected 14 transcripts genes from the top 50 list (Additional file 3) for evaluated their expression in the porcine milk exosomes by qPCR. The results showed that they were all detected in the sample (Fig. 2b).

Proteome sequencing and data analysis

Following separation by SDS-PAGE, in-gel digestion was performed and peptides were analyzed by mass spectroscopy. The four groups of P130340_6, P130340_8, P130340_10 and P130340_13 (6, 8, 10, and 13 in Fig. 1b) were corresponding to 43–55, 35–40, 28–35 and 20–25 kDa, respectively, which were treated identically, since they displayed a relatively high gray density in the gel. With a false discovery rate (FDR) setting ≤1.2%, 307,390 total spectras were detected, which only 18,638 spectras could be mapped using the Mascot software, and 2,313 peptides represent 639 proteins were ultimately identified from the sample (*Sus_scrofa*, Table 3 and Additional file 4), and which number of protein matched with a given quality match check criterion with at least possessing one unique peptide can be considered as a reliable protein. Of these, 571 proteins were present in the *Sscrofa* 10.2 database, 66 were novel candidate proteins and two were putative proteins (Additional file 4).

Most of the novel proteins (44) and the two putative proteins were not highly abundant, whereas most of high abundance proteins were known proteins. Analysis of protein and peptide length distribution after digestion revealed that most were between 8 and 54 amino acids, and the majorities were between 9 and 25 residues, with the highest proportion (12%) comprising 13 amino acids (Additional file 5: Figure S1). Analysis of the peptide and spectrogram distribution showed that lots of proteins were represented by between 1 and 10 unique peptides, and one unique peptide was the predominantly case (Additional file 6: Figure S2). In the sequence coverage range of 0% to 20%, 473 proteins were identified (77.02%, Additional file 7: Figure S3e), and the sequence coverage was increased as the number of identified proteins decreased (Additional file 7: Figure S3a, b, c, d, e).

Identification by Western blotting

Based on the above results, we randomly selected 10 proteins to confirm their presence in porcine milk exosomes. Specifically, EGF, TGFβ-3, MSTN, CTGF, IGFBP-7, PDGFA, HTRA3, THBS1, β-actin and lactoferrin (LTF) were all successfully detected (Fig. 3).

COG annotation of mRNAs and proteins

The Cluster of Orthologous Groups of proteins (COG) database was used for protein orthologous classification, and all proteins in this database are assumed to be derived from a common protein ancestor. COG analysis showed that proteins from porcine milk exosomes were connected with multiple biological processes (Fig. 4 and Additional file 8). Interestingly, proteins involved in DNA or RNA synthesis and transport particularly abundant. Furthermore, five proteins were related to intracellular trafficking, secretion, and vesicular transport, with some in the high

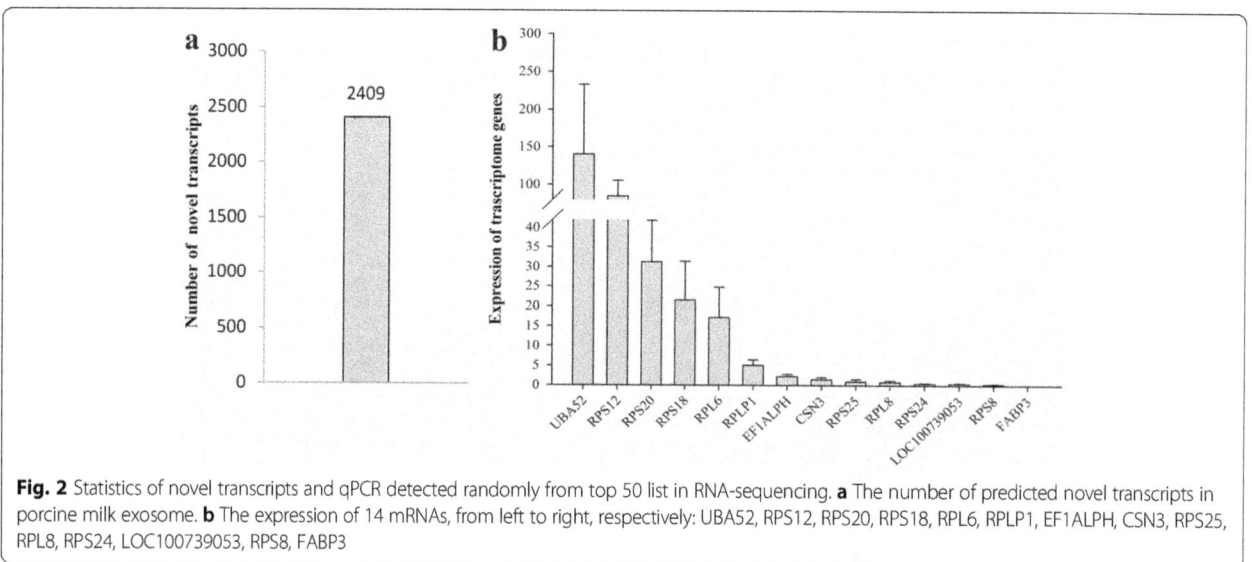

Fig. 2 Statistics of novel transcripts and qPCR detected randomly from top 50 list in RNA-sequencing. **a** The number of predicted novel transcripts in porcine milk exosome. **b** The expression of 14 mRNAs, from left to right, respectively: UBA52, RPS12, RPS20, RPS18, RPL6, RPLP1, EF1ALPH, CSN3, RPS25, RPL8, RPS24, LOC100739053, RPS8, FABP3

Table 3 Proteins identified in this study

Sample	Total spectra	Identified spectra	Identified peptides	Identified proteins	FDR (%)	Unknown protein	Putative protein
P130340_10	13,430	672 [5.0037%]	92	40	1.05	4	
P130340_13	17,281	790 [4.5715%]	108	44	0.90	2	
P130340_6	17,783	721 [4.0544%]	303	94	1.13	12	
P130340_8	15,140	596 [3.9366%]	180	59	1.03	4	
Sus_scrofa	307,390	18,638 [6.0633%]	2,313	639	1.04	66	2

abundance P130340_13 (Additional file 9: Figure S4b) and P130340_8 groups (Additional file 9: Figure S4d). Additionally, 10 conserved proteins were involved in cell cycle control, cell division and chromosome partitioning. Similarly, enriched 6 COG Ontology in mRNAs, including 31 genes related to intracellular trafficking and secretion and 18 mRNAs of Cell division and chromosome partitioning / Cytoskeleton (Fig. 5 and Additional file 10).

Go analysis of mRNAs and proteins

GO annotation was performed using DAVID version 6.7 (http://david.abcc.ncifcrf.gov) with a standard Benjamini < 0.05. We selected the top 10 GO terms of Cellular Component (CC), Molecular Function (MF) and Biological Process (BP) for further analysis. For mRNA, cytoplasm genes account for a high proportion (6.3%), and specific intracellular organelle lumen, nuclear lumen genes account for ~1.9%. Predicted functions included various bindings (including adenyl ribonucleotide, magnesium ion, nuclear hormone receptor and protein kinase) and diverse enzymatic activity (including protein kinase, pyrophosphate, transcription coactivator, exonuclease, small conjugating protein ligase and NADH dehydrogenase), predicted biological processes relative to proteins (include protein metabolic, transport, modification and catabolic process) and RNA (including RNA metabolic, processing and ncRNA processing) (Table 4

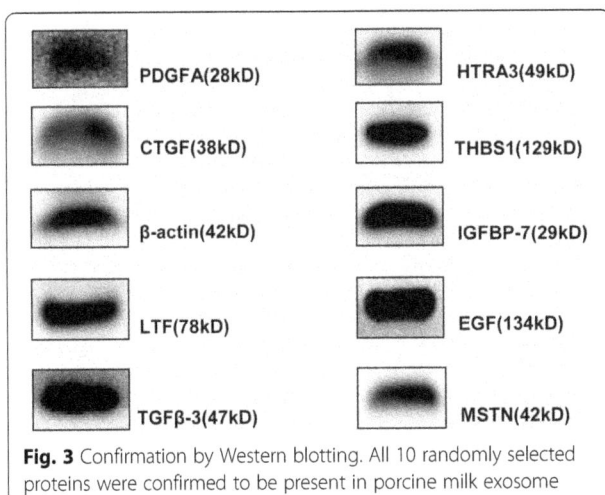

Fig. 3 Confirmation by Western blotting. All 10 randomly selected proteins were confirmed to be present in porcine milk exosome

and Additional file 10). For proteins, most of them were included in cytoplasm and cytoplasmic part, taking a proportion of 7.1%. Additionally, there were lots of specific membrane-bounded vesicle lumen, granule lumen, vesicle, lytic vacuole and reticulum lumen proteins. And major of those proteins were enriched in the molecular function in terms of diverse activity and predicted biological processes, including acute inflammatory response, complement activation, classical pathway, B cell mediated immunity, negative regulation of blood coagulation and coagulation, activation of immune response and protein maturation and processing (Table 5 and Additional file 5).

Tissues-specific analysis of mRNAs and proteins

All the known mRNAs and proteins were performed tissues-specific analysis. The results of mRNA analysis showed 8,605 of 13,895 genes were associated with 100 tissues, and were significantly correlated (Benjamini < 0.05) with 50 tissues. According to gene number, the top 5 ranking tissues were brain (3,987 genes), placenta (1,872 genes), epithelium (1,595 genes), lung (1,426 genes) and liver (1,110 genes) (Table 6 and Additional file 10). However, all the proteins were correlated with 33 tissues, and significantly correlated (Benjamini < 0.05) with only 19 tissues, including the components closely relative tissues of milk, such as plasma, blood, milk and mammary gland. More interestingly, the top five enriched tissues were liver (138 proteins), placenta (128 proteins), skin (75 proteins), lung (74 proteins) and plasma (73 proteins), and the highly correlated tissues were plasma, liver and milk (Table 7 and Additional file 8). These results suggest that mRNAs and proteins in porcine milk exosomes may have originated from multiple tissues.

KEGG pathway analysis of mRNAs and proteins

Due to the incomplete porcine bioinformatics resources in software DAVID [30], we selected the human database as reference. For mRNA, only 8,605 of 13,895 genes were enriched in 63 KEGG pathways, and the top 20 pathways were involved in various substance metabolisms, degradation, signaling pathway and some diseases pathways. Interestingly, we got 83 genes in cell cycle pathways (Fig. 6a and Additional file 10). For proteins,

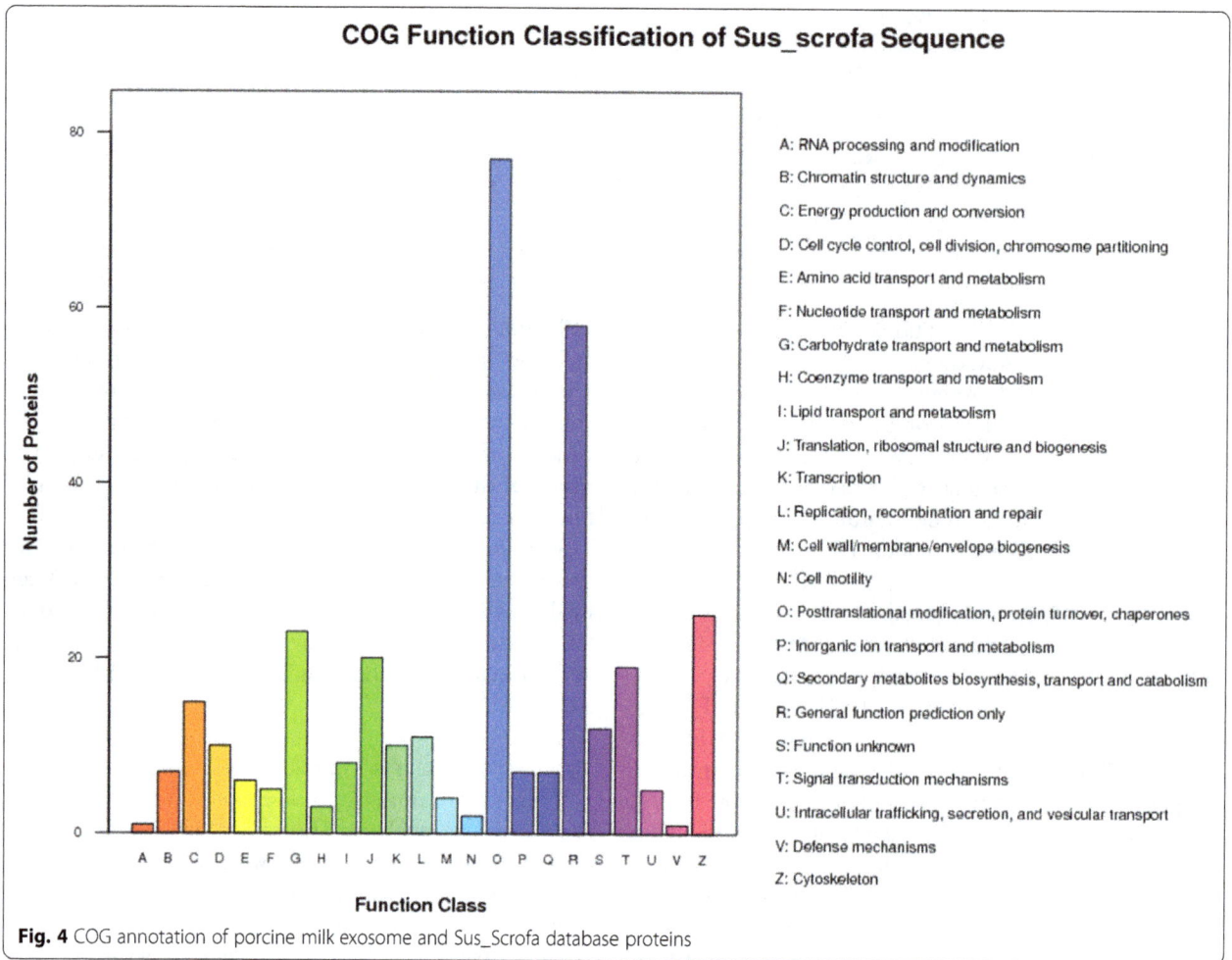

Fig. 4 COG annotation of porcine milk exosome and Sus_Scrofa database proteins

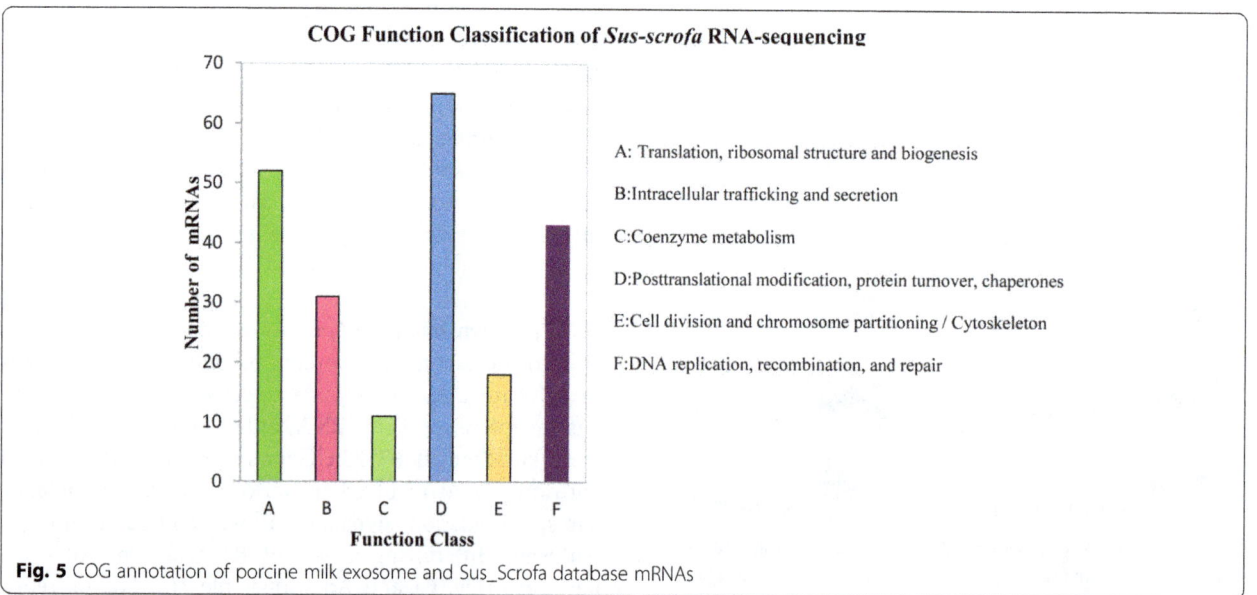

Fig. 5 COG annotation of porcine milk exosome and Sus_Scrofa database mRNAs

Table 4 GO annotation of identified mRNAs

Category	Term	Count	%	Benjamini
GOTERM_BP_5	GO:0044267 ~ cellular protein metabolic process	1,481	1.30	2.054E-42
GOTERM_BP_5	GO:0015031 ~ protein transport	531	0.47	3.262E-27
GOTERM_BP_5	GO:0016070 ~ RNA metabolic process	633	0.56	5.403E-27
GOTERM_BP_5	GO:0006396 ~ RNA processing	388	0.34	1.149E-21
GOTERM_BP_5	GO:0006464 ~ protein modification process	888	0.78	8.933E-18
GOTERM_BP_5	GO:0006886 ~ intracellular protein transport	271	0.24	1.149E-16
GOTERM_BP_5	GO:0030163 ~ protein catabolic process	404	0.36	3.782E-12
GOTERM_BP_5	GO:0043632 ~ modification-dependent macromolecule catabolic process	374	0.33	1.875E-11
GOTERM_BP_5	GO:0044257 ~ cellular protein catabolic process	390	0.34	2.34E-11
GOTERM_BP_5	GO:0034470 ~ ncRNA processing	141	0.12	4.795E-10
GOTERM_CC_5	GO:0005737 ~ cytoplasm	4,253	3.74	1.54E-126
GOTERM_CC_5	GO:0043231 ~ intracellular membrane-bounded organelle	4,485	3.94	5.61E-99
GOTERM_CC_5	GO:0044444 ~ cytoplasmic part	2,924	2.57	9.8E-86
GOTERM_CC_5	GO:0044446 ~ intracellular organelle part	2,530	2.22	5.426E-71
GOTERM_CC_5	GO:0070013 ~ intracellular organelle lumen	1,200	1.06	3.933E-66
GOTERM_CC_5	GO:0044428 ~ nuclear part	1,205	1.06	6.521E-59
GOTERM_CC_5	GO:0031981 ~ nuclear lumen	979	0.86	5.556E-53
GOTERM_CC_5	GO:0005829 ~ cytosol	879	0.77	2.958E-41
GOTERM_CC_5	GO:0005634 ~ nucleus	2,801	2.46	1.904E-33
GOTERM_CC_5	GO:0005654 ~ nucleoplasm	599	0.53	2.982E-32
GOTERM_MF_5	GO:0032559 ~ adenyl ribonucleotide binding	892	0.78	1.822E-20
GOTERM_MF_5	GO:0004672 ~ protein kinase activity	387	0.34	1.894E-13
GOTERM_MF_5	GO:0000287 ~ magnesium ion binding	294	0.26	2.071E-11
GOTERM_MF_5	GO:0016462 ~ pyrophosphatase activity	448	0.39	1.642E-08
GOTERM_MF_5	GO:0035257 ~ nuclear hormone receptor binding	64	0.06	1.128E-07
GOTERM_MF_5	GO:0003713 ~ transcription coactivator activity	147	0.13	1.259E-07
GOTERM_MF_5	GO:0004527 ~ exonuclease activity	45	0.04	1.407E-05
GOTERM_MF_5	GO:0019787 ~ small conjugating protein ligase activity	108	0.09	0.0010606
GOTERM_MF_5	GO:0019901 ~ protein kinase binding	95	0.08	0.0050729
GOTERM_MF_5	GO:0050136 ~ NADH dehydrogenase (quinone) activity	34	0.03	0.0085249

the results showed that only 426 of the 571 proteins (known proteins) were found to be enriched using KEGG pathway in the database. These 426 proteins were enriched in 20 pathways (Fig. 6b and Additional file 8), most associated with immunity and diseases.

Discussions

In the present study, we totally obtained 13,895 known genes and 2,409 putative novel genes in porcine milk exosomes. It was reported 10,948 mRNA transcripts in rats whey [21] and 19,320 transcripts in bovine milk whey exosome by mRNA microarray. Moreover, in human milk, 14,070 transcripts were found in fat globules [37]. Some of milk protein genes (CSN2, CSN3 and CSN1S1), ribosome-related proteins genes (RPS18, RPL18 and RPLP1) and other genes (e.g UBA52, FABP3

and EEF1A1) were highly expressed in the previous researches [21, 37], which were in accordance with this study (Additional file 3). Furthermore, some genes such as LALBA, TPT1, SPP1 and FASN were not found in rats whey [21], bovine milk whey exosome and human milk fat globules [37]. Additionally, the randomly selected 14 mRNAs among top 50 were further confirmed using qRT-PCR. Differences of mRNAs in milk or milk exosome exist among species, possibly indicating different functions of milk among species.

One of the aims in the present study was to explore the protein content of porcine milk exosomes using proteomics. In the present study, we observed 639 proteins (Fig. 1b) including the exosomal marker proteins CD9 [38] and CD63 [39] using Western blotting (Fig. 1a), as well as the heat shock protein family members HSPA 90

Table 5 GO annotation of identified proteins

Category	Term	Count	%	Benjamini
GOTERM_BP_5	GO:0002526 ~ acute inflammatory response	28	0.47	1.7E-17
GOTERM_BP_5	GO:0006956 ~ complement activation	15	0.25	1.06E-09
GOTERM_BP_5	GO:0016485 ~ protein processing	20	0.33	1.93E-08
GOTERM_BP_5	GO:0030193 ~ regulation of blood coagulation	13	0.22	2.19E-08
GOTERM_BP_5	GO:0006958 ~ complement activation, classical pathway	12	0.20	2.2E-08
GOTERM_BP_5	GO:0030195 ~ negative regulation of blood coagulation	11	0.18	2.31E-08
GOTERM_BP_5	GO:0051604 ~ protein maturation	21	0.35	2.32E-08
GOTERM_BP_5	GO:0050819 ~ negative regulation of coagulation	11	0.18	8.73E-08
GOTERM_BP_5	GO:0019724 ~ B cell mediated immunity	14	0.23	2.36E-07
GOTERM_BP_5	GO:0002253 ~ activation of immune response	16	0.27	2.8E-06
GOTERM_CC_5	GO:0044444 ~ cytoplasmic part	190	3.16	4.07E-16
GOTERM_CC_5	GO:0005737 ~ cytoplasm	242	4.03	1.58E-14
GOTERM_CC_5	GO:0060205 ~ cytoplasmic membrane-bounded vesicle lumen	17	0.28	8.44E-14
GOTERM_CC_5	GO:0031093 ~ platelet alpha granule lumen	16	0.27	4.9E-13
GOTERM_CC_5	GO:0016023 ~ cytoplasmic membrane-bounded vesicle	42	0.70	1.43E-09
GOTERM_CC_5	GO:0031410 ~ cytoplasmic vesicle	44	0.73	1.16E-08
GOTERM_CC_5	GO:0030141 ~ secretory granule	22	0.37	3.97E-08
GOTERM_CC_5	GO:0048770 ~ pigment granule	16	0.27	5.49E-08
GOTERM_CC_5	GO:0000323 ~ lytic vacuole	23	0.38	9.38E-08
GOTERM_CC_5	GO:0005788 ~ endoplasmic reticulum lumen	15	0.25	1.04E-07
GOTERM_MF_5	GO:0004867 ~ serine-type endopeptidase inhibitor activity	19	0.32	5.63E-09
GOTERM_MF_5	GO:0004252 ~ serine-type endopeptidase activity	20	0.33	2.61E-06
GOTERM_MF_5	GO:0008236 ~ serine-type peptidase activity	21	0.35	3.68E-06
GOTERM_MF_5	GO:0004175 ~ endopeptidase activity	31	0.52	3.8E-06
GOTERM_MF_5	GO:0008201 ~ heparin binding	15	0.25	2.18E-05
GOTERM_MF_5	GO:0005509 ~ calcium ion binding	51	0.85	2.98E-05
GOTERM_MF_5	GO:0004869 ~ cysteine-type endopeptidase inhibitor activity	7	0.12	0.010385
GOTERM_MF_5	GO:0051920 ~ peroxiredoxin activity	4	0.07	0.02113

B1, HSPA13, HSPA5, HSPA 9, HSPB1 and HSPCB (Additional file 4), which have been reported in previous exosome research [7, 8, 16]. The Western blotting results in our study confirmed that we successfully isolated porcine milk exosomes. Previous study using iTRAQ identified 2,971 milk proteins with 2,350 from exosomes, 1,012 from MFGM, and 748 from whey (FDR = 0.1%) [30], another study found 2,107 proteins in bovine milk exosomes, including the major exosomal marker proteins lactadherin/MFGE8 and TSG 101 (FDR = 0.05% for proteins and 0.2% for peptides) [29], the actin family members ACTC1, ACTN1, ACTN2, and ACTN4 are cell-specific proteins likely involved in exosome biogenesis and potentially other exosome functions [7], which were also present in porcine milk exosomes. However, xanthine oxidase (~147 kDa), Butyrophilin (~59 kDa), lactadherin/MGF8 (~47 kDa) and adipophilin/perilipin-2 (~49 kDa) and MFGM were identified in bovine milk

exosomes [29, 40], which were not detected in porcine milk exosomes. It has also been suggested that exosomes from different sources might contain different components [16] and may play tissue-specific roles in intracellular communication and immune function [41–43].

In COG ontology analysis of mRNAs and proteins, we obtained 10 conserved proteins and 18 mRNAs relative to cell cycle. Additionally, many genes and proteins involved in cell cycle and immunity related pathways by KEGG pathways analysis. Then, we randomly selected 10 proteins for Western blotting analysis. Platelet-derived growth factor (PDGF) acts as a potential binding pattern mitogen for mesenchymal cells both in vitro and in vivo [44]. Epidermal growth factor (EGF) plays an important role in regulating cell proliferation and differentiation during development [45]. Thrombospondin1 (THBS1), cysteine-rich protein 61 (Cyr61) and connective tissue growth factor (CTGF) were all involved in the

Table 6 mRNAs expressed in a tissue-specific manner

Category	Term	Count	%	Benjamini
UP_TISSUE	Epithelium	1,595	1.40	1.77E-68
UP_TISSUE	Placenta	1,872	1.65	2.06E-44
UP_TISSUE	Skin	1,079	0.95	1.66E-38
UP_TISSUE	Uterus	1,021	0.90	4.31E-35
UP_TISSUE	Brain	3,987	3.51	5.68E-34
UP_TISSUE	Lung	1,426	1.25	1.43E-28
UP_TISSUE	Cervix carcinoma	255	0.22	3.19E-19
UP_TISSUE	Muscle	492	0.43	7.94E-18
UP_TISSUE	Liver	1,110	0.98	9.72E-16
UP_TISSUE	Lymph	400	0.35	1.43E-13
UP_TISSUE	Fetal brain cortex	153	0.13	2.22E-13
UP_TISSUE	Eye	582	0.51	4.51E-13
UP_TISSUE	Platelet	338	0.30	4.58E-13
UP_TISSUE	Bone marrow	431	0.38	4.91E-13
UP_TISSUE	Cervix	308	0.27	7.3E-13
UP_TISSUE	Cajal-Retzius cell	137	0.12	3.21E-11
UP_TISSUE	Colon	642	0.56	1.35E-10
UP_TISSUE	Urinary bladder	137	0.12	1.06E-09
UP_TISSUE	B-cell lymphoma	93	0.08	7.21E-09
UP_TISSUE	Colon carcinoma	124	0.11	7.39E-09
UP_TISSUE	T-cell	199	0.17	2.83E-08
UP_TISSUE	Umbilical cord blood	197	0.17	5.14E-08
UP_TISSUE	Mammary gland	240	0.21	6.05E-08
UP_TISSUE	Ovary	452	0.40	4.35E-07
UP_TISSUE	Kidney	765	0.67	1.81E-06
UP_TISSUE	Pancreas	519	0.46	4.43E-06
UP_TISSUE	B-cell	160	0.14	6.34E-05
UP_TISSUE	Teratocarcinoma	297	0.26	0.000105
UP_TISSUE	Adipose tissue	108	0.09	0.000142
UP_TISSUE	Human small intestine	51	0.04	0.001389
UP_TISSUE	Melanoma	176	0.15	0.001846
UP_TISSUE	Leukemia	46	0.04	0.001979
UP_TISSUE	Keratinocyte	80	0.07	0.003529
UP_TISSUE	Ovarian carcinoma	90	0.08	0.00707
UP_TISSUE	Fetal kidney	106	0.09	0.007677
UP_TISSUE	Renal cell carcinoma	42	0.04	0.008965
UP_TISSUE	Umbilical vein	27	0.02	0.008996
UP_TISSUE	Pituitary	86	0.08	0.010397
UP_TISSUE	Bone	42	0.04	0.014
UP_TISSUE	Skeletal muscle	295	0.26	0.014765
UP_TISSUE	Endometrium carcinoma cell line	19	0.02	0.020668
UP_TISSUE	Embryonic kidney	48	0.04	0.021149
UP_TISSUE	Bladder	48	0.04	0.021149
UP_TISSUE	Hepatoma	128	0.11	0.021469

Table 6 mRNAs expressed in a tissue-specific manner (Continued)

Category	Term	Count	%	Benjamini
UP_TISSUE	Embryo	183	0.16	0.026251
UP_TISSUE	Mammary carcinoma	59	0.05	0.033882
UP_TISSUE	Breast	53	0.05	0.036028
UP_TISSUE	Hypothalamus	73	0.06	0.036155
UP_TISSUE	Dendritic cell	49	0.04	0.041269
UP_TISSUE	Fetal brain	391	0.34	0.041334
UP_TISSUE	Carcinoma	33	0.03	0.050043

transforming growth factor-beta (TGF-β) signaling pathway [46]. High-temperature requirement A3 (HtrA3) inhibits BMP-4, BMP-2 and TGF-β1 signaling [47]. Lactoferrin (LTF) functions in inflammation [48]. Myostatin (MSTN) was a negative regulator of myogenesis and has been implicated in the regulation of adiposity and controlling the structure and function of tendons [49]. IGFBP-7 acts through autocrine/paracrine pathways to inhibit BRAF-MEK-ERK signaling and induces senescence and apoptosis in cells containing the BRAF oncogene [50]. Additionally, IGFBP-7 inhibits cell growth and induces apoptosis in RKO and SW620 cells [51]. Confirmation of the presence of these 10 proteins in porcine milk exosomes suggests a possible function in the regulation of immunity, cell proliferation and possibly other pathways.

Table 7 Proteins expressed in a tissue-specific manner

Category	Term	Count	%	Benjamini
UP_TISSUE	Plasma	73	1.22	3.36E-52
UP_TISSUE	Liver	138	2.30	3.01E-31
UP_TISSUE	Milk	17	0.28	2.57E-16
UP_TISSUE	Fetal brain cortex	32	0.53	6.22E-15
UP_TISSUE	Cajal-Retzius cell	30	0.50	2.98E-14
UP_TISSUE	Bile	11	0.18	6.73E-10
UP_TISSUE	Placenta	128	2.13	8.25E-09
UP_TISSUE	Urine	10	0.17	1.45E-07
UP_TISSUE	Saliva	12	0.20	6.37E-07
UP_TISSUE	Skin	75	1.25	5.48E-06
UP_TISSUE	Platelet	34	0.57	7.73E-06
UP_TISSUE	Neutrophil	7	0.12	0.000105
UP_TISSUE	Pancreas	44	0.73	0.000382
UP_TISSUE	Mammary gland	24	0.40	0.000715
UP_TISSUE	B-cell lymphoma	12	0.20	0.002124
UP_TISSUE	Blood	29	0.48	0.021296
UP_TISSUE	Adipose tissue	12	0.20	0.024482
UP_TISSUE	Colon	43	0.72	0.025496
UP_TISSUE	Ovary	32	0.53	0.047235

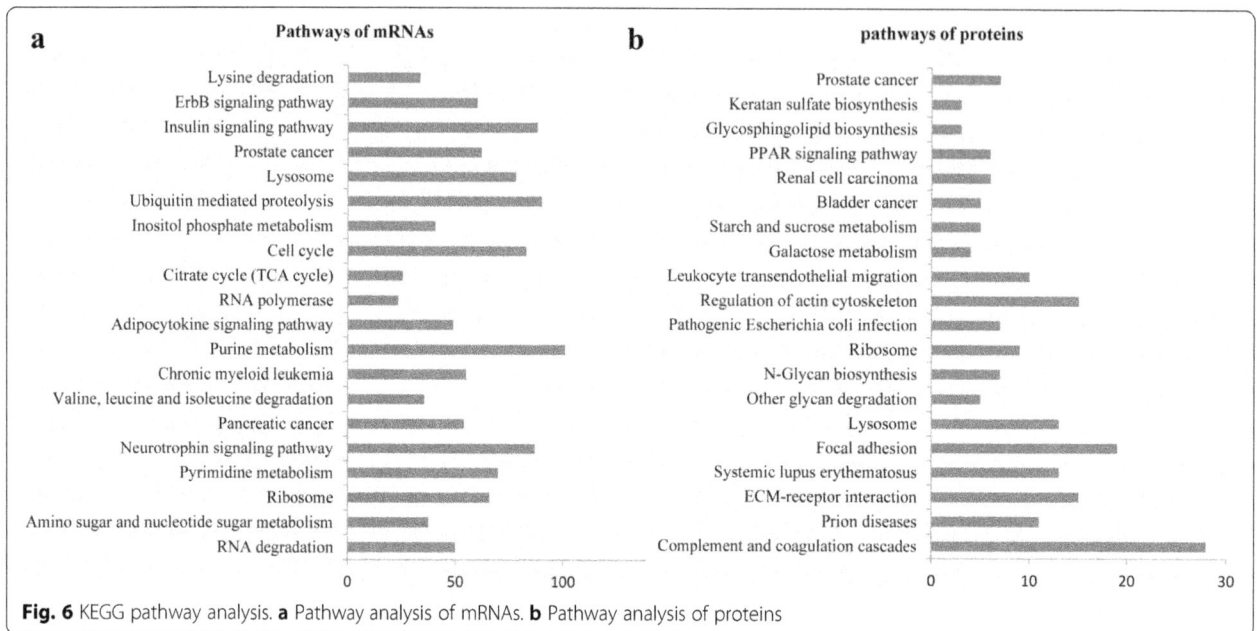

Fig. 6 KEGG pathway analysis. **a** Pathway analysis of mRNAs. **b** Pathway analysis of proteins

All previously reported exosomal proteins were cytosolic, and many of them were associated with the plasma membrane or membranes of endocytic compartments [7]. Most of the genes and proteins identified in the present study were relatived to cytoplasm or cytoplasmic GO terms (Tables 4 and 5). Analysis of GO, KEGG and COG annotations suggested that most porcine milk exosome genes and proteins might function in activation, immunity and cell cycle. KEGG analysis revealed that four pathways (ECM-receptor interaction, Focal adhesion, Regulation of actin cytoskeleton and Leukocyte transendothelial migration) were enriched in both bovine [29] and porcine milk exosomes (Fig. 6b, Additional file 8). Above results indicate a similar function in different species. Additionally, recent reports showed that the bovine milk exosomes were able to exert endocytosis and transferred their contained molecules to other cells [52]. In this study, proteins in porcine milk exosome were predicated to be involved in pathways of starch and sucrose metabolism, other glycan degradation, N-Glycan biosynthesis, galactose metabolism and glycosphingolipid biosynthesis (Fig. 6b, Additional file 8), it was deduced that porcine milk exosomes might transfer encapsulated materials, which could mediated by those proteins and played key roles in different physiological and pathological conditions. Meanwhile, the KEGG analysis of mRNAs showed lots of genes enriched in Purine metabolism, Pyrimidine metabolism, Insulin signaling pathway, Cell cycle and RNA degradation pathways, which were different to predicated pathways in the KEGG analysis of porcine milk exosomes proteins.

It is reported the viral RNA (hepatitis C virus) was able to transfer to infected cells (plasmacytoid dendritic cells)

and trigger an innate immune response, depending on membrane vesicle trafficking [53]. The glioblastoma cells derived-exosome could deliver a specific mRNA transcript to endothelia cells followed by generating functional proteins for patients [54]. When incubated with NIH-3 T3 cells, milk-derived microvesicles could transfer bovine milk related transcripts to living cells and affect the calf's gastrointestinal development and immune systems [6]. Additionally, recent reports showed that bovine milk exosomes can be uptaken by endocytosis, depending on cell exosome surface glycoproteins [52].The uptaking exosome further affected gene expression [55]. And exosome can also be incorporated into differentiated human cells with containing RNA. These data collectively indicate the exosomes could not only deliver the encapsulated miRNAs, mRNAs and proteins to recipient cells, but also make their specific functions on immunity, thereafter play a key role in different physiological and pathological conditions. Our results provided extensively mRNAs and proteins data, which are beneficial to understand how milk regulates health and development of newborns by exosomes.

Conclusions

In this study, we identified 16,304 mRNAs and 639 proteins in porcine milk exosomes by RNA-sequencing and proteomic analysis, and many of mRNAs and proteins were predicted to be involved in immunity, proliferation and cellular signaling, which would be closely associated with piglets development and healthy. These findings provided a large amount of informations and contributed to increased understanding of the role of genes and proteins in milk exosomes, and build a foundation for future studies on their physiological functions and regulatory mechanisms.

Additional files

Additional file 1: mRNA expression.

Additional file 2: Novel transcripts.

Additional file 3: Different top 50 mRNAs expression.

Additional file 4: The 639 identified proteins analysis.

Additional file 5: Figure S1. Peptide length distribution of identified proteins. a, b, c, d, and e represent the peptide length distribution of 10, 13, 6, 8, and Sus_Scrofa proteins, respectively.

Additional file 6: Figure S2. Peptide and spectrogram distribution of identified proteins. a, b, c, d, and e represent the peptide and spectrogram distribution of 10, 13, 6, 8, and Sus_Scrofa proteins, respectively.

Additional file 7: Figure S3. Distribution of protein sequences coverage. a, b, c, d, and e represent 10, 13, 6, 8, and Sus_Scrofa, respectively.

Additional file 8: Proteins bioinformatics analysis.

Additional file 9: Figure S4. COG annotation of identified proteins. a, b, c, and d represent 10, 13, 6, and 8, respectively.

Additional file 10: mRNAs bioinformatics analysis.xlsx.

Abbreviations

ACN: Acetonitrile; ACTC1: Alpha-cardiac actin one; ACTN2: Alpha-actin two; ACTN4: Alpha-actin four; AGC: Automatic gain control; BCA: Bicinchoninic acid; BMP-2: Bone morphogenetic protein 2; BMP-4: Bone morphogenetic protein 4; BP: Biological Process; CC: Cellular Component; CD24: Tetraspan-integrin complexes 24; CD63: Tetraspan-integrin complexes 63; CD9: Tetraspan-integrin complexes 24; COG: Cluster of Orthologous Groups; CSN1S1: Casein alpha s1; CSN2: Beta-casein; CSN3: κ-casein; CTGF: Connective tissue growth factor; Cyr61: Cysteine-rich protein 61; DTT: DL-Dithiothreitol; EEF1A1: Elongation factor 1 alpha-1; EGF: Epidermal growth factor; FABP3: Fatty-acid binding protein 3; FASN: Fatty acid synthase; FDR: False discovery rate; GO: Gene Ontology; HCD: High-energy collision dissociation; HRP: Horseradish Peroxidase; Hsp70: 70-kDa heat-shock protein; HSPA 9: Heat shock 70kD protein 9; HSPA 90 B1: heat shock protein 90 kDa beta,member 1; HSPA13: Heat shock protein 70 kDa family, member 13; HSPA5: Heat shock 70 kDa protein 5; HSPB1: Heat shock 27 kDa protein 1; HSPCB: Heat shock 90kD protein beta; HTRA3: High-temperature requirement A3; IGFBP-7: Insulin-like growth factor binding protein 7; iTRAQ: Isobaric tag for relative and absolute quantitation; KEGG: Kyoto Encyclopedia of Genes and Genomes; LALBA: Lactalbumin alpha; LC-MS/MS: Liquid chromatography with tandem mass spectrometric; LTF: Lactoferrin; MF: Molecular Function; MFG: Milk fat globules; MFGE8: Milk fat globule-EGF factor 8; MFGM: Milk fat globule membrane; MGF: Mammary gland factor; MHC-class II: Major histocompatibility complex class II; miRNAs: MicroRNAs; mRNAs: Messager RNA; MSTN: Myostatin; NADH: Nicotinamide adenine dinucleotide; ncRNA: Noncoding RNA; PDGF: Platelet-derived growth factor; PDGFA: Platelet-derived growth factor A chain; PMSF: Phenylmethanesulfonyl fluoride; QC: Quality control; RIPA: Radio-immunoprecipitation assay; RNA-Seq: RNA-sequencing; RPL18: 60S ribosomal subunit protein L18; RPLP1: Ribosomal phosphoprotein P1; RPS18: Ribosomal protein S18; rRNA: Ribosomal RNA; SPP1: Secreted phosphoprotein 1; TGFB-3: Transforming growth factor-beta 3; TGF-β1: Transforming growth factorβ1; THBS1: Thrombospondin1; TPT1: Translationally controlled tumor protein 1; TSG 101: Tumor susceptibility gene 101; UBA52: Ubiquitin A-52 ribosomal protein

Acknowledgements
None.

Funding
This work was supported by grants from the Key Project of Guangdong Provincial Nature Science Foundation (S2013020012766), National Nature Science Foundation of China, The National Key Research and Development Program of China (2016YFD0500503) National Basic Research Program of China (973 Program, 2013CB127304), Hainan Provincial research institutes technology development key Project of China (grant number KYYS-2015-02) and Open subject of Hainan Provincial Key Laboratory of Tropical Animal Breeding and Disease Research (HNXMSYS201501). We thank the breeding farm of the Livestock Research Institute (Guangzhou, China) for providing milk samples.

Authors' contrib
TC, QX, JS carried out the proteins and mRNA sequencing and data analysis, and participated in drafted the manuscript. XC, RS carried out the Western blotting and q-PCR. GS, SW participated in the sample collected. XZ, NW performed the raw data analysis. RY, QJ and YZ conceived of the study, and participated in its design and coordination and helped to draft the manuscript. All authors read and approved the final manuscript.

Competing interests
The authors declare that they have no competing interests.

Consent for publication
Not applicable.

Author details
[1]National Engineering Research Center For Breeding Swine Industry, Guandong Provincial Key Laboratory of Agro-Animal Genomics and Molecular Breeding, Guandong Province Research Center of Woody Forage Engineering and Technology, South China Agricultural University, 483 Wushan Road, Guangzhou 510642, China. [2]Institute of Animal Science and Veterinary Medicine, Hainan Academy of Agricultural Sciences, Haikou 571100, China.

References

1. Gartner LM, Morton J, Lawrence RA, Naylor AJ, O'Hare D, Schanler RJ, Eidelman AI. Breastfeeding and the use of human milk. Pediatrics. 2005; 115(2):496–506.
2. Armogida SA, Yannaras NM, Melton AL, Srivastava MD. Identification and quantification of innate immune system mediators in human breast milk. In: Allergy and asthma proceedings: 2004: OceanSide publications, inc, vol. 2004. p. 297–304.
3. Kramer MS, Chalmers B, Hodnett ED, Sevkovskaya Z, Dzikovich I, Shapiro S, Collet J-P, Vanilovich I, Mezen I, Ducruet T. Promotion of breastfeeding intervention trial (PROBIT): a randomized trial in the Republic of Belarus. JAMA. 2001;285(4):413–20.
4. Kosaka N, Izumi H, Sekine K, Ochiya T. microRNA as a new immune-regulatory agent in breast milk. Silence. 2010;1(1):7.
5. Høst A, Koletzko B, Dreborg S, Muraro A, Wahn U, Aggett P, Bresson J, Hernell O, Lafeber H, Michaelsen K. Dietary products used in infants for treatment and prevention of food allergy. Joint statement of the European Society for Paediatric Allergology and Clinical Immunology (ESPACI) Committee on hypoallergenic formulas and the European Society for Paediatric Gastroenterology, Hepatology and nutrition (ESPGHAN) Committee on nutrition. Arch Dis Child. 1999;81(1):80–4.
6. Hata T, Murakami K, Nakatani H, Yamamoto Y, Matsuda T, Aoki N. Isolation of bovine milk-derived microvesicles carrying mRNAs and microRNAs. Biochem Biophys Res Commun. 2010;396(2):528–33.
7. Théry C, Zitvogel L, Amigorena S. Exosomes: composition, biogenesis and function. Nat Rev Immunol. 2002;2(8):569–79.
8. Théry C, Regnault A, Garin J, Wolfers J, Zitvogel L, Ricciardi-Castagnoli P, Raposo G, Amigorena S. Molecular characterization of dendritic cell-derived exosomes selective accumulation of the heat shock protein hsc73. J Cell Biol. 1999;147(3):599–610.
9. Raposo G, Nijman HW, Stoorvogel W, Liejendekker R, Harding CV, Melief C, Geuze HJ. B lymphocytes secrete antigen-presenting vesicles. J Exp Med. 1996;183(3):1161–72.
10. Blanchard N, Lankar D, Faure F, Regnault A, Dumont C, Raposo G, Hivroz C. TCR activation of human T cells induces the production of exosomes bearing the TCR/CD3/ζ complex. J Immunol. 2002;168(7):3235–41.
11. Raposo G, Tenza D, Mecheri S, Peronet R, Bonnerot C, Desaymard C. Accumulation of major histocompatibility complex class II molecules in mast cell secretory granules and their release upon degranulation. Mol Biol Cell. 1997;8(12):2631–45.

12. Wolfers J, Lozier A, Raposo G, Regnault A, Théry C, Masurier C, Flament C, Pouzieux S, Faure F, Tursz T. Tumor-derived exosomes are a source of shared tumor rejection antigens for CTL cross-priming. Nat Med. 2001;7(3):297–303.

13. Van Niel G, Raposo G, Candalh C, Boussac M, Hershberg R. Cerf–Bensussan N, Heyman M: intestinal epithelial cells secrete exosome–like vesicles. Gastroenterology. 2001;121(2):337–49.

14. Caby M-P, Lankar D, Vincendeau-Scherrer C, Raposo G, Bonnerot C. Exosomal-like vesicles are present in human blood plasma. Int Immunol. 2005;17(7):879–87.

15. Ogawa Y, Kanai-Azuma M, Akimoto Y, Kawakami H, Yanoshita R. Exosome-like vesicles with dipeptidyl peptidase IV in human saliva. Biol Pharm Bull. 2008;31(6):1059–62.

16. Admyre C, Johansson SM, Qazi KR, Filén J-J, Lahesmaa R, Norman M, Neve EP, Scheynius A, Gabrielsson S. Exosomes with immune modulatory features are present in human breast milk. J Immunol. 2007;179(3):1969–78.

17. Chen T, Xi Q-Y, Ye R-S, Cheng X, Qi Q-E, Wang S-B, Shu G, Wang L-N, Zhu X-T, Jiang Q-Y. Exploration of microRNAs in porcine milk exosomes. BMC Genomics. 2014;15(1):100.

18. Zhou Q, Li M, Wang X, Li Q, Wang T, Zhu Q, Zhou X, Wang X, Gao X, Li X. Immune-related microRNAs are abundant in breast milk exosomes. Int J Biol Sci. 2012;8(1):118.

19. Chen X, Gao C, Li H, Huang L, Sun Q, Dong Y, Tian C, Gao S, Dong H, Guan D. Identification and characterization of microRNAs in raw milk during different periods of lactation, commercial fluid, and powdered milk products. Cell Res. 2010;20(10):1128–37.

20. Gu Y, Li M, Wang T, Liang Y, Zhong Z, Wang X, Zhou Q, Chen L, Lang Q, He Z. Lactation-related MicroRNA expression profiles of porcine breast milk Exosomes. PLoS One. 2012;7(8):e43691.

21. Izumi H, Kosaka N, Shimizu T, Sekine K, Ochiya T, Takase M. Time-dependent expression profiles of microRNAs and mRNAs in rat milk whey. PLoS One. 2014; 9(2):e88843.

22. Izumi H, Kosaka N, Shimizu T, Sekine K, Ochiya T, Takase M. Bovine milk contains microRNA and messenger RNA that are stable under degradative conditions. J Dairy Sci. 2012;95(9):4831–41.

23. Izumi H, Kosaka N, Shimizu T, Sekine K, Ochiya T, Takase M. Purification of RNA from milk whey. In: Circulating MicroRNAs edn: Springer; 2013. p. 191–201.

24. Izumi H, Tsuda M, Sato Y, Kosaka N, Ochiya T, Iwamoto H, Namba K, Takeda Y. Bovine milk exosomes contain microRNA and mRNA and are taken up by human macrophages. J Dairy Sci. 2015;98(5):2920–33.

25. Lässer C, Alikhani VS, Ekström K, Eldh M, Paredes PT, Bossios A, Sjöstrand M, Gabrielsson S, Lötvall J, Valadi H. Human saliva, plasma and breast milk exosomes contain RNA: uptake by macrophages. J Transl Med. 2011;9(1):9.

26. Campanella C, Bavisotto CC, Gammazza AM, Nikolic D, Rappa F, David S, Cappello F, Bucchieri F, Fais S: Exosomal heat shock proteins as new players in tumour cell-to-cell communication. 2014.

27. Keller S, Ridinger J, Rupp A-K, Janssen JW, Altevogt P. Body fluid derived exosomes as a novel template for clinical diagnostics. J Transl Med. 2011;9(86):240.

28. Torregrosa Paredes P, Gutzeit C, Johansson S, Admyre C, Stenius F, Alm J, Scheynius A, Gabrielsson S. Differences in exosome populations in human breast milk in relation to allergic sensitization and lifestyle. Allergy. 2014;69(4):463–71.

29. Reinhardt TA, Lippolis JD, Nonnecke BJ, Sacco RE. Bovine milk exosome proteome. J Proteome. 2012;75(5):1486–92.

30. Reinhardt TA, Sacco RE, Nonnecke BJ, Lippolis JD. Bovine milk proteome: quantitative changes in normal milk exosomes, milk fat globule membranes and whey proteomes resulting from Staphylococcus aureus mastitis. J Proteome. 2013; 82:141–54.

31. Chen C, Ai H, Ren J, Li W, Li P, Qiao R, Ouyang J, Yang M, Ma J, Huang L. A global view of porcine transcriptome in three tissues from a full-sib pair with extreme phenotypes in growth and fat deposition by paired-end RNA sequencing. BMC Genomics. 2011;12(1):448.

32. Li R, Yu C, Li Y, Lam T-W, Yiu S-M, Kristiansen K, Wang J. SOAP2: an improved ultrafast tool for short read alignment. Bioinformatics. 2009;25(15):1966–7.

33. Mortazavi A, Williams BA, McCue K, Schaeffer L, Wold B. Mapping and quantifying mammalian transcriptomes by RNA-Seq. Nat Methods. 2008;5(7):621–8.

34. Ye R-S, Xi Q-Y, Qi Q, Cheng X, Chen T, Li H, Kallon S, Shu G, Wang S-B, Jiang Q-Y. Differentially expressed miRNAs after GnRH treatment and their potential roles in FSH regulation in porcine anterior pituitary cell. PLoS One. 2013;8(2):e57156.

35. Stoscheck CM. [6] Quantitation of protein. Methods Enzymol. 1990;182:50–68.

36. Théry C, Boussac M, Véron P, Ricciardi-Castagnoli P, Raposo G, Garin J, Amigorena S. Proteomic analysis of dendritic cell-derived exosomes: a secreted subcellular compartment distinct from apoptotic vesicles. J Immunol. 2001;166(12):7309–18.

37. Maningat PD, Sen P, Rijnkels M, Sunehag AL, Hadsell DL, Bray M, Haymond MW. Gene expression in the human mammary epithelium during lactation: the milk fat globule transcriptome. Physiol Genomics. 2009;37(1):12–22.

38. Guescini M, Genedani S, Stocchi V, Agnati LF. Astrocytes and Glioblastoma cells release exosomes carrying mtDNA. J Neural Transm. 2010;117(1):1–4.

39. Mathivanan S, Simpson RJ. ExoCarta: a compendium of exosomal proteins and RNA. Proteomics. 2009;9(21):4997–5000.

40. Mather IH. A review and proposed nomenclature for major proteins of the milk-fat globule membrane. J Dairy Sci. 2000;83(2):203–47.

41. Théry C, Ostrowski M, Segura E. Membrane vesicles as conveyors of immune responses. Nat Rev Immunol. 2009;9(8):581–93.

42. Mathivanan S, Ji H, Simpson RJ. Exosomes: extracellular organelles important in intercellular communication. J Proteome. 2010;73(10):1907–20.

43. Bobrie A, Colombo M, Raposo G, Théry C. Exosome secretion: molecular mechanisms and roles in immune responses. Traffic. 2011;12(12):1659–68.

44. Gilbertson DG, Duff ME, West JW, Kelly JD, Sheppard PO, Hofstrand PD, Gao Z, Shoemaker K, Bukowski TR, Moore M. Platelet-derived growth factor C (PDGF-C), a novel growth factor that binds to PDGF α and β receptor. J Biol Chem. 2001;276(29):27406–14.

45. Riese DJ, Stern DF. Specificity within the EGF family/ErbB receptor family signaling network. BioEssays. 1998;20(1):41–8.

46. Zhou Z-Q, Cao W-H, Xie J-J, Lin J, Shen Z-Y, Zhang Q-Y, Shen J-H, Xu L-Y, Li E-M. Expression and prognostic significance of THBS1, Cyr61 and CTGF in esophageal squamous cell carcinoma. BMC Cancer. 2009;9(1):291.

47. Tocharus J, Tsuchiya A, Kajikawa M, Ueta Y, Oka C, Kawaichi M. Developmentally regulated expression of mouse HtrA3 and its role as an inhibitor of TGF-β signaling. Develop Growth Differ. 2004;46(3):257–74.

48. Britigan BE, Serody JS, Cohen MS. The role of lactoferrin as an anti-inflammatory molecule. Lactoferrin edn: Springer; 1994. p. 143–56.

49. Hickford J, Forrest R, Zhou H, Fang Q, Han J, Frampton C, Horrell A. Polymorphisms in the ovine myostatin gene (MSTN) and their association with growth and carcass traits in New Zealand Romney sheep. Anim Genet. 2010;41(1):64–72.

50. Wajapeyee N, Serra RW, Zhu X, Mahalingam M, Green MR. Oncogenic BRAF induces senescence and apoptosis through pathways mediated by the secreted protein IGFBP7. Cell. 2008;132(3):363–74.

51. Ruan W, Xu E, Xu F, Ma Y, Deng H, Huang Q, Lv B, Hu H, Lin J, Cui J. IGFBP7 plays a potential tumor suppressor role in colorectal carcinogenesis. Cancer Biol Ther. 2007;6(3):354–9.

52. Wolf T, Baier SR, Zempleni J: The Intestinal Transport of Bovine Milk Exosomes Is Mediated by Endocytosis in Human Colon Carcinoma Caco-2 Cells and Rat Small Intestinal IEC-6 Cells. J Nutr 2015:jn218586.

53. Dreux M, Garaigorta U, Boyd B, Décembre E, Chung J, Whitten-Bauer C, Wieland S, Chisari FV. Short-range exosomal transfer of viral RNA from infected cells to plasmacytoid dendritic cells triggers innate immunity. Cell Host Microbe. 2012;12(4):558–70.

54. Skog J, Würdinger T, van Rijn S, Meijer DH, Gainche L, Curry WT, Carter BS, Krichevsky AM, Breakefield XO. Glioblastoma microvesicles transport RNA and proteins that promote tumour growth and provide diagnostic biomarkers. Nat Cell Biol. 2008;10(12):1470–6.

55. Lonnerdal B, Du X, Liao Y, Li J. Human milk exosomes resist digestion in vitro and are internalized by human intestinal cells. FASEB J. 2015;29(1 Supplement):121.123.

4

Botanical ethnoveterinary therapies used by agro-pastoralists of Fafan zone, Eastern Ethiopia

Teka Feyera[1], Endalkachew Mekonnen[2], Befekadu Urga Wakayo[1] and Solomon Assefa[3*]

Abstract

Background: In Ethiopia, plant based remedies are still the most important and sometimes the only source of therapeutics in the management of livestock diseases. However, documentation of this indigenous knowledge of therapeutic system still remains at a minimum level. The aim of this study was, thus, to document the traditional knowledge of botanical ethnoveterinary therapies in the agro-pastoral communities of Fafan Zone, Eastern Ethiopia.

Methods: The study employed a cross-sectional participatory survey. Purposive sampling technique was applied to select key respondents with desired knowledge in traditional animal health care system. Data were gathered from a total of 24 (22 males and 2 females) ethnoveterinary practitioners and herbalists using an in-depth-interview complemented with group discussion and field observation.

Results: The current ethnobotanical survey indicated that botanical ethnoveterinary therapies are the mainstay of livestock health care system in the studied communities. A total of 49 medicinal plants belonging to 21 families, which are used by traditional healers and livestock raisers for the treatment of 29 types of livestock ailments/health problems, were identified in the study area. The major plant parts used were leaves (43%) followed by roots (35%). In most cases, traditional plant remedies were prepared by pounding the remedial plant part and mixing it with water at room temperature.

Conclusion: The various types of identified medicinal plants and their application in ethnoveternary practice of Fafan zone agro pastoralists indicate the depth of indigenous knowledge in ethnobotanical therapy. The identified medicinal plants could be potentially useful for future phytochemical and pharmacological studies.

Keywords: Ethnoveterinary, Medicinal plants, Livestock diseases, Fafan zone, Agro-pastoralist

Background

Livestock production is an integral part of the Ethiopian agricultural sector that approximately shares 40% of the national agricultural output [1]. Previously, it was reported that Ethiopia has the largest livestock population in Africa [2]. However, due to the prevailing animal diseases, the economic benefits gained from this sector still remain marginal. Animal diseases are among the principal causes of poor livestock performance and cause of high economic losses in the country [3, 4].

Conventional veterinary service is still less developed in the country, which is characterized by lack of adequate animal health infrastructure, veterinary clinics, and veterinarians. Furthermore, most modern drugs are expensive and not affordable to the majority of Ethiopian farmers and pastoralists [5, 6]. The majority of livestock raisers in Ethiopia are far away from the sites of animal clinic stations [7]. These factors make Ethiopian livestock raisers rely on endogenous ethnoveterinary knowledge and practices (mainly botanical products) for the management of diseases of their domestic animals. The traditional remedies are socially acceptable, inexpensive and locally available [8, 9].

However, very little of the ethnoveterinary knowledge of Ethiopian famers and pastoralists in relation to the

* Correspondence: Solomon.assefa@aau.edu.et
[3]Department of Pharmacology and Clinical Pharmacy, School of Pharmacy, Addis Ababa University, Addis Ababa, Ethiopia

use of medicinal plants is so far properly documented and analyzed [5, 6, 10]. It is estimated that up to 90% of current livestock diseases are managed through the use of traditional medicines [11]. WHO stated: the use of natural products in control of animal and human diseases are considerably effective [12].

In most scenarios, the traditional medical knowledge in Ethiopia is passed verbally from generation to generation. In addition, valuable information can be lost whenever a traditional medical practitioner passes without conveying his/her knowledge on traditional medicinal plants. Similarly, ethnoveterinary practice in the country is being affected by acculturation and depletion of plants as a result of population pressure, drought, environmental degradation, deforestation and over exploitation of the medicinal plants [13, 14]. Consequently, there is a pressing need to document medicinal plants used and the associated indigenous knowledge by conducting ethnobotanical studies [15, 16].

Compared to the multiethnic cultural diversity and the diverse flora of Ethiopia, the studies conducted on the traditional ethnoveterinary medicinal plants in Ethiopia are very limited [17]. In recent years, few ethnoveterinary surveys have been conducted in different areas of the country [10, 17–28]. As it is factual throughout the country, in Ethiopian Somali Regional State (ESRS), ethnoveterinary knowledge is believed to be rich and worth documenting. However, there is gap of information on the level, scope, role and limitations of plant based remedies in the traditional animal healthcare system. Thus, this ethnobotanical survey was initiated in view of documenting the indigenous knowledge associated with utilization of botanical ethnoveterinary therapies for the management of livestock ailments among the agro-pastoralist communities of Fafan Zone, Eastern Ethiopia.

Methods
Study area
The study area covers the Babile district and part of Jigjiga district, found in Fafan zone of ESRS (Fig. 1). The zone is situated in the northern part of ESRS. The total land coverage of the zone is 40, 861 km², of which the rangeland extends over 36, 629 km². About 52.6%, 31% and 7% of the landscape of the zone can be categorized as flat to gentle slopes, hills and steep slope, respectively. Fafan zone comprises pastoralism, agro-pastoralism and sedentary production systems. Agropastoralism (95%) is the dominant production system in the zone [29].

The zone geographically lies between 8° 44′ N to 11° 00′ N latitude and 40° 22′ E to 44° 00′ E longitude. The altitude of the zone ranges from 500 to 1650 m above sea level. The mean minimum and maximum temperature ranges from 16 to 20 °C and

28–38 °C, respectively [30]. The rainfall distribution in the zone is very erratic with a mean annual rainfall of 600 to 700 mm [31].

Study design
A cross-sectional, participatory study was employed to collect ethnoveterinary information from traditional healers in Fafan zone of ESRS between April, 2014 and August, 2015. Indigenous ethno-botanical knowledge, resources and their applications were the main study parameters.

Sampling procedure
A purposive snowball sampling technique was used to select study participants i.e. ethnopractitioners. This approach aids in acquiring the desired quality and quantity of information on traditional animal health care systems [32]. Ultimately, a total sample of 24 (22 males and 2 female) key respondents were selected.

Ethnobotanical data collection
Ethnobotanical data were mainly gathered through repeated field trips and investigations, with individual interviews, group discussion, and field observations using the same format used by [33] and [34]. Participant interviews were conducted using semi-structured questionnaires prepared in English and administered in local language (*Somali*) with the help of competent local translators. Data collected comprise: indications, local name, parts used in traditional remedies, mode of preparation (dosage), and route of administration of each medicinal plant against livestock diseases. Moreover, manner of indigenous knowledge transfer was recorded.

Plant specimen collection and identification
Ensuing interviews with selected key respondents, a field trip was arranged to identify and collect specimen of reported indigenous medicinal plants from their natural vegetation for further botanical identification. Botanical identification of plant specimens was conducted using herbarium materials and taxonomic keys described in various volumes on the Flora of Ethiopia [35, 36]. For each plant species, voucher specimens were given a collection number and deposited in the National Herbarium, Addis Ababa University.

Enumeration of documented plants
A list of plants and plant products traditionally used to manage animal health problems in the agro-pastoralist communities of Fafan zone was documented. The documentation compiled their scientific and vernacular names, family names, disease and ill-health conditions treated, target type of livestock and the preparation forms of

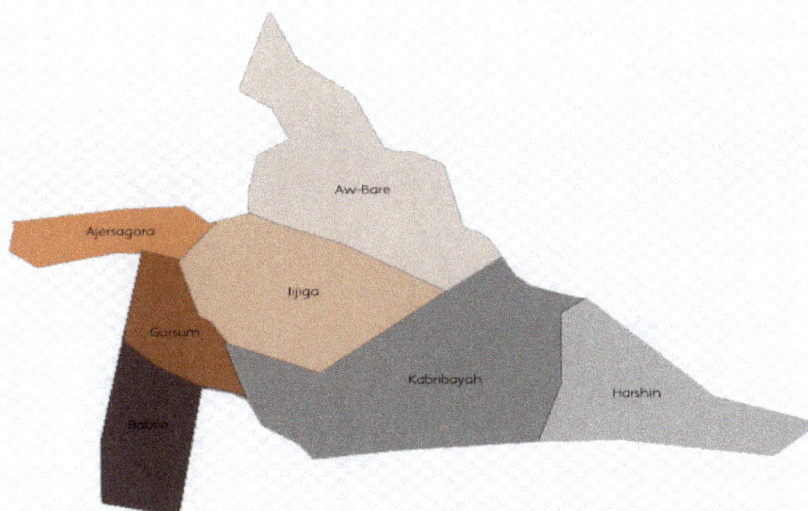

Fig. 1 Map of the study area. © User: AlaskaLava / Wikimedia Commons / https://commons.wikimedia.org/wiki/File:Fafan_Zone.png#filelinks / CC-BY-SA-4.0

different remedies (Table 2). The names of plants were arranged according to their alphabetical order.

Data analysis
Microsoft Excel spreadsheet software was employed for organizing and analyzing the collected ethnobotanical data. Descriptive statistical methods (percentage and frequency) were used to summarize data on reported medicinal plants and associated indigenous knowledge.

Results and discussion
Socio-demographic characteristics and experience of ethnoveterinary practitioners
Majority of the ethnoveterinary practitioners surveyed in Fafan zone were rural residents and males. Other studies have similarly shown that practice of Traditional Medicine in Ethiopia is largely dominated by men [25, 37]. Majority of the participants have been practicing ethnoveterinary medicine for ≥10 years. Ethno-veterinary knowledge of the traditional healers was usually obtained from family members or religious institutions (Islamic madrasas) which are passed through generation with word of mouth (Table 1). The way traditional veterinary medicine is acquired by the practitioners is largely similar to traditional human medicine. The traditional healers claimed that there is a considerable overlap in the utilization of some of the reported herbs against both human and livestock diseases. It was also interesting to note that most of the sampled ethnoveterinary practitioners were also traditional healers for several human ailments.

Table 1 Socio-demographic features and ethnoveterinary experiences of participants (n = 24)

Characteristics	Category level	Frequency	Percentage (%)
Sex	Male	22	91
	Female	2	9
Age	25–40	3	12
	41–55	9	38
	56–70	12	50
Residence	Rural	21	88
	Urban	3	12
Educational status	Formal	5	21
	Religious	18	75
	Illiterate	1	4
Level of ethnoveterinary practice experience (years)	< 10	2	9
	10–20	6	25
	21–30	10	41
	>30	6	25
Source of ethnoveterinary healing knowledge	Religious institution	7	29
	Family members or decedents	11	46
	Close friends and colleagues	4	16
	Other senior traditional healers	2	9
Mode of ethnoveterinary service delivery	Always charging	3	12
	Sometimes charging	12	50
	Free (not charging)	9	38

Table 2 List of traditional medicinal plants used to treat different livestock ailments among the agro-pastoralist communities of Fafan Zone

Scientific name	Family	Vernacular name	Part (s) used	Indication	Method of preparation and application	Livestock species treated	Voucher number
Abutilon anglosomaliae Cufod.	Malvaceae	Balanbaal	Leaf	Non-specific external wound	Grounded leaves are applied to wound and washed later	All Livestock	TF-05
Abutilon bidentatum Hochst. ex A.Rich.	Malvaceae	Maran	Root	Hyena/Jackal bite wound	Crushed root is applied to affected area	Cattle	TF-25
			Leaf	Helminthiasis, Abdominal pain andSnake bite	Decoction drenched orally	Cattle, sheep and goat	
Acacia mellifera (Vahl) Benth.	Mimosaceae	Bilcin	Bark and Root	Retained placenta	Crushed root and bark concocted with Acacia oerfota root is administered vaginally to clean uterus	Camel	TF-06
			Bark	Infertility	Bark placed in vagina to kill semen from previous unsuccessful mating	Cattle	
Acacia oerfota (Forssk.) Schweinf.	Mimosaceae	Gumar	Bark	Infertility	Bark placed in vagina to kill semen from previous unsuccessful mating	Cattle	TF-34
				Sudden sickness	Bark crushed, mixed with water and drenched orally	Camel	
Acacia tortilis (Forssk.) Galasso&Banfi	Mimosaceae	Madheedh	Gum	Non-specific external wound	Gum is applied to wound topically	All Livestock	TF-39
Adenium aculeatum (Forsk)	Apocynaceae	Dhalaandhux	Stem/Root	Ringworm	Crushed root or stem dispersed in water is applied to lesions	Cattle	TF-20
			Stem/Root	Coughing/Pasteurellosis	Decocted and drenched orally	Goat and Sheep	
Adenium obesum (Forssk) Roem. & Schult.	Apocynaceae	Aboobo wan Aad, Aboobo-gunweyn	Stem	Mange infestation	Inside of the stem which has been fermented for two days is applied to mange lesions	Camel	TF-37
Boscia minimifolia Chiov.	Capparaceae	Meygaag	Bark and Leaf	Bloat	Crushed bark and leaf mixed with water is drenched orally	Cattle	TF-31
Carullum speciosa N.E.Br.	Asclepiadaceae	Udaabeys	Leaf/Stem	Ringworm	Leaves/stem juice is applied to lesions	Cattle	TF-17
			Leaf	Eye injury or infection	Powdered leaves mixed with oil is applied locally as ointment	Cattle, sheep and goat	
Catha edulis (Vahl) Forssk. ex Endl.	Celastraceae	Jaad, qat	Leaf	Helminthiasis/Diarrhoea	Crushed leaves mixed with water is used as oral drench or mixed with feed and fed	Sheep and goat	TF-28
Celosia polystachina	Amaranthaceae	laaleys	Leaf	Non-specific external wound	Crushed leaves mixed with oil is applied to wound	Cattle	TF-22
Cissus quadrangularis L.	Vitaceae	Gaad	Aerial part	Tick infestation and external wound	Crushed aerial part mixed with water is applied topically	Cattle and Camel	TF-02
			Leaf	Mastitis, Helminthiaisis and Leach infestation	Crushed leaf mixed with water is drenched orally	Cattle and camel	

Table 2 List of traditional medicinal plants used to treat different livestock ailments among the agro-pastoralist communities of Fafan Zone (*Continued*)

Scientific name	Family	Local name	Aerial part	Black leg	Decoction drenched orally	Cattle	TF-08
Cistanche phelypae L. Cout.	Orobanchaceae	Qoodho-dameer	Leaf and root	Trypanosomiasis	Chopped, mixed with water and drenched orally	Camel	TF-08
Commiphora erlangeriana Engl.	Bursuraceace	Dhunkaal	Bark	Tick infestation	Bark crushed, mixed with water, left overnight and used as wash	Cattle, camel, sheep and goat	TF-03
Commiphora erythrea (Ehrenb.) Engl.	Burseraceae	Xagar	Leaf/Gum	Mange infestation and ring worm	Cooked gum with animal's urine is applied to the lesion; Leaf and gum burnt and applied to lesion	Camel	TF-14
Commiphora ogadensis Chiov.	Burseraceae	Xagar-madow	Gum	Ringworm	Gum mixed with water is applied to the lesions	Cattle (Calf) and camel	TF-11
Commiphora serrulata Engl.	Burseraceae	Mukh	Leaf	Orf	Leaf concocted with *C.dtangularis* and mixed with animal urine is cooked and applied to the lesions	Sheep and goat	TF-38
Crabbea velutina S. Moore	Acanthaceae	Gheg-maanyo	Leaf	Hyena/Jackal wounds	Grounded leaves applied to wound and washed after three days	Donkey	TF-23
Crotalaria albicaulis Franch.	Fabaceae	Gabal-daye	Leaf	Trypanosomiasis	Leaf extracted withwater and concocted with leaf of*C.phelypaef* is drenched orally	Camel	TF-12
Cucumella kelleri (Cogn.) C.Jeffrey	Cucurbitaceae	Afgub, uneexo	Root	Infertility	Root is inserted into vagina with *Acacia oerfota* to attract bull	Camel	TF-40
Cucumis prophetarum **L.**	Cucurbitaceae	Qalfoon-idaad	Root	Infertility	Root inserted into vagina with *A.oerfota* to attract bull	Cattle and Camel	TF-26
			Fruit	Swellings	Fruit is made warm and bandaged to affected area	All livestock	
				Retained placenta	Crushed and used to wash uterus	Cattle, sheep and goat	
Cucumis pustulatus Hook. f.	Cucurbitaceae	Qalfoon	Fruit/Seed	Non-specific external wound	Fruit pulp and seed applied to wound	All Livestock	TF-41
Cyphostemma cyphopetalum (Fresen) Desc. ex Wild & R.B.Drumm.	Vitaceae	Carmo, carmo-waraaboz	Root	Non-specific external wound	Crushed root is applied topically as paste	Cattle, camel, sheep and goat	TF-49
Cyphostemma serpens (Hochst. ex A.Rich.) Desc.	Vitaceae	Carom	Root	Non-specific external wound	Powder of dried and crushed root is applied	All Livestock	TF-10
Dichrostachys cinerea Wight et Arn.	Mimosaceae	Warsamays	Stem	Hyena/Jackal bite wounds	Burned stem is applied to wound	All Livestock	TF-46
Echidnopsis dammaniana Sprenger	Asclepiadaceae	Riyo-dararis	Stem	Lice infestation and Snake bite	Crushed stem mixed with water is used as wash; Crushed and applied to affected area	Cattle (Calf)	TF-45
Entada leptostachya Harms	Mimosaceae	Gacma-dheere	Root	Coughing	Grounded root mixed with water is given intranasal; or mixed with feed and fed	Goat	TF-09
Euphorbia hirta L.	Euphorbiaceae	Caraba-nadh	Latex	Non-specific external wound	Latex/juice is applied to wound	All Livestock	TF-44

Table 2 List of traditional medicinal plants used to treat different livestock ailments among the agro-pastoralist communities of Fafan Zone *(Continued)*

Scientific name	Family	Local name	Part used	Ailment	Application	Livestock	Code
Euphorbia longispina Chiov.	Euphorbiaceae	Qabo	Latex	Non-specific external wound	Latex is applied to wound	All Livestock	TF-43
Euphorbia schizacantha Pax	Euphorbiaceae	Qabo-yare	Whole plant	Non-specific external wound	Whole plant crushed, dried and used as powder. Juice also applied to the affected area	Cattle and camel	TF-42
Indigofera amorphoides Jaub. & Spach	Fabaceae	Meydhax-dheere	Root	Tick and Lice infestation	Crushed (broken) root is applied to ticks/lice	Cattle, sheep and goat	TF-18
			Whole plant	Helminthiasis	Decoction drenched orally	Sheep and goat	
Ipomoea cicatricosae L.	Convolvulaceae	Weylo-wad	Root	Joint diseases	Crushed root is applied topically	Cattle	TF-48
Jatropha spicata Pax	Euphorbiaceae	Mawe	Root	Non-specific external wound	Crushed root is applied topically to wound	All livestock	TF-15
			Seed	Indigestion (impaction)	Seed decocted and drenched orally	Cattle, sheep and goat	
Justica generifolia	Acanthaceae	Buuxiso	Leaf	Non-specific external wound	Crushed leaves is applied to wound	Cattle	TF-32
Kleinia abyssinica (A.Rich.) A.Berger,	Asteraceae	Godor-cad	Rhizome	Sexual impotency	Fresh rhizome is given to bulls to enhance libido	Cattle	TF-35
Lycium shawii Roem. & Schult.	Convolvulaceae	Surad	Root	Non-specific external wound /thorns	Crushed root applied near to site of embedded thorns	Camel	TF-29
Moringa borziana Mattei	Moringaceae	Mawe	Root	Coughing	Crushed root mixed with boiled water is drenched orally	Sheep and goat	TF-21
Pergularia daemia (Forssk.) Chiov.	Asclepiadaceae	Gees-riyaad	Leaf	Non-specific external wound	Leaf juice is applied to affected area	Cattle	TF-16
Psilotrichum gnaphalobryum (Hochst) Schintz	Amaranthaceae	Booga-dhaye	Leaf	Non-specific external wound	Crushed leaves concocted with *Ipomoea cicatricose* is applied to wound	Donkey	TF-47
Pupalia lappcea L. Juss.	Amaranthaceae	Maro-boob, dhegmaanyo	Leaf, fruit or root	Retained placenta, painful joints and wound	Juice or paste is applied to lesion or affected area	Cattle, sheep and goat	TF-04
Salvadora persica L.	Salvadoraceae	Caday	Root	Non-specific external wound	Crushed root is applied topically	Cattle	TF-27
Sarcostemma andongense Hiern	Asclepiadaceae	Xangey-dhurwaa	Leaf	Snake bite	Leaf juice is applied orally	All livestock	TF-30
Schinus molle L.	Anacardiace	Mirmiri	Leaf	Tick infestation	Crushed leaves rubbed on to ticks	Cattle and sheep	TF-01
			Leaf	Eye injury/infection	Leaf Juice is applied topically	Cattle and sheep	
			Bark	Helminthiasis	Water extract of the bark is applied orally	Sheep and goat	
	Convulvolaceae	Nagadh	Whole plant		Crushed whole plant is applied topically		TF-33

Table 2 List of traditional medicinal plants used to treat different livestock ailments among the agro-pastoralist communities of Fafan Zone *(Continued)*

Seddera pedunculatae (Balf.f.) Verdc				Dermatophilosis (skin infection)		Cattle and camel	
Solanium dubium fresen	Solanaceae	UUruudhi, Xunboox	Fruit	Non-specific external wound	Fruit juice is applied topically	Camel	TF-36
Solanum incanum L.	Solanaceae	Waniiye, xunboox, kiriiri	Fruit/Leaf	Tick infestation	Fruit/leaf sap concocted with leaf of *Schinusmolle* is applied on tick infested area	Cattle and camel	TF-07
			Seed	infertility	Seed inserted into vagina to attract bull	Cattle	
			Leaf	Ring worm and swollen joints	Crushed parts extracted in water is applied locally	Cattle and camel	
			Fruit	Coughing/pneumonia/mastitis	Fruit sap is applied orally/nasally or locally	Goat	
Solanum jubae Bitter	Solanaceae	Kiriiri, xunboox	Seeds, fruit, and root	Joint disease and Snake bite	Powder of dried and crushed parts is applied topically to the affected area	Cattle	TF-24
Withnia somnifera (L.) Dunal	Solanaceae	Guryo-fan	Leaf	Urinary abnormalities	Leaf concocted with *Cissusquadrangularis* and drenched orally	Cattle and camel	TF-13
Zanthoxylum chalybeum Engl.	Rutaceae	Geed-dixri	Fruit	Helminthiaisis	Powder of Crushed fruit mixed with water is applied orally as drench	Sheep	TF-19

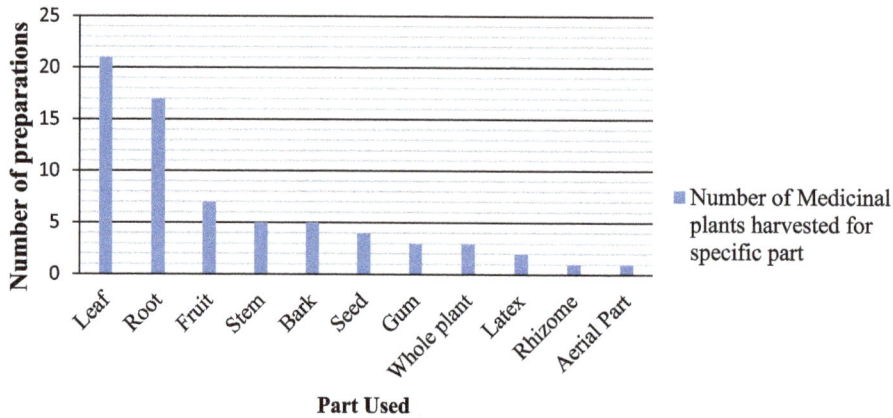
Fig. 2 Proportion of plant parts used for preparation of botanical remedies

Documented medicinal plants

The present study showed that the agro-pastoralist communities in Fafan Zone of ESRS use a variety of medicinal plant species to treat a range of livestock health problems. A total of 49 medicinal plants were reported for the treatment of different livestock ailments. The reported medicinal plants are botanically categorized under 21 plant families (Table 2).

Data from the present study showed that Mimosaceae (5 species), and Solanaceae, Bursuraceace, Asclepiadaceae and Euphorbiaceae (4 species each) took the superior share of the reported plant families, followed by Vitaceae, Amaranthaceae, Cucurbitaceous and Convulvolaceae (3 species each). In agreement with this study, Solanaceae, Bursuraceace and Cucurbitaceous have also been reported to be dominant families in other parts of the country [25, 38–40]. The fact that Solanaceae, Bursuraceace, Mimosaceae, Asclepiadaceous and Euphorbiaceae contributed relatively higher number of medicinal plants might be attributed to better abundance of species in the study area belonging to these families.

Parts used, mode of preparation and routes of administration

This study revealed that the most frequently used part of plants was leaf (43%) followed by root (35%) (Fig. 2). Other parts of the plant reported to be used were fruit (14%), stem (10%), bark (10%), seed, gum, latex, rhizome and aerial parts of the plants. Moreover, the entire plant was used in some cases (6%). In consonant with the present study, studies conducted elsewhere in Ethiopia indicated that leaves were the most frequently used plant part to treat livestock ailments [10, 22, 5, 20]. A study conducted by Poffenberger et al. [41] indicated that collection of leaves for traditional remedies poses no significant threat to the survival of plants in comparison with other parts; such as roots, stem, bark and whole plant. On contrary, harvest involving roots, rhizomes, bulb, bark and stem have a serious threat on the survival of the mother plant in its habitat. In this regard, the present study indicated that root was the second commonly utilized part of the medicinal plant, which shows the presence of high risk on the survival of those reported plants in the study area.

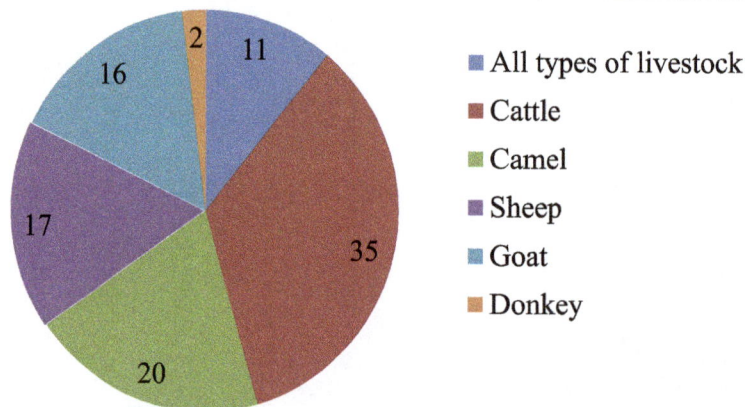
Fig. 3 Number of medicinal plants used in different livestock categories in Fafan zone, the area

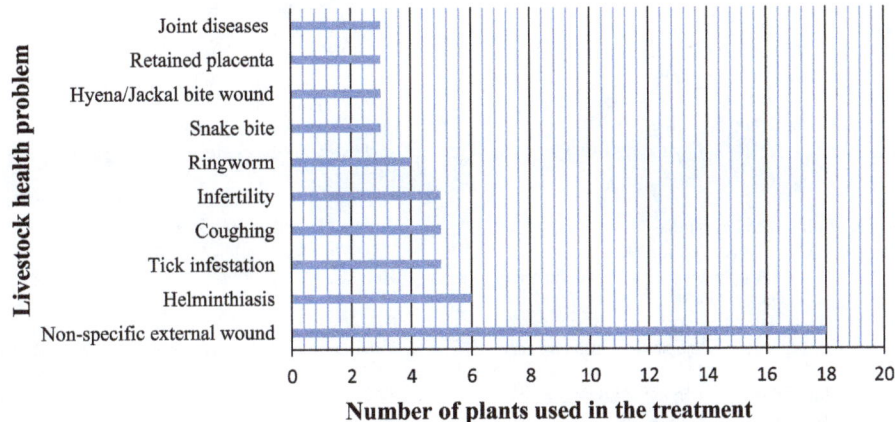

Fig. 4 Livestock health Problems against which three or more medicinal plants have been prescribed

In this study, majority (84%) of traditional remedies were prepared using a single medicinal plant. Single plant species based preparations also accounted for majority (65%) of traditional remedies in Afar [5]. However, single plant based preparations were reported at lower frequency from other parts of Ethiopia [22, 42].

In most cases, traditional plant remedies were prepared by pounding the remedial plant part and mixing it with water at room temperature. This is in line with the report of other studies [39, 40]. Some of the plants are prepared and administered in the form of topical route of administration without mixing using water. Topical applications of paste (poultice), sap, and other formulations were reported by other investigators to be common in traditional veterinary practice [18].

Types of livestock and major livestock health problems treated

The therapeutic indication of medicinal plant based remedies in Fafan zone covered all livestock species (Fig. 3) and around 29 distinct disease problems. Medicinal plant remedies were more frequently indicated for diseases affecting cattle and camels, followed by small ruminant and equine diseases. This variation is probably a reflection of the abundance and value of different livestock species in the study area rather than the therapeutic range of medicinal plants themselves.

Traditional medicinal plant remedies were prescribed against 29 different types of livestock ailments/health problems (Fig. 4). This study generally revealed that most of the traditional medicines used in the area are used for the management of skin diseases and removal of ecto-parasites. Unspecified wounds were reported to be the indication of majority of medicinal plants (18) (Fig. 4), followed by helminthiasis (6), tick infestation, respiratory disorders characterized by coughing and infertility (5). Out of the 29 animal health problems

reported to be treated by ethnobotanical remedies, 15 (51.7%) are treated by only one medicinal plant species.

Conclusions

The study suggests that the agro-pastoralist communities of the study area largely depend on ethnoveterinary medicinal plants for the treatment of different animal ailments. In total, 49 medicinal plants were reported to have been used by the ethnoveterinary practitioners and livestock raisers. Leaf followed by root was the most frequently used plant part in the preparation of ethnobotanical remedies. The identified medicinal plants could be potentially useful for future phytochemical and pharmacological studies. Thus, further studies on biological activity, phytoconstituents and safety profile of the reported medicinal plants is warranted.

Abbreviation
ESRS: Ethiopian Somali Regional State

Acknowledgments
The authors would like to thank the Directorate of Research, Publication and Technology Transfer- Jigjiga University for funding this research. The authors also highly acknowledge the contribution made by local administration and the study participants.

Funding
This study was financially supported by Jigjiga University.

Authors' contributions
TF conceived, designed and coordinated the study including the process of earning fund. EM and BUW participated in data collection, analysis and drafting the manuscript. SA finalized and submitted the manuscript for publication. All the authors revised and approved the final manuscript.

Consent for publication
Not applicable.

Competing interests
The authors declare that they have no competing interests.

Author details

[1]Department of Veterinary Clinical Studies, College of Veterinary Medicine, Jigjiga University, Jigjiga, Ethiopia. [2]Department of Basic Sciences, College of Medicine and Health Sciences, Jigjiga University, Jigjiga, Ethiopia. [3]Department of Pharmacology and Clinical Pharmacy, School of Pharmacy, Addis Ababa University, Addis Ababa, Ethiopia.

References

1. EARO. National Animal Health Research Programme Strategy. Addis Ababa: Ethiopian Agricultural Research Organization (EARO); 1999.
2. Abebe D, Ayehu A. Medicinal plants and enigmatic practices of northern Ethiopia. Addis Ababa: B. S. P. E; 1993. p. 1–200.
3. Wondimu T, Asfaw Z, Kelbessa E. Ethnobotanical study of medicinal plants around "Dheeraa" town Arsi zone, Ethiopia. J Ethnopharmacol. 2007;112:152–61.
4. Zerhiun W, Mesfin T. The status of the vegetation in the Lake region of the Rift Valley of Ethiopia and possibilities of its recovery. SINET: Ethiopian J of Sci. 1990;392:97–120.
5. Giday M, Teklehaymanot T. Ethnobotanical study of plants used in management of livestock health problems by afar people of Ada'ar district, afar regional state, Ethiopia. J Ethnobiol Ethnomed. 2013;9:8.
6. Yineger H, Yewhalaw D. Traditional medicinal plant knowledge and use by local healers in Sekoru District, Jimma zone, southwestern Ethiopia. J Ethnobiol Ethnomed. 2007;3:24. doi:10.1186/1746-4269-3-24.
7. Nasir T, Mohammed A, Nandikola J. Ethnobotanical survey of medicinal plants in the Southeast Ethiopia used in traditional medicine. Spatula. 2011;1:153–8.
8. Kibebew F, Zewdu M, Demissie A. The status of availability of data of oral and written knowledge and traditional health Care in Ethiopia. In: Conservation and sustainable use of medicinal plants in Ethiopia. Addis Ababa: Institute of Biodiversity Conservation and Research; 2001. p. 107–19.
9. Teklehaymanot T, Giday M. Ethnobotanical study of medicinal plants used by people in Zegie peninsula, northwestern Ethiopia. J Ethnobiol Ethnomed. 2007;3–12. doi:10.1186/1746-4269-3-12.
10. Giday M, Ameni G. An ethnobotanical survey on plants of veterinary importance in two woredas of southern Tigray, northern Ethiopia. SINET: Ethiopian J Sci. 2003;26:123–36.
11. Endashaw B. Study on actual situation of medicinal plants in Ethiopia. Prepared for JAICAF (Japan Association for International Collaboration of Agric and Forestry). 2007. http://www.jaicaf.or.jp/publications/ethiopia_ac.pdf. Accessed 17 Mar 2017.
12. Hoff W. Traditional practitioners as primary health care workers. Geneva: WHO; 1995. p. 141.
13. Gessese B. Review of land degradation and land management in Ethiopia up to 2008/09. In: Edwards S, editor. Ethiopian environment review; 2010. p. 187–214.
14. Berhan G, Dessie S. Medicinal plants in Bonga Forest and their uses. In Biodiversity newsletter I Addis Ababa:IBCR. 2002; 9-10.
15. Pankhurst R. The status and availability of oral and written knowledge on traditional health care in Ethiopia. In: Proceedings of the National Workshop on biodiversity conservation and sustainable use of medicinal plants in Ethiopia. Addis Ababa: IBCR; 2001. p. 92–106.
16. Hamilton AC. Medicinal plants and conservation: issues and approaches. UK: International plant conservation unit, WWF-UK, Pandahouse, Catteshall Lane; 2003.
17. Gidey A (2009). A study on traditional medicinal plants of Central Tigray. A senior essay submitted to the department of biology in partial fulfillment of the requirements for bachelor degree, Mekelle University, Ethiopia.
18. Sori T, Bekana M, Adunga G, Kelbesa E. Medicinal plants in ethnoveterinary practices of borana pastoralists, southern Ethiopia. Int J Appl Res Vet Med. 2004;2:220–5.
19. Yirga G. Assessment of indigenous knowledge of medicinal plants in central zone of Tigray, northern Ethiopia. African J Plant Sci. 2010;4(1):006–11.
20. Yirga G, Teferi M, Gidey G, Zerabruk S. An ethnoveterinary survey of medicinal plants used to treat livestock diseases in Seharti-Samre district, northern Ethiopia. African J Plant Sci. 2012;6(3):113–9.
21. Kebu B, Ensermu K, Zemede A. Indigenous medicinal utilization, management and threats in Fentale area, eastern Shewa, Ethiopia. Ethiop J Biol Sci. 2004;3(1):37–58.
22. Hunde D, Asfaw Z, Kelbessa E. Use and management of ethnoveterinary medicinal plants by indigenous people in 'Boosat', Welenchiti area. Ethiopian J Biol Sci. 2004;3:113–32.
23. Yineger H, Kelbessa E, Bekele T, Lulekal E. Ethnoveterinary medicinal plants in Bale Mountains National Park, Ethiopia. J Ethnopharmacol. 2007;112:55–70.
24. Yineger H, Yewhalaw D, Teketay D. Plants of veterinary importance in southwestern Ethiopia: the case of Gilgel Ghibe area. Forests Trees Livelihoods. 2008;18(2):165–81.
25. Yigezu Y, Berihun D, Yenet W. Ethnoveterinary medicines in four districts of Jimma zone, Ethiopia: cross sectional survey for plant species and mode of use. BMC Vet Res. 2014;10:76. doi:10.1186/1746-6148-10-76.
26. Tekle Y. Study on ethno veterinary practices in Amaro special district southern Ethiopia. Med Aromat Plants. 2015;4:186. doi:10.4172/2167-0412.1000186.
27. Melaku A. Ethnoveterinary practices and potential herbal materials for the treatment of ticks in North Gondar. Global Veterinaria. 2013;11(2):186–90.
28. Birhanu T, Gadisa M, Gurmesa F, Abda S. Survey on Ethno-Veterinary Medicinal Plants in Selected Woredas of East Wollega Zone, Western Ethiopia. J. Biol. Agric. Healthcare. 2014;4(17):97-105.
29. IPS (Industrial Project Service). Resource potential assessments and project identification study of Somali region. Vol.3. Agricultural Resources.Industrial projects service; 2002. p. 401.
30. National Meteorological Services Agency (NMSA). Annual Climatic Bulletin. Addis Ababa: Ministry of Water Resources; 2004. p. 16.
31. Milkessa W, Kurtu MY. Small scale dairy development project (SDDP). Studies and planning for the establishment of small scale dairy production. Alemay: Alemaya University of Agriculture; 1997. p. 44.
32. Russell B. Research methods in anthropology: qualitative and quantitative methods. 3rd edition, ISBN-10: 0759101485 (ISBN-13: 978–0759101487). California: Altamira Press; 2002.
33. Alexiades MN. Selected guidelines for Ethnobotanical research: a field manual. In: Advances in economic botany volume 10. The New York Botanical Garden: Bronx; 1996.
34. Martin GJ. Ethnobotany. A methods manual. London: WWF for Nature International, Chapman and Hall, London; 1995. p. 265–70.
35. Edwards S, Tadesse M, Demissew S, Hedberg I (Eds): Flora of Ethiopia and Eritrea. Volume 2, part 1. Magnoliaceae to Flacourtiaceae. The National Herbarium, Addis Ababa, Ethiopia, and Department of Systematic Botany, Uppsala, Sweden; 2000.
36. Hedberg I, Edwards S, Nemomissa S (Eds): Flora of Ethiopia and Eritrea. Volume 4, part 1. Apiaceae to Dipsacaceae. The National Herbarium, Addis Ababa, Ethiopia, and Department of Systematic Botany, Uppsala, Sweden; 2003.
37. Addis G, Abebe D, Urga K. A survey on traditional medicinal plants in Shirka district Arsi zone, Ethiopia. EPJ. 2001;19:30–47.
38. Gebrezgabiher G, Kalayou S, Sahle S. An ethnoveterinary survey of medicinal plants in woredas of Tigray region, northern Ethiopia. Int J Biodivers Conserv. 2013;5:89–97.
39. Tamiru F, Terfa W, Kebede E, Dabessa G, Roy RK, Sorsa M. Ethnoknowledge of plants used in veterinary practices in Dabo Hana District, West Ethiopia. J Med Plants Res. 2013;7:2960–71.
40. Lulekal E, Kelbessa E, Bekele T, Yineger H. An ethnobotanical study of medicinal plants in Mana Angetu District, southwestern Ethiopia. J Ethnobiol Ethnomed. 2008;4:10.
41. Poffenberger M, McGean B, Khare A, Campbell J. Community Forest economy and use pattern: participatory rural Apraisal method in South Gujarat, India. Field method manual volume II. Society for promotion of Wastelands Development: New Delhi; 1992.
42. Giday M, Asfaw Z, Woldu Z. Ethnomedicinal study of plants used by Sheko ethnic group of Ethiopia. J Ethnopharmacol. 2010;132:75–85.

5

Nontuberculosis mycobacteria are the major causes of tuberculosis like lesions in cattle slaughtered at Bahir Dar Abattoir, northwestern Ethiopia

Anwar Nuru[1,5]* , Aboma Zewude[1], Temesgen Mohammed[1], Biniam Wondale[1], Laikemariam Teshome[2], Muluwork Getahun[4], Gezahegne Mamo[3], Girmay Medhin[1], Rembert Pieper[6] and Gobena Ameni[1]

Abstract

Background: The main cause of bovine tuberculosis (bTB) is believed to be *Mycobacterium bovis* (*M. bovis*). Nontuberculosis mycobacteria (NTM) are neglected but opportunistic pathogens and obstacles for bTB diagnosis. This study aimed to isolate and characterize the mycobacteria organisms involved in causing TB-like lesions in cattle in northwestern Ethiopia.

Results: A total of 2846 carcasses of cattle were inspected for TB lesions. Ninety six tissues (including lymph nodes such as submandibular, retropharyngeal, tonsilar, mediatinal, bronchial and mesenteric, and organs such as lung, liver and kidney) with suspicious TB lesion(s) were collected and cultured on Lowenstein-Jensen medium. Twenty one showed culture growth, of which only 17 were identified containing acid fast bacilli (AFB) by Ziehl–Neelsen staining. Among the 17 AFB isolates 15 generated a polymerase chain reaction product of 1030 bp by gel electrophoresis based on the 16S ribosomal RNA gene amplification. No *M. tuberculosis* complex species were isolated. Further characterization by Genotype Mycobacterium CM assay showed 6 isolates identified as *M. peregrinum*. Eight isolates represented by mixed species, which includes *M. fortuitum-peregrinum* (3 isolates), *M. gordonae-peregrinum* (3 isolates) and *M. fortuitum-gordonae-peregrinum* (2 isolates). One NTM could not be interpreted.

Conclusion: A significant number of NTM species were isolated from TB-like lesions of grazing cattle slaughtered at Bahir Dar Abattoir. Such finding could suggest the role of NTM in causing lesions in cattle. Further investigations are recommended on the pathogenesis of the reported NTM species in cattle, and if they have public health significance.

Keywords: Cattle, Nontuberculosis mycobacteria, TB-like lesion

Background

Bovine tuberculosis (bTB), which is primarily caused by *Mycobacterium bovis* (*M. bovis*) is an endemic disease of cattle in Ethiopia and distributed in almost all parts of the country. Although its current prevalence rate, at a national level, is unknown, previous Ethiopian studies have shown that the average herd prevalence of bTB in smallholder farms is 21.1% [1–4] and intensive dairy production systems is 49.3% [5, 6]. Other Ethiopian studies [7–15] undertaken at abattoirs have reported bTB in cattle based on TB-like lesion with an estimated average prevalence of 5.57%. In addition, *M. bovis* was also recovered from TB lesions in cattle, spoligotyped, and their strain types were identified and reported by previous studies in Ethiopia [7, 9, 16–19]. Infection with *M. bovis* can be transmitted from cattle to humans, mainly through the consumption of contaminated milk and meat [20], although there is no evidence that this has happened in Ethiopia, where raw milk and meat consumption is widely habituated. Although the *M. tuberculosis* complex (MTBC) species are identified as

* Correspondence: hamduanwar@yahoo.com
[1]Aklilu Lemma Institute of Pathobiology, Addis Ababa University, P.O. Box 1176, Addis Ababa, Ethiopia
[5]College of Veterinary Medicine and Animal Sciences, University of Gondar, P.O. Box 346, Gondar, Ethiopia

strict pathogens of TB in human and animals, other mycobacteria species collectively referred to as nontuberculosis mycobacteria (NTM) also play a significant role as a source of infections [21]. However, there have been no studies to date conducted to identify the specific species of NTMs that are causing TB lesions in cattle in northwest Ethiopia. Presently fast, easy and sensitive molecular tools are available for the detection and identification of MTBC and NTM [22]. Thus, identification of mycobacteria is required using these molecular tools to guide therapy and for epidemiological purposes.

In the present study NTMs were predominantly isolated and characterized from TB-like lesions of cattle by molecular tools such as mycobacterium genus typing and Genotype Mycobacterium CM assay. *M. peregrinum* was the most dominant NTM species recovered from 6 isolates. Eight isolates represented by mixed species such as *M. fortuitum-peregrinum* (3 isolates), *M. gordonae-peregrinum* (3 isolates) and *M. fortuitum-gordonae-peregrinum* (2 isolates). One NTM could not be interpreted even if it had a band pattern of 1,2,3 and 10, and no MTBC species were identified.

Methods
Description of the study area and setting
The study was conducted in cattle slaughtered at Bahir Dar Abattoir, which is located in Bahir Dar City of Amhara Regional State, northwest Ethiopia. Currently, Bahir Dar Abattoir is the only licensed slaughter house in Bahir Dar City, which fulfils the daily beef requirements of over 200,000 inhabitants of the city, peri-urban areas and its neighboring rural villages. Cattle slaughtered at the abattoir were mainly of the Zebu type and originated from different districts of Amhara Region and the neighboring Oromia Region (Amhara and Oromia regions are among the nine ethnically based regional states of Ethiopia, and have the largest number of livestock and human population compared to other regions).

Sample collection and processing
A total of 2846 cattle slaughtered from October 2014 to December 2015 at Bahir Dar Abattoir were thoroughly inspected for TB lesions. Parotid, mandibular, retropharyngeal, tonsilar, left and right bronchial, cranial and caudal mediastinal, brochial, tracheobronchial and mesenteric lymph nodes, and organs including the lungs, liver and kidneys were examined. The seven lobes of the two lungs were inspected externally and palpated. Each lobe was sectioned into approximately 2 cm thick slices to identify the lesions. Similarly, lymph nodes sliced into sections of a similar thickness and inspected for the presence of visible lesions. The animal was classified as having lesion when gross lesion(s) suggestive of bTB were found in any of the tissues examined. Each specimen was processed and cultured for the isolation of mycobacteria following standard

procedure described by OIE [23]. In brief, the tissue samples were manually dissected in to small pieces and homogenized using a pestle and mortar. The homogenate decontaminated by an equal volume of 4% NaOH and concentrated by centrifugation at 3000×g for 15 min. The sediment was neutralized with 2 N HCl using phenol red as an indicator, and inoculated onto Lowenstein Jensen (LJ) glycerol and LJ pyruvate solid media slants. The culture media were incubated at 37 °C for 8 weeks, and considered negative if no visible growth was detected after the eighth week of incubation. Ziehl–Neelsen (ZN) staining microscopic examination was performed to select acid fast bacilli (AFB) positive isolates. Presumptive mycobacterial colonies were heat-killed at 85 °C for 45 min by mixing ~2 loop-full of cells in 200 μl distilled H$_2$O for further molecular activities.

Mycobacterium genus typing
Multiplex polymerase chain reaction (mPCR) using six oligonucleotide primers was performed as described previously [24]. Primer pairs included were MYCGEN-F 5′-AGA GTT TGA TCC TGG CTC AG-3′, MYCGEN-R 5′-TGC ACA CAG GCC ACA AGG GA-3′, which amplify a specific PCR product from the 16S rRNA gene of all know mycobacteria were used. MYCAV-R 5′-ACC AGA AGA CAT GCG TCT TG-3′ and MYCINT-F 5′-CCT TTA GGC GCA TGT CTT TA-3′ which amplify the hyper variable region of the 16S rRNA gene of *M. intracellulare* (MYCINT-F) and *M. avium* (MYCAV-R), respectively. Two primers (TB1-F 5′-GAA CAA TCC GGA GTT GAC AA-3′) and (TB1-R 5′-AGC ACG CTG TCA ATC ATG TA-3′), which target for the MPB70 gene were used to specify *M. tuberculosis* complex from the mycobacteria.

Amplification was done as recommended. In each run *M. avium* and *M. bovis* were included as a positive control with sterile water (H$_2$O Qiagen) as a negative control. The PCR products were electrophoresed in 1.5% agarose gel, and the final image visualized under ultraviolet light.

The GenoType® mycobacterium common Mycobacteria (CM) assay
The GenoType® Mycobacterium CM assay (Hain Lifescience, Nehren Germany) was used to analyze NTM isolates at the species level, and the procedure described in the manual enclosed in the kit was followed to conduct the test. The assay involved DNA amplification targeting the 23S rRNA gene region, as recommended. Followed by the reverse hybridization to specific oligonucleotide probes immobilized on membrane strips, which was conducted on a shaking TwinCubator (Hain). The final result was interpreted based on the presence and absence of bands, and compared with the evaluation sheet provided with the kit. *M. tuberculosis* H37Rv, *M.*

fortuitum and *M. abscessus* were used as appositive control while H_2O Qiagen as a negative control.

Ethical considerations

The study was approved by Ethical Review Board (Ref. number IRB/05-02/2013) of the Aklilu Lemma Institute of Pathobiology, Addis Ababa University. Study permission also obtained Amhara Region Bureau of Agriculture Department of Animal Agency, and Municipality Office of Bahir Dar City.

Results

Description of the study animals and tissues

The vast majority of studied cattle was male (88.7%, 2524/2846) and zebu breed (99.9%, 2842/2846). Seventy nine carcasses had lesion(s) suspected of bTB resulting in an overall animal level prevalence of 2.78% (79/2846). The animal level prevalence was defined as the number of cattle positive for TB-like lesion(s) per 100 cattle examined. From 79 positive cattle a total of 96 different tissues having TB-like lesions were collected, processed and cultured onto LJ media. Of which 21 showed culture growth, and only 17 colonies were identified containing mycobacteria by ZN staining with an overall AFB positivity of 17.7% (17/96). The 17 mycobacterial isolates were detected only from 12 slaughtered cattle, and the largest proportion was observed in the retropharyngeal lymph nodes (75%) followed by submandibular and the kidney tissues (each with 50% proportion). The type and number of tissues identified with suspicious TB lesion(s), and their corresponding AFB positivity are indicated in Table 1.

Identification and speciation of nontuberculosis mycobacteria

Among the 17 AFB positive mycobacterial isolates 15 generated a PCR product of 1030 bp by gel electrophoresis (Fig. 1), and consequently identified as NTM.

Further characterization of the 15 NTM by using Genotype Mycobacterium CM assay revealed that 14 isolates identified at the species level and 1 NTM could

Table 1 Cattle tissues identified with suspicious tuberculosis lesions and mycobacteria

Type of tissue	Number of tissue with lesion(s)	[a]AFB positive, +n(%)
Submandibular lymph node	2	1(50.0)
Retropharyngeal lymph node	4	3(75.0)
Tonsilar lymph node	3	1(33.3)
Mediastinal lymph node	12	1(8.33)
Bronchial lymph node	10	1(10.0)
Mesenteric lymph node	43	3(7.00)
Lung tissue	6	2(33.3)
Liver tissue	14	4(28.6)
Kidney tissue	2	1(50.0)
Total	96	17(17.7)

[a]Acid fast bacilli; +number of tissues positive for AFB; The 96 tissues suspected of having tuberculosis lesion(s) were identified from 79 of 2846 cattle slaughtered at Bahir Dar Abattoir, northwestern Ethiopia. Of which 17 tissues identified containing mycobacteria by Ziehl-Neelsen staining and were recorded from carcasses of only 12 cattle

not be interpreted even if it has a band pattern of 1, 2, 3 and 10 (Fig. 2). Among the 14 isolates with defined NTM species, 6 isolates were recognized as *M. peregrinum*, and the remaining 8 represented mixed species including *M. fortuitum-peregrinum* (3 isolates), *M. gordonae-peregrinum* (3 isolates), and *M. fortuitum-gordonae-peregrinum* (2 isolates).

Discussion

The overall prevalence of bTB from gross suspected TB lesion(s) in the present study was 2.78% which is comparable to 2.7% reported by Bekele and Belay [10], but lower than other findings ranging from 3.5% to 10.2% [7–9, 11–15, 25]. These variations could be explained by many factors including differences in the disease status in the animal populations, the sample size and the type of production system from where the slaughtered cattle were originated. Breed of animals that are slaughtered in

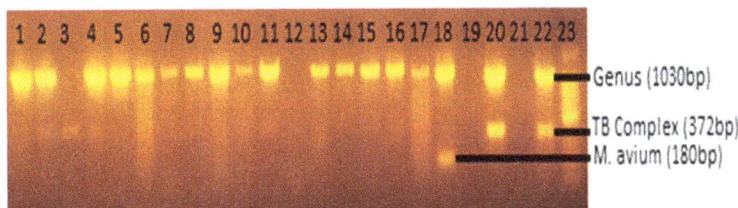

Fig. 1 Gel electrophoresis of PCR products from AFB isolated from cattle tissue containing TB-like lesion(s). The Seventeen acid fast bacilli positive tuberculosis lesions were identified from 79 tissues of 2846 cattle slaughtered at Bahir Dar Abattoir, northwest Ethiopia. Lanes 1-17 = test isolates, Lane 18 = *M. avium* (positive control), Lane 19 = missed out, Lane 20 = *M. bovis* (positive control), Lane 21 = Qiagen H_2O (negative control), Lane 22 = *M. tuberculosis* (positive control), and Lane 23 = 100 bp DNA ladder

HAIN
LIFESCIENCE

GenoType Mycobacterium CM/AS 96

VER 1.0
00299-0507-03-3

22 04 2016
dd mm yyyy

species

#						
1	C_1	Neg. control		NC		H_2O Qiagen
2	C_2	Pos. control		P_1		1,2,3,4	M. avium
3	S_1	2568 RRM		1		1,2,3,14	M. peregrinum
4	S_2	2557 Mes		2		1,2,3,14	M. peregrinum
5	S_3	2457 Lun		3		1,2,3,14	M. peregrinum
6	S_4	3059 Mes		4		1,2,3,8,10,14	Mixed-1*
7	S_5	3059 Liv		5		1,2,3,8,10,14	Mixed-1*
8	S_6	2973 RRM		6		1,2,3,7,8,10,14	Mixed-3***
9	S_7	3006 Liv		7		1,2,3,7,14	Mixed-2**
10	S_8	2682 Lun		8		1,2,3,7,14	Mixed-2**
11	S_9	3009 RRM		9		1,2,3,7,14	Mixed-2**
12	C_3	Pos. control		P_2		1,2,3,5,6,10	M. abscessus
13	C_4	Pos. control		P_3		1,2,3,10,16	M. tub. complex
14	S_{10}	3389 Liv		10		1,2,3,10	Unknown
15	S_{11}	2772 Mes		11		1,2,3,14	M. peregrinum
16	S_{12}	3059 Kid		12		1,2,3,14	M. peregrinum
17	S_{13}	2437 Liv		13		1,2,3,8,10,14	Mixed-1*
18	S_{14}	2393 BR		14		1,2,3,7,8,10,14	Mixed-3***
19	S_{15}	3059 Liv		15		1,2,3,14	M. peregrinum

* M. gordonae-peregrinum complex
** M. fortuitum-peregrinum complex
*** M. fortuitum-gordonae-peregrinum complex

LOT _____ HYB 30 min   STR 15 min   SUB 5 min _____

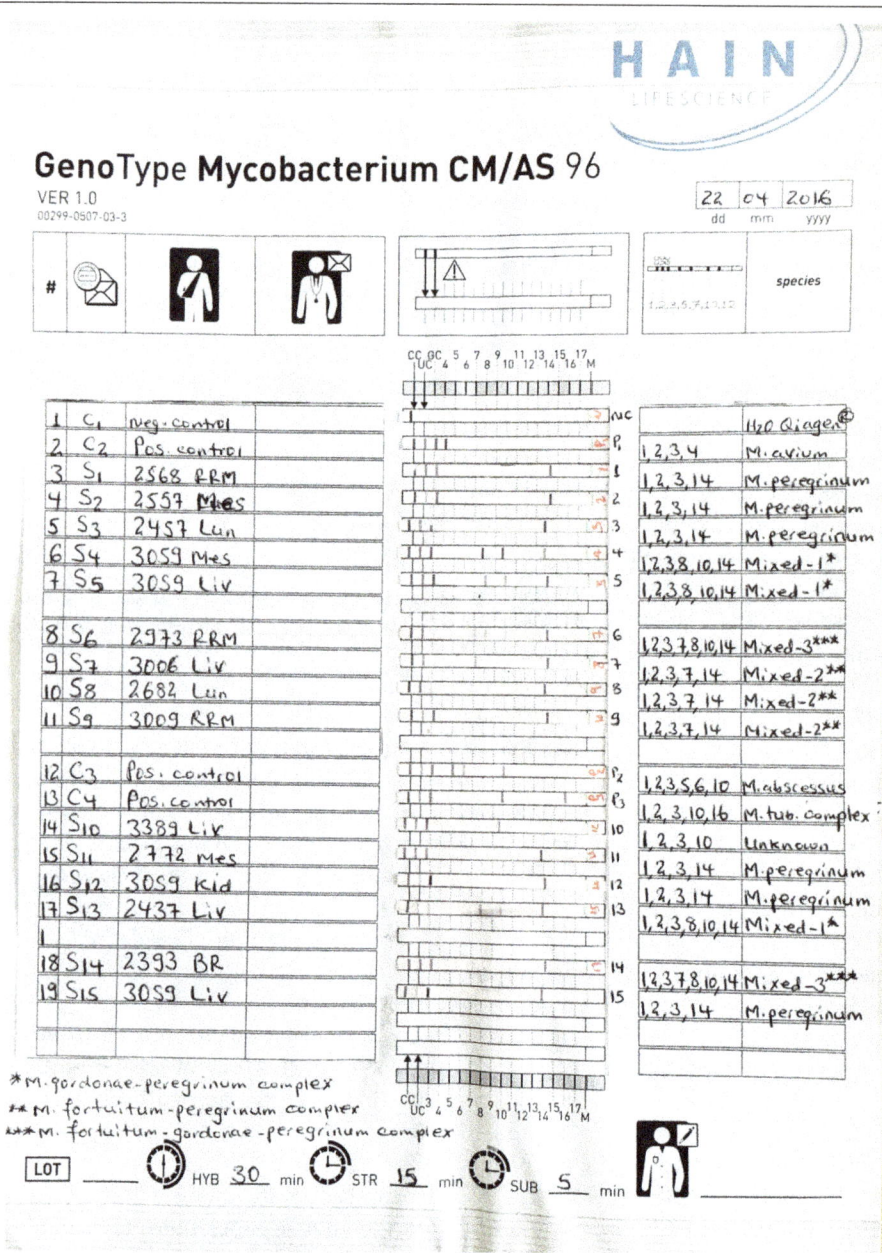

Fig. 2 Nontuberculosis mycobacteria species identified from cattle tissue containing TB-like lesion(s). *Mixed-1: *M. gordonae-peregrinum*; **Mixed-2: *M. fortuitum-peregrinum*; ***Mixed-3: *M. fortuitum-gordonae-peregrinum*; fifteen of the 17 isolates with acid fast bacilli showed bands at 1030 bp by Gel electrophoresis and identified as nontuberculosis mycobacteria (NTM). Further characterization by GenoType® mycobacterium CM showed 14 of the 15 NTMs defined at the species level and the remaining 1 NTM (Sample code: S_{10}) could not be interpreted

the abattoirs and subjective differences in identifying TB lesions could also be considered for the disparities observed. The low prevalence of bTB in this study could be explained by the fact that the vast majority of cattle in the current study were Zebu and from non-intensive smallholder farms as well as most of the cattle were originated from northwest Ethiopia, where the overall prevalence of bTB was reported very low [10]. Moreover, the TB-like lesions might not always be of mycobacterial origin, rather they could also be caused by other granuloma forming organisms like *Nocardia* and *Corynebacterium* species [26], parasites and other non-specific reactions [27, 28].

The overall culture yield of AFB from visible lesions in the present study was 17.7%, which is slightly higher than 11% reported previously in Ethiopia [7], but lower proportion when it is compared to 38.1% recorded in Jimma Municipality Abattoir, southwest Ethiopia [10]. The observed differences could also be attributed to the subjective differences in identifying TB lesions, which were subjected to ZN staining microscopic examination across the study sites.

Different NTM species were identified in this study from isolates with positive AFB, notably *M. fortuitum*, *M. gordonae* and *M. peregrinum*. The NTM species such as *M. fortuitum* and *M. gordonae* are so ubiquitous that they have previously been recovered from cattle in Ethiopia [7, 29], and human, animals and the environment elsewhere in Africa [28, 30–32]. *Mycobacterium peregrinum*, which is a rapidly growing, ubiquitous and an opportunistic but potentially pathogenic NTM [33] was isolated more frequently in this study. Similar previous studies in Ethiopia [7] and Zimbabwe [34] have also reported *M. peregrinum* from cattle as well. Moreover, miscellaneous human infections, more specifically skin and lung infections were also found associated with *M. peregrinum* in Japan [35] and Brazil [36], respectively. The high rate of *M. peregrinum* isolation from lesions in the present study can suggest that this species of NTM is abundant and has high pathogenicity to cause infection in cattle in the study area as compared to other NTM species including *M. fortuitum* and *M. gordonae*. However, the role of these NTMs in TB disease causation in cattle and their zoonotic implication is not known in our cases, and these will be the objective for further investigations. Moreover, *M. fortuitum* and *M. gordonae* have been reported to elicit reactions to purified protein derivative bovine based skin tests in cattle [37]. As a result the isolation of these species in the present study emphasized further studies as mycobacteria other than *M. bovis* may interfere with current bTB diagnostic tests and ensuing in false positive test results [38].

Conclusion

This study has isolated NTMs, notably *M. fortuitum, M. gordonae and M. peregrinum* from TB-like lesions of grazing cattle, and these findings suggest an important role of NTM in causing lesions in cattle. However, the pathogenesis of NTM species in cattle, the epidemiology (including sampling from environmental sources such as water and soil), their interactions with bTB and the zoonotic link between animal and humans is not known and needs further studies.

Abbreviations

AFB: Acid fast bacilli; Bp: Base pair; bTB: Bovine tuberculosis; CIDT: Comparative intradermal tuberculin; CM: Common mycobacteria; LJ: Lowenstein Jensen; MTBC: Mycobacterium tuberculosis complex; NTM: nontuberculosis mycobacteria; PCR: Polymerase chain reaction; PPD-B: Purified protein derivative bovine; TB: Tuberculosis; ZN: Ziehl–Neelsen

Acknowledgements

This study was jointly funded by the National Institute of Health (NIH, USA) through its H3Africa Program (Grant number: U01HG0074720I), Addis Ababa University through its Thematic Research Program, and University of Gondar. Tiru Alem, Gashaw Yitayew, Aschalew Admasu, and staff of Aklilu Lemma Institute of Pathobiology and Bahir Dar Animal Diseases Investigation and Diagnostic Laboratory were acknowledged for their technical support.

Funding

This study was jointly funded by the National Institute of Health (NIH, USA) through its H3Africa Program (Grant number: U01HG0074720I), Addis Ababa University through its Thematic Research Program, and University of Gondar.

Authors' contributions

AN participated in the design of the study, data collection, laboratory work, statistical analysis, interpretation of the data and drafted the manuscript. AZ participated in the identification of suspected TB lesions at the abattoir and interpretation of the data. TM participated in performing multiplex PCR and gel electrophoresis, and interpretation of the data. BW participated in performing multiplex PCR and gel electrophoresis, and interpretation of the data. MG participated in performing NTM species identification by GenoType Mycobacteria CM assay and interpretation the data. GM participated in the design of the study, interpretation of the data and reviewed the draft manuscript. GM participated in the design of the study, statistical analysis and reviewed the draft manuscript. RP participated in the design of the study and reviewed the draft manuscript. GA participated in the design of the study, interpretation of the data and reviewed the draft manuscript. All authors read and approved the final manuscript.

Consent for publication

Not applicable

Competing interests

The authors declare that they have no competing interests.

Author details

[1]Aklilu Lemma Institute of Pathobiology, Addis Ababa University, P.O. Box 1176, Addis Ababa, Ethiopia. [2]Animal Diseases Investigation and Diagnostic Laboratory, Amhara Region Bureau of Agriculture, P. 0. Box 70, Bahir Dar, Ethiopia. [3]College of Veterinary Medicine and Agriculture, Addis Ababa University, P.O. Box 34, Debre Zeit, Ethiopia. [4]Ethiopian Public Health Institute, P. O. Box181689, Addis Ababa, Ethiopia. [5]College of Veterinary Medicine and Animal Sciences, University of Gondar, P.O. Box 346, Gondar, Ethiopia. [6]J. Craig Venter Institute, 9704 Medical Center Drive, Rockville, MD, USA.

References

1. Mamo G, Abebe F, Worku Y, Hussein N, Legesse M, Tilahun G, et al. Bovine tuberculosis and its associated risk factors in pastoral and agro-pastoral cattle herds of afar region, Northeast Ethiopia. J Vet Med Anim Health. 2013;5(6):171–9.
2. Romha G, Gebre egziabher G, Ameni G. Assessment of bovine tuberculosis and its risk factors in cattle and humans, at and around Dilla town, southern Ethiopia. Anim Vet Sci. 2014;2(4):94–100.
3. Zeru F, Romha G, Berhe G, Mamo G, Sisay T, Ameni G. Prevalence of bovine tuberculosis and ssessment of cattle owners' awareness on its public health implication in and around Mekelle, northern Ethiopia. J Vet Med Anim Health. 2014;6(6):159–67.
4. Nuru A, Mamo G, Teshome L, Zewdie A, Medhin G, Pieper R, Ameni G. Bovine tuberculosis and its risk factors among dairy cattle herds in and around Bahir Dar City. Northwest Ethiopia Ethiop Vet J. 2015;19(2):27–40.
5. Firdessa R, Tschopp R, Wubete A, Sombo M, Hailu E, Erenso G, et al. High prevalence of bovine tuberculosis in dairy cattle in Central Ethiopia: implications for the dairy industry and public health. PLoS One. 2012;7(12):e52851.

6. Tigre W, Alemayehu G, Abetu T, Ameni G. Preliminary study on the epidemiology of bovine tuberculosis in Jimma town and its surroundings, southwestern Ethiopia. Afr J Microbiol Res. 2012;6(11):2591–7.

7. Berg S, Firdessa R, Habtamu M, Gadisa E, Mengistu A, Yamuah L, et al. The burden of Mycobacterial disease in Ethiopian cattle: implications for public health. PLoS One. 2009;4(4):e5068.

8. Demelash B, Inangolet F, Oloya J, Asseged B, Badaso M, Yilkal A, et al. Prevalence of bovine tuberculosis in Ethiopian slaughter cattle based on post-mortem examination. Trop Anim Health Prod. 2009;41(5):755–65.

9. Ameni G, Desta F, Firdessa R. Molecular typing of *Mycobacterium bovis* isolated from tuberculosis lesions of cattle in north eastern Ethiopia. Vet Rec. 2010;167:138–41.

10. Bekele B, Belay I. Evaluation of routine meat inspection procedure to detect bovine tuberculosis suggestive lesions in Jimma municipal abattoir, south West Ethiopia. Global Veterinaria. 2011;6:172–9.

11. Ewnetu L, Melaku A, Birhanu A. Bovine tuberculosis prevalence in slaughtered cattle at Akaki municipal abattoir based on meat inspection methods. Global Veterinaria. 2012;9(5):541–5.

12. Mekibeb A, Fulasa TT, Firdessa R, Hailu E. Prevalence study on bovine tuberculosis and molecular characterization of its causative agents in cattle slaughtered at Addis Ababa municipal abattoir, Central Ethiopia. Trop Anim Health Prod. 2013;45(3):763–9.

13. Biru A, Ameni G, Sori T, Desissa F, Teklu A, Tafess K. Epidemiology and public health significance of bovine tuberculosis in and around Sululta District, Central Ethiopia. Afr J Microbiol Res. 2014;8(24):2352–8.

14. Zeru F, Romha G, Ameni G. Gross and molecular characterization of mycobacterium tuberculosis complex in Mekelle town municipal abattoir. Northern Ethiopia Global Veterinaria. 2013;11(5):541–6.

15. Terefe D. Gross pathological lesions of bovine tuberculosis and efficiency of meat inspection procedure to detect-infected cattle in Adama municipal abattoir. J Vet Med Anim Health. 2014;6(2):48–53.

16. Ameni G, Aseffa A, Engers H, Young D, Gordon S, Hewinson G. Vordermeier. High prevalence and increased severity of pathology of bovine tuberculosis in Holsteins compared to zebu breeds under field cattle husbandry in central Ethiopia. Clin Vaccine Immunol. 2007;14:1356–61.

17. Biffa D, Skjerve E, Oloya J, Bogale A, Abebe F, et al. Molecular characterization of Mycobacterium Bovis isolates from Ethiopian cattle. BMC Vet Res. 2010;6:28.

18. Firdessa R, Berg S, Hailu E, Schelling E, Gumi B, Erenso G, et al. Mycobacterial lineages causing pulmonary and extrapulmonary tuberculosis, Ethiopia. Emerg Infect Dis. 2013;19(3):460–3.

19. Ameni G, Tadesse K, Hailu E, Deresse Y, Medhin G, Aseffa A, et al. Transmission of mycobacterium tuberculosis between farmers and cattle in central Ethiopia. PLoS One. 2013;8(10):e76891.

20. de la Rua-Domenech R. Human *Mycobacterium bovis* infection in the United Kingdom: incidence, risks, control measures and review of the zoonotic aspects of bovine tuberculosis. Tuberculosis. 2006;86:77–109.

21. van Ingen J, Boeree MJ, Dekhuijzen PN, van Soolingen D. Environmental sources of rapid growing nontuberculous mycobacteria causing disease in humans. Clin Microbiol Infect. 2009;15:888–93.

22. Ting SW, Chia CL, Hsin CL. Current situations on identification of nontuberculous mycobacteria. J Biomed Lab Sci. 2009;21:1–6.

23. OIE. Bovine tuberculosis. Manual of diagnostic tests and vaccines for terrestrial animals (mammals, birds and bees), vol. Chapter 2.4.7; 2009. p. 683.

24. Wilton S, Cousins D. Detection and identification of multiple mycobacterial pathogens by DNA amplification in a single tube. PCR Methods Appl. 1992;1(4):269–73.

25. Beyi AF, Gezahegne KZ, Mussa A, Ameni G, Ali MS. Prevalence of bovine tuberculosis in dromedary camels and awareness of pastoralists about its zoonotic importance in eastern Ethiopia. J Vet Med Anim Health. 2014;6(4):109–15.

26. Grist A. Bovine meat inspection – anatomy, physiology and disease conditions. 2nd ed. Nottingham: Nottingham University Press; 2009.

27. Shitaye JE, Getahun B, Alemayehu T, Skoric M, Treml F, Fictum P, et al. A prevalence study of bovine tuberculosis by using abattoir meat inspection and tuberculin skin testing data, histopathological and IS6110 PCR examination of tissues with tuberculous lesions in cattle in Ethiopia. Veterinarni Medicina. 2006;51:512–22.

28. Diguimbaye-Djaibe C, Vincent V, Schelling E, Hilty M, Ngandolo R, Mahamat HH, et al. Species identification of non-tuberculous mycobacteria from humans and cattle of Chad. Schweiz Arch Tierheilkd. 2006;148(5):251–6.

29. Ameni G, Vordermeier M, Firdessa R, Aseffac A, Hewinson G, Gordon SV, et al. Mycobacterium tuberculosis infection in grazing cattle in central Ethiopia. Vet J. 2011;188(3):359–61.

30. Katale BZ, Mbugi EV, Botha L, Keyyu JD, Kendall S, Dockrell HM, et al. Species diversity of non-tuberculous mycobacteria isolated from humans, livestock and wildlife in the Serengeti ecosystem, Tanzania. BMC Infect Dis. 2014;14:616.

31. Egbe NF, Muwonge A, Ndip L, Kelly RF, Sander M, Tanya V, et al. Abattoir-based estimates of mycobacterial infections in Cameroon. Sci Rep. 2016;6:24320.

32. Malama S, Munyeme M, Mwanza S, Muma JB. Isolation and characterization of non tuberculous mycobacteria from humans and animals in Namwala District of Zambia. BMC Research Notes. 2014;7:622.

33. Kamijo F, Uhara H, Kubo H, Nakanaga K, Hoshino Y, Ishii N, Okuyama R. A case of mycobacterial skin disease caused by *Mycobacterium peregrinum*, and a review of cutaneous infection. Case Rep Dermatol. 2012;4(1):76–9.

34. Padya L, Chin'ombe N, Magwenzi M, Mbanga J, Ruhanya V, Nziramasanga P. Molecular identification of mycobacterium species of public health importance in cattle in Zimbabwe by 16S rRNA gene sequencing. Open Microbiol J. 2015;9:38–42.

35. Sawahata M, Hagiwara E, Ogura T, Komatsu S, Sekine A, Tsuchiya N, et al. Pulmonary mycobacteriosis caused by Mycobacterium Peregrinum in a young, healthy man. Nihon Kokyuki Gakkai Zasshi. 2010;48(11):866–70.

36. Wachholz PA, Sette CS, do Nascimento DC, Soares CT, Diório SM, Masuda PY. *Mycobacterium peregrinum* skin infection: Case Report J Cutan Med Surg. 2015; [Epub ahead of print].

37. Bercovier H, Vincent V. Mycobacterial infections in domestic and wild animals due to Mycobacterium Marinum, M. Fortuitum, M. Chelonae, M. Porcinum, M. Farcinogenes, M. Smegmatis, M. Scrofulaceum, M. Xenopi, M. Kansasii, M. Simiae and M. Genavense. Rev - Off Int Epizoot. 2001;20:265–90.

38. Thacker TC, Robbe-Austerman S, Harris B, Van Palmer M, Waters WR. Isolation of mycobacteria from clinical samples collected in the United States from 2004 to 2011. BMC Vet Res. 2013;9:100.

6

Gross and histopathological evaluation of human inflicted bruises in Danish slaughter pigs

Kristiane Barington[1]* (iD), Jens Frederik Gramstrup Agger[2], Søren Saxmose Nielsen[2], Kristine Dich-Jørgensen[1] and Henrik Elvang Jensen[1]

Abstract

Background: Human inflicted bruises in slaughter pigs are hampering animal welfare, are an infringement of the animal protection act, and are a focus of public attention. The aim of the present study was to evaluate the gross appearance of human inflicted bruises in slaughter pigs and to compare the inflammatory changes in two lesions as a basis for estimating the age of lesions in the same pig.

Pigs with human inflicted bruises slaughtered at two major slaughterhouses in Denmark from November 2013 to May 2014 were evaluated. After slaughter, the bruises were examined grossly and skin and underlying muscle tissue from two similar but separate bruises (a and b) on each pig were sampled for histology.

Results: Skin and muscle tissue from 101 slaughter pigs were subjected to gross evaluation. Eighty-one of these were also subjected to histological evaluation. Most frequently (51 out of 101 pigs, 50 %), bruises had a tram-line pattern due to blunt trauma inflicted with long objects such as sticks. Other bruises reflected the use of tattoo-hammers, plastic paddles, double U profiles and chains. Histological evaluation of two bruises from a pig with multiple lesions was found insufficient to assess the overall age of the lesions as substantial variation in the inflammatory response between bruises was present.

Conclusions: Grossly, the pattern of bruises often reflected the shape of the object used for inflicting the lesions. When determining the age of multiple bruises on a pig more than two lesions should be evaluated histologically.

Keywords: Age assessment, Bruise, Forensic, Pig

Background

During recent years, human inflicted bruises in pigs have received increased attention [1–7]. Estimation of the age of such bruises is crucial in order to determine in whose custody the pig was when the lesions were inflicted [2]. The age determination of the lesions is based on a histological evaluation of skin and underlying muscle tissue [2]. Histological evaluation of skin and underlying muscle tissue is able to determine if bruises are 1 to 3 h or 4 to 10 h old based on experimental bruises in pigs [4]. Estimation of the age of bruises was primarily based on

the infiltration of neutrophils in the subcutaneous tissue and of macrophages in the underlying muscle tissue.

At gross evaluation, human inflicted bruises on slaughter pigs are defined by being multiple, having a uniform pattern often identifying the object used to inflict, and are always placed on the back or upper sides of the animals [2]. Moreover, based on histological evaluation, more than 90 % of bruises on slaughter pigs are assumed to be inflicted in less than 8 h prior to slaughter, i.e., when the pigs are managed around transport to slaughter [2]. Therefore, multiple bruises on a slaughter pig are most likely of the same age (applied during a short period, i.e., minutes); and an overall assessment of the age in relation to the time of slaughter is stated [2]. However, a comparison of the inflammatory changes between individual bruises on the

* Correspondence: krisb@sund.ku.dk
[1]Department of Veterinary Disease Biology, Faculty of Health and Medical Sciences, University of Copenhagen, Ridebanevej 3, DK-1870 Frederiksberg C, Denmark

same animal has not previously been done, and the impact on the age determination is unknown.

As part of a study requested by the Danish Veterinary and Food Administration [1] skin and muscle tissue from human inflicted bruises on Danish slaughter pigs were sampled prospectively at two major slaughterhouses. The aim of the present study was to evaluate the gross appearance of bruises in slaughter pigs. Moreover, the inflammatory changes and the estimated age of each of two bruises sampled from pigs with multiple bruises were also compared.

Methods

Routine monitoring of human inflicted bruises occur at the slaughter lines in all abattoirs in Denmark. All pigs detected with human inflicted bruises slaughtered at two major slaughterhouses in Denmark from November 2013 to May 2014 were included. All pigs were slaughter pigs, i.e., around 5 to 6 months old and had a mean carcass weight of 83.25 ± 6.58 (SD) kg. The animals were all transported directly from farms to the slaughter houses within an average transportation time of 2.13 ± 3.33 (SD) h. After arrival, the animals were herded together by the use of noise making plastic paddles and were slaughtered within 3.03 ± 3.44 h (SD). After slaughter, all animals were closely examined for the presence of human inflicted bruises. Bruises are defined as being inflicted by humans if they are multiple, having a uniform pattern often reflecting the object used for infliction and are localized on the back or upper sides of the slaughter pigs [2]. After slaughter, skin and underlying muscle tissue from two separate bruises (a and b) on each pig were sampled by veterinarians employed at the slaughterhouses and immersion-fixed in 10 % neutral buffered formalin for at least 5 days (Fig. 1). The remaining skin was stored at -18 °C before being submitted for gross evaluation at the University of Copenhagen.

Pathological evaluation

At gross evaluation, the pattern and dimensions of bruises were registered. For histology, a single tissue section of each of the immersion-fixed skin and muscle tissues (a and b) were cut at 4-5 μm and stained with hematoxylin and eosin [8]. Dermis, subcutis and muscle tissue were assessed for hemorrhage, hyper-leukocytosis (capillary pavement of leukocytes), and infiltration of neutrophils and macrophages. Hemorrhage and hyper-leukocytosis were recorded as present or absent in all tissue layers. In the dermis, infiltration of neutrophils and macrophages was registered as present or absent. In the subcutis and muscle tissue, infiltration of neutrophils and macrophages was scored on a semi-quantitative scale: 0) absence of neutrophils or macrophages, 1) 1-10 neutrophils or macrophages, 2) 11-30 neutrophils or macrophages, 3) >30 neutrophils or macrophages. The scoring was carried out by using a 40x objective in the area with the highest density of neutrophils and macrophages. In the muscle tissue, the percentile area of necrotic muscle fibers was

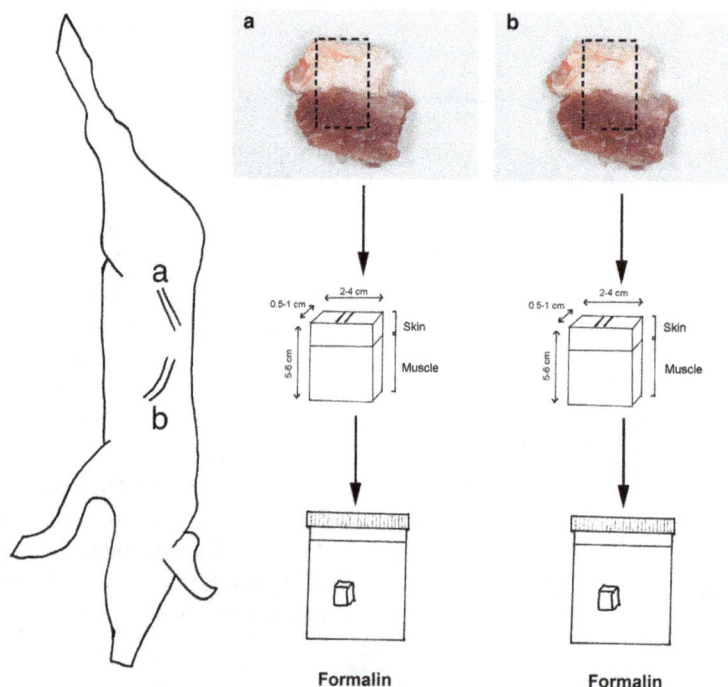

Fig. 1 Drawing illustrating a carcass of a slaughter pig with multiple bruises. Two bruises are marked (**a** and **b**). Skin and underlying muscle tissue from bruises **a** and **b** were sampled at the slaughter houses and immersion-fixed in 10 % buffered formalin

scored according to the following scale by using a 10x objective: 0) no necrosis: absence of necrotic muscle fibers, 1) minor necrosis: <12.5 %, 2) moderate necrosis: 12.5-50 %, 3) severe necrosis: >50 %. Moreover, in the muscle tissue, the localization of neutrophils and macrophages was recorded as predominantly being present in the interstitial spaces (>50 % of the leukocytes) or as intracellular infiltrations (>50 % of the leukocytes) in the necrotic muscle fibers.

Age estimation

The age of a bruise was determined as a time interval based on the infiltration of neutrophils and macrophages in the subcutis and muscle tissue, respectively, according to the results from experimental porcine bruises [4]. This was done by combining the age interval given by the neutrophil score together with the age interval given by the macrophage score (Table 1). In case of overlap between the age intervals, the two intervals were combined so that the narrowest age interval was obtained for each bruise. If the age interval obtained by the neutrophil score and the macrophage score did not overlap, the highest score (based on infiltration of either neutrophils or macrophages) was used to determine the age of the bruise. A score of zero was considered as inconclusive.

Based on the age interval, bruises were assigned to one of the following four categories: 1) inconclusive (neutrophils and macrophages were absent), 2) <4 h, 3) >4 h, and 4) age intervals overlapping 4 h (e.g., 1 to 8 h, 2 to 10 h).

Statistics

Agreement between observations in bruises a and b was determined for each of the histological parameters and for the age of the bruises (categories 1 to 4) by calculating Cohen's kappa using the fmsb function [9] in R version 3.2.2 [10].

Results

Skin from 101 slaughter pigs was submitted for gross evaluation, and seven patterns of bruises were recognized either occurring alone or in combination (Table 2).

Tramline bruises: Bruises with a tramline pattern were characterized by two longitudinal, parallel lines of hemorrhage (width from 0.1 to 1.2 cm) separated by apparently normal skin (width from 0.2 to 3 cm). This pattern was compatible of being inflicted with a stick and was present on 51 pigs (Table 2 and Fig. 2a).

Tattoo-hammer bruises: Uniform bruises consistent with being inflicted by the back of the head of a tattoo-hammer were found on 21 pigs (Table 2). This type of pattern consisted of a short (4.5 to 7 cm) tramline pattern or a confluent accumulation of hemorrhage (Fig. 2b). On 3 out of the 21 pigs, a second hemorrhage was seen perpendicular to the first hemorrhage, i.e., compatible of being caused by the shaft of the tattoo-hammer (Figs. 2b and c).

Paddle formed bruises: Bruises reflecting the handle of a plastic paddle ("pig-paddle") were seen on 8 pigs. The bruises appeared as oblique rows of parallel hemorrhages similar to the shape of the handle of a plastic paddle, which is a handling device normally used when herding pigs (Figs. 2d and e). Sometimes, bruises with a tramline pattern were present in proximity to the handle shaped bruises consistent with the shaft of the paddle.

Double U profile bruises: Bruises inflicted by double U profiles were present on 6 pigs (Table 2 and Figs. 2f and g). The bruises consisted of three parallel lines of hemorrhages (width from 0.1 to1.5 cm) separated by apparently normal skin (width from 0.3 to 0.8 cm).

Circle formed bruises: On 3 pigs, multiple circular hemorrhages were seen (Table 2 and Fig. 2h). The bruises consisted of a small circle of hemorrhage (diameter from 0.6 to 1 cm) within a larger circle (diameter from 2.3 to 2.5 cm). The small circles were occasionally seen as a confluent hemorrhage, and both small and large circles were sometimes open, i.e., not complete. The tool used for applying the circle formed bruises is unknown.

Table 1 Estimation of the age of bruises localized on the back or upper side of slaughter pigs

Cell type and tissue	Score	Bruise age
Neutrophils, subcutis	1	1 to 3 h
Neutrophils, subcutis	2	1 to 8 h
Neutrophils, subcutis	3	4 to 10 h
Macrophages, muscle	1	2 to 9 h
Macrophages, muscle	2	2 to 10 h
Macrophages, muscle	3	4 to 10 h

The age of bruises was based on the degree of infiltration by neutrophils and macrophages (cell type) in the subcutis and the muscle tissue, respectively [4]. The age of a bruise was stated by combining the age intervals given by the neutrophil score and the macrophage score

Table 2 The pattern of bruises reflecting the object used for infliction on 101 slaughter pigs

Pattern	Number of pigs (n = 101)	
Tramline	51	(50 %)
Tattoo-hammer	21	(21 %)
Paddle	8	(8 %)
Double U profile	6	(6 %)
Circle	3	(3 %)
Chain	2	(2 %)
Other	10	(10 %)

Six patterns of bruises were recognized. In 13 % of pigs the object used to inflict bruises could not be identified

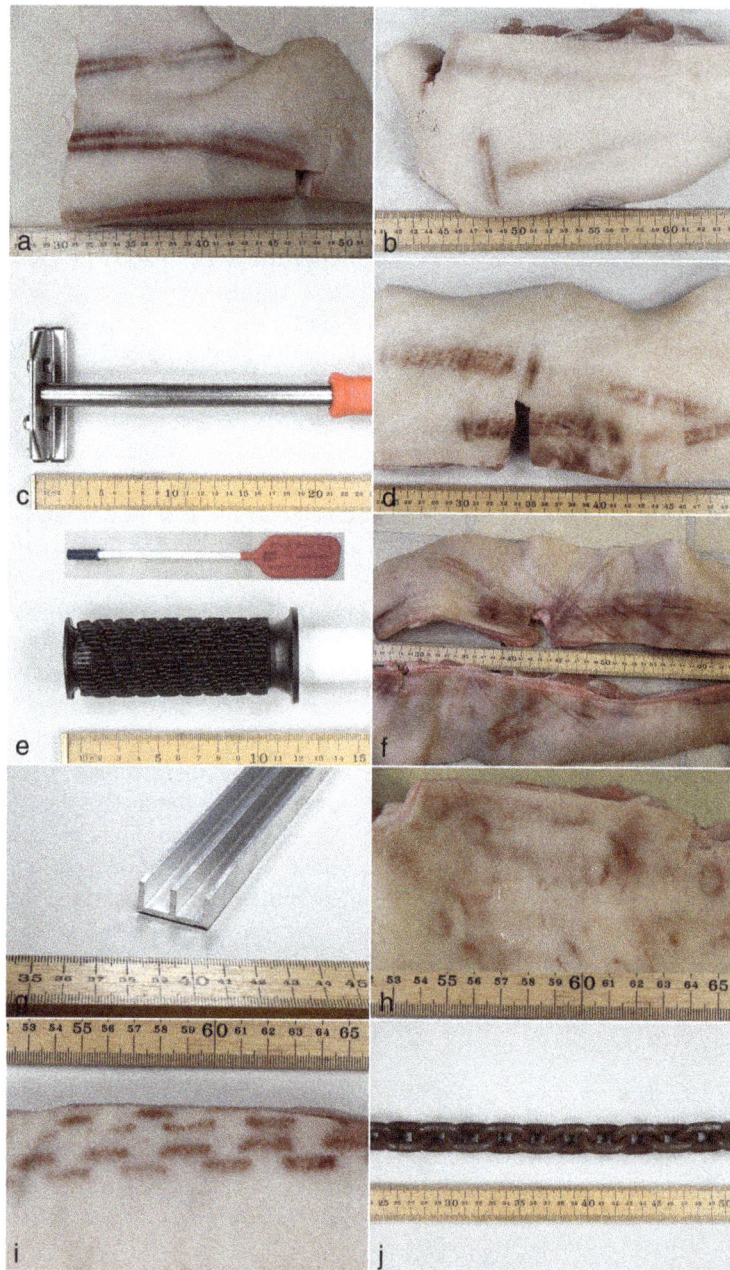

Fig. 2 The patterns of bruises on pig skin and objects clearly used to inflict lesions. **a** Bruises with a tramline pattern compatible of being inflicted with a stick. **b** Bruises consistent with being inflicted by the back of the head of a tattoo-hammer (see **c**). **c** The back of a tattoo-hammer. **d** Bruises reflecting the handle of a plastic paddle (see **e**). **e** Handle of a plastic paddle and the entire plastic paddle (inset). **f** Bruises inflicted by a double U profile (see **g**). **g** Double U profile. **h** Multiple circle formed bruises consisting of a small circle of hemorrhage within a larger circle. The object used to inflict these bruises is unknown. **i** Bruises reflecting strikes using a chain (see **j**). **j** Chain. Ruler on figures is in cm

Chain shaped bruises: Bruises reflecting strikes using a chain were seen on 2 pigs (Table 2 and Figs. 2i and j). This type of bruise consisted of up to four rows of parallel hemorrhages with a length from 1.4 to 2.3 cm and a width from 0.2 to 0.5 cm.

Other bruises: On 10 pigs, bruises with other patterns were present (Table 2). The tools used to inflict these bruises could not be identified, and the shape of the lesions differed.

In total, skin and muscle tissue from 81 of the pigs (i.e., bruises a and b) were available for histological evaluation. From the remaining 20 pigs, the tissues were excluded from histological evaluation due to two reasons: missing samples ($n = 15$) and putrefaction of samples due to an insufficient volume of formalin ($n = 5$).

Cohen's kappa value, p-value and interpretation of agreement for each of the histological parameters are presented

Table 3 Diagnostic agreement (estimated as Cohen's kappa) between two samples (a and b) of bruises from each of 81 slaughter pigs with multiple bruises

Tissue layer	Variable	Kappa	Lower95	Upper95	Level of agreement	P-value
Dermis	Hyper-leukocytosis	0.58	0.29	0.88	Moderate	0.0037
	Leukocytes	0.43	0.23	0.64	Moderate	0.0002
	Hemorrhage	0.29	-0.18	0.76	Not significant	0.1508
Subcutis	Neutrophils	0.30	0.14	0.46	Fair	0.0000
	Hyper-leukocytosis	0.29	-0.18	0.76	Not significant	0.5000
	Macrophages	0.52	0.23	0.80	Moderate	0.0034
	Hemorrhage	-0.04	-0.77	0.70	Not significant	0.5395
Muscle	Necrosis	0.14	-0.19	0.48	Not significant	0.2146
	Neutrophils	0.32	0.12	0.53	Fair	0.0015
	Macrophages	0.17	-0.06	0.40	Not significant	0.0806
	Leukocytes in the interstitial space	0.43	0.15	0.71	Moderate	0.0077
	Leukocytes present intramuscularly	0.40	0.18	0.62	Moderate	0.0008
	Hyper-leukocytosis	0.00	-1.95	1.95	Not significant	0.5000
	Hemorrhage	0.34	0.13	0.54	Fair	0.0012

Limits of 95 % confidence interval not including zero and a P-value for kappa below 0.05 means there is some level of agreement between the test results of tissue samples a and b within the same pig. The level of agreement (fair, moderate) depends on the kappa value
The agreement was based on histological variables within the dermis, subcutis and underlying muscle tissue of bruise a and bruise b

in Table 3. A fair agreement (lowest level) was found between the age estimation of bruise a and bruise b ($\kappa = 0.24$, 95 % confidence interval = [0.08 to 0.40], $p = 0.0008$). The distribution of the age is shown in Table 4. In 48 % of pigs, bruise a and bruise b were estimated to the same age category (Table 5).

Discussion

In approximately half of the slaughter pigs, the bruises had a tram-line pattern similar to the pattern previously described in pigs, cattle and humans beaten with sticks such as broom handles or other wood and metal rods [2, 4, 6, 11–14]. Moreover, in a study from Brazil, bruises in pigs with a rectangular shape were associated to be inflicted by sticks and other solid handling devices when herding the pigs [5]. The pattern of some bruises clearly revealed the equipment used to inflict the bruises, e.g., the back of a tattoo-hammer, the handle of a plastic paddle or a chain, which are all tools or parts of equipment that are normally present in a pig production environment. Tattoo-hammers are normally used for applying an identification number on

the pigs at the farm, while plastic paddles are used for herding pigs together on farms and at slaughter houses.

Histologically, the degree of inflammatory changes and the age assessment varied between bruise a and bruise b as none of the kappa values exceeded 0.6. The variation in the inflammatory response between the two bruises is in accordance with human cases, in which bruises with and without inflammatory changes coexisted although they were known to have been established 30 h or more before death [15]. In addition, the degree of inflammatory changes has been seen to vary between experimental bruises of the same age on the same pig [4]. In the present study, the exact age of the bruises was unknown. However, due to the uniform pattern of bruises on each pig, it is reasonable to assume that they were inflicted almost simultaneously (i.e., within minutes) at a given time before slaughter. In forensic cases of multiple bruises on a pig, all lesions are presumed to have been inflicted almost at the same time (within minutes) [2]. Therefore, one overall assessment of the

Table 4 The pattern of the estimated age of two bruises (a and b) on each of 81 slaughter pigs with multiple bruises

Age	Bruise a		Bruise b	
Inconclusive	5	(6 %)	7	(9 %)
<4 h	30	(37 %)	36	(44 %)
>4 h	15	(19 %)	13	(16 %)
Overlapping 4 h	31	(38 %)	25	(31 %)

The number and percentage of bruises in each of the four age categories are presented

Table 5 Agreement and difference in the estimated age of bruise a and bruise b on each of 81 pigs

		Bruise a			
		Inconclusive	<4 h	>4 h	Overlapping 4 h
Bruise b	Inconclusive	1	3	0	3
	<4 h	1	20	3	12
	>4 h	0	3	6	4
	Overlapping 4 h	3	4	6	12

age is made for all bruises based on the examination of an unspecified number of samples of skin and underlying muscle tissue [2]. However, due to the difference regarding the age estimation and the variation in the inflammatory response in bruise a and bruise b, histological evaluation of skin and muscle from two bruises is insufficient in order to determine an overall state of lesions and thereby a common age of bruises on a pig, despite being inflicted almost simultaneously.

Conclusions

Grossly, the pattern of bruises in slaughter pigs often reflected the shape of the object used for inflicting the lesions. Most frequently, bruises had a tram-line pattern due to blunt trauma inflicted with long objects such as sticks. Histological evaluation of two bruises was insufficient to determine the age (i.e., at what time before slaughter the pigs were beaten) as substantial variation in the inflammatory response between the two bruises (a and b) was present.

Acknowledgements
We wish to thank the meat inspection personal at the two slaughterhouses for collecting material and information on transportation of pigs. For technical assistance we would like to acknowledge Dennis Brok, Betina Gjedsted Andersen, and Elisabeth Wairimu Petersen.

Funding
The study was funded by the University of Copenhagen and the Danish Veterinary and Food Administration.

Authors' contribution
KB contributed to the conception and design of the study, the acquisition of data, the gross and histological analysis and interpretation of data, and drafted the manuscript. JFA and SSN contributed to the conception and design of the study, the statistical analysis and interpretation of data and revised the manuscript. KDJ contributed to the histological analysis of data and revised the manuscript. HEJ contributed to the conception and design of the study, the gross and histological analysis and interpretation of data, and revised the manuscript. All authors read and approved the final manuscript.

Authors' information
Not applicable.

Competing interests
The authors declare that they have no competing interest.

Consent for publication
Not applicable.

Author details
[1]Department of Veterinary Disease Biology, Faculty of Health and Medical Sciences, University of Copenhagen, Ridebanevej 3, DK-1870 Frederiksberg C, Denmark. [2]Department of Large Animal Sciences, Faculty of Health and Medical Sciences, University of Copenhagen, Frederiksberg C, Denmark.

References

1. Agger JF, Jensen HE, Barington K., Nielsen SS, 2015. Prospektivt studium af slagmærker hos svin [Prospective study of bruises in pigs]. Danish Veterinary and Food Administration. 2015. http://www.foedevarestyrelsen.dk/SiteCollectionDocuments/Dyresundhed/Oms%C3%A6tnings-gruppen/Rapport%20om%20prospektivt%20studium%20af%20slagm%C3%A6rker%20hos%20svin%20JF%20Agger%20et%20al%20%2023-03-2015%20(3).pdf. In Danish. Accessed 29 Feb 2016

2. Barington K, Jensen HE. Forensic cases of bruises in pigs. Vet Rec. 2013;173:526–30.

3. Barington K, Jensen HE. Experimental animal models of bruises in forensic medicine – A review. SJLAS. 2015;41:14.

4. Barington K, Jensen HE. A novel, comprehensive and reproducible porcine model for the timing of bruises in forensic pathology. Forensic Sci Med Pathol. 2016;12:58–67.

5. Dalla Costa OA, Faucitano L, Coldebella A, Ludke JV, Peloso JV, Dalla Roza D, da Costa MJR P. Effect of the season of the year, truck type and location on truck on skin bruises and meat quality in pigs. Livest Sci. 2007;107:29–36.

6. Jensen HE, Dahl-Pedersen K, Krarup C, Selmer O, Elverstad K, Kristensen SS, Lassen SJ. Skader efter stump vold hos svin – Patologiske og forensiske aspekter [Lesions in pigs due to blunt trauma – pathological and forensic aspects]. Dansk Veterinærtidsskrift. 2009;22:13–5. In Danish.

7. Nielsen SS, Michelsen AM, Jensen HE, Barington K, Opstrup KV, Agger JF. The apparent prevalence of skin lesions suspected to be human-inflicted in Danish finishing pigs at slaughter. Prev Vet Med. 2014;117:200–6.

8. Grizzle WE, Fredenburgh JL, Myers RB, Billings PE, Spencer LT, Bancroft JD, Horobin RW, Gamble M. Chapter 4-9. In: Bancroft JD, Gamble M, editors. Theory and practice of histological techniques. 6th ed. Philadelphia: Churchill Livingstone Elsevier; 2008. p. 53–134.

9. Functions for Medical Statistics Book with some Demographic Data. https://cran.r-project.org/web/packages/fmsb/index.html. Accessed 29 Feb 2016.

10. The R Project for Statistical Computing. https://www.r-project.org/. Accessed 29 Feb 2016.

11. Armstrong EJ. Distinctive patterned injuries caused by an expandable baton. Am J Forensic Med Pathol. 2005;26:186–8.

12. McNally PW, Warriss PD. Recent bruising in cattle at abattoirs. Vet Rec. 1996; 138:126–8.

13. Saukko P, Knight B. The pathology of wounds. In: Saukko P, Knight B, editors. Knight's Forensic Pathology. 3rd ed. London: Arnold; 2004. p. 136–73.

14. Weeks CA, McNally PW, Warriss PD. Influence of the design of facilities at auction markets and animal handling procedure on bruising in cattle. Vet Rec. 2002;150:743–8.

15. Byard RW, Wick R, Gilbert JD, Donald T. Histological dating of bruises in moribund infants and young children. Forensic Sci Med Pathol. 2008;4:187–92.

7

Acute-onset high-morbidity primary photosensitisation in sheep associated with consumption of the Casbah and Mauro cultivars of the pasture legume Biserrula

Jane C. Quinn[1,2], Yuchi Chen[1], Belinda Hackney[2], Muhammad Shoaib Tufail[1,2], Leslie A. Weston[2] and Panayiotis Loukopoulos[1,2*] (iD)

Abstract

Background: Primary photosensitisation (PS) subsequent to ingestion of the pasture legume *Biserrula pelecinus L.* (biserrula) has recently been confirmed in grazing livestock. Given the potential utility of this pasture species in challenging climates, a grazing trial was undertaken to examine if both varieties 'Casbah' and 'Mauro' were able to cause photosensitisation in livestock, and if this could be mitigated by grazing in winter, or in combination with other common pasture species.

Results: A controlled grazing trial was undertaken in winter in Australia with plots containing a dominant pasture of *Biserrula pelecinus L. cv.* 'Casbah' or 'Mauro', or mixed biserrula/perennial ryegrass populations. A photosensitisation grading system was established. 167 prime meat ewe lambs were introduced to the plots and monitored twice daily. Mild clinical signs were observed at 72 h on pasture. All animals were removed from biserrula dominant stands at this point. Four animals grazing 'Casbah' dominant pasture rapidly proceeded to severe photosensitisation in the following 12 h. Animals remaining on mixed biserrula/ryegrass stands did not exhibit severe PS but showed an 89% incidence of mild to moderate photosensitisation over the following 14 days. Animals on mixed lucerne showed significantly lower PS score than animals grazing biserrula varieties of any composition. The trial was halted at 14 days as only plots with low biserrula proportion still contained unaffected animals.

Necropsy revealed severe multifocal erythematous ulcerations and alopecia of the ear pinnae, severe bilateral periorbital and conjunctival oedema and variably severe subcutaneous facial oedema. No evidence of hepatopathy was present. A diagnosis of acute unseasonal primary photosensitisation caused by biserrula ingestion with no other underlying pathology was confirmed.

Conclusions: We report an unseasonal outbreak of acute photosensitisation in sheep grazing *Biserrula pelecinus L cvs.*'Casbah' and 'Mauro' with exceedingly high morbidity. A grading system is also proposed as a tool for objective and consistent clinical appraisal of future PS outbreaks. This finding expands our definition of seasonal and temporal risk periods for biserrula photosensitisation, and is the first to identify that both commercial cultivars of biserrula can cause primary photosensitisation in sheep.

Keywords: Photosensitisation, Primary, Unseasonal, *Biserrula pelecinus L.*, Legume, Sheep

* Correspondence: ploukopoulos@csu.edu.au
[1]School of Animal and Veterinary Sciences, Charles Sturt University, Wagga Wagga, NSW 2650, Australia
[2]Graham Centre for Agricultural Innovation; Charles Sturt University and NSW Department of Primary Industries, Wagga Wagga, NSW 2650, Australia

Background

Biserrula pelecinus is a self-regenerating annual legume highly suited to low to medium rainfall zones [16]. It is able to adapt to a broad range of soil types with higher productivity compared to other legumes [20]. *Biserrula pelecinus* L. var. 'Casbah' was commercialised in 1997, and has been widely used by farmers in Australian mixed farming systems as a break crop, especially in South Australia and New South Wales [14, 23–26]. Anecdotal reports of cases of photosensitisation in livestock associated with biserrula ingestion had been noticed since this legume was commercialised in 1997 as a highly productive pasture with multiple agronomic advantages [17]. However, it was only in the recent past that a direct connection between exposure to biserrula pasture and onset of primary photosensitisation was confirmed by detailed diagnostics [4].

Photosensitisation (PS) is a skin disorder caused by photodynamic pigments activated by long-wavelength ultraviolet or visible light in exposed areas of skin, resulting in dermatitis [1]. In livestock, areas commonly affected include those lacking protective fleece, hair coat or skin pigmentation such as the muzzle, face, ears, eyes, mammary glands and genitalia [1].

In this study, we present an outbreak with exceedingly high morbidity in sheep grazing the Mediterranean pasture legume *Biserrula pelecinus L.* during its vegetative stage, in which animals exhibited an unexpected acute primary photosensitisation during mid-winter. Dissimilar to the current report, all but one [4] of the previous reports of PS outbreaks were reported in spring, a season classically associated with presentation of photosensitivity in livestock. Winter is not traditionally a time when photosensitisation outbreaks present to the livestock veterinarian. There are a number of reasons why this is the case: 1) lack of significant UV exposure due to cloudy conditions and short day length; 2) lack of actively growing plants, the phase during which toxic secondary compounds are commonly generated; 3) increased skin cover by hair or fleece for seasonal protection against colder conditions.

Furthermore, the clinical course and the speed of onset of clinical photosensitisation subsequent to exposure or ingestion of the presumptive or established aetiologic agent has not been clearly defined or evaluated. This is the first report of primary photosensitisation in sheep resulting from ingestion of both commercially available varieties of Biserrula, namely 'Casbah' and 'Mauro', in mid-winter, the presentation of which was acute, severe and with exceedingly high morbidity. We also define a clinical grading scale that is capable of differentiating mild, moderate and severe PS as a tool for objective and consistent clinical appraisal of PS outbreaks in the future, and describe the clinical course in relation to the

severity of PS and the causative agent. This report also identifies that mixed pastures are not a guaranteed protection against incidence of PS in animals grazing this pasture species. This report highlights the need to consider photosensitisation as a differential diagnosis, even during winter months, when cases of acute dermatitis are rarely reported in grazing livestock.

Methods

Pastures

Two varieties of *Biserrula pelecinus* L., *cv.* 'Casbah' and 'Mauro', were used under commercial conditions in an established long-term cropping paddock (−35.03°, 147.34°) to be available for 2015 spring grazing. The study site was owned and managed by the Charles Sturt University Farm, and no additional permission was required for experimentation on this site. Planting occurred in a paddock that was previously sown with dual purpose wheat in as per usual commercial practice. The site contained up to 10% infestation of annual ryegrass (*Lolium rigidum*) that exhibited resistance to multiple classes of post-emergent herbicides. A very low infestation of common pasture weeds was also present (<5%). All trial plots were sown according to standard commercial practices on 16 May 2015. Prior to sowing, plots were treated with a pre-emergent herbicide and then the stubble burned and light tillage performed. Trial sites were sown with 10 kg/ha of commercially available scarified biserrula seed of the two biserrula varieties 'Casbah' and 'Mauro' with 8 kg/ha 'biserrula special' inoculant (Alosca Technologies, Australia) for rhizobial establishment. Plots were treated post-sowing with Talstar insecticide (bifenthrin) at 100 ml/ha on 21 May 2015 to prevent insect herbivory post emergence. In order to establish pastures containing mixed populations of biserrula and ryegrass as a mitigating fodder, plots were oversown with annual ryegrass (*Lolium rigidum*) at a rate of 120 kg/ha ("medium biserrula" plots) or 40 kg/ha ("low biserrula"). Control plots ("high biserrula") contained no or very low contamination with other pasture species (<10%) and were considered biserrula-dominant. All plots were over-sown in triplicate using a random mixed block design. Plots ranges in size from 0.21 ha (plot 2, 6 & 7) to 0.38 ha in size (plots 1, 13, 17, 18). The plot design and pasture composition is shown in Fig. 1.

Animals

One hundred sixty seven 'Primeline' meat ewe lambs were sourced at 10 months old from a local producer. Prior to purchase for this study, animals were maintained according to standard industry practices on a local property and were grazing mixed annual ryegrass pasture. Inclusion criteria for the trial included a healthy

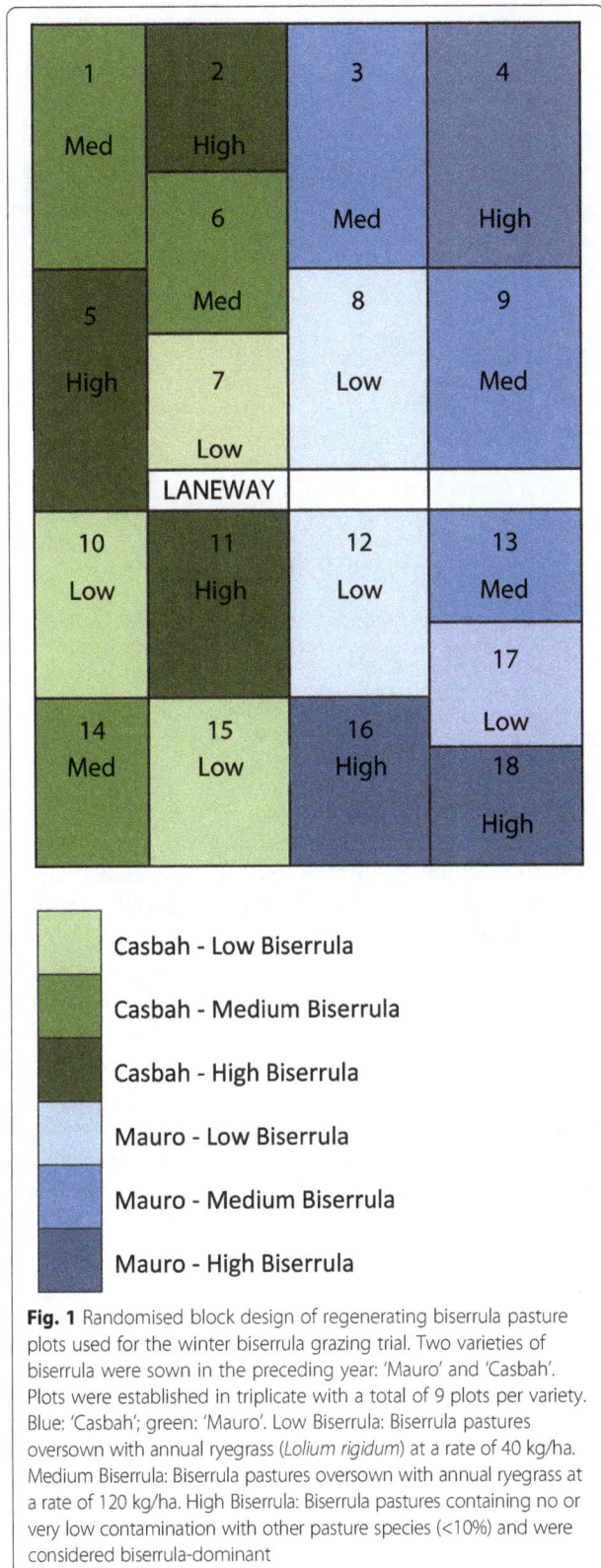

Fig. 1 Randomised block design of regenerating biserrula pasture plots used for the winter biserrula grazing trial. Two varieties of biserrula were sown in the preceding year: 'Mauro' and 'Casbah'. Plots were established in triplicate with a total of 9 plots per variety. Blue: 'Casbah'; green: 'Mauro'. Low Biserrula: Biserrula pastures oversown with annual ryegrass (*Lolium rigidum*) at a rate of 40 kg/ha. Medium Biserrula: Biserrula pastures oversown with annual ryegrass at a rate of 120 kg/ha. High Biserrula: Biserrula pastures containing no or very low contamination with other pasture species (<10%) and were considered biserrula-dominant

appearance at yarding with no visible pathology. All animals were naive for exposure to biserrula pasture. Prior to entry, animals were yarded and weighed (mean entry weight, 35.1 kg). All lambs were drenched prior to entry to pasture according to manufacturers recommendations (Ivomec®, Merial Australia). Water was provided ad libitum via automatic drinkers to all plots. As a control, an established mixed lucerne (*Medicago sativa*)/perennial ryegrass (*Lolium perenne*)/subclover (*Trifolium subterraneum*) paddock adjacent to the sown biserrula trial site was used as an industry-standard comparison pasture.

Photosensitisation grading system

To quantify clinical signs and severity of photosensitisation a de novo photosensitisation clinical scoring and grading system was designed (Table 1). The presence and severity of PS associated lesions in the face, eyes, ears and body were evaluated separately for each of these regions and scored o a scale of 0 to 5, as shown in Table 1. The four individual scores were then added to produce a PS grade. PS was defined as being mild, moderate or severe if having a composite PS score of <7, <12 or ≥12 respectively.

Grazing trial

A total of 167 lambs was randomly assigned to pasture in July (mid-winter) 2015 at matched stocking densities using restricted randomisation by weight (6–7 animals per plot; approximately 14 DSE/ha) (Fig. 1), 53 grazing in plots containing the biserrula variety 'Mauro' and 65 grazing on the variety 'Casbah'. A further 49 lambs were grazing the non-PS mixed lucerne pasture. The *'Mauro'* dominant pasture contained 88% biserrula by composition (*n* = 18 animals); while the biserrula *'Casbah'* dominant pasture contained 81% biserrula by composition (*n* = 21 animals). Medium biserrula pastures of both varieties contained between 52 and 75% biserrula by composition ('Mauro' medium, n = 18 animals; 'Casbah' medium, *n* = 23 animals). Low biserrula pastures contained <50% biserrula species ('Mauro' low, *n* = 17 animals; 'Casbah' low, n = 21 animals). Animals were monitored twice daily for behavioural or physical changes (shade-seeking behaviour, reduction in grazing or general activity, physical isolation, lameness) and scored every 3 days for clinical signs of photosensitisation.

All lambs were introduced to pastures on 9 July 2015 with all remaining animals removed from trial plots on 23 July 2015. Animals exhibiting clinical signs of photosensitisation were removed from the trial on the date of observation and scored for clinical signs of PS. Withdrawn animals were removed to a non-photosensitising pasture (lucerne/ryegrass/subclover) with full availability of shade and water. Once less than 50% of the original pen cohort were remaining in any plot, all members of

Table 1 Proposed photosensitisation clinical grading system. Lesions were defined as being mild, moderate or severe if having a composite score of <7, <12 or ≥12 respectively

Score	Lesion description			
	Face and muzzle	Eyes	Ears	Fleece/body
0	No apparent lesions	No apparent lesions	No apparent lesions	No apparent lesions
1	Mild cutaneous oedema and erythema	Mild serous blepharitis	Drooping of ears with mild oedema.	Mild erythema of exposed areas.
2	Cutaneous oedema and erythema; mild to mderate aural and facial oedema	Serous exudation, mild to moderate perorbital oedema and conjunctival erythema	Mild aural pitting oedema	Marked erythema of exposed areas.
3	Severe cutaneous erythema, crusting and black discolouration; moderate aural and facial oedema	Serous exudation, possible crusting; marked palpebral and conjunctival erythema and oedema; mild to moderate periorbital oedema	Marked aural pitting oedema, curling of ear ends, some flaking or other lesions possible	Multifocal, possibly multifocally extensive, patchy fleece loss, erythema of underlying skin.
4	Severe cutaneous erythema, crusting and black discolouration; severe aural and facial oedema	Severe perorbital, palpebral and conjunctival oedema. Possible corneal opacity or ulceration or opacity, serous exudate, eyelid crusting	Marked aural pitting oedema, curling of ear ends, other dermal lesions present. Skin flaking and multifocal necrosis; some tissue loss from rubbing may be evident; moderate serous exudation	Marked focal or multifocal fleece loss and dermatitis of exposed skin.
5	Multifocal irregularly shaped cutaneous necrosis and exudation; severe dermatitis including secondary lesions; severe facial oedema	Severe periorbital, palpebral and conjunctival oedema; eyes closed; corneal ulceration and/or opacity	Marked aural pitting oedema, curling of ear ends and tissue loss, other dermal lesions present including flaking and multifocal necrosis; abundant serous exudation	Widespread fleece loss, severe dermatitis of exposed skin. Significant fleece loss in dorsum and flanks.

that cohort were removed from the trial. Any animal with a PS score > 11 was selected for euthanasia. All animals were carefully monitored both within and after the trial for increasing signs of PS. PS was observed to resolve as soon as animals were withdrawn from biserrula pastures. No other clinical intervention was required. This trial was approved and compliant with requirements of the Charles Sturt Institutional Animal Care and Ethics Committee (Protocol 13/018).

Statistical analysis

Photosensitisation scores were compared between cohorts using ANOVA and general mixed linear models using SPSS™ (IBM, Version 20), with variety, plot, composition and time as fixed factors. Significance was defined as $p < 0.05$.

Results

Clinical course of the photosensitisation outbreak

Initially, animals grazing biserrula 'Casbah' dominant pastures were observed to exhibit some reduced grazing behaviour compared to other plots, showed less apparent activity within the plots and some individuals exhibited overt shade-seeking behaviour such as standing in a close line, nose to tail, or lying close to troughs or fence poles.

The first clinical signs of photosensitisation, ranging from mild to severe, were observed in 5 animals after only 3 days on biserulla dominant pasture. Three animals were located in the plot containing the highest

contribution of biserrula 'Casbah' (plot 4, 98% composition, PS scores 2.5, 2.5 and 15), one was in the second highest contribution 'Casbah' plot (8, 92% composition, PS score 2.5), and one in a dominant 'Mauro' plot (plot 7, 90% composition, PS score 4). All five lambs exhibited swelling and drooping of the pinnae and blepharitis and showed shade-seeking behaviour. No PS was observed in any animal on mixed lucerne pasture at this timepoint.

All those animals identified to be showing clinical signs of photosensitisation on day 3 of the grazing period were removed from the trial. Despite removal, two of the animals (animals 1 and 2) showed clinical signs which either remained severe (PS score 10.5) or increased in severity (PS score 4 increasing to 7). The most severely affected animal (animal 1) (Plot 4, high Casbah plot, PS score 15) was selected for necropsy, while animal 2 was closely observed and was observed to resolve over the following 7 days. Whilst oedema of the pinnae and muzzle and blepharitis were observed in other cases, extensive facial oedema was only apparent in animal 1 (Fig. 2). Venous blood was collected for haematology and biochemistry, faecal samples for analysis of parasite burden, and the animal was submitted for full diagnostic work up.

Clinical photosensitisation (mild, moderate or severe) was observed in 100% ($n = 39$) of animals grazing either 'Mauro' or 'Casbah' dominant plots by day 6, facilitating complete withdrawal of all animals on biserrula dominant pastures at this time. Comparison between biserrula

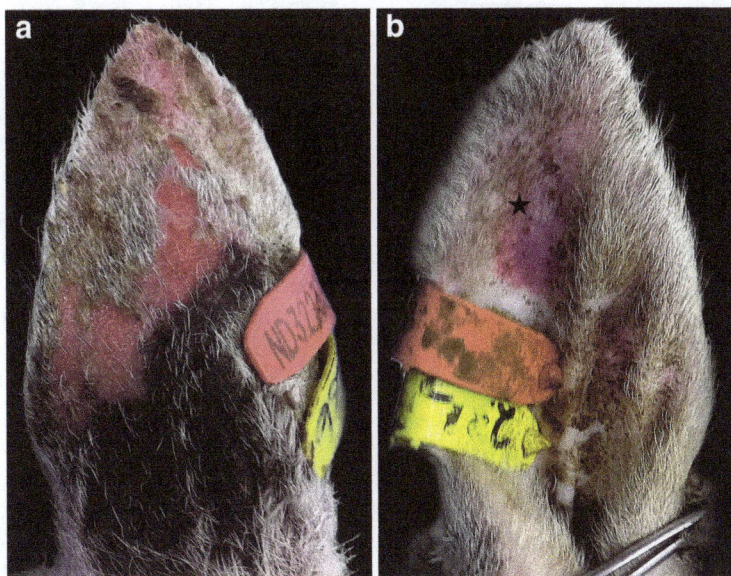

Fig. 2 Skin lesions associated with primary photosensitisation caused by ingestion of biserrula. **a** External aspect of left ear pinna: multifocal to coalescing erosion, ulceration and erythema; (**b**) Inner surface of the tip of the ear: erythema and alopecia (star). Skin covered by ear tags was not affected

dominant pastures at day 6 showed that the animals on 'Casbah' had significantly higher PS score compared to the animals on 'Mauro' at this time point (p = <0.01). No animals showed signs of icterus.

In total, 107/118 (91%) animals on biserrula pastures showed some clinical signs of photosensitisation during the 14 day monitoring period. On day 14, the majority of animals showed composite PS scores ranging from 0 to 7.5 with most within the 0–5 range, indicating mild photosensitisation was apparent in the majority of animals that had remained on the plots until day 14 (n = 68; mean PS score 2.4). Only four moderately affected were identified on day 14 of the grazing trial, all were present in plots containing a medium composition of either *B. pelecinus* L. 'Casbah' or 'Mauro' (Plot 3, Casbah medium: PS scores 9 and 10; Mauro medium: plot 5, PS scores 9 and 11).

Effect of pasture and biserulla composition and biserulla variety on photosensitisation severity

Mean PS score for animals grazing mixed lucerne pastures at day 14 was 0 (zero) with animals on these pastures showing significantly lower PS score than animals grazing biserulla varieties of any composition ($p \geq 0.0001$). Exit weight was not found to be statistically significant between the different pastures by variety or composition.

At day 14, biserulla variety was found to have a significant effect on severity of PS, with 'Casbah' showing significantly higher PS values than 'Mauro' ('Casbah' mean PS score: 3.09 ± 0.30, n = 46; 'Mauro' mean PS

score: 1.97 ± 0.35, n = 34, F = 5.776; p = 0.019). Composition was also shown to exert a significant effect, with the medium biserulla composition mean PS score being significantly higher than the low biserulla low composition mean PS score (2.89 ± 0.34 vs: 2.31 ± 0.33, F = 6.332, p = 0.001).

Clinical pathology findings

Complete blood count of animal 1 was consistent with the presence of a mild inflammatory process, showing mild leucocytosis due to neutrophilia and monocytosis (white blood cell count 17.4×10^9/L, reference range $4.1–13.0 \times 10^9$/L; neutrophils 11.5×10^9/L, reference range $0.5–9.3 \times 10^9$/L; monocytes 1.1×10^9/L, reference range $0.0–0.7 \times 10^9$/L), likely attributable to the aural dermatitis. Serum biochemistry showed mildly elevated AST (157 U/L, reference range 87–156 U/L) and CK (600 U/L, reference range 91–472 U/L), without elevation of GGT (52 U/L, reference range 35–61 U/L) or bilirubin (2 μmol/L, reference range 4–15 μmol/L). These changes were unremarkable. Faecal egg count was negative.

Gross pathology findings

Necropsy of animal 1 revealed multifocal to coalescing erythematous ulcerations, hair loss and crusting on the external aspect of both ear pinnae. The areas of ear skin which had been protected from UV exposure by the animal's identification tag were unaffected (Fig. 2a, b). Moderate erythematous and alopecic patches were present on the internal aspect of the left ear pinna.

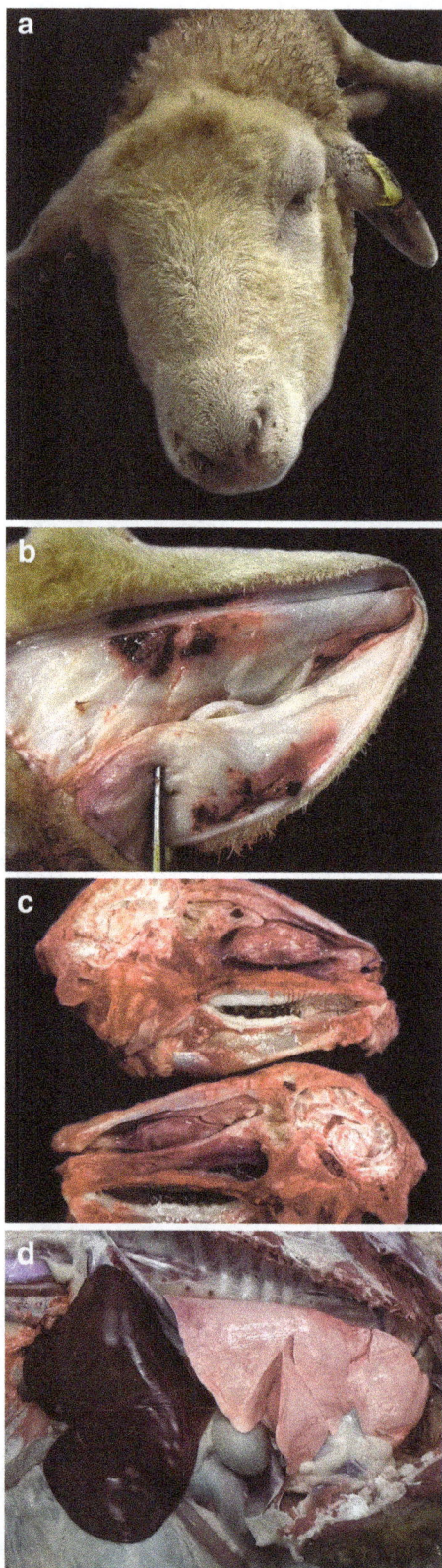

Fig. 3 a Severe bilateral periorbital and conjuctival oedema and variably severe subcutaneous facial oedema. **b** Severe focally extensive haemorrhage in the nasal subcutaneous tissues. **c** Severe narrowing of the nasal cavity due to oedema. **d** No significant changes were observed in the liver and other internal organs

Severe bilateral periorbital and conjunctival oedema and variably severe subcutaneous facial oedema were noted, the latter ranging from moderate to severe in the occipital region, severe in the nasal region, to exceedingly severe in the mandibular region (Fig. 3a). In the nasal subcutaneous tissues, there was severe focally extensive haemorrhage (Fig. 3b). Severe blepharitis was observed bilaterally but no gross corneal damage was observed. The nasal mucosa was extensively congested and the submucosa was oedematous and congested, both changes sparing the ethmoid area (Fig. 3c). No significant changes were observed in other organs, including the liver (Fig. 3d).

Histopathology findings

Histopathological examination of ear lesions from animal 1 revealed similar changes. These included exceedingly severe neutrophilic and eosinophilic epidermitis; dermatitis with intradermal nodular pustules containing neutrophils, eosinophils, macrophages and necrotic keratinocytes; severe haemorrhage and oedema; and individual keratinocyte to transmural epidermal necrosis. In the most severe lesions, moderate multifocal coagulative necrosis of sebaceous glands was noted (Fig. 4). The liver showed minimal periportal and occasionally periacinar lymphoplasmacytic aggregation. The heart, lungs, kidneys and brain showed no noteworthy changes.

Discussion

Previous reports of photosensitisation in lambs grazing biserrula identified dermal lesions of the face and ears [4, 15, 18] suggesting established or resolving lesions. Kessell et al. (2015) [4] also presented a PS outbreak due to biserrula Casbah in winter, affecting 25% of animals. However, the high morbidity, the speed of onset (4 days on pasture) and severity of presentation of PS described in our study have not been reported for this pasture species previously [2, 4, 18–20]. This study is the first to define the speed of onset of clinical photosensitisation subsequent to animals ingesting biserrula (<72 h). This study also identifies a change in behaviour in otherwise subclinically affected animals that could be used to predict the onset of clinical cases. Together these findings identify unique aspects of this clinical entity.

Prior to this experiment, it was not known whether both varieties of biserrula commercially available in Australia, 'Casbah' and 'Mauro', were able to induce photosensitisation. Previous reports had identified the

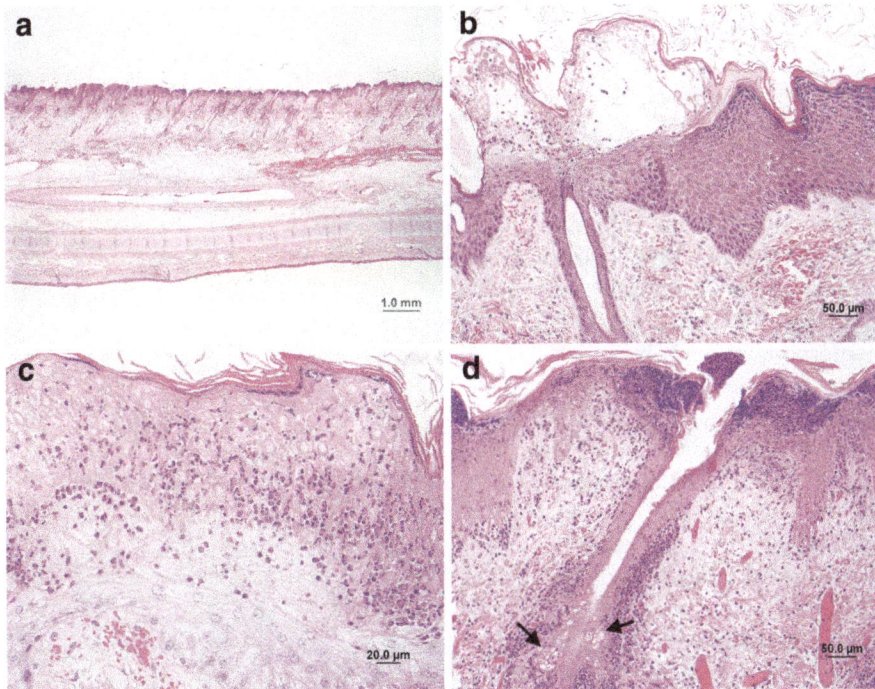

Fig. 4 Photomicrographs of the alopecic and oedematous areas of the ear pinnae: (**a**) Low magnification showing severe haemorrhage and oedema in the dermis. H&E, objective × 1.25. **b** Micropustules containing neutrophils, eosinophils and necrotic debris in the stratum corneum and upper stratum granulosum on the external aspect of the ear pinna. Mild individual cell keratinocyte necrosis, mild acanthosis and haemorrhage in the upper dermis are also shown. H&E, × 200. **c** Moderate zonal necrosis of the outer layers of the epidermis, which is infiltrated by numerous eosinophils and neutrophils, at places forming small intraepidermal pustules. Keratinocytes show moderate individual cell necrosis. H&E, × 400. (**d**) Severe epidermal necrosis with obliteration of the follicular epithelial structure and sebaceous gland necrosis (arrows) in the most severely affected section of the ear pinna. Intraepidermal nodular pustules formed by dead and viable neutrophils; eosinophils can be seen multifocally. The dermis is infiltrated by eosinophils and neutrophils, particularly adjacent to necrotic hair follicles. H&E, × 200

variety 'Casbah' to be phytophototoxic, but it was not known if 'Mauro' was also able to exert this effect. The current report identified that both varieties can induce photosensitisation in sheep grazing on them with the vast majority of animals showing clinical signs on either variety at both low and high composition densities. Such high morbidity rates have also not been reported previously nor has the severity of clinical presentation been evaluated systematically in prior outbreaks [4, 13–16]. Together our data suggest that a low composition pasture of *B. pelecinus L* cv. 'Mauro' might be the least photosensitising option available to producers, although neither variety is inert in its effect at a composition above 25% of total pasture.

Clinical photosensitisation occurs in three forms. Type I (primary PS) results from the direct ingestion, or exposure by dermal contact to, photodynamic compounds found within certain plant species, including biserulla, *Froelichia humboldtiana*, and alfalfa [2–5]. Type II, not reported in sheep to date [6], is associated with congenitally abnormal porphyrin metabolism. Type III (hepatogenous) is the most common form of PS in livestock, and is caused by impaired liver function resulting in

failure to excrete circulating phylloerythrin, a natural breakdown product of chlorophyll [7–12], which in turn causes phototoxic damage to the dermal and subdermal layers of exposed skin.

This study, and previous case studies and reports on biserulla, showed no evidence of hepatopathy [4, 13–16]. It is therefore suggested that biserulla cultivars contain photodynamic agents that lead to the onset of primary photosensitisation, occurring when the plant reaches show reproductive maturity [2, 4, 13, 15, 18, 21]. Although photoactive metabolites involved in this syndrome have not all been fully structurally elucidated, bioassay-guided isolation and analysis by NMR and UV spectroscopy along with mass spectrometry have revealed structural features of at least three bioactive metabolites present in the shoot extracts of both biserulla cultivars (LW, JQ unpublished data) and in seasonal equivalence to spring – early summer in NSW Australia. Photodynamic molecules such as these are polycyclic ring structures with conjugated bonds, generating potential electronic cycling and free radicals following exposure to UV irradiation, thereby leading to photosensitisation [2, 4]. At this time, investigation and structural elucidation is underway to 1) identify the photodynamic constituent(s) present in fresh *B. pelecinus* L. foliage and foliar extracts and 2) to investigate the potential to screen and select for less photocytotoxic genotypes for use in Mediterranean climates, including Australia.

The current study identifies the earliest phenological presentation of clinical photosensitisation to occur well prior to the reproductive phase of the plant's life cycle, indicating that the currently unidentified photodynamic compound that causes photosensitisation may exist across all phenological stages of biserrula growth.

A diagnosis of photosensitisation is based on history, clinical signs and the exclusion of all other possible dermatopathies [1, 22]. Furthermore, the acute onset of symptoms in winter, when extensive cloud cover was present, in conjunction with the consumption of a known primary photosensitising plant species, largely rules out sunburn as a differential diagnosis in this case.

In this case, the eosinophilic infiltration of the epidermis and dermis could be a tissue response to necrotic keratinocytes, but it could also be a reaction to the deposited photodynamic substances in the skin, either through ingestion or possibly also by dermal contact in affected animals. Eosinophilic inflammation has not been reported as a histopathological finding in primary photosensitisation previously [1], and may represent an undetermined component involved in the aetiopathogenesis. The high morbidity rate of primary photosensitisation observed in the feeding trial (91% overall), combined with the pathologic findings in this case are highly suggestive of the presence of photosensitising compound(s) in sufficient quantity to cause severe dermatitis even in seasons of low ambient temperatures and lower UV exposure, conditions which are not usually associated with outbreaks of photosensitisation in livestock.

Conclusion

The high incidence of primary photosensitisation combined with the pathologic findings strongly suggest the presence of photosensitising agent(s) in sufficient quantity in both biserrula cultivars. Severe dermatitis in grazing sheep was observed unexpectedly during late winter growth conditions in central NSW, when low ambient air and soil temperatures and variable UV exposure were experienced.

Finally, the design and introduction of a quantitative grading system, such as the one proposed and applied in this report, to evaluate the severity of photosensitisation lesions will facilitate the objective evaluation, monitoring and comparison of photosensitisation cases in the future.

Abbreviations
AST: Aspartate aminotransferase; CK: Creatine kinase; DSE: Dry sheep equivalent; EDTA: Ethylenediaminetetraacetic acid; GGT: Gamma-glutamyl transferase; H&E: Haematoxylin and eosin; ha: Hectare; PS: photosensitisation

Acknowledgments
The authors thank the staff of the Charles Sturt University Veterinary Diagnostic Laboratory for their excellent service and James Stephens for animal assistance..

Funding
JCQ and BH are both supported by Meat and Livestock Australia (MLA). This trial was funded by Meat and Livestock Australia and Australian Wool Innovations under the project code B.PSP.0013 'Pasture legumes in the mixed farming zones of WA and NSW: shifting the baseline'. Additional funding was provided by a Research Centre Fellowship awarded to JQ by the Graham Centre for Agricultural Innovation, an alliance between NSW Department of Primary Industries and Charles Sturt University. YC is funded by a PhD Scholarship from the Graham Centre for Agricultural Innovation, Charles Sturt University.

Authors' contributions
JCQ conducted the grazing trial, designed the clinical grading system and finalised the manuscript. YC reviewed the literature, drafted the manuscript and prepared the photos. BH carried out agronomic analysis of pasture compositions and assisted with the management of the trial. MST assisted with the management of the trial and animal care post trial. PL performed the necropsy, histology and diagnostic interpretation. LW provided comment on the manuscript and provided information on putative causal compounds. All authors reviewed and approved the final manuscript.

Consent for publication
Not applicable.

Competing interests
The authors declare that they have no competing interests.

References
1. Mauldin EA, Peters-Kennedy J. Integumentary System. In: Maxie MG, editor. Jubb, Kennedy, and Palmer's Pathology of Domestic Animals, vol. 1. 6th ed. St. Louis; 2016. p. 577–80.
2. Quinn JC, Kessell A, Weston LA. Secondary plant products causing photosensitization in grazing herbivores: their structure, activity and regulation. Int J Mol Sci. 2014;15:1441–65.
3. Santos DS, Silva CCB, Araújo VO, de Fátima Souza M, Lacerda-Lucena PB, Simões SVD, Riet-Correa F, Lucena RB. Primary photosensitization caused by ingestion of Froelichia Humboldtiana by dairy goats. Toxicon. 2017;125:65–9.
4. Kessell AE, Ladmore GE, Quinn JC. An outbreak of primary photosensitisation in lambs secondary to consumption of Biserrula Pelecinus (biserrula). Aust Vet J. 2015;93:174–8.
5. Puschner B, Chen X, Read R, Affolter VK. Alfalfa hay induced primary photosensitization in horses. Vet J. 2016;211:32–8.
6. Agerholm JS, Thulstrup PW, Bjerrum MJ, Bendixen C, Jørgensen CB, Fredholm MA. Molecular study of congenital erythropoietic porphyria in cattle. Anim Genet. 2012;43:210–5.
7. Sargison ND, Baird GJ, Sotiraki S, Gilleard JS, Busin V. Hepatogenous photosensitisation in Scottish sheep casued by Dicrocoelium Dendriticum. Vet Parasitol. 2012;189:233–7.
8. Haydardedeoğlu NO. Hepatogenous photosensitization in Akkaraman lambs: special emphasis to oxidative stress and thrombocytopenia. Ankara Üniversitesi Veteriner Fakültesi Dergisi. 2013;60:116–22.
9. Minervino AHH, Júnior RAB, Rodrigues FAML, Ferreira RNF, Reis LF, Headley SA, et al. Hepatogenous photosensitization associated with liver copper accumulation in buffalos. Res Vet Sci. 2010;88:519–22.
10. Jesse FF, Ramanoon SZ. Hepatogenous photosensitization in cattle - a case report. Vet World. 2012;5:764–6.
11. Glastonbury JRW, Doughty FR, Whitaker SJ, Sergant E. A syndrome of hepatogenous photosensitisation, resembling geeldikkop, in sheep grazing Tribulus Terrestris. Aust Vet J. 1984;61:314–6.

12. Scheie E, Ryste EV, Flåøyen A. Measurement of phylloerythrin (phytoporphyrin) in plasma or serum and skin from sheep photosensitised after ingestion of Narthecium Ossifragum. N Z Vet J. 2003;51:99–103.

13. Revell C, Revell D. Meeting "duty of care" obligations when developing new pasture species. Field Crops Res. 2007;104:95–102.

14. Hackney B. Hardseeded annual legumes – an on-demand break option with significant benefit to the mixed farming zone. Grains Research and Development Corporation Update papers. 2015; 2. https://grdc.com.au/ resources-and-publications/grdc-update-papers/tab-content/grdc-update-papers/2015/02/hardseeded-annual-legumes; Accessed 18/7/2017.

15. Loi A, Revell C, Casbah NB, Biserrula M. Persistent pasture legumes for Mediterranean farming systems. Genetic Resources of Mediterranean Pasture and Forage Legumes. Dordrecht: Department of Agriculture Farmnote. 2005:90–5.

16. Salam KP, Murray-Prior R, Bowran D, Salam MU. Cadiz and Casbah pastures in Western Australia: Breeders' expectation, Farmers' evaluation and achieved adoption. Ext Farm Sys J. 2009;5:103–12.

17. Ghamkhar K, Revell C, Erskine W. Biserrula Pelecinus L. - genetic diversity in a promising pasture legume for the future. Crop Pasture Sci. 2012;63:833.

18. Hogg H. Photosensitisation in sheep grazing Biserrula. Western Australian Department of Agriculture and Food Farmnote. 2010;396:4.

19. Thomas DT, Milton JTB, Revell CK, Ewing MA, Lindsay DR. Individual and socially learned preferences for biserrula (Biserrula Pelecinus L.) in sheep. Grass Forage Sci. 2015;70:374–80.

20. Carr SJ, Loi A, Howieson JH. Attributes of Biserrula pelecinus L.(biserrula): A new pasture legume for sustainable farming on acidic sandy soils in Mediterranean environments. Cahiers Options Mediterraneennes. 1999;39:87–90.

21. Swinny E, Revell CK, Campbell N, Spadek E, Russo C. Search of photosensitising compounds in the annual forage legume Biserrula Pelecinus L. Crop Pasture Sci. 2015;66:1–6.

22. Chapman RE, Bennett JW, Carter NB. Erythemal response of biologically denuded sheep to sunlight and the effects on skin structure and wool growth. Aust J Biol Sci. 1984;37:217–35.

23. Loi A, Howieson JH, Carr SJ. Register of Australian herbage plant varietys. Biserrula Pelecinus L.(biserrula) cv. Casbah. Aus J Exp Agr. 2001;41:841–2.

24. Salam KP, Murray-prior R, Bowran D. A 'Dream'pasture and its comparison with two existing annual pasture legumes for Western Australia: a farmers' eye view. Livest Res Rural Dev. 2010;22:1.

25. Nichols P, Loi A, Nutt BJ, Evans PM, Craig AD. New annual and short-lived perennial pasture legumes for Australian agriculture-15 years of revolution. Field Crops Res. 2007;104:10–23.

26. Boschma SP, Crockerb GJ, Lodge GM. Evaluation of pasture legumes in northern new South Wales, Australia. N Z. J Agr Res. 2011;54:203–13.

8

Ultrasonographic findings in goats with contagious caprine pleuropneumonia caused by *Mycoplasma capricolum* subsp. *capripneumoniae*

Mohamed Tharwat* and Fahd Al-Sobayil

Abstract

Background: In goats, contagious caprine pleuropneumonia (CCPP) is a cause of major economic losses in Africa, Asia and in the Middle East. There is no information emphasising the importance of diagnostic ultrasound in goats with CCPP caused by *Mycoplasma capricolum* subsp. *capripneumoniae* (*Mccp*). This study was designed to describe the ultrasonographic findings in goats with CCPP caused by *Mccp* and to correlate ultrasonographic with post-mortem findings. To this end, 55 goats with CCPP were examined. Twenty-five healthy adult goats were used as a control group.

Results: Major clinical findings included harried, painful respiration, dyspnoea and mouth breathing. On ultrasonography, a liver-like echotexture was imaged in 13 goats. Upon post-mortem examination, all 13 goats exhibited unilateral pulmonary consolidation. Seven goats had a unilateral hypoechoic pleural effusion. At necropsy, the related lung was consolidated and the pleural fluid appeared turbid and greenish. Pleural abscessiation detected in five goats was confirmed post-mortem. Twenty-eight goats had a bright, fibrinous matrix extending over the chest wall containing numerous anechoic fluid pockets with medial displacement and compression of lung tissue. Echogenic tags imaged floating in the fluid were found upon post-mortem examination to be fibrin. In two goats, a consolidated right parenchyma was imaged together with hypoechoic pericardial effusions with echogenic tags covering the epicardium. At necropsy, the right lung was consolidated in three goats and fibrin threads were found covering the epicardium and pericardium.

Conclusions: In goats with CCPP, the extension and the severity of the pulmonary changes could not be verified with clinical certainty in most cases, whereas this was possible most of the time with sonography, thus making the prognosis easier. Ultrasonographic examination of the pleurae and the lungs helped in the detection of various lesions.

Keywords: Contagious caprine pleuropneumonia, CCPP, Goat, Mycoplasma, Ultrasonography

Background

In many parts of the world, goats are considered important domestic animals. They are kept as a source of meat, milk, cheese, and fibre. Goats are also wonderful to raise purely for enjoyment, as a hobby or for show. Goats can survive in harsh environments in which other livestock species would perish. In addition, they are able to live and reproduce in icy mountainous areas as well as in the hot,

dry desert. Therefore, improved goat husbandry will help maximize human food supplies from marginal agricultural lands under restrictive climatologic circumstances [1].

Among the important goat diseases, mycoplasmal infections result in significant losses in many countries, and morbidity and mortality can reach 100% [1]. Of these mycoplasmal infections, contagious caprine pleuropneumonia (CCPP), occurring in many countries in Asia and Africa, is a severe contagious respiratory disease of goats [2]. It is characterised by fever, high morbidity, and high mortality. Respiration is accelerated and painful, coughing is frequent, and, in the terminal

* Correspondence: mohamedtharwat129@gmail.com
Department of Veterinary Medicine, College of Agriculture and Veterinary Medicine, Qassim University, Buraydah, Saudi Arabia

stages, the animal is unable to move, standing base wide with neck extended. The gross lesions of the disease are typically limited to the thoracic cavity and characterised by fibrinous pleuropneumonia, lung hepatisation, and accumulation of pleural fluid [3–5]. The pneumonia may often be unilateral [3, 6]. Rapid and inexpensive detection of CCPP is carried out using a *Mccp* capsular polysaccharide-specific antigen detection latex agglutination test (LAT) [7–10]. The World Organization for Animal Health (OIE) has recommended the LAT for confirmation of clinical cases of goats with CCPP [2].

The type, severity, and extent of lung disease cannot always be determined by physical examination alone, which may lead to misinterpretation of respiratory symptoms and ineffective therapy [11]. Because the waves are incapable of penetrating gas-filled structures, physiologically normal lung tissue cannot be examined by ultrasound; however, sonography is suitable for the detection of a number of respiratory pathologies [12–17].

In sheep, accurate identification and distribution of pleural and superficial lung pathology necessitate ultrasonographic examination. Ultrasonographic examination of the chest allows critical evaluation of the pleurae and establishment of a definitive diagnosis in most diseased sheep [18–21]. In goats, however, only one report of thoracic osteosarcoma was found in the veterinary literature [22]. The present study was designed to describe the ultrasonographic findings in goats with CCPP. The clinical and post-mortem findings are also described.

Methods

Animals, history and physical examination

Fifty-five goats (mean age 2.5 ± 1.1 years; mean body weight 26.4 ± 10.1 kg) were examined in the Veterinary Teaching Hospital, Qassim University, Saudi Arabia, between February 2010 and August 2015. The goats had been admitted because of weight loss, anorexia and respiratory signs which included dyspnoea, polypnea, cough and nasal discharges. Twenty-five healthy adult goats (mean age 2.8 ± 0.9 years; mean body weight 31.0 ± 12.7 kg) were used as a control group.

The diseased goats were enrolled in the study in situ based on a positive serological LAT (CapriLAT, product code: RAI 6224, lot number: MccpLAT304141, Animal Health and Veterinary Laboratories Agency, Surrey, United Kingdom) that confirmed the detection of *Mccp* as the causative agent of CCPP [7–10]. The control goats were enrolled based on a negative result of the LAT. The owners of the goats (both sick and controls) provided informed consent for their animals to participate in the study, and the owners of the controls gave permission for the healthy animals to be euthanised.

Ultrasonographic examination

A real-time, B-mode ultrasound machine equipped with a 7.5 MHz-sector transducer (SSD-500, Aloka, Tokyo, Japan) was used to image the thorax and heart in the non-sedated diseased goats. Parallel, the lungs and the heart was scanned in the control animals. Firstly, the two sides of the thorax of each animal were clipped and the skin was shaved. The thoracic ultrasonography was carried out in the goats as in that reported for sheep [19, 20], and the echocardiography was conducted as has been recently reported [23]. As reported by Buczinski et al. [24], criteria for lung consolidation, pleural fluid accumulation, fibrinous pleurisy and pericardial fluid accumulation was defined as follows. Lung consolidation was defined as the ability to observe the abnormal lung parenchyma as a heterogenous hypoechoic to echoic area. Pleural fluid accumulation was diagnosed if disruptions between the parietal and the visceral pleura were observed during examination. Fibrinous pleurisy was defined if fibrinous matrix was observed during examination extending over the chest wall and containing numerous fluid pockets, and the pleural line was serrated with an irregular shape. Pericardial fluid accumulation was diagnosed if

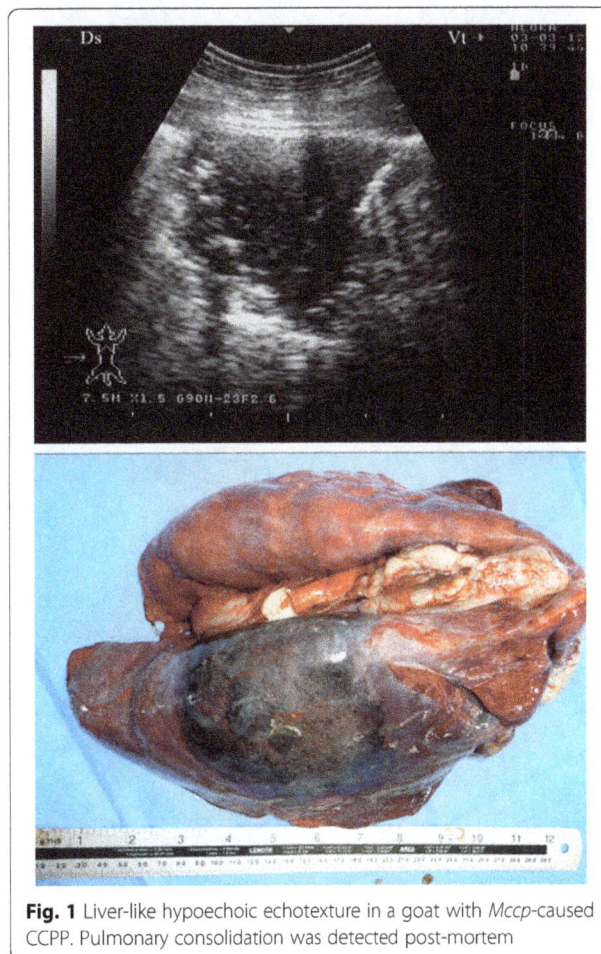

Fig. 1 Liver-like hypoechoic echotexture in a goat with *Mccp*-caused CCPP. Pulmonary consolidation was detected post-mortem

disruptions between the parietal and the visceral pericardium were observed during examination.

Postmortem examination

Both the diseased and control goats were euthanized by throat cutting without breaking the neck and thoroughly examined post-mortem. If present, the lung consolidation, pleural fluid accumulation, fibrinous pleurisy, pericardial fluid accumulation were recorded and described.

Results

Of the 47 females and eight males, 45 were local goat breeds (Ardi) and the remaining 10 were Syrian goats. On initial examination, the diseased goats had a mean internal body temperature of 39.8 ± 1.7 °C, a mean pulse of 122 ± 23 beats per minute and a mean respiratory rate of 45 ± 9 breaths per minute. Harried, painful respiration was detected in 48 goats, 15 had dyspnoea and 33 animals displayed open mouth breathing. Spontaneous coughing was detected in 39 goats and seven had coughing upon stimulation. Eighteen goats were admitted in a recumbent position. Surprisingly, nine goats had no history of coughing at all and did not cough upon stimulation. Fifty-one of the 55 diseased goats were admitted in a depressed and poor body condition. Upon percussion of the lungs, 51 goats exhibited a unilaterally reduced volume and four had increased resonance. Upon auscultation of the lungs, 24 cases exhibited unilaterally mild to severe increased vesicular breath sounds, nine had rough breath sounds, nine had splashing and pleuritic friction sounds, six had an absence of lung sounds, and wheezing was detected in seven cases. Some goats had a combination of these findings. In all 55 diseased goats, the LAT was positive for the presence of *Mccp*.

Imaging of the lungs in the control goats revealed normally aerated lungs characterised by the uppermost hyperechoic linear image with numerous reverberation artefacts running regularly and parallel below this line. Both pleural leaves appeared as a broad, smooth, hyperechoic line between the surface of the lungs and the musculature of the thoracic wall moving synchronously with respiration. It was not possible to differentiate the parietal and visceral pleurae. The motion of the lungs synchronous with respiration was visible. No pleural

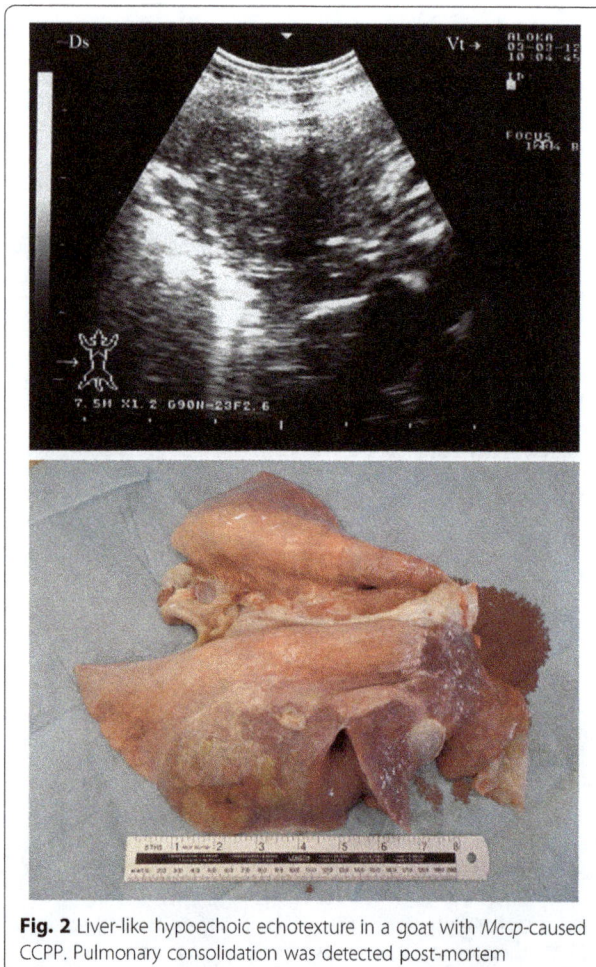

Fig. 2 Liver-like hypoechoic echotexture in a goat with *Mccp*-caused CCPP. Pulmonary consolidation was detected post-mortem

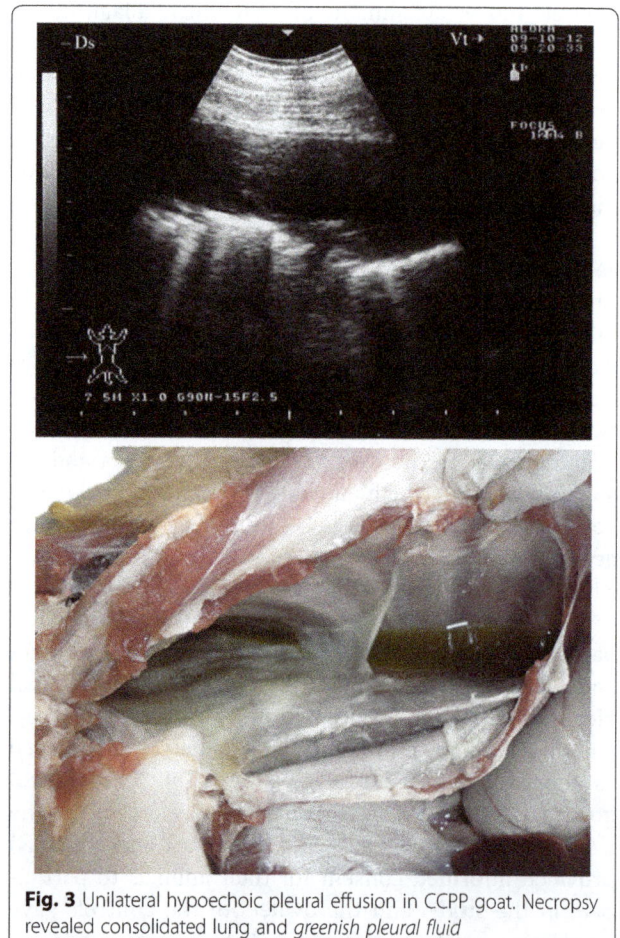

Fig. 3 Unilateral hypoechoic pleural effusion in CCPP goat. Necropsy revealed consolidated lung and *greenish pleural fluid*

fluid was visualised in any of the control goats. Upon post-mortem examination, none of the control goats displayed any pulmonary abnormality.

In 13 of the diseased goats, a non-ventilated lung parenchyma with a liver-like echotexture was imaged. Depending on the degree of atelectasis, the ventilated lung deep to the consolidation could be identified by the weak, defined, and blurry reverberation artefacts. The extensive hypoechoic zones in the cranioventral lung fields and the cranioventral portions of the main lobes were confirmed to be consolidated lung tissue upon post-mortem examination (Figs. 1 and 2). The post-mortem examination showed all 13 goats to have unilateral pulmonary consolidation: nine at the diaphragmatic lobes (Fig. 1) and four at the diaphragmatic together with the anterior lobes (Fig. 2). Remarkably, 11 goats had consolidation in the right lung and only two in the left lung.

Ultrasonography of the thorax in seven goats revealed a unilateral pleural effusion that appeared hypoechoic (Fig. 3). The visceral pleura appeared broader and more hyperechoic than normal due to acoustic enhancement by the pleural exudates. Upon necropsy, the related lung

was consolidated and the pleural fluid appeared turbid and greenish. Five other goats displayed a free hypo-and anechoic fluid with echogenic foci. At post-mortem examination, it was confirmed to be pus (Fig. 4).

Sonograms obtained from 28 goats with marked fibrinous pleurisy revealed a bright fibrinous matrix extending over the chest wall containing numerous anechoic fluid pockets with medial displacement and compression of lung tissue (Figs. 5, 6, 7, 8, 9). Due to acoustic enhancement, the surface of the displaced lung lobes had the appearance of broad hyperechoic lines. Echogenic tags that were imaged floating in the fluid were found to be fibrin upon post-mortem examination. The pleural fluid was a clear yellow in eight goats (Fig. 5), turbid and yellowish in five (Fig. 6), reddish in eight (Fig. 7) and dark red in the remaining seven goats (Figs. 8 and 9).

In two goats, a consolidated right parenchyma was imaged together with hypoechoic pericardial effusions and echogenic tags covering the epicardium (Fig. 10). Upon necropsy, the right lung was consolidated in three goats and fibrin threads were found covering the epicardium and pericardium.

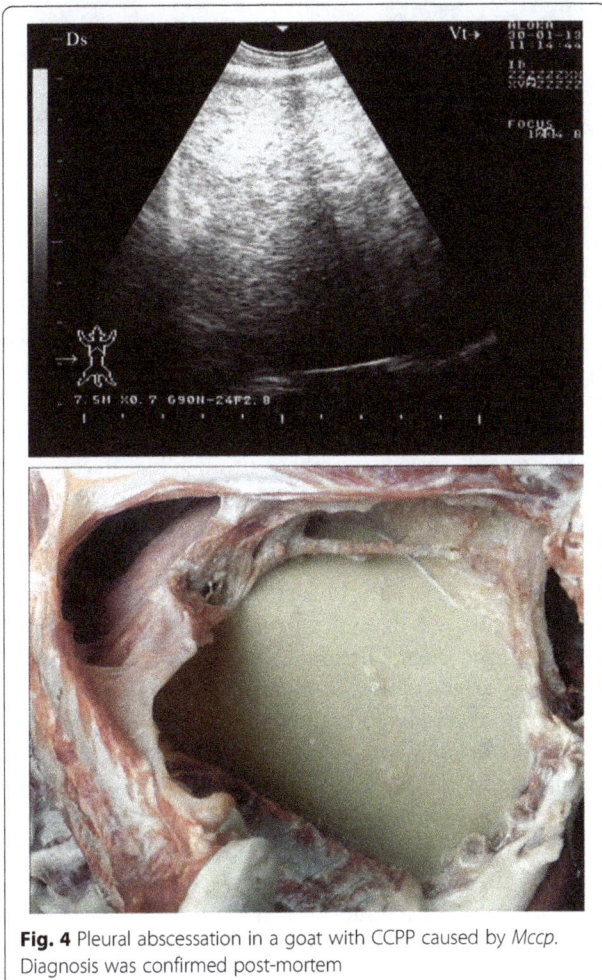

Fig. 4 Pleural abscessation in a goat with CCPP caused by *Mccp*. Diagnosis was confirmed post-mortem

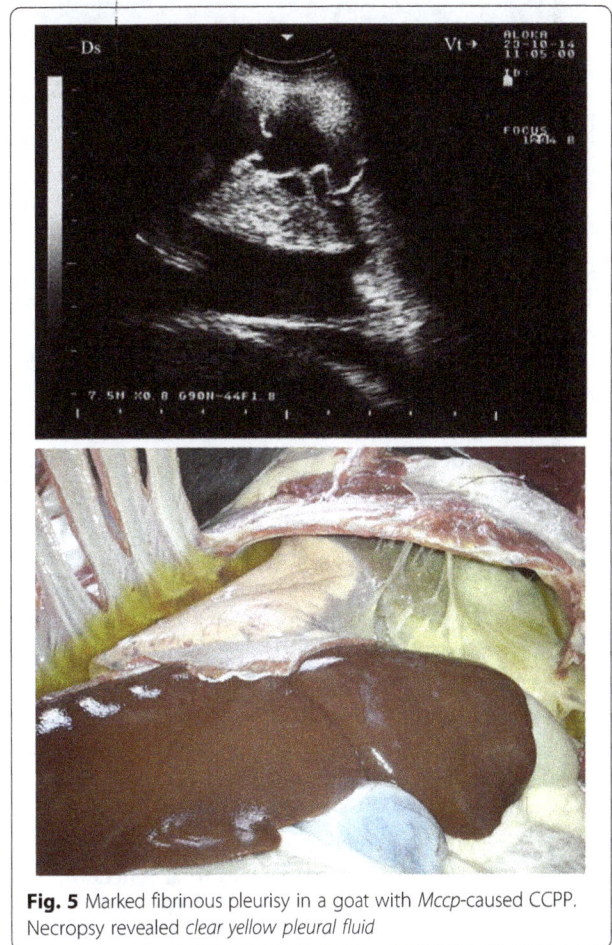

Fig. 5 Marked fibrinous pleurisy in a goat with *Mccp*-caused CCPP. Necropsy revealed *clear yellow pleural fluid*

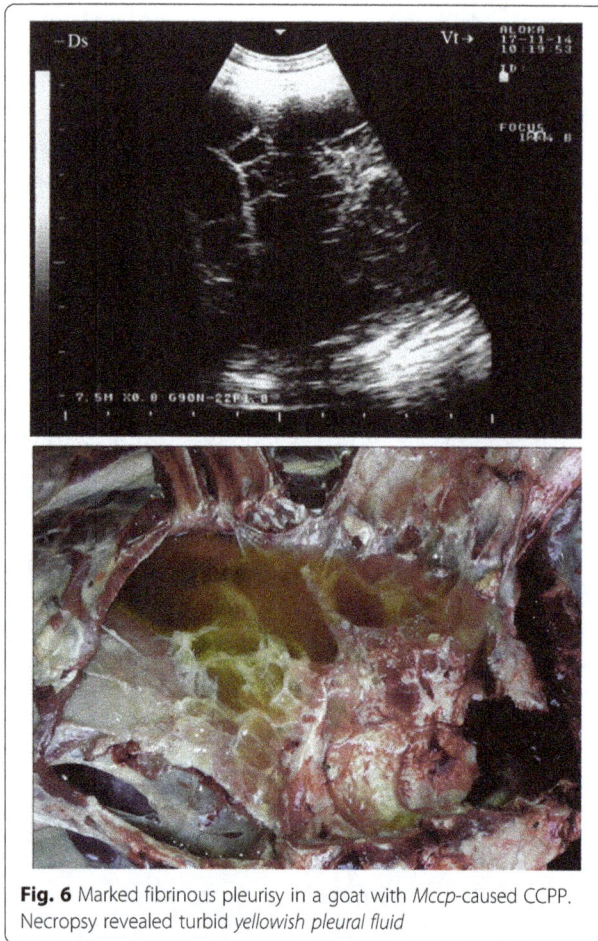

Fig. 6 Marked fibrinous pleurisy in a goat with *Mccp*-caused CCPP. Necropsy revealed turbid *yellowish pleural fluid*

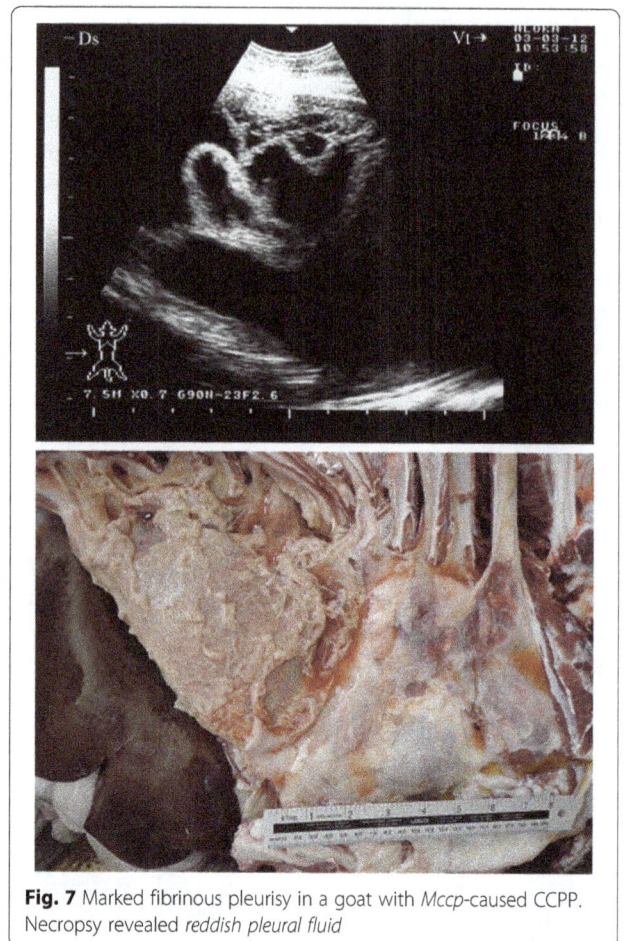

Fig. 7 Marked fibrinous pleurisy in a goat with *Mccp*-caused CCPP. Necropsy revealed *reddish pleural fluid*

Discussion

To the author's knowledge, there is no information emphasising the importance of diagnostic ultrasound in goats with CCPP caused by *Mccp*. A knowledge of the pathological lung changes is necessary in order to evaluate the diagnostic value of ultrasonography. In this study, post-mortem examinations were therefore undertaken to evaluate the diagnostic value of the imaging technique.

In human medicine, conventional radiographs are still the first diagnostic imaging choice for thoracic examination [25]. Similarly, in veterinary medicine, radiography is superior to ultrasonography in the identification of diffuse diseases of lung parenchyma such as pulmonary emphysema, oedema, interstitial pneumonia and diffuse neoplastic or granulomatous processes [26–28]. However, unlike radiography, ultrasonography requires no special restrictions or health and safety procedures [19].

As shown in the present study, in a healthy animal, pulmonary air content complicates the ultrasonographic assessment of lung parenchyma. Due to total reflection, the intercostal transmission of ultrasound where the lungs are air-filled extends only to the visceral pleura and ends at

the air-filled alveoli [29]. A useful ultrasonographic assessment of the lung tissue can be achieved when the pulmonary air content is reduced and the lung has the appearance of liver. The irregularity of the visceral pleural surface can be a first sign of consolidation [28].

In the consolidated lung areas in cattle and horses, there are still varying numbers of air-filled alveoli that form hyperechoic zones [12, 30, 31]. In correlation to the size of the compressed lung area and the duration of the existence of a consolidation, a reduction of these hyperechoic zones can be detected as hyperechoic reflective bands [12, 32, 33]. In calves, Rabeling et al. [34] have reported consolidations as echogenic regions with comet-tail artefacts. In the present study, the consolidations were always hypoechoic and homogenous, possibly due to the accumulation of exudate, blood and mucous. Similar findings have been reported previously [16, 30]. In horses, consolidation was observed most often cranioventrally, whereby the right lung was usually more severely affected [30]. In this study, consolidation was observed mostly caudodorsally in the right lung.

In bovines, the most profound lung changes can be observed in most cases in the cranial and above all the

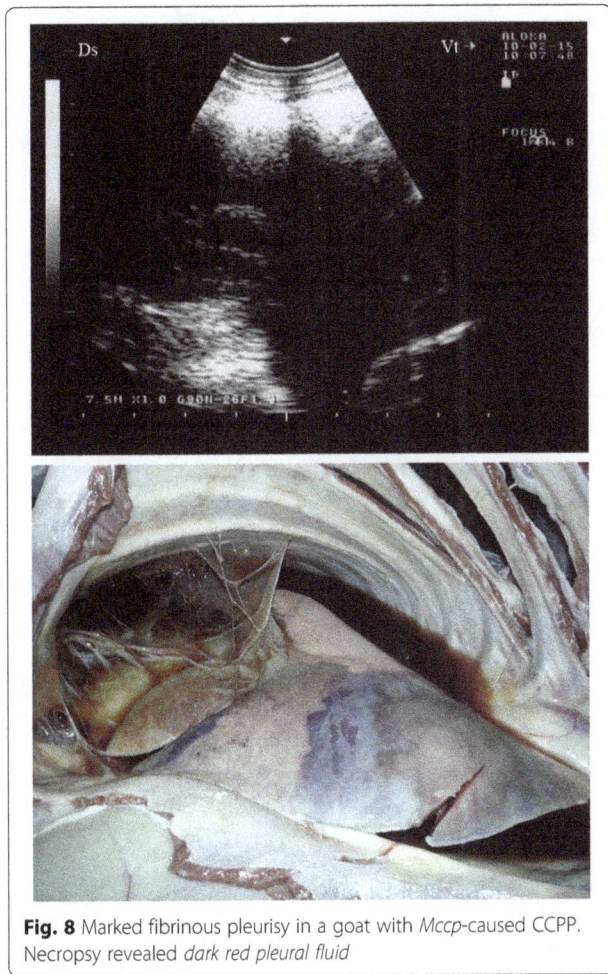

Fig. 8 Marked fibrinous pleurisy in a goat with *Mccp*-caused CCPP. Necropsy revealed *dark red pleural fluid*

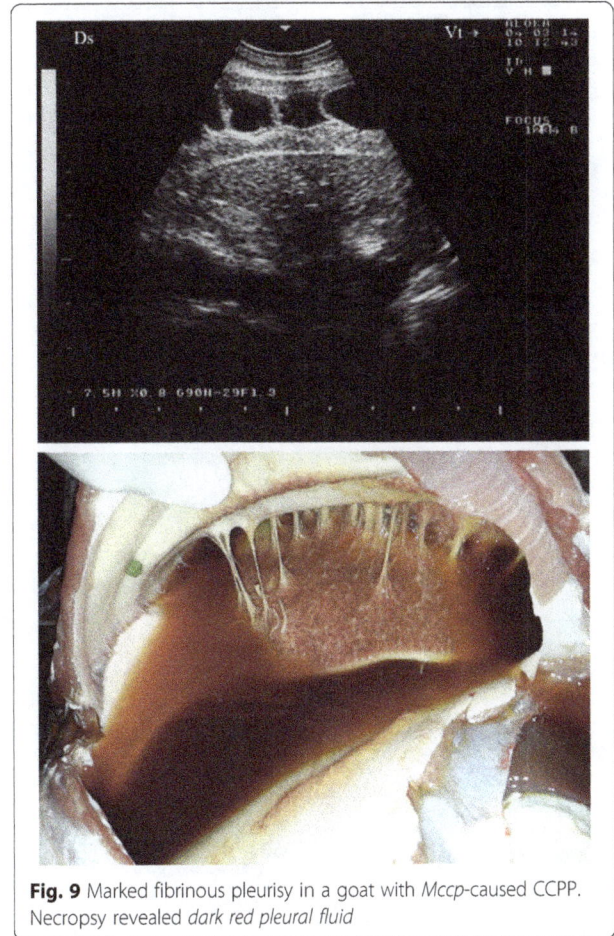

Fig. 9 Marked fibrinous pleurisy in a goat with *Mccp*-caused CCPP. Necropsy revealed *dark red pleural fluid*

cranioventral lung areas. Therefore, it is advisable to begin the sonographic examination in that location [31]. In contrast, in this study, most lung changes were detected in the caudodorsal lung areas.

In this study, ultrasonography allowed the pleural effusion to be visualised in a much more definitive manner and was able to qualify the nature and the extent of the effusion. Pleural effusion appears as an anechoic space between the lung, thoracic wall, diaphragm and heart, with acoustic enhancement deep to the lesion and often with septa floating within it [28]. The parital and visceral pleurae were also separated from one another; similar findings were reported by Reef et al., [26]. Within the pleural cavity, anechoic fluid represents transudate, and increased echogenicity points toward an increased cell count or total protein concentration [12, 26, 28]. This feature is, however, unreliable and must always be confirmed by thoracocentesis [35].

In practice, ultrasonography is mostly used for diagnostic purposes in the case of pleural effusion. The images are often characteristic, and ultrasound-guided aspiration provides a safe way to obtain liquid for analysis. Another

use for such technology is the identification of the area to be drained so as to provide relief to the patient. It is also an objective tool for monitoring patients secondary to therapy [16]. Ultrasonography has been most helpful in the definitive diagnosis of superficial lung abscesses where the anechoic areas containing multiple hyperechoic dots were bordered distally by a broad hyperechoic capsule [21]. The present ultrasonographic findings of pleural abscesses were similar to the ultrasonographic description of suppurative pleuropneumonia in a ram [35–37].

This study has 2 limitations; first was using the LAT in detecting goats with CCPP. A conclusive diagnosis should have included culture or a molecular confirmation of the pathogen. Excluding co-infections with bacteriology was a second limitation. We aimed to assist filed veterinarians in early detection and isolation of suspected cases until the results of culture or molecular confirmation arrive. However, the causative agent, *Mccp*, is very difficult to cultivate in vitro. This may be attributed to its fastidiousness and/or misuse of antimicrobials [38]. In addition, current serological tests used for detection of *Mccp* antibodies include the growth inhibition test, complement fixation test (CFT), indirect haemagglutination assay, competitive

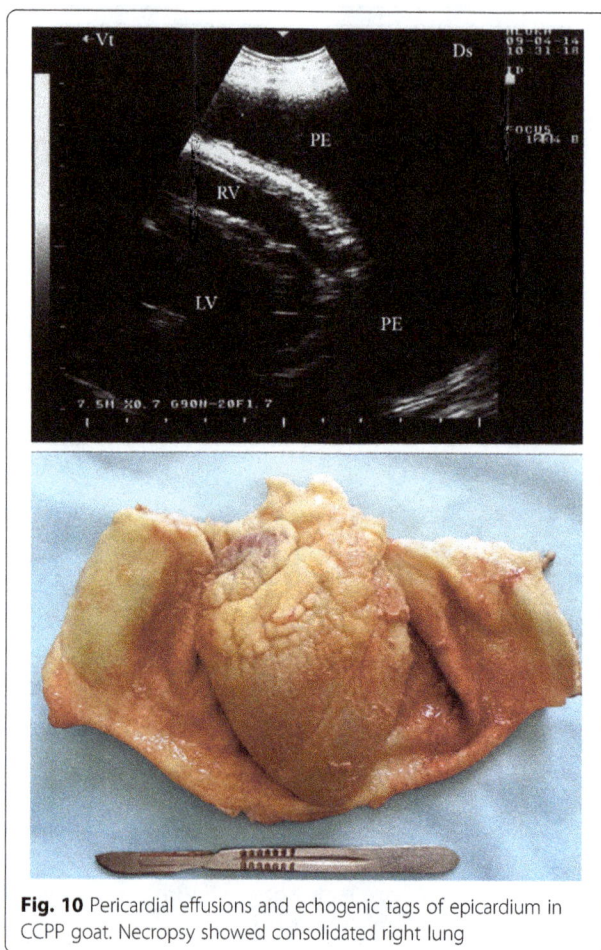

Fig. 10 Pericardial effusions and echogenic tags of epicardium in CCPP goat. Necropsy showed consolidated right lung

enzyme-linked immunosorbent assay and LAT (c-ELISA). The first two tests are ranked the least sensitive for confirmation of clinical CCPP cases (+); the third and fourth are suitable (++), and the fifth (+++) is the recommended [2]. The LAT is considered the simplest, quick and excellent procedure for the "*in field* or in situ" diagnosis of CCPP either in whole blood or serum [39]. Moreover, the LAT is reported to be more sensitive than the CFT [40] and also easier to perform than the c-ELISA [41].

Conclusions

In this study, some ultrasound images in goats with CCPP were characteristic of pathologic lesions in the chest and can help clinicians with their diagnosis by allowing visualisation of the lesion itself or serving as a guide for aspiration. In current veterinary farm practice, however, in which radiographic examination is impossible, ultrasonography is an available diagnostic tool that is quickly implemented and non-invasive.

Acknowledgements
The authors would like to thank N. Peachy, English professor, Deanship for Educational Services, Qassim University, for language revising.

Funding
No funding was received for this study.

Authors' contributions
MT initiated and planned the study. MT and FA carried out the experimental work. MT wrote the manuscript and made the figures. FA read, revised and approved the manuscript.

Consent for publication
Not applicable.

Competing interests
The authors declare that they have no competing interests.

References
1. DaMassa AJ, Wakenell PS, Brooks DL. Mycoplasmas of goats and sheep. J Vet Diagn Investig. 1992;4:101–13.
2. World Organization for Animal Health. Contagious caprine pleuropneumonia. In: terrestrial manual, world assembly of delegates of the OIE. Paris: Office International Des Epizooties; 2014. Chapter 2.7.6. p. 1–15.
3. Thiaucourt F, Bölske G, Leneguersh B, Smith D, Wesonga H. Diagnosis and control of contagious caprine pleuropneumonia. Rev Sci Tech. 1996;15:1415–29.
4. World Organization for Animal Health. Contagious caprine pleuropneumonia. In: manual of diagnostic tests and vaccines for terrestrial animals (mammals, birds and bees), vol. Vol. II. 5th ed. Paris: Office International Des Epizooties; 2004. Chapter 2.4.6. p. 623–34.
5. Hernandez L, Lopez J, St-Jacques M, Ontiveros L, Acosta J, Handel K. Mycoplasma mycoides subsp. capri associated with goat respiratory disease and high flock mortality. Can Vet J. 2006;47:366–9.
6. Thiaucourt F, Bölske G. Contagious caprine pleuropneumonia and other pulmonary mycoplasmoses of sheep and goats. Rev Sci Tech. 1996;15:1397–414.
7. Rurangirwa FR, McGuire TC, Kibor A, Chema S. A latex agglutination test for field diagnosis of contagious caprine pleuropneumonia. Vet Rec. 1987;121:191–3.
8. March JB, Harrison JC, Borich SM. Humoral immune responses following experimental infection of goats with Mycoplasma capricolum subsp. capripneumoniae. Vet Microbiol. 2002;84:29–45.
9. March JB, Gammack C, Nicholas R. Rapid detection of contagious caprine pleuropneumonia using a mycoplasma capricolum subsp. capripneumoniae capsular polysaccharide-specific antigen detection latex agglutination test. J Clin Microbiol. 2000;38:4152–9.
10. Ozdemir U, Ozdemir E, March JB, Churchward C, Nicholas RA. Contagious caprine pleuropneumonia in the Thrace region of Turkey. Vet Rec. 2005; 156:286–7.
11. Divers TJ. Respiratory diseases. In: Divers TJ, Peek SF, editors. Rebhun's diseases of dairy cattle. 2nd ed. Philadelphia: Saunders; 2007. p. 79–129.
12. Braun U, Pusterla N, Fluckinger M. Ultrasonographic findings in cattle with pleuropneumonia. Vet Rec. 1997;141:12–7.
13. Mannison P. Diagnostic ultrasound in small animal practice. Oxford: Blackwell Scientific Publications; 2006. p. 170–87.
14. Tharwat M, Oikawa S. Ultrasonographic characteristics of abdominal and thoracic abscesses in cattle and buffaloes. J Vet Med A. 2007;54:512–7.
15. Tharwat M, Oikawa S. Ultrasonographic evaluation of cattle and buffaloes with respiratory disorders. Trop Anim Health Prod. 2011;43:803–10.
16. Babkine M, Blond L. Ultrasonography of the bovine respiratory system and its practical application. Vet Clin North Am Food Anim Pract. 2009;25:633–49.
17. Buczinski S, Tolouei M, Rezakhani A, Tharwat M. Echocardiographic measurement of cardiac valvular thickness in healthy cows, cows with bacterial endocarditis, and cows with cardiorespiratory diseases. J Vet Cardiol. 2013;15:253–61.
18. Scott P. The role of ultrasonography as an adjunct to clinical examination in sheep practice. Ir Vet J. 2008;61:474–80.
19. Scott P, Sargison ND. Ultrasonography as an adjunct to clinical examination in sheep. Small Rum Res. 2010;92:108–19.
20. Scott P, Collie D, McGorum B, Sargison N. Relationship between thoracic auscultation and lung pathology detected by ultrasonography in sheep. Vet J. 2010;186:53–7.
21. Scott PR. Antibiotic treatment response of chronic lung diseases of adult sheep in the United Kingdom based upon ultrasonographic findings. Vet Med Intern. 2014:537501. doi:10.1155/2014/537501.
22. Braun U, Schwarzwald CC, Forster E, Becker-Birck M, Borel N, Ohlerth S. Extraskeletal osteosarcoma of the thorax in a goat: case report. BMC Vet Res. 2011;7:55.

23. Leroux AA, Moonen ML, Farnir F, Sandersen CF, Deleuze S, Salciccia A, Amory H. Two-dimensional and M-mode echocardiographic reference values in healthy adult Saanen goats. Vet Rec. 2012;170:154

24. Buczinski S, Forté G, Francoz D, Bélanger AM. Comparison of thoracic auscultation, clinical score, and ultrasonography as indicators of bovinerespiratory disease in preweaned dairy calves. J Vet Intern Med. 2014;28:234–42.

25. Mathis G. In: Mathis G, editor. Berlin: Springer; 1996.

26. Reef VB, Boy MG, Reid CF, Elser A. Comparison between diagnostic ultrasonography and radiography in the evaluation of horses and cattle with thoracic disease: 56 cases (1984–1985). J Am Vet Med Assoc. 1991;198:2112–8.

27. Braun U. Pleura, Lunge und Mediastinum. In: Braun U, editor. Atlas und Lehrbuch der Ultraschalldiagnostik beim Rind. Berlin: Parey; 1997. p. 115–41.

28. Flock M. Diagnostic ultrasonography in cattle with thoracic disease. Vet J. 2004;167:272–80.

29. Banholzer P. Thoraxwand, pleura und lunge. In: Kremer H, Dobrinski W, editors. Sonographische Diagnostik. Berlin: Urban & Schwarzenberg; 1993. p. 307–15.

30. Reef VB. Thoracic Ultrasonography: Noncardiac Imaging. In: Reef VB, Saunders WB, editors. Equine Diagnostic Ultrasound. Philadelphia: Saunders; 1998. p. 187–213.

31. Scott PR. Ultrasonographic examination of the bovine thorax. Cattle Practice. 1998;6:151–3.

32. Reef VB. Equine pediatric ultrasonography. Compound Contin Educ. 1991;13:1277–85.

33. Jung C, Bostedt H. Thoracic ultrasonography technique in newborn calves and description of normal and pathological findings. Vet Radiol Ultrasound. 2004;45:331–5.

34. Rabeling B, Rehage J, Dopfer D, Scholz H. Ultrasonographic findings in calves with respiratory disease. Vet Rec. 1998;143:468–71.

35. Lorenz J. Ultraschalldiagnostik. In: Ferlinz R, editor. Diagnostik in der Pneumologie. Stuttgart: Thieme; 1992. p. 104–21.

36. Braun U, Flukiger M, Sicher D, Theil D. Suppurative pleuropneumonia and a pulmonary abscess in a ram: ultrasonographic and radiographic findings. Scheizer Archiv fur Tierheilkunde. 1995;137:272–8.

37. Scott PR, Gessert ME. Ultrasonographic examination of the ovine thorax. Vet J. 1998;155:305–10.

38. Awan MA, Abbas F, Yasinzai M, Nicholas RAJ, Babar S, Ayling RD, Attique MA, Ahmad Z. Prevalence of Mycoplasma capricolum subspecies capricolum and Mycoplasma putrefaciens in goats in Pishin district of Balochistan. Pak Vet J. 2009;29:179–85.

39. March JB, Gammack C, Nicholas RAJ. Rapid detection of contagious caprine pleuropneumonia using a Mycoplasma capricolum subsp. capripneumoniae capsular polysaccharide-specific antigen detection latex agglutination test. J Clin Microbiol. 2000;38:4152–9.

40. Houshaymi B, Tekleghiorghis T, Worth DR, Miles RJ, Nicholas RAJ. Investigations of outbreaks of contagious Caprine Pleuropneumonia in Eritrea. Trop Anim Health Prod. 2002;34:383–9.

41. Thiaucourt F, Bölske G, Libeau G, Le Goff C, Lefevre PC. The use of monoclonal antibodies in the diagnosis of contagious caprine pleuropneumonia (CCPP). Vet Microbiol. 1994;41:191–203.

9

Development and test of a visual-only meat inspection system for heavy pigs in Northern Italy

Sergio Ghidini[1]*[ID], Emanuela Zanardi[1], Pierluigi Aldo Di Ciccio[1], Silvio Borrello[2], Giancarlo Belluzi[3], Sarah Guizzardi[2] and Adriana Ianieri[1]

Abstract

Background: There is a general consensus in recognizing that traditional meat inspection is no longer able to address the hazards related to meat consumption. Moreover, it has been shown that invasive procedures, such as palpation and incision, can increase microbial contamination in carcasses. For these reasons, legislations all over the world are changing meat inspection techniques, moving towards visual-only techniques. Hence, there was also the need to test visual-only inspection in pigs in Italy.

Results: A protocol for visual-only post-mortem inspection was produced together with a 24-class scheme used to record pathological lesions. A list of guidelines needed for univocal interpretation and classification of lesions was developed. To record lesions at the slaughtering line, a light instrument that is resistant to the slaughter environment was designed and then produced in collaboration with an electro-medical company. Six contracted veterinarians were chosen and trained. They performed visual-only post-mortem inspections on 231.673 heavy pigs in three different slaughterhouses of Northern Italy. Visual-only inspection was compared to traditional inspection on 38.819 pig carcasses. No relevant differences were found between the two systems.

Conclusions: The comparison between traditional and visual-only inspection showed that visual-only inspection can be adopted in pig slaughterhouse. The analysis of the performance of the veterinarians stressed the importance of standardization and continuous education for veterinarians working in this field.

Background

Veterinary inspection has been performed for more than a century in slaughterhouses, and it has been effective in protecting consumers against classical hazards such as *Mycobacterium bovis* and parasites. However, there is a consensus around the idea that traditional inspection methods in slaughterhouses no longer cope with the hazards that pose the highest foodborne risks today, such as Salmonella and Yesinia. In industrialised countries, classical diseases are now more effectively controlled with eradication plans [1]. Back in 2011, EFSA [2] stated that the traditional inspection system in swine is not targeted to the main hazards deriving from meat consumption. These hazards are no longer detectable by classical meat inspection because they are no longer caused by pathogens associated with specific lesions and are sometimes related to chemicals. Moreover, procedures such as palpation and incision of the viscera by veterinarians can lead to cross contamination of the carcasses [3].

Considering this evidence, in 2014, the European Commission amended EU Regulation 854/2004 via EU Regulation 219 [4], which laid down specific rules for the organisation of official controls on products of animal origin intended for human consumption [5]. In particular, the regulation stated that starting in June 2014, post-mortem inspection in domestic swine should only be visual and that the official veterinarians shall proceed with additional post-mortem inspection procedures using incision and palpation of the carcass and offal when, in his or her opinion, clinical signs and

* Correspondence: sergio.ghidini@unipr.it
[1]Department of Food and Drug, Parma University, Via Del Taglio, 10, 43126 Parma, Italy

lesions may indicate a possible risk to public health, animal health or animal welfare.

A classification of pig producers as a function of their risk level could help the official veterinarian choose the inspection method [6]. Such a classification should be possible using the food chain information (FCI) module. However, FCI proved to be inefficient in providing such information [7]. In fully integrated chains, it is certainly easier to get more information regarding the farm of origin. Such additional information can be useful for a classification of the farms based on risk.

In Italy, pork production shows a variety of organisational structures and farm size patterns. In 2012, the national pig population was approximately 8.600.000 animals (Eurostat). Southern Italy is characterized by a large number of small-scale farms and many low productivity slaughterhouses, producing a total of 5.700.000 carcasses per year (2012). The North of Italy, where approximately 9.300.000 carcasses are produced per year, is characterized by large-scale indoor intensive farms and high production slaughterhouses (up to 500 carcasses/h). A peculiar feature of swine production in the North of Italy is that there is a very high degree of integration between farmers and meat producers because the majority of swine production in this area processes Protected Designation of Origin products (PDOs). The animals, therefore, share the same genetics, breeding techniques, and feeding schemes, and they have to be born in the North of Italy. In addition, the weight and age of the animals are quite constant since they have to fulfil the requirements of the Parma Ham disciplinary of production. In fact, the animals have to be slaughtered at a minimum age of 9 months and usually weigh approximately 160 kg at the time of slaughtering, with a very small dispersion around the mean because there are economic penalties for lighter and heavier animals [8].

Given this scenario, pig production in the North of Italy can be considered almost fully integrated. Therefore, the holdings in which pigs are raised in this area are fully controlled. When categorizing the holdings according to the risks they pose to public health, they fall into a low-risk class. For this reason, it was considered feasible to test visual-only inspection in this area.

In Italy, there are no data on possible applications of a visual-only inspection system in pigs. In addition, consistent data on post-mortem lesions for pigs at the slaughterhouse are lacking. There have been some local projects in Northern Italy, but the obtained data are not homogenous and comparable. Moreover, in their review, Stark el al. [9] highlighted "a substantial lack of suitable and accessible published data on the frequency of occurrence of many diseases and conditions affecting food animals in Europe." In this context, the Italian Ministry of Health, on behalf of the National Committee for Food Safety, financed a project to study new inspection systems for both the South and the North of Italy.

To fulfil the needs of the high productivity slaughterhouses of the North of Italy, which are characterized by a high working speed, a visual-only inspection system was designed. The system was then tested in three slaughterhouses in the North of Italy to obtain data on the prevalence of post-mortem lesions in pigs dedicated to the production of PDO products. The visual system was then compared to the "traditional" inspection using invasive procedures.

Methods
Study area and population
The Parma Ham Consortium of production limits the area of origin of the animals dedicated to Parma Ham production (and other PDO products) to the following regions: Emilia-Romagna, Veneto, Lombardy, Piedmont, Molise, Umbria, Tuscany, Marche, Abruzzo and Lazio [8]. These regions represent the whole north and a large part of the centre of Italy. The pigs belong to the Large White, Landrace, Duroc breeds and their hybrids. They must be slaughtered at a minimum age of 9 months. At this age, they reach an average weight of 160 kg.

In 2013, 4199 farms in this area produced and then sent to slaughter 8.071.726 animals for transformation into PDO products. Pigs are usually sent to the slaughterhouses in batches of approximately 120 animals. The whole animals are slaughtered in 65 slaughterhouses. All the slaughterhouses have the possibility to buy animals from all the PDO regions mentioned above. Eighteen of these slaughterhouses, which are in only two regions (Emilia-Romagna and Lombardy), process 93% of all the animals [10]. For logistical convenience, the present study was performed in 3 of the 18 slaughterhouses (2 in Lombardy and one in Emilia-Romagna) that share the same layout and slaughtering technique and that are very similar in size and processing speed.

Animal selection
Only heavy pigs following the Parma Ham disciplinary (therefore of national origin) were considered in this study. In the slaughterhouses, no further selection of the animals was performed so that all the animals of the Parma Ham area could have the same probability of being chosen for the study. To minimise the influence of the distance between the farm and the slaughterhouse, the sampling times were homogeneously distributed between the different working days of the week and the working hours of the day.

The study was designed to achieve relative standard errors of the prevalence of lesions lower than 1% for lesions with a prevalence higher than 5% and lower than 10% for lesions with a prevalence as low as 0.1%. Using

FAO [11] formulae, we aimed to inspect 200.000 pig carcasses. The study lasted from January to August 2013.

Visual inspection protocol

A new protocol of visual-only inspection for pigs was developed based on EU Regulation 854/2004 because there were no visual-only inspection protocols at the time of this study. To give an operative tool to veterinarians, the anatomical structures to be inspected were re-arranged into three main groups (carcass, red offal, green offal), which resembles the way organs are found at the end of a slaughtering line.

Together with the veterinary service in the Emilia-Romagna and Lombardy regions and the Italian Ministry of Health, a 24-class scheme (Table 1) was developed. The scheme was designed to be easily adopted in high production slaughterhouses, shared at national level and comparable with schemes adopted by the Food Safety and Inspection Service in the USA [12] and the Food Standard Agency in the UK [13]. A list of guidelines needed for univocal interpretation and classification of lesions was developed (Table 1).

Recording system

An electro-medical company (Omicron T S.R.L., Napoli, Italy) was commissioned to design a light tablet (Fig. 1). The tablet had to record lesions on the slaughtering line and be resistant to the slaughterhouse environment.

The instrument weights 420 g, and it is 24 cm wide, 25 cm height and 1 cm thick. It can be connected to a computer via a mini USB port, which is used for both data downloading and charging. On the front panel, it has 24 square buttons (2 cm on each side), representing the lesions in Table 1. Two larger buttons (2 cm high and 4 cm wide) are used for normal animals and to record a change of batch. A vibration is emitted when a button is pushed. In case of a mistake or a change in diagnosis, the operator can change his decision within 2 s, after which the decision is automatically confirmed by a flashing LED light. The data of each working day are then saved in a file and transferred to the central unit, which handles the database.

Software and data analysis

Software was developed with the help of Omicron T S.R.L., (Napoli, Italy). The software had to build a database, starting from the data recorded on the tablets, and then handle a database of at least 400.000 inspected carcasses. The database system used by this software is MySql Server (Oracle, CA). At present, this software is able to extract the data from the database by using five filters: date, type of farm, distance from slaughterhouse, breeding farm code and veterinarian. In the future, the software could be implemented with other filters if necessary. The results of the queries were exported to MS Excel-compatible datasheets, and MS Excel was used for data elaboration. The mean data were compared using Student's t-test.

Personnel

Six veterinarians experienced in meat inspection of pigs were contracted to perform visual-only post-mortem inspection in the slaughterhouses. First, they were trained to use the recording system and then to handle it in operating conditions. Before collecting the data, each veterinarian was trained in the slaughterhouse for a period of about one month (approximately 5000 carcasses). After this period, their results were analysed and they were given further training on the classification of lesions, following the previously developed guidelines. The veterinarians then inspected approximately 40,000 carcasses each to achieve the target of 200,000 carcasses that was previously set. The contracted veterinarians were regularly rotated between the three slaughterhouses.

Place of work

The three slaughterhouses had a capacity varying from 380 to 450 carcasses per hour. In these slaughterhouses, the contracted veterinarians were placed before the official colleagues performing traditional inspections to prevent the contracted veterinarians from diagnosing lesions by relying on cuts made by the official colleagues. To minimise mutual influence, the contracted and official veterinarians were always the maximum possible distance apart in the slaughtering environment (never <5 m) .

Visual-only vs. traditional inspection comparison

In the last period of the study, the developed recording system was also given to official veterinarians, and the data from visual-only (performed by the contracted veterinarians) and traditional inspections (performed by the official veterinarians) of the same pigs were compared on 38.919 pig carcasses. In this period, the work was conducted only in one slaughterhouse to minimise environmental effects. Furthermore, because a different tool to record lesions was already in use in the chosen slaughterhouse, the official veterinarians working there were already trained to perform post-mortem inspections while recording data on an electronic device.

The study was submitted to the Institutional Review Board of The University of Parma that gave a favourable opinion since compliant with ethical principles.

Results

Overall, 231.673 carcasses were inspected by means of a visual-only post-mortem inspection. The carcasses composed 1.832 batches (mean of 126 animals/batch) and came from 323 different farms. A batch is defined as a

Table 1 Lesion classification and the guidelines adopted to record the data

Apparatus	Lesion	Guideline
Respiratory	Pneumonia	Detect both pneumonia and outcomes of pneumonia. Detect pneumonia when an entire lobe is interested, or when not involving the entire lobe, it involves two contralateral lobes. Always consider specific pneumonia. Consider lung abscesses (even one) as pneumonia.
	Pleuropneumonia	Is recognized when adhesions are present on the carcass. Is recognised when fibrin is present on the visceral layer of the pleura.
Digestive	Hepatitis	Hepatitis and outcomes of hepatitis. The presence of fibrin on the capsule should not be classified as hepatitis (classified as peritonitis).
	Hepatosis/hepatic dystrophies	Steatosis and necrosis are to be classified only in cases involving at least an entire lobe or parts of several lobes.
	Peritonitis/perihepatitis	
	Enteritis	Haemorrhagic or necrotic. Thickening of the small intestine.
Reproductive-Urinary	Nephritis	Nephritis and glomerulonephritis.
	Nephrosis	Cystitis and hydronephrosis.
	Cryptorchidism	
Cardio Circulatory	Myocarditis	Involvement of pericarditis. Do not classify degenerative processes in the absence of inflammation as myocarditis.
	Pericarditis	
Integumentary	Dermatitis	Recognized when there is a thickening of the skin. Detect when lesions exceed 50% of the body surface and not when confined to the abdominal region and chest. Detect carcasses massively affected by bites of ectoparasites as dermatitis.
	Erysipelas	Detect whenever the typical skin lesions are encountered.
Locomotor	Arthritis	
	Muscle colour alteration (PSE/DFD)	PSE / DFD
	Oedema/emaciation	
Other (carcass)	Jaundice	
	Abscesses	Detect all abscesses that are not located in the lung or in the liver. Also detect phlegmons as abscesses.
	Neoplasms / tumours	
	Biliary or faecal contamination	Both faecal and bile contamination. In addition, the residual presence of parts of the rectal mucosa is considered contamination.
	Trauma	Skin
		Bruises and injuries due to mismanagement during loading / unloading (bruises and haematomas). Wounds from intraspecific fights and numerous injuries that get to in the derma, possibly infected.
		Skeletal muscle
		Splay-leg animals (open). Do not report results of old injuries.
	Lymphadenopathy	Mesenteric lymph nodes, lung, and generally an increase in the volume of lymph nodes in the carcass.
	Splenomegaly	Detect when affecting more than 50% of the organ.
	Petechial haemorrhages	

group of animals from one farm delivered on one day, usually transported by a single truck. In Table 2, the number and percentage of each lesion detected in each slaughterhouse and an estimate of the prevalence for each lesion. Table 3 presents the results of the comparison between traditional and visual-only inspections. Table 4 shows the total variability achieved and the variability within each lesion (standard deviation and variation coefficient).

Discussion

The majority of lesions were at the respiratory level (Table 2). In fact, more than 20% of the animals had pneumonia or pleuropneumonia. This result is not surprising because intensively bred, fat animals nine months in age were inspected. Furthermore, these data are consistent with those coming from international literature. For instance, in a review of post-mortem data in pig slaughterhouses of New Zealand from 2000 to 2010,

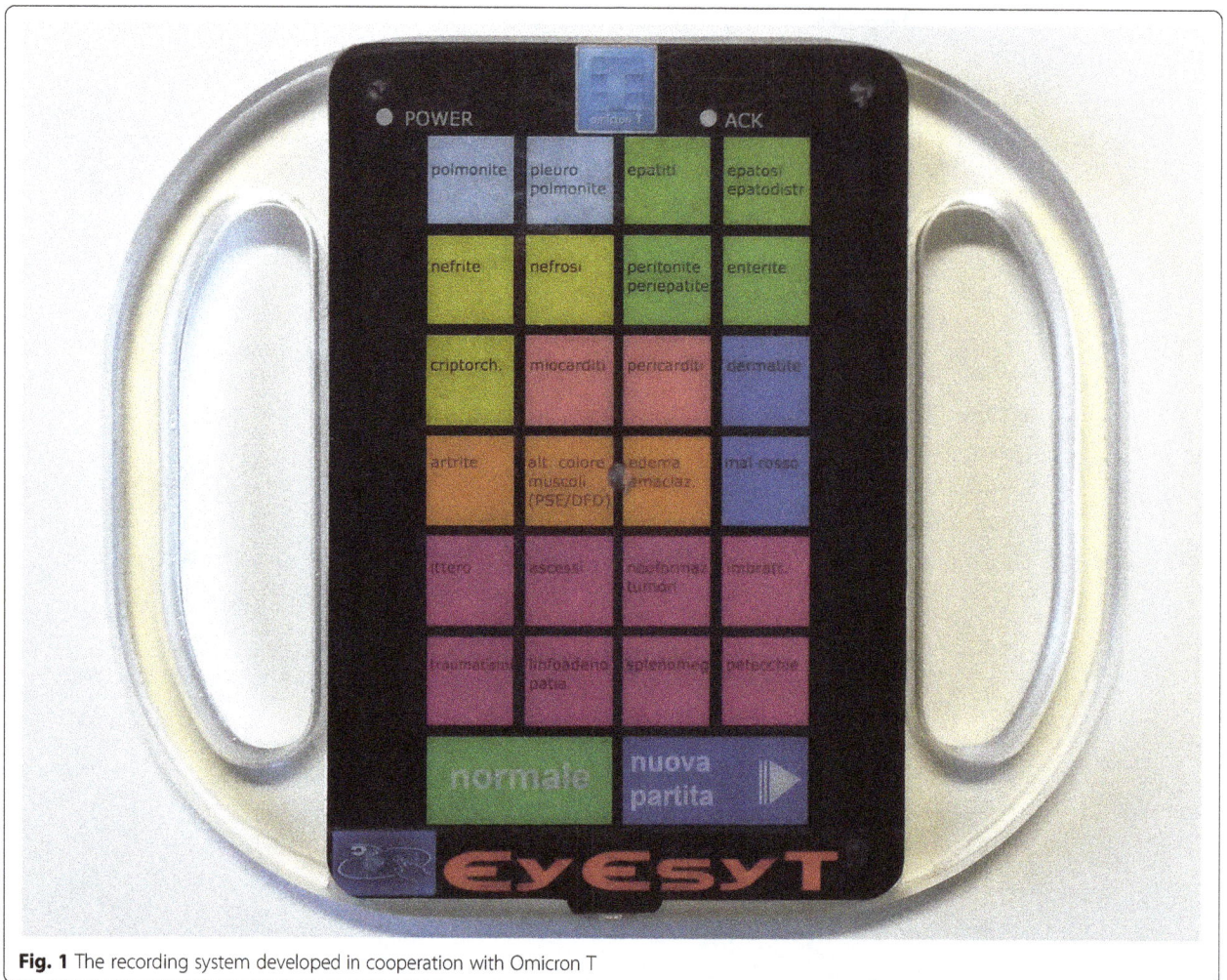

Fig. 1 The recording system developed in cooperation with Omicron T

Neumann et al. [14] found a prevalence of pleurisy, pneumonia and pleuropneumonia of approximately 16%. This prevalence is slightly lower than the one found in the present study, which can be explained by the lower age and weight of their animals at the time of slaughter.

For heavy pigs from Northern Italy, Merialdi et al. [15], found a prevalence of respiratory lesions of up to 40%, which is even higher than the prevalence in the present study. However, the focus of this previous study was different, and the researchers probably included all minimal lung lesions. In the present study, pneumonia was considered only if the lesion (Table 1) intersected a whole lobe. They found a prevalence of milk spot lesions near 10%, while in the present study, the prevalence of hepatic lesions was 16%. Milk spot lesions composed the majority of hepatic lesions in the present study, but the fact that all hepatic lesions were not classified in more detail can explain the difference in results.

According to European Union Regulation (EC) No. 854/2004, erysipelas should be detected ante-mortem, and the slaughtering must be deferred. Nevertheless,

erysipelas can be undiagnosed ante-mortem because the typical lesions become evident only after scalding and bristle removal. In this case, swine carcasses affected by erysipelas must either undergo skin removal or be destroyed depending on the disease stage. Occasional cases of erysipelas were recorded during post-mortem inspection, but the number was very low. In all of these cases, the carcasses were destroyed.

No large differences were detected between the three slaughterhouses. In particular, as could be expected due to the homogeneity of the animals, no relevant differences in lesions related to animal health were found. Only a relevant difference in biliary or faecal contamination was found. In particular, one slaughterhouse showed an prevalence of carcass contamination (3.6%) that was much higher than that of the other two slaughterhouses (2.2% and 2.5%). The slaughtering lines of the three plants did not have relevant technological differences. The two slaughterhouses with lower incidences had a visual inspection of carcasses for faecal or biliary contamination, defined as a critical control point in their

Table 2 Number and percentage of each lesion detected during the work in the three slaughterhouses and an estimate of the prevalence of each lesion

	Slaughterhouse 1		Slaughterhouse 2		Slaughterhouse 3		Tot.		
	number	%	number	%	number	%	number	Prevalence %	standard error
Pneumonia	5100	5.40	8840	8.99	1911	4.91	15,851	6.43	0.050
Pleuropneumonia	15,242	16.14	12,654	12.87	6756	17.35	34,652	15.46	0.074
Hepatitis	21,972	23.27	10,535	10.71	5594	14.37	38,101	16.12	0.075
Hepatosis/hepato-dystrophies	625	0.66	3537	3.60	618	1.59	4780	1.95	0.028
Peritonitis/perihepatitis	355	0.38	730	0.74	71	0.18	1156	0.43	0.013
Enteritis	206	0.22	514	0.52	137	0.35	857	0.36	0.012
Nephritis	234	0.25	261	0.27	113	0.29	608	0.27	0.011
Nephrosis	134	0.14	137	0.14	224	0.58	495	0.29	0.011
Cryptorchidism	139	0.15	140	0.14	115	0.30	394	0.20	0.009
Myocarditis	11	0.01	3	0.00	4	0.01	18	0.01	0.002
Pericarditis	3341	3.54	3345	3.40	1059	2.72	7745	3.22	0.036
Dermatitis	832	0.88	1120	1.14	858	2.20	2810	1.41	0.024
Erysipelas	29	0.03	115	0.12	291	0.75	435	0.30	0.011
Arthritis	0	0.00	8	0.01	0	0.00	8	0.00	0.000
Muscle colour alteration (PSE/DFD)	7	0.01	2	0.00	4	0.01	13	0.01	0.002
Oedema/emaciation	22	0.02	12	0.01	4	0.01	38	0.02	0.003
Jaundice	79	0.08	11	0.01	9	0.02	99	0.04	0.004
Abscesses	571	0.60	865	0.88	422	1.08	1858	0.86	0.019
Neoplasms / tumours	25	0.03	7	0.01	9	0.02	41	0.02	0.003
Biliary or faecal contamination	2126	2.25	3582	3.64	956	2.46	6664	2.78	0.034
Trauma	405	0.43	1316	1.34	674	1.73	2395	1.17	0.022
Lymphadenopathy	138	0.15	47	0.05	25	0.06	210	0.09	0.006
Splenomegaly	254	0.27	173	0.18	176	0.45	603	0.30	0.011
Petechial haemorrhages	24	0.03	1	0.00	0	0.00	25	0.01	0.002
Tot.	51,871	54.94	47,955	48.77	20,235	51.45	119,965	51.72	0.102
Animals	94,411		98,333		38,929		231,590		

self-control plan, while the third slaughterhouse did not. This difference probably resulted in the operators paying greater attention during the evisceration phases.

No differences in trauma lesions were found between the slaughterhouses. The relatively low number of cases (2395, 1.03%) shows that the operators pay attention to animal welfare and handling during transportation and ante-mortem care.

Overall, the kidney conditions of the animals were good, and nephritis or nephrosis lesions were detected in less than 0.3% of the cases.

Dermatitis lesions were found in approximately 1.4% of cases. This figure is much lower than the data recorded by Neuman et al. [14], who found mange lesions in 3.6% of the animals. Still, the data can be considered comparable because dermatitis in the present study was recorded only when the lesion involved more than 50% of the whole skin surface (Table 1).

Regarding peritonitis/perihepatitis, enteritis, cryptorchidism, pericarditis, abscesses and splenomegaly, it is almost impossible to compare these data with international literature since these data are scarce.

Myocarditis, arthritis, muscle colour alteration, oedema/emaciation, jaundice, neoplasms/tumours, lymphadenopathy and petechial haemorrhages cannot be considered since their prevalence was lower than 0.1%, and at this level, the relative standard error of the estimate is too high to make reliable conclusions.

Visual vs. traditional inspections

As a whole, the visual-only inspection showed greater efficiency than the traditional inspection in detecting lesions (Table 3). In fact, the visual-only inspection detected lesions in 52% of the animals, while the traditional inspection detected lesions in only 42% of the animals. There was a large difference in the sensitivity in

Table 3 Results of the comparison between traditional and visual-only inspections

	Traditional	%	Visual-only	%	δ % over traditional	Relative δ % over traditional
Pneumonia	2709	6.96	1911	4.91	−2.05	−29.5
Pleuropneumonia	4150	10.66	6756	17.35	6.69	62.8
Total respiratory	6859	17.62	8667	22.26	4.64	
Hepatitis	6566	16.87	5594	14.37	−2.50	−14.8
Hepatosis/hepato-dystrophies	1	0.00	618	1.59	1.58	61,700
Peritonitis/perihepatitis	66	0.17	71	0.18	0.01	7.58
Enteritis	38	0.10	137	0.35	0.25	261
Total digestive	6671	17.14	6420	16.49	−0.65	
Nephritis	35	0.09	113	0.29	0.20	223
Nephrosis	163	0.42	224	0.58	0.16	37.4
Cryptorchidism	40	0.10	115	0.30	0.19	188
Total reproductive-urinary	238	0.61	452	1.16	0.55	
Myocarditis	1	0.00	4	0.01	0.01	300
Pericarditis	575	1.48	1059	2.72	1.24	84.2
Total cardio-circulatory	576	1.48	1063	2.73	1.25	
Dermatitis	520	1.34	858	2.20	0.87	65.0
Erysipelas	148	0.38	291	0.75	0.37	96.6
Total tegumentary	668	1.72	1149	2.95	1.24	
Arthritis	0	0.00	0	0.00	0.00	
Muscle colour alteration (PSE/DFD)	0	0.00	4	0.01	0.01	
Oedema/emaciation	3	0.01	4	0.01	0.00	33.3
Total locomotor	3	0.01	8	0.02	0.01	
Jaundice	4	0.01	9	0.02	0.01	125
Abscesses	454	1.17	422	1.08	−0.08	−7.05
Neoplasms / tumours	3	0.01	9	0.02	0.02	200
Biliary or faecal contamination	685	1.76	1161	2.98	1.22	69.5
Trauma	103	0.26	674	1.73	1.47	554
Lymphadenopathy	6	0.02	25	0.06	0.05	317
Splenomegaly	111	0.29	176	0.45	0.17	58.6
Petechial haemorrhages	1	0.00	0	0.00	0.00	−100
Total other	1367	3.51	2476	6.36	2.85	
Tot lesions	16,382		20,235			
Tot Animals	38,929		38,929			
% lesions	42.09		51.98			

pneumonia and pleuropneumonia detection probably because official veterinarians performing traditional inspections did not undergo training for lesion classification before the trial. As a matter of fact, if we consider respiratory lesions (pneumonia and pleuropneumonia) together, the difference is much lower and not statistically relevant. In synthesis, comparable numbers of respiratory diseases were detected by both systems, but the lesion classifications were different.

In addition, the difference in hepatitis detection ability was not statistically relevant, but it is not surprising that traditional liver palpation leads to more sensitivity in this area.

As a whole, almost the same sensitivity was noticed in detecting lesions in red and green offal, while visual-only inspection showed greater sensitivity in detecting lesions on the carcass. The slaughtering line was working at 380 pigs per hour, meaning that there was less than 10 s to perform a whole post-mortem inspection. If the veterinarian had to perform invasive actions, the time available for looking at the whole carcass was probably too short.

Table 4 Means, standard deviations and percent variation coefficients of lesion detection achieved by contracted veterinarians in the preliminary phase, when they inspected 5000 carcass each (not included in the global database), and the comparison period at the end of the study, after the guidelines were applied

	Preliminary period			Final period			Δ v.c. after training
	mean	st. dev.	v. c.	mean	st. dev.	v. c.	
Pneumonia	8.58	8.88	103.52	4.69	3.83	81.66	−21.85
Pleuropneumonia	10.86	7.21	66.39	17.29	1.78	10.31	−56.08
Hepatitis	17.00	9.10	53.54	14.56	2.81	19.26	−34.28
Hepatosis/hepato-dystrophies	1.72	2.61	151.90	1.36	2.17	159.73	7.83
Peritonitis/perihepatitis	0.55	0.54	97.81	0.20	0.17	83.45	−14.36
Enteritis	0.30	0.27	90.53	0.38	0.22	59.68	−30.84
Nephritis	0.18	0.07	40.29	0.26	0.15	59.44	19.14
Nephrosis	0.16	0.16	97.47	0.55	0.25	45.50	−51.98
Cryptorchidism	0.14	0.03	24.34	0.30	0.06	20.80	−3.55
Myocarditis	0.02	0.02	150.55	0.01	0.02	113.93	−36.62
Pericarditis	2.96	1.02	34.43	2.71	0.50	18.61	−15.82
Dermatitis	0.96	0.67	69.44	1.97	1.11	56.40	−13.04
Erysipelas	0.06	0.09	154.20	0.71	0.68	96.10	−58.10
Arthritis	0.01	0.01	167.33	0.00	0.00	0.00	−167.33
Muscle colour alteration (PSE/DFD)	0.00	0.01	154.92	0.01	0.01	115.72	−39.20
Oedema/emaciation	0.02	0.02	89.57	0.01	0.01	120.94	31.36
Jaundice	0.04	0.03	88.06	0.03	0.03	96.50	8.44
Abscesses	0.74	0.28	38.07	1.09	0.35	31.87	−6.20
Neoplasms / tumours	0.01	0.01	77.46	0.04	0.08	200.00	122.54
Biliary or faecal contamination	2.44	0.90	36.85	3.10	0.88	28.44	−8.41
Trauma	0.56	0.71	125.58	1.58	1.13	71.25	−54.33
Lymphadenopathy	0.14	0.19	143.46	0.06	0.04	62.70	−80.77
Splenomegaly	0.24	0.19	82.18	0.46	0.24	50.97	−31.21
Petechial haemorrhages	0.02	0.02	150.55	0.00	0.00	0.00	−150.55
Total	47.66	19.02	39.91	51.37	5.51	10.73	−29.19

The analytical results agree with an assessment of risk associated with changes in meat inspections conducted by the Danish Agriculture and Food Council in 2014 [16], which found higher sensitivity for visual inspections than traditional inspections. Hill in 2013 [17], Mousing in 1997 [18] and Blagojevich in 2015 [19] also stressed that switching to visual inspection in pigs does not imply an increase in risk, even if the pigs are raised outdoors. Figure 2 graphically represents the differences between the two inspection systems.

Pre- and post-training evaluation of veterinarians

To conduct this analysis, it was postulated that on a very large number of inspected animals, each operator should obtain the same mean data. This approach was only possible in field conditions. In such a scenario, the deviation from the median is a good parameter to define how good the inspector is compared to other colleagues. Obviously, such a system is most reliable and meaningful for the

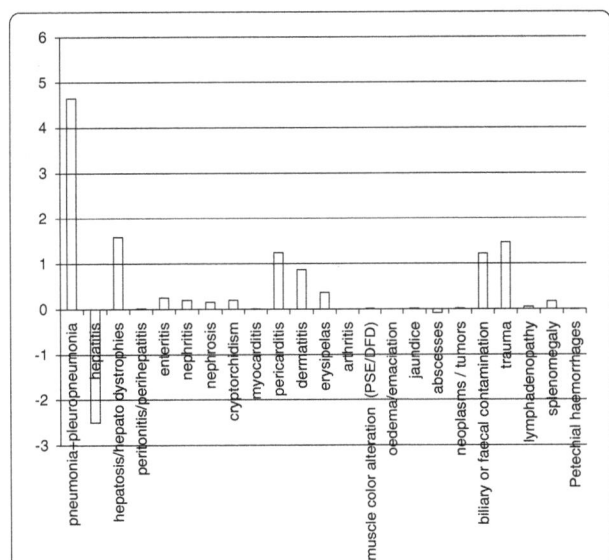

Fig. 2 Percent differences between visual and traditional inspections (positive values represent greater sensitivity of visual inspection, and negative values lower represent lower sensitivity)

most common lesions, and it is not reliable for more "exotic" lesions.

After setting guidelines and training, a generally low variation was achieved at the end of the study period, as shown by a decrease in the variation coefficient of almost every lesion category (Table 4). The decrease was present in common lesions and in the total number of lesions, showing that training is crucial to obtain homogenous judgements by veterinarians. This approach was not successful for detecting and classifying every lesion (e.g., hepatosis and nephritis), but one of the functions of such an instrument is the ability to address future training actions.

Official vs. contracted veterinarians

The same principle used for evaluating pre- and post-training performance was adopted to compare the performance of official and contracted veterinarians. Following this principle, the official veterinarians that inspected a low number of animals were excluded from this analysis. The classification and recording of lesions can be extremely useful because these data can be used for epidemiological purposes, for farming suggestions and even for farm classification. However, such a system can be effective only if the inspector's judgements are repeatable and reliable. As much as possible, the inspections have to be independent of the individuals conducting the inspections. Moreover, these judgements have extremely important economic relevance since different condemnation rates of single organs or whole carcasses imply different costs both for slaughterers and for famers.

From the data in Table 5, it is clear that the trained contracted veterinarians achieved a globally lower variability than the official colleagues. The fact that the official veterinarians were not trained to apply the guidelines can easily explain the difference. The data

Table 5 Means, standard deviations and percent variation coefficients of lesion detection achieved by the official veterinarians performing traditional inspection and by the contracted veterinaries performing visual-only inspection in the comparison period

	Official veterinarians			Contracted veterinarians			
	mean	st. dev.	v. c.	mean	st. dev.	v. c.	δ over official
Pneumonia	5.82	4.39	75.57	4.69	3.83	81.66	6.09
Pleuropneumonia	10.70	2.86	26.76	17.29	1.78	10.31	−16.45
Hepatitis	14.96	8.80	58.83	14.56	2.81	19.26	−39.56
Hepatosis/hepato-dystrophies	0.00	0.00	300.00	1.36	2.17	159.73	−140.27
Peritonitis/perihepatitis	0.11	0.26	239.75	0.20	0.17	83.45	−156.30
Enteritis	0.04	0.07	181.11	0.38	0.22	59.68	−121.43
Nephritis	0.05	0.06	123.58	0.26	0.15	59.44	−64.15
Nephrosis	0.21	0.32	152.40	0.55	0.25	45.50	−106.91
Cryptorchidism	0.06	0.09	161.92	0.30	0.06	20.80	−141.13
Myocarditis	0.00	0.00	300.00	0.01	0.02	113.93	−186.07
Pericarditis	1.81	1.20	66.13	2.71	0.50	18.61	−47.52
Dermatitis	0.80	1.65	205.77	1.97	1.11	56.40	−149.37
Erysipelas	0.39	0.32	82.58	0.71	0.68	96.10	13.52
Arthritis	0.00	0.00	0.00	0.00	0.00	0.00	0.00
Muscle colour alteration (PSE/DFD)	0.00	0.00	0.00	0.01	0.01	115.72	115.72
Oedema/emaciation	0.01	0.01	217.58	0.01	0.01	120.94	−96.64
Jaundice	0.01	0.01	198.29	0.03	0.03	96.50	−101.79
Abscesses	0.97	0.55	56.44	1.09	0.35	31.87	−24.57
Neoplasms / tumours	0.02	0.05	254.11	0.04	0.08	200.00	−54.11
Biliary or faecal contamination	1.96	0.64	32.94	3.10	0.88	28.44	−4.50
Trauma	0.10	0.21	205.51	1.58	1.13	71.25	−134.26
Lymphadenopathy	0.01	0.02	198.62	0.06	0.04	62.70	−135.92
Splenomegaly	0.14	0.19	135.20	0.46	0.24	50.97	−84.24
Petechial haemorrhages	0.00	0.00	300.00	0.00	0.00	0.00	−300.00
Total	38.15	16.30	42.71	51.37	5.51	10.73	−31.98

Nine official veterinarians conducted inspections during the study, but two of these veterinarians were excluded in this evaluation since they inspected less than 1000 carcasses

demonstrate that it is essential to reach a high level of standardisation, which can be achieved only through the adoption of strict operative guidelines and training veterinarians to adopt and follow these guidelines. The training should be aimed towards reaching a lower variability in judgement by understanding and following the guidelines.

Conclusions

The data derived from local projects on post-mortem lesions in slaughterhouses in Northern Italy were not homogenous and comparable.

For the first time, a classification of lesions was developed and shared with the Ministry of Health and the two most productive regions in the swine sector. Moreover, a relevant dataset of these lesions and instruments able to further expand this database were built.

In industrial high-speed slaughtering lines of pigs, visual inspection was shown to be comparable to traditional inspection and was even more sensitive for some lesions.

Post-mortem inspection is a human judgement and is therefore prone to large error. To minimize error and to achieve a high level of standardization, it is necessary to develop operative guidelines. In addition, training the operators involved is crucial for obtaining consistent data. Only with reliable data can post-mortem inspection reports be used for several purposes, such as epidemiological studies or the classification of farms based on risk. It is therefore important to have the same classification and guidelines, and the veterinarians involved in meat inspection should undergo continuous education.

Acknowledgements
The authors want to acknowledge the veterinary services of the Emilia Romagna and Lombardy regions and all the slaughterhouse personnel who were involved in the project.

Funding
The project was financed by the Italian Ministry of Health.

Authors' contributions
SG, EZ, PADC and AI designed the study, followed the work and analysed the data. SB, GB and SG contributed to the study design and its implementation in the field scenario. All authors read and approved the final manuscript.

Consent for publication
Not applicable

Competing interests
The authors declare that they have no competing interests.

Author details
[1]Department of Food and Drug, Parma University, Via Del Taglio, 10, 43126 Parma, Italy. [2]Italian Ministry of Health, Via Giorgio Ribotta, 5, 00144 Rome, Italy. [3]Italian Ministry of Health, Viale Tanara 31/A, 43100 Parma (PR), Italy.

References
1. Huey R. Toroughly modern meat inspection. Vet Rec. 2014;170(3):68–70.
2. EFSA. Scientific opinion on the public health hazards to be covered by inspection of meat (swine). EFSA J. 2011;9(10):2351. [198 pp.]
3. Buncic S, Nychas GJ, Lee Michael RF, Koutsoumanis K, Hebraud M, Desvaux M, Chorianopoulos N, Bolton D, Blagojevic B, Antic D. Microbial pathogen control in the beef chain: recent research advances. Meat Sci. 2014;97(3):288–97.
4. Commission Regulation (EU) No 219/2014 amending Annex I to Regulation (EC) No 854/2004 of the European Parliament and of the Council as regards the specific requirements for post-mortem inspection of domestic swine. Official Journal of the European Union. L 69/99. 206. 08 March 2014;
5. Regulation (EC) No. 854/2004. Official Journal of the European Union. L155. 206. 29 April 2004;
6. Heinonen M, Grohn YT, Saloniemi H, Eskola E, Tuovinen V. K. The effects of health classification and housing and management of feeder pigs on performance and meat inspection findings of all-in-all-out swine-finishing herds. Prev Vet Med. 2001;49(1–2):41–54.
7. Felin E, Jukola E, Raulo S, Heinonen J, Fredriksson-Ahomaa M. Current food chain information provides insufficient information for modern meat inspection of pigs. Prev Vet Med. 2016;127:113–20.
8. Prosciutto di Parma. Disciplinary of Production. 2015. Available at http://www.prosciuttodiparma.com/pdf/it_IT/disciplinare.28.11.2013.it.pdf. Accessed 7 June 2015.
9. Stärk KDC, Alonso S, Dadios N, Dupuy C, Ellerbroek L, Georgiev M, Hardstaff J, Huneau-Salaün A, Laugier C, Mateus A, Nigsch A, Afonso A, Lindberg A. Strengths and weaknesses of meat inspection as a contribution to animal health and welfare surveillance. Food Control. 2014;39:154–62.
10. Istituto Parma Qualità. Rapporto sull'attività dei servizi coordinati di controllonell'anno 2013. 2014. Available at http://www.google.it/url?sa= t&rct=j&q=&esrc=s&source=web&cd=8&ved=0CEYQFjAHahUKEwjjnq_ nj5bGAhUKXBQKHX0IAHg&url=http%3A%2F%2Fineqweb.it%2Findex. php%2Ffiles%2F7dcee9883e139530df552ecdb3da8e948927fc1b. pdf%3Faction%3Ddownload&ei=uBaBVePiCoq4Uf2QgMAH&usg= AFQjCNFOijufCr_9C9ZZ1-3zfONrRpOFwg&sig2=IRajHTlzIlxPfzi5yrPPVQ&bvm= bv.96041959.d.d24&cad=rja. Accessed 5 June 2015.
11. FAO. The Epidemiological Approach to Investigate Disease Problems. 2013. Available at http://www.fao.org/wairdocs/ilri/x5436e/x5436e06.htm#4. 4%20sample%20sizes. Accessed on November 2012;
12. Food Safety and Inspection Service. FSIS Directive 6100.2 17/09/2007. 2007. Available at http://www.fsis.usda.gov/OPPDE/rdad/FSISDirectives/6100.2.pdf;
13. Food Standard Agency. Trial of visual inspection of fattening pigs from non-controlled housing conditions. 2013. Available at https://www.food.gov.uk/ science/research/choiceandstandardsresearch/meatcontrolsprojects/ fs145003/#.UqhAieKmYxA
14. Neumann EJ, Hall WF, Stevenson MA, Morris RS, Ling Min Than J. Descriptive and temporal analysis of post-mortem lesions recorded in slaughtered pigs in New Zealand from 2000 to 2010. N Z Vet J. 2014; 62(3):110–6.
15. Merialdi G, Dottori M, Bonilauri P, Luppi A, Gozio S, Pozzi P, Spaggiari B, Martelli P. Survey of pleuritis and pulmonary lesions in pigs at abattoir with a focus on the extent of the condition and herd risk factors. Vet J. 2012;193:234–9.
16. Danish Agriculture and Food Council. Assessment of risk associated with a change in meat inspection, 2014. Available at http://lf.dk/Aktuelt/ Publikationer/Svinekod.aspx#. Accessed on November 2015;
17. Hill A, Adam Brouwer A, Donaldson N, Lambton S, Buncic S, Griffiths I. A risk and benefit assessment for visual-only meat inspection of indoor and outdoor pigs in the United Kingdom. Food Control. 2013;30:255–64.
18. Mousing J, Kyrval J, Jensen T. K, Aalbaek B, Buttensch J, Svensmark B, Willeberg P. Meat safety consequences of implementing visualpostmortem meat inspection procedures in Danish slaughter pigs. Vet Rec. 1997;140(8):472–77.
19. Blagojevic B, Dadios N, Reinmann K, Guitian J, Stärk KDC. Green offal inspection of cattle, small ruminants and pigs in the United Kingdom: impact assessment of changes in the inspection protocol on likelihood of detection of selected hazards. Res Vet Sci. 2015;100:31–8.

10

Biosecurity survey in relation to the risk of HPAI outbreaks in backyard poultry holdings in Thimphu city area, Bhutan

Tenzin Tenzin[1*], Chador Wangdi[2] and Purna Bdr Rai[3]

Abstracts

Background: A questionnaire survey was conducted to assess the biosecurity and other practices of backyard poultry holdings and knowledge and practices of poultry keepers following an outbreak of highly pathogenic avian influenza (H5N1) virus in poultry in Thimphu city area, Bhutan.

Results: The study identified 62 backyard poultry holdings in 12 settlement areas, and the owners were subsequently interviewed. The birds are kept in a low-input low-output system, fed locally available scavenging feed base, and supplemented with food scraps and some grain. Although the birds are housed at night in a small coop to protect them against theft and predators, they are let loose during the day to scavenge in the homestead surroundings. This invariably results in mixing with other poultry birds within the settlement and wild birds, creating favorable conditions for disease spread within and between flocks. Moreover, the poultry keepers have a low level of knowledge and awareness related to the importance of biosecurity measures, as well as veterinary care of the birds and reporting systems. Of particular concern is that sick birds within backyard holdings may not be detected rapidly, resulting in silent spread of disease and increased risk of humans contacting the virus (e.g. HPAI) from infected poultry. Nevertheless, all the respondents have indicated that they know and practice hand washing using soap and water after handling poultry and poultry products, but rarely use face-masks and hand gloves while handling poultry or cleaning poultry house.

Conclusions: This study highlights the importance of educating poultry keepers to improve the housing and management systems of poultry farming within the backyard holdings in the Thimphu city area in order to prevent future disease outbreaks.

Keywords: Backyard poultry holdings, Biosecurity, Knowledge, Attitudes, Practices, HPAI and H5N1 virus, Bhutan

Background

Highly pathogenic avian influenza (HPAI), subtype H5N1, was first reported in Southeast Asia in late 2003, and then spread rapidly with outbreaks being reported in 63 countries across Asia, Europe, Africa, and the Middle-East [1]. Since the emergence of HPAI virus in poultry in 2003, there has been 856 laboratory-confirmed human cases officially reported to World Health Organization (WHO) from 16 countries, including 452 deaths up to 3 October 2016 [2]. The outbreaks have had serious economic impact to the affected countries, with millions of birds either killed by the disease or mandatory culled in an effort to limit the spread of virus [3, 4]. Although, different countries have implemented various strategies aimed at preventing and mitigating infection within poultry with varying degree of success, in some countries, the virus remains entrenched within poultry populations [1, 5]. One of the factors responsible for outbreaks and the persistence of the virus in domestic poultry populations is cited to be the widespread practice of small holder backyard poultry farming and associated live bird markets [5–7]. This is mainly because basic biosecurity measures are rarely implemented in backyard poultry farming systems allowing HPAI to circulate within poultry populations resulting in a perpetual virus source to other poultry flocks [5, 8, 9]. Therefore, one of the most effective forms of protection against HPAI and other poultry diseases is biosecurity, which is principally the implementation of

* Correspondence: tenzinvp@gmail.com
[1]Disease Prevention and Control Unit, National Centre for Animal Health, Department of Livestock, Thimphu, Bhutan
Full list of author information is available at the end of the article

measures to prevent the introduction of infectious agents into the farm/environment (bio-exclusion) or containment measures to prevent spread of infectious agents from exiting in the event of outbreaks (bio-containment) [9–12].

In Bhutan, the poultry farming system comprise of both commercial and backyard holdings but backyard farming is predominant in the country. The first outbreak of HPAI (H5N1) virus was reported in February 2010 in a backyard poultry holding in the southwest Bhutan, near the border with India [13, 14]. Since then, at least seven separate outbreaks of HPAI (H5N1) have been confirmed at 21 locations in six districts with outbreaks reported every year in 2011, 2012 and 2013. The most recent outbreak occurred on 03 April, 2015 in a backyard poultry holding in Thimphu city area (capital of Bhutan) [15]. To our knowledge, no studies have been conducted to understand the biosecurity practices of backyard poultry holdings in Thimphu or elsewhere in the country. Therefore, it is important to understand the types of backyard poultry holdings and biosecurity practices in farms for better preparedness planning. In this context, we conducted a rapid biosecurity survey among the backyard poultry holdings in Thimphu city, as a part of rapid risk assessment following an outbreak of HPAI in one of the backyard holdings in Thimphu.

The main objectives of this study were to (1) identify backyard poultry holdings in Thimphu city area that have potential risk of possible outbreaks in future, (2) generate baseline information about flock characteristics and assess basic biosecurity practices, and (3) understand poultry keepers' knowledge in relation to poultry keeping and personal hygiene practices to prevent incursion and transmission of HPAI.

Methods

Study area

This survey was conducted in Thimphu City Area, which is the capital of Bhutan (Fig. 1). The city is located at 27°28′00″N, 89°38′30″E at an altitude of about 2300 m above sea level. Thimphu city covers an area of 26 km^2 with an estimated population of 93,270. Backyard poultry keeping is practiced in 12 areas within the city inhabited by people who work for Thimphu City Corporation and public work department (PWD) as daily wage laborers' or on contract system. The first outbreak of avian influenza A (H5N1) was reported in a backyard poultry holding at Changedaphu (Kalabazar) area in January 2012 [16]. The second outbreak also occurred in a backyard holding at Motithang city camp area on 3 April 2015, which is about 3 km away from 2012 outbreak area (Fig. 1). Following this outbreak, we formed a rapid response team (RRT) to implement the containment activities and have identified 12 settlement areas within the city, where poultry birds are reared as backyard free-ranging system (Fig. 1, Table 1).

Questionnaire design

A questionnaire consisting of closed questions was designed to collect information on various aspects of backyard chicken keeping and the owners' knowledge and practices in relation to avian influenza as summarized in Table 1. The questionnaire was piloted with three poultry owners prior to the actual survey and was modified to improve clarity and interpretation.

Data collection

Owing to the lack of a proper sampling frame, a purposive sampling was used to recruit backyard poultry keepers within the 12 identified settlement. After visiting each settlement, a door-to-door survey was conducted using a rolling sample method in which the first selected household that owned poultry provides information about the next household that owned poultry in the area. In this way, 62 poultry birds owning household (HHs) within the 12 settlement area were selected and interviewed. When the poultry owning HHs was not available during the first visit, we revisited the HHs in the evening after owners returned home after the work, ensuring all the poultry owning HHs were interviewed. One adult person from each selected household/family was interviewed face-to-face. The selected person was informed about the purpose of the survey by explaining that the data collected will be used for understanding the backyard poultry keeping practices and to strengthen backyard poultry biosecurity in the city. All the identified poultry owners ($n = 62$) agreed and consented to be interviewed. Since the questionnaire survey was conducted by the rapid response team as part of an emergency response during the door-to-door surveillance and awareness education campaign during the time of HPAI outbreak in one of city areas, no formal ethical approval was necessary. The interview was carried out from 24 to 27 April 2015. In addition, when a survey team come across any sick birds oropharyngeal and cloacal swabs were collected ($n = 10$) and rapid antigen detection test was performed at the site. The samples were then referred to the laboratory to carry out RT-PCR test to detect H5N1 virus, but none of the samples tested positive to avian influenza A virus and H5N1 virus strain. Awareness education related to poultry bird management, risk of disease spread, biosecurity practices and public health risk of bird flu were provided to the respondents and to the community at the time of interview.

Data management and analysis

Data was entered into a database developed in Epi Info V.7.1 (http://www.cdc.gov/epiinfo) (CDC, Atlanta, GA, USA). Data cleaning, management and analyses was

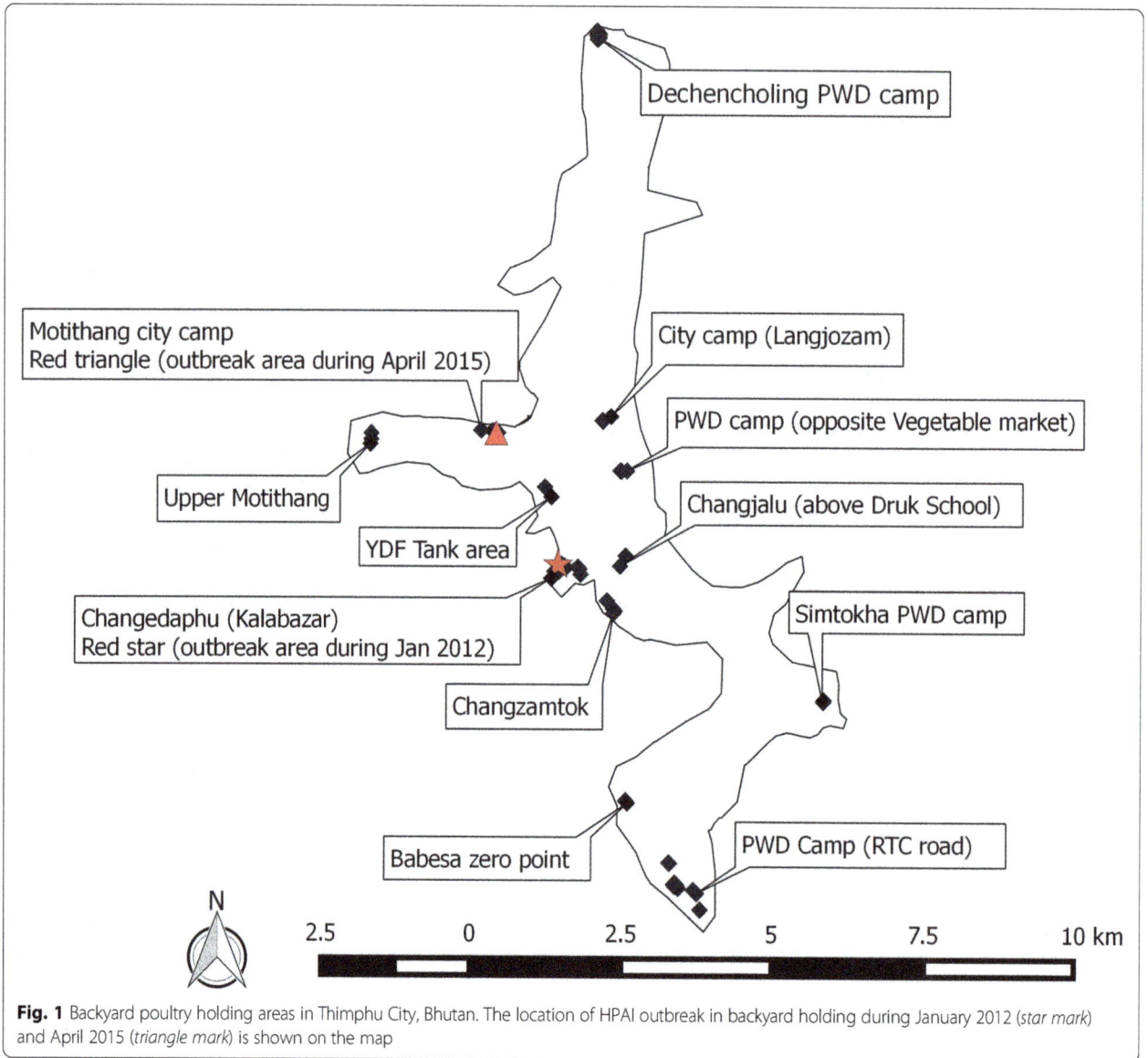

Fig. 1 Backyard poultry holding areas in Thimphu City, Bhutan. The location of HPAI outbreak in backyard holding during January 2012 (*star mark*) and April 2015 (*triangle mark*) is shown on the map

Table 1 Characteristics of backyard poultry farming and the knowledge, attitude and biosecurity practices of poultry owners addressed by the questionnaire

Items	Details
Respondents details	Name, contact detail, place of living, geo-coordinates, gender, occupation, number of people in the household, and approximate monthly family income
Poultry & husbandry characteristics	Number and category of poultry birds kept, breed, source of birds, purpose of keeping poultry by household, number of years of backyard poultry keeping by the household in the city
Biosecurity and management practices	Poultry housing type, location of coop/shed, husbandry practices (intensive/free ranging), contact with wild birds, cleaning & disinfection of poultry house/coop, water bodies near house, foot dip at the entrance to the coop/shed, type of feeds given to the chicken, feeding and watering container, poultry death and disposal system, vaccination of chicken against diseases, notification of poultry death, poultry litter management
Bird flu knowledge and practices, and personnel hygiene	Awareness of bird flu outbreak in the city, source of information, knowledge of bird flu transmission to humans, personnel hygiene (hand washing after handling of chicken and its products, and use of face mask and hand gloves while handling chicken)

carried out using Microsoft Excel 2007 (Microsoft Corp., Redmond, WA, USA) and Stata software V.13 (Stata Corp, Texas, USA). The data described in the manuscript can be requested and obtained from the corresponding author.

Results

Respondents' demographic characteristics

Table 2 shows the demographic characteristics of the respondents. Of the 62 participants, 62% (39/62) were

Table 2 Respondents' demographic, poultry characteristics and purpose of keeping poultry birds

Variables/categories	Number	Percent
Gender of respondent		
Female	39	62.9
Male	23	37.1
Occupation of respondent		
House wife	11	17.74
Work in City Corporation	30	48.39
Work in public road maintenance section	11	17.74
Private work	5	8.06
Others (hospital, forest nursery)	5	8.06
Number of people in the household		
1 to 3	11	17.74
4 to 6	38	61.29
7 to 9	9	14.52
10 to 12	4	6.45
Approximate monthly income		
Up to Nu. 5000	21	33.87
Nu. 5000 to Nu. 15,000	30	48.39
Nu. 15,000 to Nu. 25,000	7	11.29
Above Nu. 25,000	4	6.45
Sources of poultry		
Brought from villages/other areas within the country	12	19.35
Hatched within farm	49	79.03
Government poultry farm	1	1.61
Purposes of keeping poultry birds		
Egg production & family consumption	42	64.52
Egg production & sale	3	4.84
Meat purpose for family	3	4.84
Egg production & meat purpose for family	14	22.58
No. of years of poultry keeping by the HHs		
Less than 1 year	13	20.97
1 to 3 years	14	22.58
3 to 5 years	15	24.19
More than 5 years	20	32.26

female and the majority of the respondents were working for the Thimphu City Corporation and public work department's road maintenance section in the city as daily wage labourers. The family size of the respondents ranged from 2 to 12 (median 7) and the majority (82%; 51/62) of the family earned an approximate monthly income of Nu. 15,000 (US$ 250).

Poultry characteristics and purpose of keeping birds

Sixty two respondents kept a total of 562 local indigenous breed birds (chick: 333, hen: 166, cock: 41) and Hyline brown breed: 22) (Table 3). The main source of the poultry birds was from the hatching of chicks in the households (79%; 49/62) in comparison to purchase of poultry from other places. The majority (67%; 42/62) of the respondents had been keeping poultry for up to 5 years and 32% (20/62) of the owners had been engaged with poultry keeping for more than 5 years. Forty two respondents reported keeping poultry for egg production and family consumption, while 14 respondents kept poultry for both egg production and for meat for family consumption. Only six respondents reported keeping birds for egg production, sale and for meat purposes. In addition 22 respondents also reported keeping poultry birds for sale since local breed fetches higher price (Table 2).

Biosecurity and management practices

The summary of management practices of birds in relation to the biosecurity, disease prevention and control issues is presented in Table 4. Briefly, the poultry birds are housed in a coop constructed with either wooden box (64.5%; 40/62), wire mesh box (12.9%; 8/62), basket (11.3%; 7/62) or coop with wire mesh fencing (11.3%; 7/62) which are either attached to the family house (38.7%; 24/62) or located away from the house (61.3%; 38/62). There is a significant difference ($\chi^2 = 32.495$, P-value = 0.001) between the 12 settlement regarding the location of poultry house in which 17% (12/62) of poultry keepers in Changedaphu (Kalabazar) have the poultry house attached to their house whilst 20.95% (13/62) of the poultry keepers in PWD Camp-RTC road have their poultry house located away from their house.

All poultry keepers reported cleaning the poultry house daily (4.9%; 3/62), weekly (83.9%; 52/62) or monthly (11.5%; 7/62) but none of the keepers used disinfectant to clean the poultry house or surrounding. The majority (87.1%; 54/62) of the owners used the deep litter produced as fertilizer in the kitchen garden. Some poultry keepers also reported that their poultry houses and birds had access to outside people/visitors (54.8%; 34/62) and contact with wild birds (70.9%; 44/62), particularly feral pigeons since all poultry are reared as free ranging. Six of the 12 settlement locations had a small

Table 3 Number & type of poultry birds reared as backyard poultry in different areas in Thimphu City (April 2015)

Location of risk areas	HHs	Local breed (categories)				Hyline brown breed	Total birds
		Cock	Hen	Chicks	Total		
Changedaphu (Kalabazar)	15	1	28	127	156		156
PWD camp (RTC road)	15	7	28	92	127	2	129
Dechencholing city/PWD camp	9	11	34	21	66		66
Changjalu (above Druk School)	3	7	13	19	39		39
PWD camp (opposite vegetable market)	3	4	10	21	35		35
Motithang city camp	4	1	8	21	30		30
Upper Motithang	2	2	22	1	25	13	38
Changzamtok	3	3	4	16	23		23
City camp (Langjozam)	3	2	5	15	22		22
Babesa zero point area	2	0	8	0	8	7	15
Simtokha (PWD camp)	2	2	4	0	6		6
YDF Tank area	1	1	2	0	3		3
Grand Total	62	41	166	333	540	22	562

stream or river near the settlement. The majority of the poultry keepers reported feeding their poultry with left over family food and local grains to supplement the scavenging system, and only 53.2% (33/62) of the keepers used clean containers for feeding and watering birds. Other poultry keepers (46.8%; 29/62) sprayed food/grains into the household surroundings. When asked about any poultry mortality in the backyards, majority (88.7%; 55/62) of the bird keepers reported no unusual mortality during the past 2 week period. The most widely used methods for disposal of dead birds were either burial (59.7%; 37/62) or disposal into open area/bushes (40.3%; 25/62). Of the total 562 birds recorded among the 62 keepers at the time survey, 96% (540/562) of the birds were not vaccinated against poultry diseases since they did not know about the importance of vaccination or even the availability of vaccine. And, only 27.7% (17/62) of the poultry keepers understood how to seek veterinary assistance in the event of any illness in the birds, whilst the rest of the respondents (72.6%; 45/62) were not aware of how to seek assistance.

Knowledge and practice of personnel hygiene in relation to poultry diseases
When asked whether they had heard of the recent avian influenza H5N1 virus (bird flu) outbreak in one of the city camps in Thimphu, 88.7% (55/62) of the respondents had heard about the outbreak either through livestock surveillance team, news media or friends. More than half (66.1%; 41/62) of the respondents were also aware that bird flu can be transmitted from infected poultry to humans, but how it is transmitted is unknown (Table 5).

All the respondents indicated that they understood and practiced hand washing using soap and water after handling poultry and poultry products. When asked whether they used a face-mask while handling poultry or cleaning poultry house, 45 (72.6%) respondents knew the importance of use of face mask but only 22 (48.9%) used one while 23 (51.1%) of the respondents knew of but did not use (practice) face masks. Similarly, 40 (64.1%) respondents knew of the importance of use of hand gloves while handling poultry/products but only 13 (32.5%) practically used gloves while 27 (67.5%) of the respondents knew of but did not use (practice) hand gloves (Table 6). There was no significant (P value >0.05) difference between the location, occupation and income level of the respondents with the biosecurity practices and knowledge and practices of personnel hygiene such as use of a facemask and hand gloves while handling poultry and poultry products. There was also no significant (P value >0.05) difference between awareness on avian influenza of the poultry keepers with the biosecurity practices.

Discussion
To our knowledge this is the first study conducted to explore and assess the biosecurity situation of backyard poultry holdings and the owners' knowledge and practices in relation to HPAI prevention and control measures among backyard poultry keepers in Thimphu city area. The poultry keeping was found to be a secondary activity, as a means to supplement families' dietary protein and also generate some additional income for the households. Most birds were of local non-descript breed that either hatched chicks from within the household poultry birds or were bought from other families within the country. However, the result showed that backyard

Table 4 Poultry birds management practices in relation to biosecurity practices

Variables/categories	Number	Percent
Type of poultry house/coops		
Basket (made from bamboo)	7	11.3
Wire mesh box coop	8	12.9
Wooden box coop	40	64.5
Wire mesh/wooden box coop with wire mesh fencing	7	11.3
Location of poultry house		
Attached to family house	24	38.71
Outside family house (separate house)	38	61.29
Schedule of poultry house cleaning		
Daily	3	4.92
Weekly	52	83.87
Monthly	7	11.48
Disposal method of poultry litter		
Use as fertilizer in the kitchen garden	54	87.1
Dispose into open area	8	12.9
Sale	0	0
Do the people have access to poultry house		
No	28	45.16
Yes	34	54.84
Do the poultry birds come in contact with wild birds		
No	18	29.03
Yes	44	70.97
Is there water bodies near poultry house/premises		
No	26	41.94
Yes	36	58.06
Type of feed given to poultry birds		
Family food left over	9	14.52
Local feed grains (maize, wheat)	7	11.29
Family food left over & local feed grains	44	70.97
Commercial feed	2	3.23
Have clean container for feeding & watering		
No	29	46.77
Yes	33	53.23
Was there any poultry death during the past 2 weeks		
No	55	88.71
Yes	7	11.29
Way of disposal of dead birds		
Disposal into open area/bush	25	40.32
Burial	37	59.68
Sale	0	0
Consumption	0	0
Have poultry birds been vaccinated against poultry diseases		
No	61	98.39
Yes	1	1.61

flocks were reared as a free-ranging system where flocks from different households within the settlement scavenged together. During the daytime birds scavenge freely close to the homestead and have access to cheap feed, insects on the ground, water from the drain and waste water accumulation around the houses or stream. Although the nutrient requirement for the chickens may be fulfilled through scavenging feed resources, the birds were also provided feed supplementation such as grains and household family food scraps [17]. But the majority of the poultry keepers have no clean feeder and water container to feed the supplementary feed. Instead, the grains and food scraps are spread around the homestead which also attracts wild birds such as feral pigeons and other birds, providing an avenue for domestic poultry-wild bird interface for disease transmission. Although the risk of HPAI transmission from pigeons and other wild birds into poultry is unclear, there is risk of other avian diseases transmission to both poultry as well as to humans [18].

Regarding poultry housing, the birds are confined in small houses made of wood, wire mesh or are kept in a basket made of bamboo during night time and are released for free-range scavenging during day. The majority of the chicken houses were found to be poor and unhygienic state condition that did not offer adequate protection either from predators and theft or protection against diseases. Therefore, housing systems need to be improved to enhance biosecurity measures. Also the poultry houses were found to be attached to the family house in order to protect them from predators such as stray dogs or from theft. This indicates there is close interaction at the human-poultry bird interface and poses risks for disease transmission. Moreover, biosecurity measures such as disinfection, foot dip, and restriction of visitors have never been implemented in all backyard holdings surveyed in 12 settlements. Since disinfectants are often not easily available in the market it may not be practical to emphasize their use in backyard settings. The cleaning of poultry shed are mostly done on a weekly basis and the wastes products (poultry litter) are used as fertilizer in the kitchen garden. However, the use of untreated poultry manure as fertilizer may pose a risk of infection spread if the birds are infected [11, 19]. In addition, the poultry waste disposal into garden or any land may attract wild birds due to the presence of spilled feed in these wastes thereby infecting wild birds and contributing to long distance transmission [20]. This may be addressed by composting the litter before spread in the garden [11, 19]. Unfortunately, poultry keepers are not aware of and therefore do not practice this measure. In the backyard and resource-poor setting, composting is rarely applied in developing countries [11, 19].

Table 5 Knowledge about poultry diseases and personal hygiene

Variables/categories	Number	Percent
Heard of bird flu outbreak in Thimphu in the recent weeks		
No	7	11.29
Yes	55	88.71
What were the sources of information[a]		
From disease surveillance team (Livestock)	3	5.45
Television (TV)	14	25.45
Radio news	2	3.64
Print media (Kuensel)	1	1.82
Friends/neighbours	20	36.36
Health officials/clinics	15	27.27
Awareness & knowledge that bird flu can transmit to humans[a]		
No	21	33.87
Yes	41	66.13
Knowledge where to report in case of poultry bird sickness/death		
No	45	72.58
Yes	17	27.42
Where to report the sickness/deaths of poultry[b]		
Livestock Officials	11	64.71
Livestock regulatory authority	4	23.53
City officials	2	11.76

[a]Data based on the total number of person who have heard of bird flu outbreak in Thimphu (n = 55) in the previous question
[b]Data based on who have knowledge where to report in case of poultry bird sickness/death (n = 17) in the previous question

Table 6 Knowledge and practice of hand washing, using face mask and hand gloves while handling poultry & poultry products

Knowledge on the importance of hand wash as well as practice while handling poultry birds

Knowledge on hand wash	Practice hand wash		Total (percent)
	No (percent)	Yes (percent)	
No	0	0	0
Yes	0	62 (100)	62 (100)

Knowledge on the importance of using facemask as well as practice while handling poultry birds

Knowledge	Practice (using facemask)		Total (percent)
	No (percent)	Yes (percent)	
No	17	0	17 (27.42)
Yes	23 (51.11)	22 (48.89)	45 (72.58)

Knowledge on the importance of using hand gloves as well as practice while handling poultry birds

Knowledge	Practice (using hand gloves)		Total (percent)
	No (percent)	Yes (percent)	
No	22	0	22 (35.49)
Yes	27 (67.50)	13 (32.50)	40 (64.52)

This study also indicates that the majority of the poultry owners are not aware of the existence of veterinary facilities and do not know how to seek veterinary assistance in the event of illness in their chickens. They also do not have any knowledge or awareness of any legal obligation to report any unusual mortality or sickness in their flocks to the veterinary authorities. Another concern is that the majority of poultry keepers dispose of dead birds by burying them in the gardens or dispose into open area/dustbins when the mortality should be reported to the veterinary authority for postmortem examination and investigation. These are inappropriate methods of disposal since any infectious disease outbreaks in poultry, for example, Newcastle disease or HPAI could silently spread within the backyard flocks and act as a perpetual source of infection to other birds in the neighborhood/country as well as pose risk to humans. In backyard poultry farms, sickness or mortality of few number of birds are usually considered as a normal pattern and owners would not normally report these cases. This may be due to limited knowledge of the poultry keepers. The deficiency of knowledge about

health problem and relevant regulations such as reporting of any illness or mortality of birds within the flocks/ settlement indicate the importance of poultry keepers to have accessible and reliable source of information. Therefore, the veterinary and regulatory agencies should regularly educate the poultry keepers about poultry diseases and biosecurity practices.

In relation to knowledge and awareness of HPAI (bird flu), the majority of the respondents had heard of the recent outbreak of HPAI in a backyard poultry holding in one of the city camps in Thimphu. The study also revealed that poultry keepers are aware of the risk of transmission of disease from poultry to humans. These findings were expected since the current study was conducted shortly after the declaration and announcement of HPAI outbreak in Thimphu in the mass media. Also, the poultry owners have clear memory of the past outbreak containment and awareness program when H5N1 outbreak had occurred in one of the city camps in Thimphu during January 2012.

The findings of this study demonstrate that the poultry keepers are aware of the importance of hand washing with soap and water and undertake washing after handling poultry & poultry products, and after cleaning of poultry shed. This finding is consistent with other studies where hand washing was found to be the best known practice among poultry workers [21, 22]. However, a knowledge gap and practice was found amongst the poultry keepers such as wearing protective hand gloves and face masks while handling poultry or poultry litter. Although some poultry keepers knew of its importance, but were not practiced because the poultry keepers could not afford to procure it for daily use. But there was no variations in the biosecurity practices and personnel hygiene measures between different backyard holdings, occupation, monthly family income and awareness level on avian influenza of the respondents.

Conclusions

We conclude that the backyard poultry holdings in the study area have very weak biosecurity management practices and the poultry keepers have minimal knowledge and awareness related to the importance of biosecurity measures, veterinary care of the birds and reporting systems, and personnel hygiene. It is therefore important to educate the poultry keepers and improve the housing and management system of poultry farming within the backyard holdings in Thimphu city area in order to prevent future disease outbreak.

Abbreviations
HH: Households; HPAI: Highly pathogenic avian influenza; PWD: Public work department; RRT: Rapid response team; WHO: World Health Organization

Acknowledgements
We would like to thank the poultry keepers who participated in the questionnaire survey. The management of National Centre for Animal Health, Department of Livestock and Bhutan Agriculture and Food Regulatory Authority, Ministry of Agriculture and Forests, Thimphu are acknowledged for providing support to implement outbreak containment measures and surveillance.

Funding
This study was done as part of an emergency response activities to investigate and control HPAI outbreak by the RRT and was funded by the Royal Government of Bhutan. The funding body has no role in the design of the study and data collection, analysis, interpretation and writing of the manuscript.

Authors' contributions
TT and CW conceived and designed the study. TT and PBR carried out the study. TT analyzed the data and wrote the paper. TT, CW and PBR read and approved the final manuscript.

Competing interests
The authors declare that they have no competing interests.

Consent for publication
The manuscript contains no individual person's data in any form and therefore does not require consent for publication.

Ethics approval and consent to participate
Since the survey was conducted as part of an emergency response by the rapid response team (RRT) during door-to-door surveillance and awareness education campaign during the time of HPAI outbreak in Thimphu city, so no formal ethical approval was necessary. The RRT have explained the purpose of the survey and obtained verbal consent from the respondents prior to the interview, and all (n = 62) have agreed and consented to be interviewed. Moreover, the surveys were conducted anonymous and were not linked to a name or address of the respondents.

Author details
[1]Disease Prevention and Control Unit, National Centre for Animal Health, Department of Livestock, Thimphu, Bhutan. [2]Bhutan Agriculture and Food Regulatory Authority, Ministry of Agriculture & Forests, Thimphu, Bhutan. [3]Laboratory Service Unit, National Centre for Animal Health, Department of Livestock, Thimphu, Bhutan.

References
1. OIE: Update on highly pathogenic avian influenza in animals (Type H5 and H7). 2016 http://www.oie.int/animal-health-in-the-world/update-on-avian-influenza/2015/. Accessed 20 Nov 2016.
2. World Health Organization (WHO), 2016: Cumulative number of confirmed human cases for avian influenza A(H5N1) reported to WHO, 2003–2016. http://www.who.int/influenza/human_animal_interface/2016_10_03_tableH5N1.pdf?ua=1. Accessed 10 Nov 2016.
3. Rushton J, Viscarra R, Bleich EG, Mcled A. Impact of avian influenza outbreaks in the poultry sectors of five South East Asian countries (Cambodia, Indonesia, Lao PDR, Thailand, Viet Nam) outbreak costs, responses and potential long term control. Worlds Poult Sci J. 2005;61:491–514.
4. Alders R, Awuni JA, Bagnol B, Farrell P, de Haan N. Impact of avian influenza on village poultry production globally. Ecohealth. 2014;11:63–72.
5. Paul M, Wongnarkpet S, Gasqui P, Poolkhet C, Thongratsakul S, Ducrot C, Roger FO. Risk factors for highly pathogenic avian influenza (HPAI) H5N1 infection in backyard chicken farms, Thailand. Acta Trop. 2011;118:209–16.
6. Sonaiya EB. Family poultry, food security and the impact of HPAI. Worlds Poult Sci J. 2007;63:132–8.
7. Henning KA, Henning J, Morton J, Long NT, Ha NT, Meers J. Farm and flock-level risk factors associated with Highly patahogenic avian influenza outbreaks on small holder duck and chicken farms in the Mekong Delta of Viet Nam. Pre Vet Med. 2009;91:179–88.
8. Capua I, Marangon S. Control and prevention of avian influenza in an evolving scenario. Vaccine. 2007;25:5645–52.
9. Food and Agriculture Organization, (FAO). Biosecurity for Highly pathogenic avian influenza: Issues and options. FAO Animal Production and Health Paper No. 165. 2008. http://www.fao.org/3/a-i0359e.pdf. Accessed 6 Nov 2016.
10. Alhaji NB, Odetokun IA. Assessment of Biosecurity Measures Against Highly Pathogenic Avian Influenza Risks in Small-Scale Commercial Farms and Free-Range Poultry Flocks in the Northcentral Nigeria. Trans Emerg Dis. 2011;58:157–61.
11. Conan A, Goutard FL, Sorn S, Vong S. Biosecurity measures for backyard poultry in developing countries: a systematic review. BMC Vet Res. 2012;8:240.
12. Koch G, Elbers ARW. Outdoor ranging of poultry: a major risk factor for the introduction and development of high pathogenic avian influenza. NJAS. 2006;54(2):179–94.
13. Dubey S, Dahal N, Nagarajan S, Tosh C, Murugkar H, Rinzin K, Sharma B, Jain R, Katare M, Patil S, Khandia R, Syed Z, Tripathi S, Behera P, Kumar M, Kulkarni D, Krishna L. Isolation and characterization of influenza A virus (subtype H5N1) that caused the first highly pathogenic avian influenza outbreak in chicken in Bhutan. Vet Microbiol. 2012;155:100–5.
14. Pandit PS, Bunn DA, Pande SA, and Aly SS. Modeling highly pathogenic avian influenza transmission in wild birds and poultry in West Bengal, India. Scientific reports 3, 2175. 2013. doi: 10.1038/srep02175.
15. OIE 2015: Bhutan Immediate notification; 2015. http://www.oie.int/wahis_2/public%5C..%5Ctemp%5Creports/en_imm_0000017533_20150416_143425.pdf. Accessed 20 Nov 2016
16. OIE 2012: Bhutan follow up report No 7: 2012, notification http://www.oie.int/wahis_2/temp/reports/en_fup_0000012491_20121025_163148.pdf. Accessed 20 Nov 2016
17. Karabozhilova I, Wieland B, Alonso S, Salonen L, Hasler B. Backyard chicken keeping in the Greater London Urban Area: welfare status, biosecurity and disease control issues. Br Poult Sci. 2012;53:421–30.
18. Teske L, Ryll M, Rubbenstroth D, Hänel I, Hartmann M, Kreienbrock L, Rautenschlein S. Epidemiological investigations on the possible risk of distribution of zoonotic bacteria through apparently healthy homing pigeons. Avian Pathol. 2013;42:397–407.
19. Cristalli A, Capua I. Practical problems in controlling H5N1 high pathogenicity avian influenza at village level in Vietnam and introduction of biosecurity measures. Avian Dis. 2007;51(1 Suppl):461–2.

20. Otte J, Pfeiffer D, Tiensin T, Price L, Silbergeld E. Evidence-based policy for controlling HPAI in poultry: bio-security revisited. Research report 2006. https://assets.publishing.service.gov.uk/media/57a08c1ae5274a27b2000f9d/PPLPIrep-hpai_biosecurity.pdf. Accessed 2 Nov 2016.
21. Abdullahi MI, Oguntunde O. Knowledge, attitudes, and practices of avian influenza among poultry traders in Nigeria. Internet J Infect Dis. 2010;8:1–8.
22. Neupane D, Khanal V, Ghimire K, Aro AR, Leppin A. Knowledge, attitudes and practices related to avian influenza among poultry workers in Nepal:a cross sectional study. BMC Infect Dis. 2012;12:76.

11

Spatial analysis and characteristics of pig farming in Thailand

Weerapong Thanapongtharm[1,2]*, Catherine Linard[2,3], Pornpiroon Chinson[1], Suwicha Kasemsuwan[4], Marjolein Visser[5], Andrea E. Gaughan[6], Michael Epprech[7], Timothy P. Robinson[8] and Marius Gilbert[2,3]

Abstract

Background: In Thailand, pig production intensified significantly during the last decade, with many economic, epidemiological and environmental implications. Strategies toward more sustainable future developments are currently investigated, and these could be informed by a detailed assessment of the main trends in the pig sector, and on how different production systems are geographically distributed. This study had two main objectives. First, we aimed to describe the main trends and geographic patterns of pig production systems in Thailand in terms of pig type (native, breeding, and fattening pigs), farm scales (smallholder and large-scale farming systems) and type of farming systems (farrow-to-finish, nursery, and finishing systems) based on a very detailed 2010 census. Second, we aimed to study the statistical spatial association between these different types of pig farming distribution and a set of spatial variables describing access to feed and markets.

Results: Over the last decades, pig population gradually increased, with a continuously increasing number of pigs per holder, suggesting a continuing intensification of the sector. The different pig-production systems showed very contrasted geographical distributions. The spatial distribution of large-scale pig farms corresponds with that of commercial pig breeds, and spatial analysis conducted using Random Forest distribution models indicated that these were concentrated in lowland urban or peri-urban areas, close to means of transportation, facilitating supply to major markets such as provincial capitals and the Bangkok Metropolitan region. Conversely the smallholders were distributed throughout the country, with higher densities located in highland, remote, and rural areas, where they supply local rural markets. A limitation of the study was that pig farming systems were defined from the number of animals per farm, resulting in their possible misclassification, but this should have a limited impact on the main patterns revealed by the analysis.

Conclusions: The very contrasted distribution of different pig production systems present opportunities for future regionalization of pig production. More specifically, the detailed geographical analysis of the different production systems will be used to spatially-inform planning decisions for pig farming accounting for the specific health, environment and economical implications of the different pig production systems.

Keywords: Intensive pig farm, Sustainable development, Spatial distribution, Random forest, Two-part model

Background

In the recent decades, changes in the pig production sector have occurred in many countries, enabling increases in production of pig meat per capita and per farm [1, 2]. The changes to the production systems included a shift from extensive, small-scale, subsistence, mixed production systems towards more intensive, large-scale, geographically-concentrated, commercially-oriented and specialized production [1]. In Thailand, this process of intensification started in the 1960s when the first commercial pig breeds were imported from the United Kingdom by the Department of Livestock Development (DLD) and then from the United States by Kasetsart University [2]. Since then, smallholders who raise indigenous native pig breeds for both personal consumption and as a supplementary source of income have been gradually replaced by large-scale farming of

* Correspondence: weeraden@yahoo.com
[1]Department of Livestock Development (DLD), Bangkok 10400, Thailand
[2]Lutte biologique et Ecologie spatiale (LUBIES), Université Libre de Bruxelles, Brussels 1050, Belgium
Full list of author information is available at the end of the article

improved pig breeds [4, 5]. The pig revolution in Thailand corresponds to the introduction of modern technologies and farm management. The introduction of modern technology include the use of evaporated cooling animal housing, which provides temperatures ranging between 25 and 27 °C (pigs are particularly susceptible to heat stress) artificial insemination, and optimized feed ingredients and additives. These combined factors have allowed commercial farmers to raise more pigs per square meter with faster production cycles [2]. These production systems are referred to as 'intensive' in the sense that a high amount of infrastructure, technology, health care and feeds are used to increase the productivity of high-yielding animals on the farm, resulting in increased outputs (kg meat per animal space per year) [3]. In the pig sector, intensive production systems characterized by high input/output ratios generally, also correspond to large farm size. Although intensive systems could also be obtained in small-scale farming, using high inputs of manpower for example, this does not correspond to the current situation in the Asian region. The very large majority of smallholders use very low levels of inputs in their production cycle, have limited outputs in return, and can therefore be characterized as extensive. Consequently, in Asia, pig production systems are still largely classified in extensive vs. intensive by their farm size, expressed as number of head per farm. For example, following an extensive review of the farm-sizes in different countries, Robinson et al. used thresholds of 10 and 100 pigs/farm to distinguish extensive (<10), semi-intensive (10–100) and intensive (>100) pig farming systems [1].

There is a strong link between the occurrence of diseases, pig production systems and farm scales [3–5]. Smallholders pig production systems are usually linked to poor hygiene and low bio-security with few barriers to potential contacts between the pigs, humans and wildlife. This facilitates disease transmission from wildlife to pig, pig to pig and pig to human. A typical example of disease affecting smallholders in Thailand is trichinosis, a parasitic disease circulating in wild and domestic animals such as rats, pigs, and wild pigs, and occasionally infecting human through the consumption of inadequately cooked infected pork [6]. So, smallholders are characterized by endemic and parasitic diseases with a relatively limited impact. In contrast, intensive pig production systems are hosts to other types of diseases. The hygiene and bio-security can be much higher than in small-scale production systems, but the high concentrations of genetically similar animals, sharing a limited space and producing large quantities of effluent results in i) increased contact rates and pathogen transmission within and between these populations, ii) the build-up of potential pathogens in the environment and in carrier

animals e.g. older breeding stock; and iii) the emergence of new serotypes or mutations [4, 5]. For example, an atypical and highly virulent form of Porcine Reproductive and Respiratory Syndrome (PRRS) recently emerged in pig farms in China [7] and spread to many other countries throughout Asia resulting in a significant productivity impact in the pig production systems [8–12]. Swine influenza is endemic in the pig production sector, but one of the few factors positively associated with disease risk is the farm size [13]. Intensive pig production also has an indirect potential effect through the emergence of zoonotic diseases. The concurrence of several conditions such as high densities of pigs and farms, together with the immunological characteristics of pigs themselves, increase the chance of emergence and spread of some zoonotic pathogens that originate from wild animals passing to pigs (called "mixing vessel") and then on to humans [14]. For examples, pigs have been identified as mixing vessels for influenza viruses [15] – having receptors both for avian and mammalian viruses - and as intermediate hosts for Nipah viruses [16, 17]. In environmental terms, intensive pig production systems are also a serious cause of environmental pollution, both air and water, due to poor manure management [18]. Intensive pig production systems can also radically alter biodiversity of aquatic ecosystems because water polluted by manure that is rich in phosphates, nitrates, and organic matter stimulates the growth of oxygen-depleting plant life, such as blue algae, that then affects fisheries and other valuable aquatic biodiversity [18].

In Thailand, pig farming systems can be categorized into three groups: i) the farrow-to-finish production system, which includes breeding pigs, producing piglets and fattening pigs in the same farm; ii) the nursery system, which only raises breeding pigs to produce piglets; and iii) the finishing system, which raises weaners until they reach market weight [19, 20]. Nowadays, two groups of pig breeds are used in Thailand: the native breeds such as Raad or Ka Done, Puang, Hailum, Kwai, and wild pigs ([21, 22] and the main commercial breeds, including the Large White, Landrace, Duroc, and crosses of these [20]. Native pig breeds grow slowly and their reproduction rates are lower than those of commercial breeds. However, they are better adapted to hot and humid climates and to low-quality feed [21] and they apparently show higher resistance to endemic diseases such as Foot and Mouth Disease (FMD) and internal parasites [21]. In contrast, commercial pig breeds grow much faster, with comparatively higher feed conversion rates and their carcass and meat quality better meet supermarket needs for standardized products [2].

Previous studies demonstrated that farm-level characteristics (i.e. production systems), could be an important risk factor for different diseases in Thailand [2, 12, 23]. For examples, the movements of pigs between production stages provide significant opportunities for the transmission of diseases between herds or farms. Examples include Transmissible Gastroenteritis (TGE) and PRRS [5]. Purchasing feeder pigs from outside the farm increases the risk of introducing diseases such as PRRS, Classical Swine Fever (CSF), and FMD [2]. Farms with breeding sows are at a higher risk from PRRS [12]. The traditional farrow-to-finish system, with high levels of mixing between age groups, facilitates the exchange of a wide number of potential pathogens within the farm, especially enteric and respiratory diseases [23]. In terms of environmental impacts, the Thailand Pollution Control Department (PCD) reported that the high concentration of pig farms in the central plain caused significant water pollution in rivers, and consequently, PCD added pig farming to the list of regulated activities in 2001 [2, 24].

In order to reduce the adverse impacts of intensive pig farming, both in epidemiological and environmental terms, the Agricultural Standard Committee (Ministry of Agriculture and Cooperatives MOAC, Thailand), established the "Standard for Good Agricultural Practices for Pig Farms", which aimed to provide guidance to pig farmers and promote healthy and hygienic pig farming practices [25]. This document provides recommendations relating to eight topics: i) farming conditions (location, farm layout, and housing), ii) use of feed, iii) management of water, iv) overall farm management, v) animal health, vi) animal welfare, vii) the environment (in relation to proper disposal of refuse, manure, discarded carcasses, and water treatment) and viii) the keeping of records allowing tracing of animals. The standards outlined in the document are also used as guidelines for responsible agencies such as the Provincial and Regional DLD Livestock Offices to accredit and monitor pig farms [25]. However, in order to assess the epidemiological and environmental risk associated with pig farming, as well as to guide future planning, a thorough understanding of how different pig production systems are geographically distributed is needed.

Over the last few years, the DLD has been undertaking regular, detailed livestock censuses throughout Thailand, thanks to a very large network of volunteers coordinated by regional, provincial, and district veterinary officers. This study aimed to analyze these very detailed census data on pig distributions in Thailand with two objectives. First, we aimed to describe the geographical patterns and trends in pig farming in Thailand in terms of pig breeds, farming systems, and farm scales. Second, we aimed to analyse the spatial distribution of these different systems in relation to spatial factors that may influence their distribution.

Methods
Pig and human population data
Throughout the paper, we use the term of "farm" or "holder" to refer to a household keeping at least one pig.

Pig population data, both globally and for Thailand during 1964–2013 were obtained from FAOSTAT [26]. More detailed time-series data between 2004 and 2013 on the number of pigs per holder were obtained from the DLD annual census data conducted every year in January [27]. Local DLD staff and livestock volunteers conducted house-to-house census surveys and reported data through a web-based reporting system [27]. The census includes locations (owner name and address), annual counts of native pigs, breeding pigs (boars, sows and piglets), and fattening pigs per holder. The census includes annual counts of native pigs, breeding pigs (boars, sows and piglets), and fattening pigs per holder. There was no strict definition of farming systems used by the pig census so holders were allocated to different farming system according to the following rules, illustrated schematically in Fig. 1. We considered a holder to be of the farrow-to-finish system if its records showed that it was keeping all types of breeding pig (boar, sow, and piglet) as well as fattening pigs. A nursery farming system was assumed for holders keeping all types of breeding pig (but no fattening pigs), whereas a finishing system was assumed for holders keeping only fattening pigs.

Smallholders and large-scale farming systems were separated based on the number of pigs per holder, with holders raising less than 50 pigs being considered as smallholders (<5 pigs per holder for backyard and 5–50 pigs per holder for commercial) and holders with fifty or more pigs being considered as large-scale farming system (50–500 pigs per holder for small, 500–5000 pigs per holder for moderate, and >5000 pigs per holder for large) (the categories shown in Table 1). We previously indicated that farm size is strongly linked to extensive and intensive system. Here, we use the number of 50 pigs per holder with <50 to match the definition of the agricultural standard on the "Good Agricultural Practices for Pig Farms" [25], that is used for operational and management purposes in Thailand. Data on human population counts were provided by the Bureau of Registration Administration (BORA), Department of Provincial Administration [28].

Analysis
Previous studies relating livestock distributions to spatial variables have mostly employed linear regression models [29–32]. For example the global livestock distribution maps provided by the Gridded Livestock of the World 1 (GLW1) [31] and GLW2 [30] were carried out through the use of stratified linear regression models. A similar method was employed to predict the distribution of chickens, ducks and geese in China [29] and to predict the distribution of domestic ducks in Monsoon Asia [32]. A slightly different methodological approach was

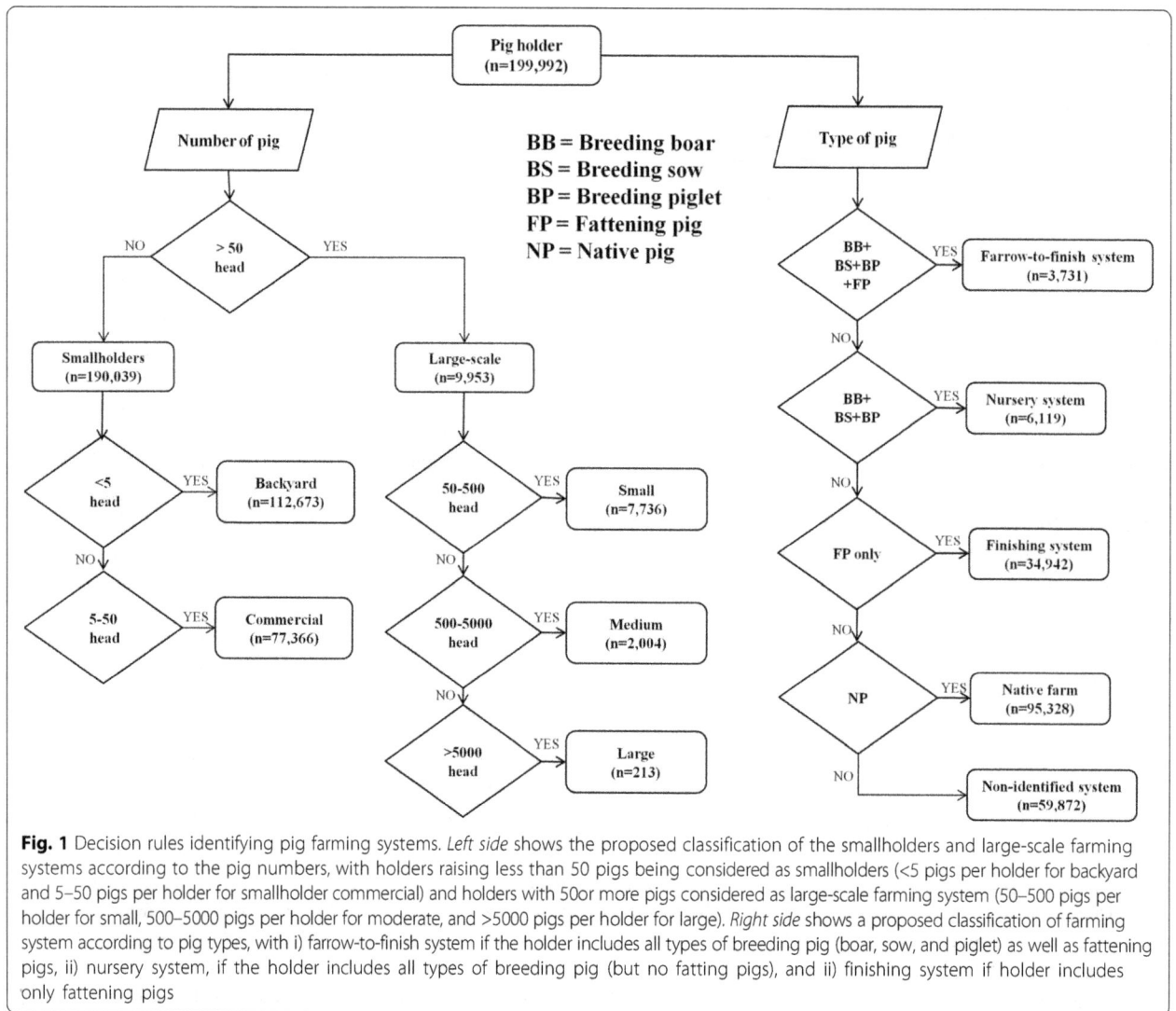

Fig. 1 Decision rules identifying pig farming systems. *Left side* shows the proposed classification of the smallholders and large-scale farming systems according to the pig numbers, with holders raising less than 50 pigs being considered as smallholders (<5 pigs per holder for backyard and 5–50 pigs per holder for smallholder commercial) and holders with 50or more pigs considered as large-scale farming system (50–500 pigs per holder for small, 500–5000 pigs per holder for moderate, and >5000 pigs per holder for large). *Right side* shows a proposed classification of farming system according to pig types, with i) farrow-to-finish system if the holder includes all types of breeding pig (boar, sow, and piglet) as well as fattening pigs, ii) nursery system, if the holder includes all types of breeding pig (but no fatting pigs), and ii) finishing system if holder includes only fattening pigs

Table 1 Criteria to discriminate pig farming systems. Criteria to discriminate pig farming systems using farm scales as defined in the Standard for Good Agricultural Practices for Pig Farm in Thailand in 2009

Categories	Definitions	Approximate number of pigs
Smallholder	Raising boar and sow or finishing pig or piglet or combination of different ages that has the livestock weight less than six units[a]	<50 head[b]
Large-scale farm		
Small	Raising boar and sow or finishing pig or piglet or combination of different phases of age that has the livestock weight between 6 and 60 units[a]	50–500 head
Medium	Raising boar and sow or finishing pig or piglet or combination of different phases of age that has the livestock weight between 60 and 600 units.	500–5000 head
Large	Raising boar and sow or finishing pig or piglet or combination of different phases of age that has the livestock weight more than 600 units	>5000 head

[a]Unit of livestock weight means net weight of boar and sow or finishing pig or piglet or combination of different ages that have total weight equal to 500 kg by assigning 170 kg for the average weight of boar or sow, 60 kg for finishing pig and 12 kg for piglet
[b]50 head calculated from (6 units × 500 kg)/60 kg

used to map the distribution of intensive poultry farming in Thailand, through the use of a simultaneous autoregressive model (SAR) that incorporates an explicit component to account for spatial autocorrelation in the linear regression modeling framework [33]. Two different approaches were used to downscale livestock distribution data in Europe: i) an expert-based suitability rule and ii) a statistical modeling approach based on multiple regression [34].

In this study, we used a Random Forest (RF) approach to quantify the association between the predictor variables and the pig population data in 2010. RF is a machine learning method, which combines the prediction of a high number of classification trees in an ensemble, non-parametric approach [35]. The RF algorithm for regression works by: i) drawing n bootstrap sub-samples from the original data; ii) growing un-pruned regression trees by randomly sampling m variables from a list of predictor variables and choosing the best split from those predictor variables for each of the bootstrap samples (i.e. each tree) and iii) generating a final predicted value by averaging the predictions of the n trees [36]. RF estimates the error rate based on the training data that are randomly sampled 36 % of the whole part at each bootstrap iteration (called as "out-of-bag", or OOB) [35, 36]. The error rate is calculated from the predictions aggregated from all bootstrapped training sets (called as the OOB estimate of error rate). The variable importance is reported by counting the number

of time each variable is selected in the different trees, so it is an absolute measure where variables importance is assessed according to their relative contribution [36]. In general, the variable importance may vary from run to run, but the ranking of the variable is generally stable, so these estimates should not necessarily be interpreted in absolute terms. Compared with other methods, RF has a high ability to model complex interactions among predictor variables [37] and was recently shown to provide highly accurate results in modeling livestock [38] and human population [39, 40].

Predictor variables used to explain the distribution of pig types and pig farming systems were according to the literature, with variables that may account for market and consumer access (travel time, human population density), local provision of feed (crop) and topographic constrains [2, 33, 41]. Six spatial covariates were included in the model in order to quantify their association with the spatial distribution of different pig production systems (Fig. 2). The covariates were: i) two variables accounting for the spatial distribution of croplands used for animal feed (the proportion of rain-fed croplands within a square kilometre and the proportion of irrigated croplands within a square kilometre); ii) two variables that account for access to urban markets; travel time to provincial capitals and to the capital city of Thailand, Bangkok (iii) human population density; and iv) elevation (to account for the observation that native breeds are usually raised on highland and commercial

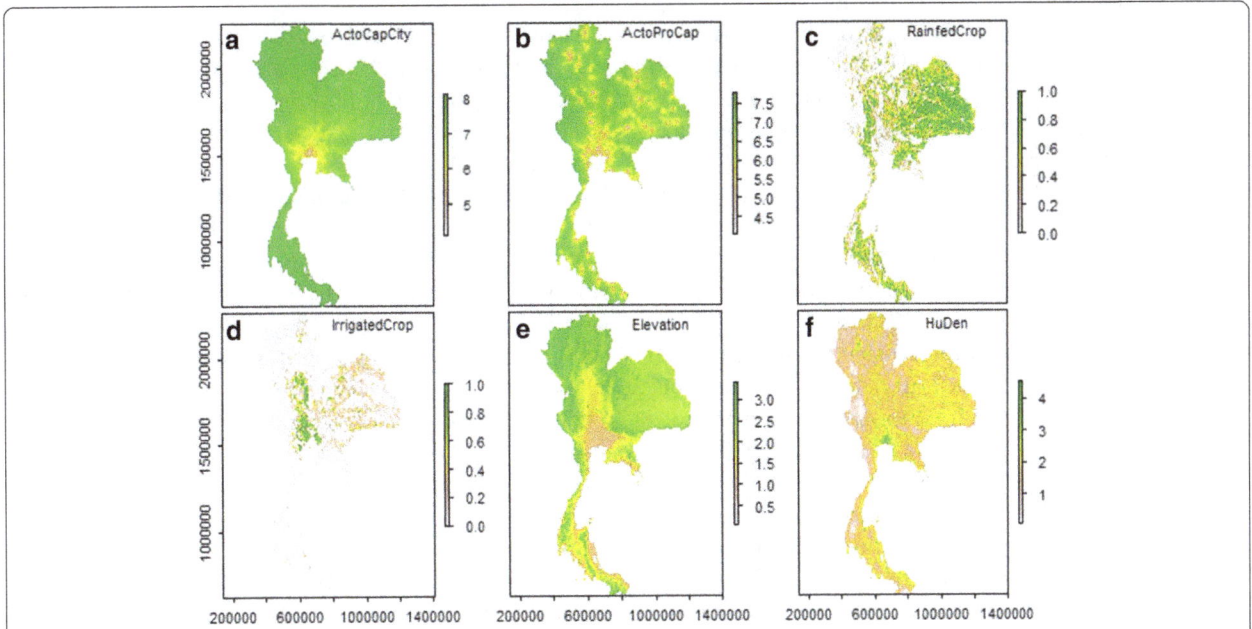

Fig. 2 Spatial datasets used as predictor variables for modeling the pig distribution in Thailand. The variables (1 km resolution) include; **a** Travel time to the capital city (Bangkok) (log10 of time) [44], **b** Travel time to the provincial capitals (log10 of time) [44], **c** rainfed croplands (proportion within a square kilometer) [43], **d** irrigated croplands (proportion within a square kilometer) [43], **e** elevation (log10 of meter) [42], and **f** human population density (log10 of number of human per a square kilometer) [39]

breeds in the plains). To ensure that predictor variables could generate results potentially comparable with other regions, these were obtained from global or regional datasets. A human population density raster map at 100 m resolution was obtained from the Worldpop project [39]. We used the SRTM elevation database with 90 m spatial resolution produced by NASA [42]. The two maps of croplands at 300 m resolution were extracted from the land cover map obtained from the GlobCover project [43], and each cropland class quantification was computed using a focal mean within 1 km. Travel time (accessibility) was estimated using a travel "friction surface" [44], which was initially created by calculating the total time needed to cross the cell of a raster grid based on ancillary data such as land cover, road type, water bodies and slope. The travelled-time maps were created from the friction surface using a cost-distance algorithm to determine the cost of travelling from each pixel to the closest point of interest; either the province capital or Bangkok. Finally, all raster maps of predictor variables were aggregated to 1 km resolution, and then averaged to sub-district unit. The data processing was implemented in ArcGIS 10.2.

The predictor variables were used to build five separate RF models with the following dependent variables i) native pig density, ii) breeding pig density, iii) fattening pig density, iv) smallholders density, and iv) large-scale pig farms density, with all densities expressed in head per square km. Exploratory data analysis indicated that there was strong over-dispersion, especially in the farm density variables, as well as zero inflation [45].

We dealt with the zero inflation through a zero-altered model (also called a hurdle model or a two-part model) [46–49] where presence/absence was modelled separately from abundance, upon presence. First, a binomial RF model was first constructed to predict zero and non-zeros observations [48, 49]. Indeed, a zero value in the census may occur for a variety of reasons: i) the absence of pig because of unsuitable conditions for farming, such as in the urban areas (structural unsuitability); ii) the conditions were suitable for pig farming but pig was absent at the specific time of the census (e.g. moved out to be slaughtered, design error); iii) a pig was present but the observer misidentified it or missed its presence (observer error); iv) conditions were suitable for pig farming but no farmer had taken it up (farmer error). The zeros due to design, observer and farmer errors are also called false zeros or false negatives and the structural unsuitability are called positive zeros, true zeros, or true negatives [45]. The keystone of the two-part model developed here is that the model does not discriminate between the four different types of zeros. In previous studies having to deal with zero-inflated data, a comparison of five regression models was studied (Poisson,

negative binomial, quasi-Poisson, the zero-altered (a two-part model) and the zero-inflated Poisson), the results showed that the zero-altered model performed best, with the highest correlations between the observed and predicted abundances [47]. Second, the non-zero observations were predicted using a quantitative RF model, where we dealt with over-dispersion through a log10 transform ($\log10(x + 1)$). All RF models were built from 500 trees, each bootstrap being predicted by four predictor variables randomly selected from the set of six. The RF models were used to derive predicted density maps both at the sub-district level (using predictors aggregated at the sub-district level) and at the 1 km pixel level (applying the RF model to the 1 km resolution predictors) and the predicted values of each map were then combined. We summed the 1 km cell values within each of the sub-district units.

Two statistical metrics were used to quantify the goodness of fit between observed and predicted densities: the correlation coefficient (COR) and the root mean square error (RMSE). A correlation coefficient provides an indication of precision, i.e. how closely the observed and predicted values agree in relative terms, with a perfect correlation equal to one [47]. RMSE depends on the sample size (n), and the discrepancy between the observed (y_i) and predicted ($\hat{y_i}$) values [47]. It provides an estimate of accuracy and is calculated as:

$$ RMSE = \sqrt{\frac{1}{n}\sum_{i=1}^{n}(\hat{y}-y_i)^2} $$

Analyses were carried out using the "randomForest" [50] and "hydroGOF" [51] packages in R for the RF model and goodness of fit estimates, respectively.

Results
The development of pig population in Thailand
The overall trend in pig production in Thailand over the past 50 years differed from the global pattern. While the global pig population has increased regularly over the past half century, the pig population in Thailand has shown a much more variable trajectory, within an overall trend of increase since the mid-1980s (Fig. 3-Top). The number of pig holders in Thailand remained fairly stable for the last 10 years 2004–2013, but showed an interesting fluctuating pattern (Fig. 3-Left bottom). Intensification of the pig sector can be quantified through the number of pigs per holder, which increased steadily during the same period (Fig. 3-Right bottom). Changes in human and pig population between 2004 and 2013, at the global level and in Thailand, are presented in Table 2. These figures show that while the total Thai population increased, the number of pig holders slightly decreased. The Thai pig population represented 0.70 % of the global

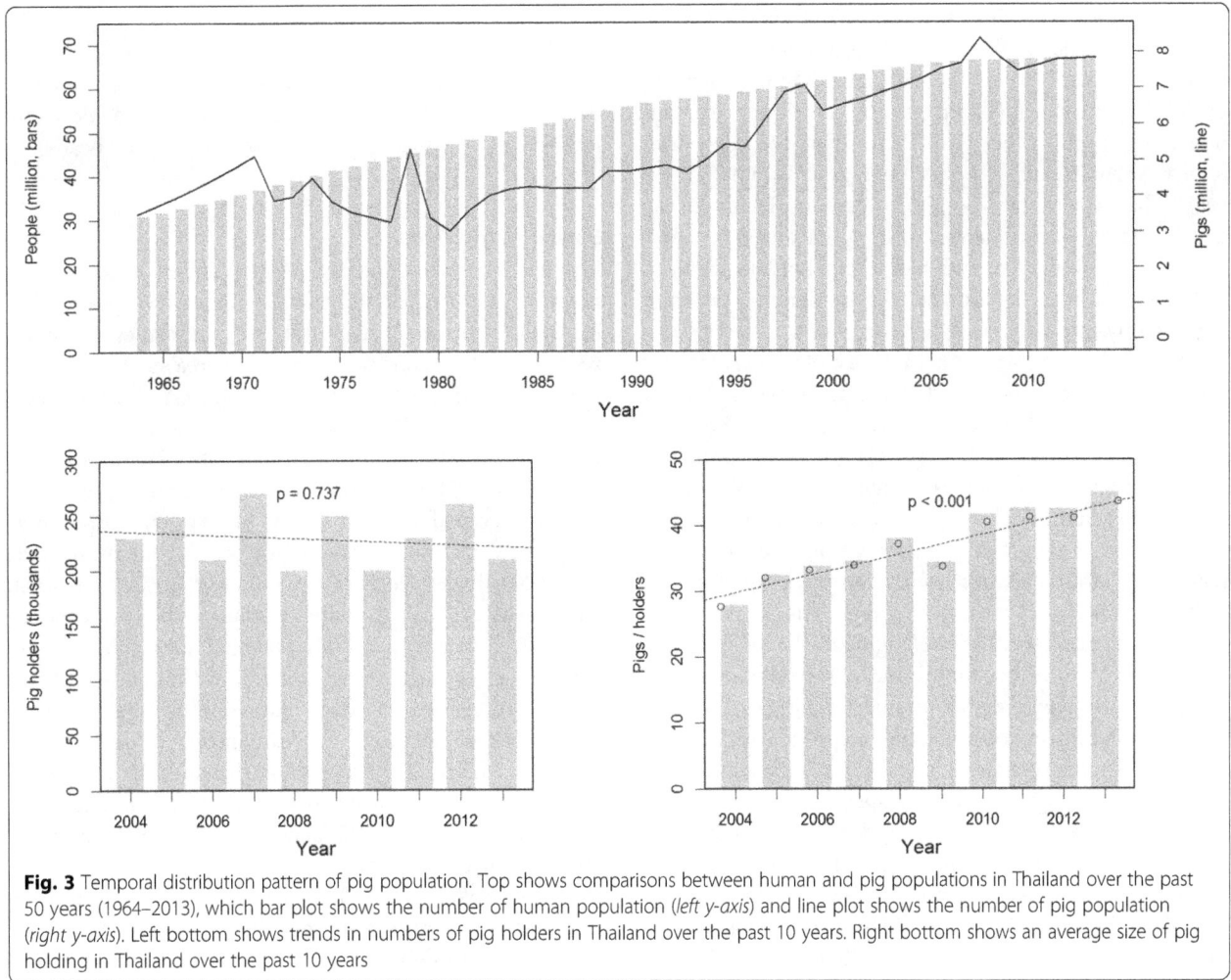

Fig. 3 Temporal distribution pattern of pig population. Top shows comparisons between human and pig populations in Thailand over the past 50 years (1964–2013), which bar plot shows the number of human population (*left y-axis*) and line plot shows the number of pig population (*right y-axis*). Left bottom shows trends in numbers of pig holders in Thailand over the past 10 years. Right bottom shows an average size of pig holding in Thailand over the past 10 years

Table 2 Trends of global and Thai pig production during 10 years. Changes in human population and pig population globally and in Thailand between 2004 and 2013

Type	2004		2013		Compound annual growth rate[a] (%)	
Human population	Person	Household	Person	Household	Person	Household
Global level (million)	6436		7162		1.19	
Thailand (million)	62	18	65	23	0.53	2.76
Pig population	Head	Holder	Head	Holder	Head	Holder
Total global level (million)	873		977		1.26	
Total Thai pigs	6,285,603	225,592	9,511,389	210,978	4.71	−0.74
Native pigs	504,075	86,622	580,069	82,083	1.57	−0.60
Breeding pigs	2,032,561	96,024	3,054,758	87,121	4.63	−1.08
Boars	137,226		126,208		−0.93	
Sows	721,341		885,928		2.31	
Piglets	1,173,994		2,042,622		6.35	
Fattening pigs	3,748,967	79,173	5,876,562	79,843	5.12	0.09

[a]Compound annual growth rate (CAGR) is a business and investing specific term for the geometric progression ratio that provides a constant rate of return over the time period

pig population in 2004 against 0.97 % of global pig population in 2013. The compound annual growth rate over that period was 4.7 %, with fattening pigs growing by 5.1 % per year, breeding pigs by 4.6 % per year and native pigs by 1.6 % per year. In contrast, the compound annual growth rate of boars decreased by 0.93 % per year. The growth rate of pig holders in Thailand was also negative, with decreases of 0.74, 1.1 and 0.60 % per year for breeding pigs and native pig holders, respectively. Meanwhile, the number of holders of fattening pigs slightly increased by 0.09 % per year.

Detailed data on pig populations by pig types, farming systems and farm scales for 2010 are presented in Table 3. There were 8.3 million head of pigs throughout the country with 5.2 million fattening pigs (62 %), 2.5 million breeding pigs (30 %) and 0.68 million native pigs (8 %). The median number of pigs per holder was five, but when broken down by pig type it was three pigs per holder for native pigs, four pigs per holder for breeding pigs, and eight pigs per holder for fattening pigs. The breakdown of commercial farms was 78 % belonging to the finishing systems, 14 % to nursery systems and 8 % to the farrow-to-finish systems. However, the number of pigs per holder of the farrow-to-finish systems (556) was much higher than that of the nursery systems and of the finishing systems with 96, and 88 pigs per holder, respectively.

In terms of farm scales, pig holders were classified as smallholders (95.02 %) and large-scale farming systems (4.98 %). Smallholders can be classified into two groups: backyards (representing 60 % of smallholders) and commercial smallholders (40 % of smallholders). 60.83 % of the backyard holders held native pigs, whereas 42.99 and 20.43 % of holders held breeding and fattening pigs, respectively (the percentages do not sum to 100 % because one backyard holder may have pigs of different types). In contrast, these proportions were 33.76, 40.75 and 36.98 % for the commercial smallholders. Even though

there were only 5 % of large-scale farming systems (50 to > 5000), they held 82 % of the total pig stock. Within the 5 % of farms classified as large-scale farming systems, 3.9 % were small (50–500 heads), 1.0 % were moderate (500–5000 heads), and 0.10 % were large (>5000 heads).

Spatial distribution

The spatial distributions of pig population in 2010 were mapped by pig type and farm size (Fig. 4). With a total of 8.3 million pigs in Thailand (Fig. 4a), the highest densities, regardless of pig type, were located in area surrounding the Bangkok Metropolitan region. The lowest densities were found within the city of Bangkok itself, in the three provinces in the lowest areas provinces of Yala, Pattani, and Narathiwat, and the western areas adjacent to Myanmar.

When considering different pig types, native pigs (Fig. 4b) were mostly found in isolated and rural areas of the Northwest (high mountains) and in the Northeast, where plateaus and arid lands dominate the landscape. In contrast, breeding pigs and fattening pigs showed very similar patterns (Figs. 4c and d), with concentrations of high densities in areas surrounding the Bangkok Metropolitan Region, two provinces in the North (Chiang Mai and Chiang Rai provinces), and three provinces at the border between Livestock Region 8 and Livestock Region 9 (Nakorn Sithammarat, Pattalung, and Songkhla provinces).

When broken down by farms size (Fig. 4e and g), smallholders appeared to be relatively homogeneously distributed throughout the country, but with lower densities in the Bangkok Metropolitan Region, the eastern region (Livestock Region 2), the western region (forested areas), and the three provinces in the south. In contrast, intensive larger farms were mostly located in the areas nearby the main cities including the areas surrounding the Bangkok Metropolitan Region, the areas nearby

Table 3 Pig production in Thailand in 2010. Pig production in Thailand in 2010 categorized by pig types, pig farming systems and pig farm scales

| Groups | Sub-groups | Total number | | Scales | | | | | | | | | |
| | | | | <5 | | 5–50 | | 50–500 | | 500–5000 | | >5000 | |
		Head	Farm	Head	Farm	Head	Farm	Head	Farm	Head	Farm	Head	Farm
Pig types	All pigs	8,346,614	199,992	285,932	112,673	1,214,288	77,366	1,250,106	7736	2,412,105	2004	3,184,183	213
	Native pigs	681,463	95,328	172,988	68,540	342,745	26,122	63,966	626	47,133	35	54,631	5
	Breeding pigs	2,517,651	83,502	108,957	48,443	490,224	31,533	421,327	2965	670,792	489	826,351	72
	Fattening pigs	5,147,500	56,884	66,397	23,027	468,176	28,610	758,085	3491	1,941,515	1629	1,913,327	127
Farming systems*	Farrow-to-finish	2,074,423	3731	145	31	45,932	1613	229,119	1654	526,654	343	1,272,573	90
	Nursery	590,136	6119	2969	708	77,941	4564	104,334	682	173,181	146	231,711	19
	Finishing	3,063,122	34,942	49,481	17,555	217,052	14,020	548,364	2009	1,381,166	1,302	867,059	56

*Farming systems (farrow-to-finish, nursery, and finishing systems) based on commercial pig breeds only

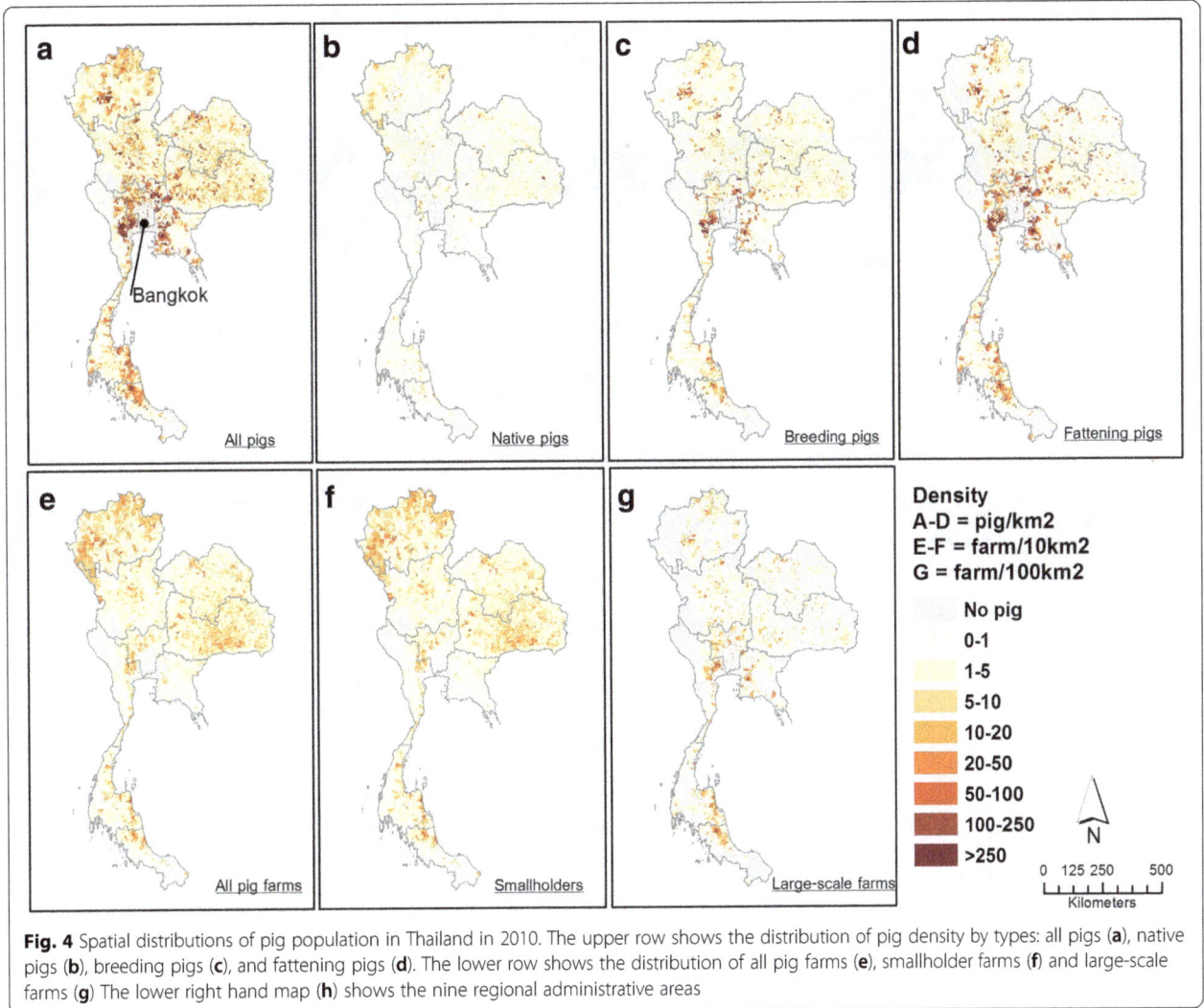

Fig. 4 Spatial distributions of pig population in Thailand in 2010. The upper row shows the distribution of pig density by types: all pigs (**a**), native pigs (**b**), breeding pigs (**c**), and fattening pigs (**d**). The lower row shows the distribution of all pig farms (**e**), smallholder farms (**f**) and large-scale farms (**g**) The lower right hand map (**h**) shows the nine regional administrative areas

Chiang Mai and Chiang Rai provinces (in the North), and the areas nearby Song Khla province (in the South).

Distribution modeling

The distribution of absences for different categories was fairly similar for different categories (Additional file 1: Figure S1) with a small minority of sub-districts with no pigs, located either in very remote and inhabited areas, or in very dense urban areas. So, we only report the results of the quantitative part of the zero-altered model, and the equivalent results for the binomial presence/absence models are presented as supplementary information. The importance of different spatial covariates in the quantitative RF models is shown in Table 4 ((Additional file 1: Table S1 for the binomial model). The strongest predictors of distribution for the different pig types and farm scales was the human population density (median variable importance of 66.9 %), followed by travel time to the capital city (median variable importance of 58.7 %), elevation (median

variable importance of 44.8 %), travel time to the provincial capitals (median variable importance of 37.6 %), rainfed croplands (median variable importance of 36.8 %), and irrigated croplands (median variable importance of 30.5 %). We obtained better accuracies (Table 4), when predictions were made directly at the sub-district level (with aggregated predictors) rather than by aggregating the results of the pixel-level predictions.

The association between the fitted functions and the predictor variables modelled by the quantitative RF model are shown in Fig. 5. The plots show that three variables, including rainfed croplands, irrigated croplands, and human population density shown a similar positive association with the predicted values for all pig farming types (Fig. 5d to f). In contrast, for two predictor variables, the travel time to the capital city and the travel time to the provincial capitals, different relationships were found according to the type of pig farming (Fig. 5a to b). Breeding pigs and fattening pigs showed a negative association with

Table 4 Important variables modeled by the quantitative Random forests and evaluation of predicted maps modeled by combined models. The variable importance (%) used to predict pig types and pig farm scales and the evaluation of the combined models. Predictor variables include, travel time to the capital city (Bangkok), travel time to the provincial capitals (Meung districts), rainfed croplands irrigated croplands, elevation, and human density)

Categories	Response variables[a]	The variable importance[b]						Evaluation			
		TCapCity	TProCap	RaCrop	IrCrop	Elev	HuDen	RMSE[c] (sub-district)	Correlation (sub-district)	RMSE (pixel)	Correlation (pixel)
Pig types (heads/km2)	Native pigs	41.97	27.54	29.44	23.86	34.64	66.89	0.12	0.94	1.19	0.78
	Breeding pigs	58.74	38.84	36.76	30.46	44.82	75.38	0.23	0.91	1.32	0.79
	Fattening pigs	61.16	37.57	47.71	33.07	63.81	63.55	0.31	0.87	1.28	0.83
Pig farm scales (farms/10 km2)	SM	100.27	46.86	62.10	50.90	77.83	148.82	0.14	0.95	1.43	0.80
	LF	21.85	21.68	25.33	15.39	36.33	62.09	0.07	0.92	0.74	0.74

[a]Response variables include: number of native pigs, number of breeding pigs, number of fattening pig, number of smallholders (SM), and number of large-scale farming systems (LF)
[b]Predictor variables include: travel time to the capital city (TCapCity), travel time to the provincial capitals (TProCap), rainfed croplands (RaCrop), irrigated croplands (IrCrop), elevation (Elev), and human density (HuDen)
[c]RMSE stands for root mean square error

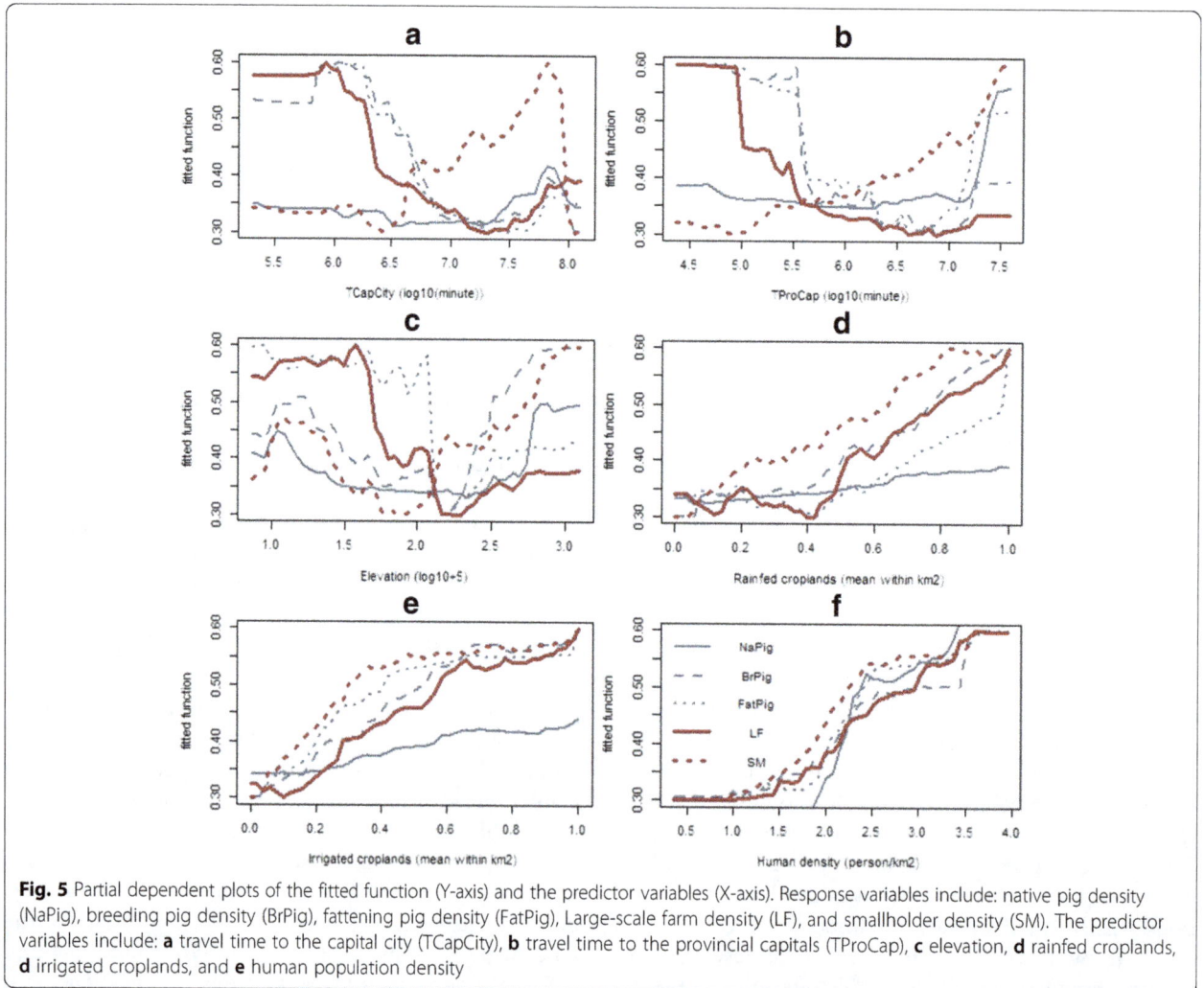

Fig. 5 Partial dependent plots of the fitted function (Y-axis) and the predictor variables (X-axis). Response variables include: native pig density (NaPig), breeding pig density (BrPig), fattening pig density (FatPig), Large-scale farm density (LF), and smallholder density (SM). The predictor variables include: **a** travel time to the capital city (TCapCity), **b** travel time to the provincial capitals (TProCap), **c** elevation, **d** rainfed croplands, **d** irrigated croplands, and **e** human population density

those predictors, whereas native pigs showed an inverse positive association. The same contrasting pattern with these predictors was found for the farm scale categories, where large-scale production systems showed a negative association with travel time to the capital city and travel time to the provincial capitals, whereas smallholders showed a positive association. Regarding elevation (Fig. 5c), fattening pigs and large-scale production systems showed correlation with low elevation, while smallholders, native pigs, and breeding pigs showed both a negative and positive association. In the binomial presence/absence model, such inverse associations were not apparent ((Additional file 1: Figure S2), as there was much more similarity between the presence/absence distributions of the different categories (e.g. probability of absence was predicted to be positively associated with elevation in all categories).

Discussion

Pig populations in Thailand over the past 50 years showed an initial decline and then increased from the mid-1980s. Interestingly, although the overal mean number of pig owners has remained fairly stable, it has shown strong fluctuations around the mean. This phenomon has been described as the "pork cycle" or "hog cycle" [41] The "pork cycle" in Thailand has been characterized typically by a 32 month cyle, with 16 months of loss and 16 months of profit [41]. The cycles have been attributed to interactions between economic and animal health factors [41]. A large proportion of the stock is held by smallholders who quickly adapt to changing market-prices on the markets. When the prices are high, many smallholders start producing pork. After the time required bringing them to slaughter weight, the supply increases, but the demand remains the same resulting in a drop in price. Smallholders then start losing money and stop raising pigs. This gradually reduces the amount of pork on the market, and the prices revert to higher levels. These fluctuations are not absorbed by export and imports, because majority of the Thai pork production, especially that of smallholders, serves domestic markets. In addition, outbreaks of pig diseases, such as virulent strains of PRRS [11, 12], can have an influence on pork production and market prices, and contribute to trigger or amplify these fluctuations. Interestingly, although the total number of pig holders did not vary so much in time over this study period, the average number of pigs per holding showed an increasing trend, which confirms that Thailand is still intensifying its pig production, and this will impact over longer time period on the number of smallholders, who will gradually either, quit pork production, or move to large-scale pig farming.

The spatial distribution of large-scale farming systems currently largely corresponded to the distribution of commercial pig types (breeding pigs and fattening pigs), in the suburban areas surrounding the main cities, particularly around the Bangkok Metropolitan region. Both large-scale farming systems and commercial pig breeds showed negative associations with travel time to the capital city and to provincial capitals (Meung districts), and with elevation, and a positive association with human density, rainfed croplands and irrigated croplands. This indicates that most of the intensive pig farms and commercial pig breeds were located in suburban areas in lowlands, conveniently placed to transport produce to consumption markets such as provinces capital and the Bangkok Metropolitan region, and at the same time to access a local supply chain of pig feed ingredients [52]. A similar spatial pattern has been observed in a similar study carried out for poultry [33]. The major pig production provinces were previously within 60–150 km of Bangkok but this catchment has now expanded to 250 km because of the rapid increase in demand for livestock products and improvements in transport [2]. In contrast to the spatial distribution of large-scale farming systems, smallholders, comprising 95 % of all farms types, were distributed in the more rural parts of the country. They showed positive associations with all predictor variables, suggesting that the smallholders were more likely to be located in highland, remote, rural areas to supply the rural and local markets. However, the basic requirement of a supply chain for pig feed, suggested by the positive association with croplands, is also necessary for smallholders, a pattern observed also in poultry [33]. So, smallholders and large-scale farms showed inverse relationships for travel times to cities and elevation, but similar relationships with human population density and crop-related variables. These results makes perfect sense, since smallholders need local consumers (hence a positive association with human population density) and local feed supply (hence the positive association with cropland), and are not affected by difficulties of transport encountered in high elevation areas. Conversely, large-scale farming required similar conditions in terms of consumer and feed, but rely so strongly on large consumption centers that market access quantified through travel time become an inverse relationship, i.e. high numbers of farms for short travel time to the province capital or to Bangkok. One should note that the travelled-time layer used here may not adequately reflect the travel constraints of trucks, as it also included landcover in its design. However, we feel that this probably would have a marginal effect on our results given the scale of our analysis that include a very wide gradient of travel time ranging from Bangkok to the remote mountainous areas. We also used a somewhat arbitrary classification threshold for smallholders, matching operational definitions already used in Thailand, but not necessarily matching home-consumption vs. commercial destination of outputs. However, a backyard vs. commercial divide would probably show similar patterns, i.e. a strong association between backyard producers and remote and rural areas.

All types of pigs were found in smallholders, but the majority of native pig farms were of the smallholders (99.3 %). As shown in the distribution map, native pigs were found throughout the country, but with high-density locations in northern Thailand. The native Thai pigs in the northern highland region are raised by smallholders in the hill tribe communities and are important in relation to local customs and religion, where animals are sacrificed for special celebrations such as New Year and weddings [21, 22]. However, it is noteworthy that 0.7 % of the native pigs were still raised in intensive farms; by 626 small farms, 35 moderately sized and five large farms (>5000 head). This is linked to the increasingly popularity of consumption of wild pigs in restaurants, to which some large-scale producers have responded by increasing their production of native pigs [21].

Most of the pig farming systems in Thailand belonged to the finishing systems (78 %) followed by the nursery systems (14 %) and the farrow-to-finish systems (8 %). The average number of pigs per holder in the farrow-to-finish systems (556) was much higher than that of the nursery systems (96) and the finishing systems (88). The farrow-to-finish systems handle all pig production stages. Consequently, they need to have a high level of specialization and a long experience in using modern technologies to increase productivity [41]. They control the entire production chain by adjusting both the number and quality of pigs raised and fattened [2]. In contrast, owners of the finishing systems need to purchase feeder pigs from external sources, which is a more risky strategy; exposing them to fluctuations in supply, unreliable genetic background, and to poor overall quality and health of the animals [2].

The different geographical patterns of large-scale commercial and smallholder production offer opportunities to their future sustainable developments, since better and more sustainable modes of production could be applied to both modes of production.

Small holders pig farmers could integrate pig farming with a combination of other livestock, crops, vegetables and fruit production as an integrated organic farming [53, 54]. The combination of different vegetable and animal products could also cover the family's consumption needs, and reduce dependency on the sale of products, thus protecting themselves from price fluctuations [54]. In addition, the combination of different farming activities can facilitate synergistic interactions [53]. Pig waste can be used to produce biogas for the household as well as organic fertilizer for plants [55–57]. In turn, crop products and residues can be used as animal feed. So,

rather than the waste from pig production becoming a source of air and water pollution, it can be better treated by i) using simple biofilters such as rice straw, coconut husks, wood shavings, rattan strips and oil palm [58]; ii) decomposing the waste using the Effective Microorganisms (EM) [59]; and iii) biodigestion to produce biogas in simple containers [55]. Better knowledge of basic of bio-security could be encouraged to protect smallholder farms from harmful agents. In economic terms, the demand for organic farming products is growing in Thailand and this may present new market opportunities for smallholders [60]. Farmers could also work together under co-operatives in order to increase their negotiation power with buyers. This could potentially lead to more sustainable agriculture, environmental protection and animal welfare for this sector, which could be favored through incentives in some particular regions of Thailand.

In the commercial sector, the concept of "Area Wide Integration" (AWI) could be applied in some areas in Thailand, and be geographically informed by the results of this study. The concept of AWI [4, 18] for the most commercially oriented farming involves integrating a particular livestock activity with other forms of crop farming in a specific geographic area not used for other types of livestock production and away from urban development. Within such areas, facilities involved in the production cycle, such as feed mills, slaughterhouses and processing plants can be established, which can greatly enhance bio-security by securing the area as a "pig-zone". Proper practice can be carried out within the area, such as farm management, distribution of cropping land, utilization of manure for biogas production and composting.

The results of this study indicate that intensified pig farms are already mostly located in suburban areas in lowlands, in areas that area already conveniently placed to transport pig products to the main markets, and with good access to pig feed ingredients. However, proper geographical planning, accounting for different aspects such as a health, environmental and economic sustainability, remains to be carried out [61] in order better to refine the definition of potentially suitable regions for long-term, more sustainable large-scale and small-scale pig raising in Thailand.

One of limitations of this quantitative assessment is that the identification of the farming system was made through the farm size and composition at the time of the census. Since farming systems were not defined prior to the census, farms were simply classified according to the number of pigs of different types. So, farms where some particular types of pigs would not have been present or raised at the time of the census could be misclassified. For the intensification level, we used the number of pigs on a farm as a classifier, with a threshold of

50 head, to separate smallholders and large-scale production systems. This may not reflect perfectly the actual level of inputs (i.e. intensification) and level of productivity in those farms. A more comprehensive assessment of inputs and productivity would be difficult to implement in census studies, but could rather be the focus of specific surveys, which may be stratified according to the categories outlined here. However, using this threshold to distinguish between both types of farming systems also has advantages including: i) the bio-security systems, enforced by the regulation on the "Good Agricultural Practice for Pig Farm" for the farms keeping more than 50 pigs [25], are differentiable; and ii) the results of the study can be used to support the strategy of the government directly. However, future study on pig systems in Thailand should consider collecting more detailed data on the pig production systems, such as information on inputs (feed, energy, manpower) outputs (volumes, quality), bio-security and disease prevention practices. These data should not only allow a finer definition of the systems within Thailand, but also facilitate the comparison with data from other countries that could be pursued in the future. We used RF to investigate the relationship between the spatial predictor variables and the pig count, as the method was recently shown to clearly outperform other regression-based techniques in large-scale livestock modeling [38]. However, our primary objective of the RF model was not to optimize the predictive power of our model, but rather to quantify how different spatial factors rank against each other's in best predicting different categories, and to provide a detailed view of their influence on the fitted values, and we feel that it was helpful in this regard too. For example, the possibility to plot the profiles corresponding to different predictor variables (Fig. 5) allows investigating these with great details, and to show some fairly complex patterns (compare e.g. Fig. 5 A large farms LF versus small farms SM). In comparison, a multiple regression, for example, would provide only coefficients that allows to give the overall direction of the association (positive or negative), but would have more difficulties in handling non-linear relationships, or to account for the multiple interactions between variables, which is one of the strength of machine learning techniques (Random Forest or BRT). A limitation, however, is the lack of formal tests allowing quantifying the significance of a particular variable in terms of hypothesis testing. A formal comparison of different modeling options goes beyond the scope of this paper, and several alternatives such as General Additive Models (GAMs) could have been used as alternatives, but we felt that RF provided a good trade-off between the details of the information it provides and ease of implementation.

Conclusions

Detailed census data and spatial modeling has enabled the geographical and functional characterization of pig farming systems in Thailand. They highlight a process of intensification of the production, with increasing numbers of pigs per owner over time, large-scale pigs farms concentrated around the capital city to supply its demand, with a tendency of being located increasingly far from the center. Their distribution mostly corresponds to that of breeding and fattening pigs of improved breeds. In contrast, smaller-scale producers are distributed in more rural regions, and more strongly concentrated around local province capitals. These historical developments have not resulted from any specific planning in the past, and have resulted in a present distribution that may not be optimal in terms of environment and health impacts, for example. As the sector is still expanding, future developments may benefit from spatially-informed planning accounting for the specific health, environment and economical implications of the different pig production systems recognizing their specificities. This could be achieved, for example, through the promotion of sustainable intensification of small-scale producers to limit their potential local environmental impact, and by the implementation of AWI for the most intensive production sector in geographically limited parts of the country. Defining these areas geographically could be the scope of follow-up works using multiple-criteria decision analysis tools such as to incorporating environment, heath and economic spatial criteria in the decision-making.

Abbreviations
AWI: Area Wide Integration; BORA: Bureau of Registration Administration; BRT: Boosted regression tree; COR: Correlation coefficient; CSF: Classical Swine Fever; DLD: Department of Livestock Development; EM: Effective Microorganisms; FMD: Foot and Mouth Disease; GAMs: General Additive Models; GLW1: Gridded Livestock of the World 1; GLW2: Gridded Livestock of the World 2; MOAC: Ministry of Agriculture and Cooperatives; OOB: Out of bag; PCD: Pollution Control Department; PRRS: Porcine Reproductive and Respiratory Syndrome; RF: Random forest; RMSE: Root mean square error; SAR: Simultaneous autoregressive model; TGE: Transmissible Gastroenteritis

Acknowledgements
We thank staff of Department of Livestock Development (DLD) composed of District Livestock Offices, Provincial Livestock Offices, and Center for Information Technology for animal census data; Ministry of Transportation for geodata; Department of Provincial Administration, Ministry of Interior for population data. We also thank the colleagues of Lutte biologique et Ecologie spatiale (LUBIES), ULB, Belgium, for assistance and suggestion. TPR is funded by the CGIAR Research Programmes on the Humidtropics; Climate Change, Agriculture and Food Security (CCAFS) and Agriculture for Nutrition and Health (A4NH).

Funding
Part of this work was supported through the NIH NIAID grant (1R01AI101028-01A1).

Authors' contributions

WT and MG conceived and designed the study. WT generated the raw data and performed statistical analysis with contributions from MG and CL. WT drafted the paper, which MG, MV and TR critically reviewed and revised the paper. PC, SK, AG, and ME supported raw data. All authors read and approved the final manuscript.

Authors' information

Not applicable.

Competing interests

The authors declare that they have no competing interests.

Consent for publication

Publication was approved by the Research Committee of the Bureau of Disease Control and Veterinary Services, Department of Livestock Development (Permit Number: 0601/814).

Author details

[1]Department of Livestock Development (DLD), Bangkok 10400, Thailand. [2]Lutte biologique et Ecologie spatiale (LUBIES), Université Libre de Bruxelles, Brussels 1050, Belgium. [3]Fonds National de la Recherche Scientifique (FNRS), Brussels 1050, Belgium. [4]Faculty of Veterinary Medicine, Kasetsart University, Kampangsaen Campus, Nakornpatom 73140, Thailand. [5]Research Unit of Landscape Ecology AND Plant Production Systems (EPSPV), University of Brussels, 1050 Brussels, Belgium. [6]Department of Geography and Geosciences, University of Louisville, Louisville 40292, USA. [7]Centre for Development and Environment (CDE), Country office in the Lao PDR, Vientiane 6101, Lao PDR. [8]Livestock Systems and Environment (LSE), International Livestock Research Institute (ILRI), Nairobi 30709, Kenya.

References

1. Robinson T, Thornton P, Franceschini G, Kruska R, Chiozza F, Notenbaert A, et al. Global livestock production systems. Rome: Food and Agriculture Organization of the United Nations (FAO) and International Livestock Research Institute (ILRI); 2011.

2. Poapongsakorn N, NaRanong V. Annex IV: Livestock Industrialization Project: Phase II - Policy, Technical, and Environmental Determinants and Implications of the Scaling-Up of Swine, Broiler, Layer and Milk Production in Thailand [Internet]. IFPRI-FAO project entitled Livestock Industrialization, Trade and Social-Health-Environment Impacts in Developing Countries; 2003. Available from: http://www.fao.org/wairdocs/lead/x6170e/x6170e39. htm#TopOfPage. Accessed 19 Aug 2014.

3. Svendsen J, Svendsen LS. Intensive (commercial) systems for breeding sows and piglets to weaning. Livest Prod Sci. 1997;49:165–79.

4. Cameron RDA. A review of the industrialisation of pig production worldwide with particular reference to the Asian region [Internet]. Anim. Prod. Health. 2000. Available from: http://www.fao.org/Ag/againfo/themes/ en/pigs/production.html.. Accessed 20 Dec 2014.

5. Otte J, Roland-Holst D, Pfeiffer D, Soares-Magalhaes R, Rushton J, Graham J, et al. Industrial livestock production and global health risks. -Poor Livest. Policy Initiat. Living Livest. Res. Rep. Rome: Food and Agriculture Organization of the United Nations; 2007.

6. Kaewpitoon N. Food-borne parasitic zoonosis: distribution of trichinosis in Thailand. World J Gastroenterol. 2008;14:3471.

7. Tong G-Z, Zhou Y-J, Hao X-F, Tian Z-J, An T-Q, Qiu H-J. Highly pathogenic porcine reproductive and respiratory syndrome. China Emerg Infect Dis. 2007;13:1434–6.

8. An T-Q. Highly pathogenic porcine reproductive and respiratory syndrome virus. Asia Emerg Infect Dis. 2011;17:1782–4.

9. Feng Y, Zhao T, Nguyen T, Inui K, Ma Y, Nguyen TH, et al. Porcine respiratory and reproductive syndrome virus variants, Vietnam and China, 2007. Emerg Infect Dis. 2008;14:1774–6.

10. Ni J, Yang S, Bounlom D, Yu X, Zhou Z, Song J, et al. Emergence and pathogenicity of highly pathogenic Porcine reproductive and respiratory syndrome virus in Vientiane, Lao People's Democratic Republic. J Vet Diagn Invest. 2012;24:349–54.

11. Nilubol D, Tripipat T, Hoonsuwan T, Kortheerakul K. Porcine reproductive and respiratory syndrome virus, Thailand, 2010–2011. Emerg Infect Dis. 2012; 18:2039–43.

12. Thanapongtharm W, Linard C, Pamaranon N, Kawkalong S, Noimoh T, Chanachai K, et al. Spatial epidemiology of porcine reproductive and respiratory syndrome in Thailand. BMC Vet Res. 2014;10:174.

13. Mastin A, Alarcon P, Pfeiffer D, Wood J, Williamson S, Brown I, et al. Prevalence and risk factors for swine influenza virus infection in the English pig population. PLoS Curr. 2011;3, RRN1209.

14. FAO. World Livestock 2013-Changing disease landscape. Rome: FAO; 2013.

15. Ma W, Kahn RE, Richt JA. The pig as a mixing vessel for influenza viruses: human and veterinary implications. J Mol Genet Med. 2009;3: 158–66.

16. Pulliam JRC, Epstein JH, Dushoff J, Rahman SA, Bunning M, Jamaluddin AA, et al. Agricultural intensification, priming for persistence and the emergence of Nipah virus: a lethal bat-borne zoonosis. J R Soc Interface. 2012;9:89–101.

17. WHO. Nipah Virus (NiV) Infection [Internet]. Glob. Alert Response GAR; 2014. Available from: http://www.who.int/csr/disease/nipah/en/. Accessed 8 Nov 2014.

18. The World Bank. Managing the Livestock Revolution Policy and Technology to Address the Negative Impacts of a Fast-Growing Sector -Report No. 32725-GLB. The International Bank for Reconstruction and Development,The World Bank, Washington, DC, USA; 2005.

19. Aksornphan P, Isvilanonda S. Profit efficiency of standardized pig production in Thailand. Kasetsart Univ. J. Econ. 2009;16:26–38.

20. Sakpuaram T, Kasaemsuwan S, Udomprasert P. Swine industry farms in Thailand. ACIAR WORKING PAPER NO. 53. Canberra: Australian Centre for International Agricultural Research; 2002.

21. Rattanaronchart S. Present situation of Thai native pigs. Chiangmai: Department of Animal Science, Faculty of Agriculture, Chiangmai University; 1994.

22. Charoensook R, Knorr C, Brenig B, Gathphayak K. Thai pigs and cattle production, genetic diversity of livestock and strategies for preserving animal genetic resources. Maejo IntJSciTechnol. 2013;7:113–32.

23. Thanawongnuwech R, Suradhat S. Taming PRRSV: revisiting the control strategies and vaccine design. Virus Res. 2010;154:133–40.

24. Tapinta S, Boonrat P, Buanak S, Songkamilin A, Laisood J. Guildines for environmental management in pig farm. Pollution Control Department, Bangkok, Thailand; 2014.

25. Viriyapak C, Mahantachaisakul C, Bunrung P, Tangjaipatana A, Chomchai S, Tantasuparuk W, et al. Good Agricultural Practice for Pig Farm B.E.2551 (2008). : National Bureau of Agricultural Commodity and Food Standards Ministry of Agriculture and Cooperatives, Bangkok, Thailand; 2008.

26. FAOSTAT. Statistic Division, Food and Agriculture organizaiton of the united nations [Internet]. 2014. Available from: http://faostat3.fao.org/download/Q/ QL/E. Accessed 5 Nov 2014.

27. DLD. Animal statistics, Infomation Technology Center, Department of Livestock Development [Internet]. 2014. Available from: http://ict.dld.go.th/ th2/index.php/th/report/11-report-thailandlivestock-livestock. Accessed 5 Nov 2014.

28. BORA. Official statistics registration systems, Department of Provincial Administration [Internet]. 2014. Available from: http://stat.bora.dopa.go.th/ stat/. Accessed 5 Nov 2014.

29. Prosser DJ, Wu J, Ellis EC, Gale F, Van Boeckel TP, Wint W, et al. Modelling the distribution of chickens, ducks, and geese in China. Agric Ecosyst Environ. 2011;141:381–9.

30. Robinson TP, Wint GRW, Conchedda G, Van Boeckel TP, Ercoli V, Palamara E, et al. Mapping the global distribution of livestock. PLoS One. 2014;9:e96084. Baylis M, editor.

31. Robinson TP, Franceschini G, Wint W. The Food and Agriculture Organization's grided of the world. Vet Ital. 2007;43:745–51.

32. Van Boeckel TP, Prosser D, Franceschini G, Biradar C, Wint W, Robinson T, et al. Modelling the distribution of domestic ducks in Monsoon Asia. Agric Ecosyst Environ. 2011;141:373–80.

33. Van Boeckel TP, Thanapongtharm W, Robinson T, D'Aietti L, Gilbert M. Predicting the distribution of intensive poultry farming in Thailand. Agric Ecosyst Environ. 2012;149:144–53.

34. Neumann K, Elbersen BS, Verburg PH, Staritsky I, Pérez-Soba M, de Vries W, et al. Modelling the spatial distribution of livestock in Europe. Landsc Ecol. 2009;24:1207–22.

35. Breiman L. Random Forests. Mach. Learn. [Internet]. The Netherlands: Kluwer Academic Publishers; 2001. p. 5–32. Available from: http://link.springer.com/article/10.1023%2FA%3A1010933404324. Accessed 30 Oct 2013.

36. Liaw A, Wiener M. Classification and regression by randomForest. R News. 2002;2/3:18–22.

37. Cutler DR, Edwards TC, Beard KH, Cutler A, Hess KT, Gibson J, et al. Random forests for classificaiton in ecology. Ecology. 2007;88:2783–92.

38. Nicolas G, Robinson TP, Wint GRW, Conchedda G, Cinardi G, Gilbert M. Using random forest to improve the downscaling of global livestock census data. PLoS One. 2016;11:e0150424. Bond-Lamberty B, editor.

39. Gaughan AE, Stevens FR, Linard C, Jia P, Tatem AJ. High resolution population distribution maps for Southeast Asia in 2010 and 2015. PLoS One. 2013;8:e55882. Pappalardo F, editor.

40. Stevens FR, Gaughan AE, Linard C, Tatem AJ. Disaggregating census data for population mapping using random forests with remotely-sensed and ancillary data. PLoS One. 2015;10:e0107042. Amaral LAN, editor.

41. Maanan B. Handbook of pig farming management. Thailand: Livestock productivity institute, Kasetsart University; 2009.

42. CGIAR-CSI. SRTM 90 m DEM Digital Elevation Database [Internet]. 2014. Availablefrom: http://srtm.csi.cgiar.org/.. Accessed Aug 6 2014.

43. Arino O, Bicheron P, Achard F, Latham J, Witt R, Weber J. GLOBCOVER: the most detailed portrait of earth. ESA Bull. Bull. ASE Eur. Space Agency [Internet]. 2008. Available from: http://due.esrin.esa.int/page_globcover.php. Accessed 9 Nov 2014.

44. Nelson A. Accessibility, transport and travel time information [Internet]. 2000. Available from: http://forobs.jrc.ec.europa.eu/products/gam/. Accessed 6 Aug 2014.

45. Zuur AF, Ieno EN, Walker NJ, Saveliev AA, Smith GM. Mixed Effects Models and Extensions in Ecology with R [Internet]. New York: Springer New York; 2009. Available from: http://link.springer.com/10.1007/978-0-387-87458-6. Accessed 9 Nov 2014.

46. Cantoni E, Zedini A. A Robust Version of the Hurdle Model. Switzerland: University of Geneva; 2009.

47. Potts JM, Elith J. Comparing species abundance models. Ecol Model. 2006; 199:153–63.

48. Kuhnert PM, Martin TG, Mengersen K, Possingham HP. Assessing the impacts of grazing levels on bird density in woodland habitat: a Bayesian approach using expert opinion. Environmetrics. 2005;16:717–47.

49. Martin TG, Wintle BA, Rhodes JR, Kuhnert PM, Field SA, Low-Choy SJ, et al. Zero tolerance ecology: improving ecological inference by modelling the source of zero observations: Modelling excess zeros in ecology. Ecol Lett. 2005;8:1235–46.

50. Breiman L, Cutler A, Liaw A, Wiener M. randomForest: Breiman and Cutler's random forests for classification and regression [Internet]. CRAN; 2014. Available from: http://cran.r-project.org/web/packages/randomForest/index.html. Accessed 23 Nov 2014.

51. Zambrano-Bigiarini M. hydroGOF: Goodness-of-fit functions for comparison of simulated and observed hydrological time series [Internet]. CRAN; 2014. Available from: http://cran.r-project.org/web/packages/hydroGOF/index.html. Accessed 20 Mar 2015.

52. Kunavongkrit A, Heard T. Pig reproduction in South East Asia. Anim Reprod Sci. 2000;60–61:527–33.

53. Jordan CF. An Ecosystem Approach to Sustainable Agriculture [Internet]. Dordrecht: Springer Netherlands; 2013. Available from: http://link.springer.com/10.1007/978-94-007-6790-4. Accessed 26 Oct 2015.

54. Mongsawad P. The philosophy of the sufficiency economy : a contribution to the theory of development. Asia-Pac Dev J. 2010;17.

55. Angelidaki I, Ellegaard L. Codigestion of manure and organic wastes in centralized biogas plants: status and future trends. Appl Biochem Biotechnol. 2003;109:95–105.

56. Holm-Nielsen JB, Al Seadi T, Oleskowicz-Popiel P. The future of anaerobic digestion and biogas utilization. Bioresour Technol. 2009;100:5478–84.

57. Somanathan E, Bluffstone R. Biogas: Clean Energy Access with Low-Cost Mitigation of Climate Change. Environ. Resour. Econ. [Internet]. 2015. Available from: http://link.springer.com/10.1007/s10640-015-9961-6. Accessed 7 Oct 2015.

58. Sommer SG, Mathanpaal G, Dass GT. A simple biofilter for treatment of pig slurry in Malaysia. Environ Technol. 2005;26:303–12.

59. Jusoh MLC, Manaf LA, Latiff PA. Composting of rice straw with effective microorganisms (EM) and its influence on compost quality. Iran J Environ Health Sci Eng. 2013;10:17.

60. Sarich C. Thailand Restores Organic, Sustainable Farming Practices [Internet]. 2004. Available from: http://naturalsociety.com/peaceful-coup-thailand-restores-organic-agriculture-sustainable-farming-practices/. Accessed 23 Nov 2014.

61. Gerber PJ, Carsjens GJ, Pak-uthai T, Robinson TP. Decision support for spatially targeted livestock policies: diverse examples from Uganda and Thailand. Agric Syst. 2008;96:37–51.

12

A large-scale study of a poultry trading network in Bangladesh: implications for control and surveillance of avian influenza viruses

N. Moyen[1]* 🆔, G. Ahmed[3], S. Gupta[2,3], T. Tenzin[3,4], R. Khan[3], T. Khan[3], N. Debnath[3], M. Yamage[3], D.U. Pfeiffer[1,5] and G. Fournie[1]

Abstract

Background: Since its first report in 2007, avian influenza (AI) has been endemic in Bangladesh. While live poultry marketing is widespread throughout the country and known to influence AI dissemination and persistence, trading patterns have not been described. The aim of this study is to assess poultry trading practices and features of the poultry trading networks which could promote AI spread, and their potential implications for disease control and surveillance.

Data on poultry trading practices was collected from 849 poultry traders during a cross-sectional survey in 138 live bird markets (LBMs) across 17 different districts of Bangladesh. The quantity and origins of traded poultry were assessed for each poultry type in surveyed LBMs. The network of contacts between farms and LBMs resulting from commercial movements of live poultry was constructed to assess its connectivity and to identify the key premises influencing it.

Results: Poultry trading practices varied according to the size of the LBMs and to the type of poultry traded. Industrial broiler chickens, the most commonly traded poultry, were generally sold in LBMs close to their production areas, whereas ducks and backyard chickens were moved over longer distances, and their transport involved several intermediates. The poultry trading network composed of 445 nodes (73.2% were LBMs) was highly connected and disassortative. However, the removal of only 5.6% of the nodes (25 LBMs with the highest betweenness scores), reduced the network's connectedness, and the maximum size of output and input domains by more than 50%.

Conclusions: Poultry types need to be discriminated in order to understand the way in which poultry trading networks are shaped, and the level of risk of disease spread that these networks may promote. Knowledge of the network structure could be used to target control and surveillance interventions to a small number of LBMs.

Keywords: Poultry network, Bangladesh, Surveillance, Avian influenza

* Correspondence: nmoyen3@rvc.ac.uk
[1]Department of Pathobiology and Population Sciences, Royal Veterinary College, University of London, Hatfield, Hertfordshire AL9 7TA, UK
Full list of author information is available at the end of the article

Background

Bangladesh has a human population of more than 145 million resulting in a density of 1072 people per km², and an estimated national poultry population of 304 million resulting in a poultry density of about 2400 poultry per km² [1, 2]. About 75% of the Bangladeshis live in rural areas and depend heavily on poultry, both as a source of proteins and of income: 1.8 million people are involved in poultry farming alone [2], and, in 2013, poultry meat represented half of the country's meat production [3]. As a result, a zoonotic infectious disease affecting the Bangladeshi poultry production system could have a large impact on the country's food security, economy and public health.

The first highly pathogenic avian influenza subtype H5N1 (HPAI H5N1) outbreak in Bangladesh was reported in 2007 [4, 5]. This strain is now considered to be endemic in the country [6, 7]. Live bird trading and marketing have been shown to play a major role in the maintenance of avian influenza viruses (AIVs) within a number of poultry production systems [8, 9]. Yet, in Bangladesh, live bird trading is ubiquitous, as more than 90% of poultry are marketed through live bird markets (LBMs) [10]. Previous studies conducted in other settings have identified an association between LBM characteristics (such as the number of poultry traded, trading frequency, and the number of poultry traded with other LBMs) and the risk of dissemination of AIVs [11–13]. In addition, the networks shaped by commercial poultry movements, within which LBMs generally act as hubs, have also been shown to support the dissemination and maintenance of avian viruses such as Avian Influenza (AI) and Newcastle disease [13–15].

In Bangladesh, migratory birds have been associated with the introduction and spread of HPAI H5N1 [16], but poultry trade and trade-related activities such as the exchange of egg trays between farms, or the introduction of contaminated vehicles into farms have also been identified as potential sources of AIV infection for commercial and backyard poultry flocks [5]. The HPAI H5N1 outbreaks that occurred between 2007 and 2009 in Bangladesh were spatially clustered along the country's main highways and principal poultry trading routes, which also supports the role of poultry trading activities in the spread of viruses through transport of infected poultry or contaminated material and vehicles [16–19]. Nevertheless, so far, national poultry trading networks have not been described, nor has their potential role in virus spread been assessed. In order to address this knowledge gap, a cross-sectional survey was conducted throughout the country to assess practices of live poultry traders, and to characterise the structure of the networks resulting from the trade of live poultry.

In Bangladesh, husbandry systems and the geographic location of poultry farms are strongly associated with the type of poultry reared. The four main poultry types traded in the country are: Industrial white-feathered broiler chickens (such as Hybro-PN, Hubbard classic, Ross, Cobb 500 [10]), sonali poultry (crossbreed between a Fayoumi female and a Red Island Red male [20]), deshi (local chickens raised in backyards [21]) and ducks. The first two poultry types are raised in commercial farms, and almost 70% of the commercial flock is located in the two most densely populated divisions of Bangladesh. Whereas deshis and ducks are raised in traditional scavenging systems, and more than a third of the backyard flock is located in one rural division [10, 22–24]. We therefore hypothesised that live poultry trading practices and networks may vary according to the types of poultry. While live poultry trading networks have been described in multiple settings [14, 25–28], they have not been characterised according to poultry types. Yet, identifying the types of poultry traded through these networks may provide insights to understand how these networks may generate disease risk for poultry and human populations.

Methods

Data collection

A trader was defined as a person whose main activity is to buy poultry from other poultry traders or farmers and to sell it either to other poultry traders or consumers. In order to select the LBMs in which the poultry traders would be interviewed a multi-stage purposive sampling process was followed. Bangladesh is divided into 64 districts (second administrative division) and 490 upazilas (sub-districts, third administrative division). A cross-sectional survey was carried in 138 LBMs across 17 upazilas purposively selected, located in 17 different districts of Bangladesh in September 2014. These 17 upazilas included the 2 main urban centres of the country, Dhaka City Corporation (DCC) and Chittagong City Corporation (CCC) which represent about 15% of the total human population of the country. The 15 other upazilas were selected based on their proximity to previous reported AI outbreaks, their high poultry density and their proximity to international borders. Within each upazila, LBMs identified with the highest quantity of traded poultry by local experts were recruited. Of the recruited LBMs, 50 (36%) were located in Dhaka district, all in DCC. In the 16 other selected districts, 1 to 16 LBMs were visited (Fig. 1). Once the LBMs selected, the traders were selected either randomly or purposively depending on if they operated in the LBM permanently or not. In LBMs where less than 5 traders operated permanently, all traders were interviewed. In LBMs with 6 to 10 permanent traders, 50% of them were randomly selected, and in LBMs where more than 10 permanent

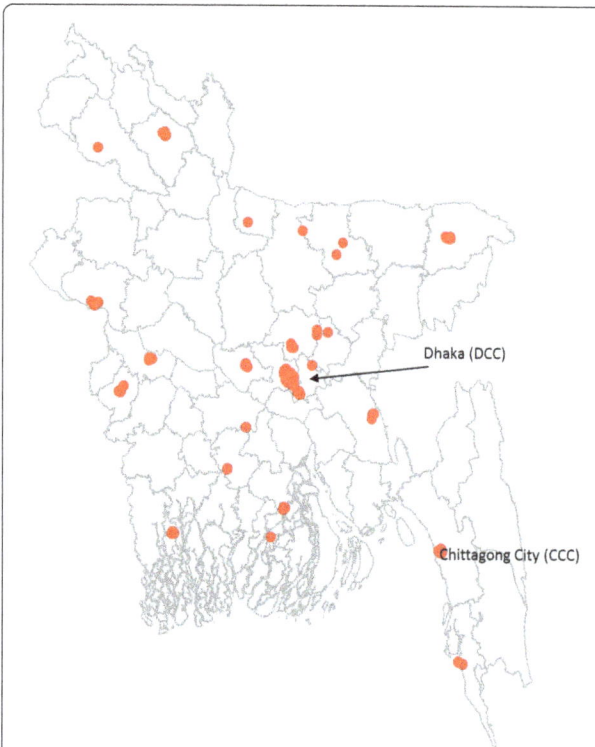

Fig. 1 Location of the 138 surveyed LBMs included in the study. One week data, collected from 849 traders in 2014

traders were present, 30% of them were randomly selected. In addition, as many traders (often referred to as middlemen) supplying permanent LBM traders as possible were also interviewed. As a result, a total of 849 poultry traders were interviewed. Verbal consent was obtained from the interviewees prior to the interviews. Standardised questionnaires were administered in Bengali by trained interviewers and Global Positioning System (GPS) coordinates of the surveyed LBMs were recorded. Informants were asked about their trading practices in the week preceding the interview: number of poultry sold to other poultry traders or consumers, number of poultry bought, types and locations from which poultry were sourced. Although traders remembered the names and towns of the LBMs they bought from, this was not the case for the farms, for which only the upazilas were given. Some traders could not identify the origin of their poultry: they reported having bought poultry from a trader who had bought from another trader. As these transactions represented only 1.2% of the total number of poultry traded through this network, and they did not affect the overall structure of the network, they were not considered in the analysis. All data was collected according to poultry types. However, for clarity, and given the small proportion of traded poultry that they represented (5.5%), spent hens, geese, pigeons and quail were grouped into an "other poultry type" category. The 4 main poultry types

considered here are: industrial white-feathered broiler chicken, sonali poultry, deshi and ducks.

Statistical analyses and social network analysis

Data was entered in Microsoft Excel 2013® (Microsoft Corp., Redmond, WA, USA), and analyses were conducted in R 3.2.5 [29].

Network terminology and metrics are defined in Table 1.

The total number of poultry traded per LBM could not be estimated because the total number of poultry traders operating permanently or not at the surveyed LBMs was unknown. We therefore used the average weekly number of poultry traded per trader in a surveyed LBM (i.e. the total number of poultry traded by all the traders interviewed in a LBM in the past week divided by the number of traders interviewed in that LBM).

The number of poultry traded per trader, the number and types of poultry sources and the number of LBMs that an interviewed trader supplied were summarised using median and inter-quartile ranges (IQR). The surveyed LBMs were then compared with respect to the practices of their traders: proportion of poultry supplied according to the type of source, number of different poultry sources per LBM, number of upazilas of origin of the poultry traded at the LBM, and distribution of transaction distances. A transaction distance was the Euclidean distance between a poultry source and a surveyed LBM. It was weighted by the number of poultry traded between these two locations. It was calculated using the GPS coordinates of the surveyed LBMs and the centroid of the upazilas in which farms and non-surveyed LBMs supplying surveyed LBMs were located, as their GPS coordinates were not available. Thus, non-surveyed LBMs and farms located in the same upazila had the same coordinates. The relative importance of the effect of within and between LBM variance on the number of poultry traded per trader was assessed using a mixed effects model with surveyed LBMs as random effects.

Weighted and directed networks were built for each poultry type. A node was defined as either the group of interviewed traders of the surveyed LBMs or a "farm upazila", and each network arc was weighted with the number of poultry traded between the considered sources and destinations. In order to assess the network connectivity, as well as the potential lower and upper bounds of potential epidemic sizes [30], the sizes of the giant weak and strong components (GWC and GSC) were calculated for each type of network described. For LBMs, the average weekly number of poultry traded by a trader was calculated as a measure of node centrality. Normalised betweenness-centrality was calculated for each node, accounting for arc strength. The directed

Table 1 Definitions of network terminology and metrics used in this study

Poultry sources	1) "Farm upazila": An upazila (third administrative division) in which farms supplied a trader in a surveyed LBM. Exact names and locations of farms were unknown, they were therefore grouped into so-called "farm upazilas". 2) A LBM (each LBM was an independent location).
Network node	A surveyed LBMs, a "farm upazila" or a non-surveyed LBM.
Network arc	Link between 2 nodes, weighted with the number of poultry traded between the considered sources and destinations.
Giant weak Component (GWC)	The largest subset of nodes in which all the nodes were connected, regardless of the direction of the arcs.
Giant Strong Component (GSC)	The largest subset of nodes in which all the nodes were connected, accounting for the direction of the arcs.
Connectedness of the network	The proportion of nodes included in the GWC [30].
Normalised betweenness-centrality	The proportion of shortest paths (i.e. geodesic distances) on which a given node lies.
Geodesic distances	Shortest path between two nodes, with the distance being calculated as the sum of the inverse of the arc strengths [64].
Input domain	Proportion of the nodes that can reach a node following network arcs.
Output domain	Proportion of the nodes that can be reached by a node following network arcs.

network's assortativity coefficient was calculated to assess the network's resilience to targeted removal of LBM nodes. Indeed, disassortative networks, in which high degree nodes are preferentially connected to low degree nodes, are less resilient to node removal then assortative networks, in which nodes are preferentially connected to other nodes with similar degree [31, 32]. While the output domain may be interpreted as the potential of a given node to spread a pathogen, and, therefore, be an indicator of the node suitability as a target for control measures, the input domain may indicate nodes which should be targeted by surveillance programs. These two metrics were calculated for each node. Finally, in order to assess the impact of targeted control measures on the spread of AIVs, the effect of targeted node removal on network connectedness, the size of maximum output and input domains was assessed. Nodes to be removed were selected according to their betweenness score, the size of their input and output domains. Removal of LBM nodes can be interpreted as the implementation of cleaning, disinfection and rest day programmes in the given LBMs [14, 15, 33]. Such measures would result in the local elimination of viruses, so that arcs leading to and starting from such a node would then not be infectious any more.

The following R packages were used for the above mentioned analyses: "lme4" [34], "sp" [35, 36], "maptools" [37], "shapefiles" [38], "sna" [39], and "igraph" [40].

Results

Traders and LBM characteristics

Half of the interviewed traders reported having traded at least 1250 (IQR: 600–2945) poultry during the previous week. The total number of poultry traded weekly by interviewed traders in surveyed LBMs is presented in an

Additional file 1. As the proportion of traders interviewed per surveyed LBM was not recorded, the total number of poultry traded in these LBMs could not be estimated. We therefore present the average number of poultry traded per trader and the proportion of trade represented by each poultry type. Broiler chickens were the main poultry type traded, they represented 42% of all the poultry sold by interviewed traders. In contrast, deshi and sonali chickens accounted for 19% and 33%, respectively, and ducks for only 0.5% of the interviewees' poultry trade. About 90% of interviewed traders sold at least 2 poultry types. Sixty-five percent of traders traded broilers, almost half traded deshi and/or sonali, and only 4.5% traded ducks (Table 2).

Broiler chickens were sold in almost all LBMs, deshi and sonalis in 70% and 75% of LBMs, respectively, and ducks in 14.5% of LBMs. In half of the surveyed LBMs, at least 85.2% of the interviewed traders traded broilers, at least 35.4% traded deshis and at least 40% sonalis (Table 2).

Using a mixed effects model, with LBMs as random effects, 65% of the variance in the number of poultry traded per trader was estimated to be due to within-LBM variance. However, when only considering the 25% largest LBMs ($n = 34$), 98% of this variance was explained by within-LBM variance.

The distributions of the number of poultry sources (i.e. "farm upazilas" or LBMs) per interviewed trader and per LBM were right-skewed. Interviewed traders reported having up to 13 different poultry sources (median: 2, IQR: 1–3) in the past week, resulting in some surveyed LBMs having at least 24 different sources, as not all sources were identified given the sampling strategy (median: 4, IQR: 2–7) (Table 3). The number of different sources was positively correlated with the number of traders interviewed per LBM (Spearman's correlation

Table 2 Number and types of poultry traded by surveyed traders and at LBMs in the week preceding the interviews

		Broiler	Sonali	Deshi	Ducks	Others	All
Trader level	Proportion of traders selling each poultry type (%).	64.9% (n = 551)	47.9% (n = 407)	46.5% (n = 395)	4.5% (n = 38)	26.9% (n = 228)	100% (n = 849)
	No. of poultry traded per week per trader interviewed[a] (median and IQR[b]).	1000 (420–2170)	650 (250–1575)	450 (200–1000)	50 (40–275)	400 (150–800)	1250 (600–2945)
	Proportion of a trader's sales represented by each poultry type[a] (median % and IQR).	77.5% (47.4–100)	40% (21–59.3)	38.5% (19.6–66.7)	5.5% (2.6–18.9)	24.6% (12.6–40)	NA
LBM level	Proportion of LBMs in which a type of poultry is sold (%).	94.9% (n = 130)	76.8% (n = 106)	71% (n = 98)	14.5% (n = 20)	48.6% (n = 67)	100% (n = 138)
	Proportion of poultry of a given type sold in each LBM (median % and IQR).	52.9% (31.8–78.8)	14.7% (1.6–33.6)	9.1% (0–23.6)	0% (0–0)	0% (0–12.3)	NA
	Proportion of traders trading each type of poultry in a given LBM (median % and IQR).	85.2% (50–100)	40% (17.6–80)	35.4% (0–75)	0 (0–0)	0 (0–50)	NA
	Average no. of poultry sold per week and per trader in a given LBM[a] (median and IQR[b]).	900 (337–2029)	296 (140–1106)	183 (82–696)	29 (14–75)	223 (71–328)	1800 (766–3303)

One-week data, collected in Bangladesh in 2014, from 849 traders in 138 LBMs
In this table "LBM" refers to the group of interviewed traders from the surveyed LBM and cannot be generalised to the entire LBM
[a]Including only the traders/LBMs which sold these types of poultry
[b]Inter-quantile range

coefficient: 0.64, p-value $< 10^{-15}$). In 20 % ($n = 25$) of the surveyed LBMs the interviewed traders were exclusively supplied by farm upazilas. These LBMs were distributed throughout the country; all of the interviewed traders in these LBMs sold broilers, 44% of them sold sonalis and 36% sold deshis, none of them sold ducks. In contrast, in 39.3% ($n = 48$) of the surveyed LBMs interviewed traders were supplied exclusively by other LBMs. Forty of those LBMs were in DCC, in 88% of them the interviewed traders sold broilers, in 73% of them they sold sonalis, in 71% of them they sold deshis, and in 6% of them they sold ducks.

In at least 70% of the surveyed LBMs, more than half of ducks and deshis supplied to the interviewed traders were supplied by other LBMs. In contrast, broilers were supplied to the interviewed traders of the surveyed LBMs equally by other LBMs and farms (Table 3).

Overall, half of the poultry supplied to the interviewed traders were sourced less than 15 km away from the surveyed LBMs, but this distance could reach up to 420 km (Fig. 2). While most poultry, regardless of their type, were sourced in the vicinity of the LBMs in which they were sold, the tail of the distribution of the distances over which they were transported varied according to

Table 3 Proportion and types of poultry sources, according to poultry type for surveyed traders and LBMs

	Broiler	Sonali	Deshi	Ducks	Others	All
No. of poultry sources/trader (median and IQR[a]).	2 (1–3)	1 (1–2)	1 (1–2)	1 (1–1)	1 (1–1)	2 (1–3)
No. of poultry sources/LBM (%).	3 (2–4)	3 (1–5)	3 (1–5)	1 (1–2)	2 (2–4)	4 (2–7)
Proportion of surveyed LBMs supplied exclusively by other LBMs (%).	34.6% (n = 48)	46.7% (n = 64)	61.4% (n = 85)	71.4% (n = 99)	37.3% (n = 51)	39.3% (n = 54)
Proportion of surveyed LBMs supplied exclusively by farm upazilas (%).	32.8% (n = 45)	24.3% (n = 34)	14.9% (n = 21)	23.8% (n = 33)	28.4% (n = 39)	20.5% (n = 28)
Proportion of surveyed LBMs supplied by other LBMs and farm upazilas (%).	32.8% (n = 45)	29% (n = 40)	23.8% (n = 33)	4.8% (n = 7)	34.3% (n = 47)	40.2% (n = 55)
Proportion of poultry supplied to a LBM by another LBM (% and IQR).	48% (0–100)	78% (11.8–100)	100% (52.8–100)	100% (12.5–100)	61.5% (0–100)	57.6% (13.5–100)
Proportion of poultry supplied to a LBM by a farm upazila (% and IQR).	51% (0–100)	19.6% (0–76.4)	0% (0–29.7)	0% (0–37.5)	38.5% (0–100)	40.7% (0–79.4)
Number of different upazilas of origin of the poultry sold at surveyed LBMs (median and IQR).	2 (1–3.5)	2 (1–3)	2 (1–3)	1 (1–2)	2 (1–3)	3 (1.3–5)

One-week data, collected in Bangladesh in 2014, from 849 traders in 138 LBMs
Only the traders or the LBMs trading the poultry type considered were included in the calculations. In this table "LBM" refers to the group of interviewed traders from the surveyed LBM and cannot be generalised to the entire LBM
[a]Inter-quantile range

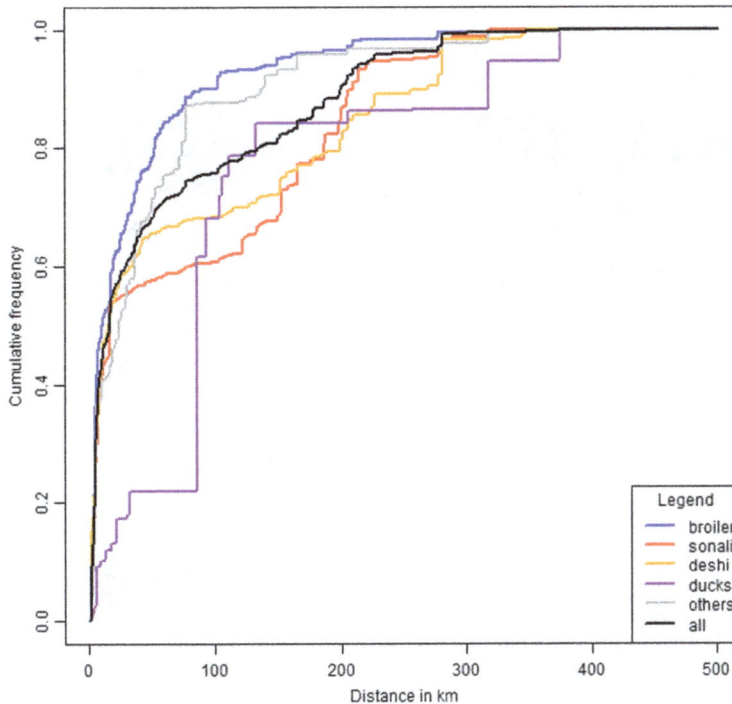

Fig. 2 Cumulative distances (distances are weighted) between poultry sources and surveyed LBMs according to poultry type. One-week data, collected in Bangladesh in 2014, from 849 traders in 138 LBMs

the poultry type considered. Broiler chickens were sourced the closest to the LBMs (25% of all broilers were transported over more than 40 km, and 3% over more than 200 km), deshis and ducks the furthest (25% of all the deshis and ducks were transported over more than 105 km and 16% over more than 200 km).

In only 41% of the surveyed LBMs did interviewed traders report supplying other LBMs, and in 24% ($n = 33$) of those LBMs, interviewed traders only supplied a unique LBM. In 4 surveyed LBMs, interviewed traders sold to more than 8 other LBMs, 3 of these LBMs were in DCC and one was in the N-E of Bangladesh (Sylhet district).

Poultry trading networks

In the networks described here, nodes were either farm-upazilas or the group of interviewed traders of each surveyed LBM, later referred to as "LBM".

As a result of the heterogeneity in trading practices, the trading networks differed according to poultry types. In the broiler trading network, 69.5% (206/296) of nodes were encompassed within the GWC. Sonali and deshi trading networks had similar levels of connectedness, 87% and 72% respectively. The duck trading network was the smallest and the least connected of the 4 trading networks represented in Fig. 3, with only 20.5% nodes encompassed within the GWC. The largest sources of sonalis for the interviewed traders were located in the centre and far north-west of Bangladesh, and deshis

were mainly sourced from the west of the country. In this study, interviewed traders did not source broilers from a specific geographical location.

When considering all poultry types, the network size and connectivity increased: it was composed of 445 nodes (Fig. 4), of which 73.2% were LBMs, and it was highly connected, with the GWC encompassing 97% ($n = 433$) of those nodes. The 12 remaining nodes were LBMs and farms in the South-East of the country (Cox's Bazar district), all encompassed within another component. This region was however connected to the GWC through other LBMs. The GSC only grouped 2 nodes, reflecting the strong directionality of poultry commercial movements through the network.

The poultry trading network described here was disassortative (–0.27), therefore unlikely to be resilient to targeted node removal. Input and output domains were right-skewed with median values of 0 (IQR: 0–0.011) and 0.004 (IQR: 0–0.012) respectively; 159 nodes had an input domain greater than 0, 324 nodes had an output domain greater than 0. Of all the LBMs included in the network, 8.9% had an input domain greater than 5%, and 2.2% had an input domain greater than 10%. Only 4% of the LBMs or the farm upazilas had an output domain greater than 5%, and 2.5% of them had an output domain greater than 10%.

Thirty-eight surveyed LBMs (28% of the total number of surveyed LBMs, and 8.6% of all the nodes of the

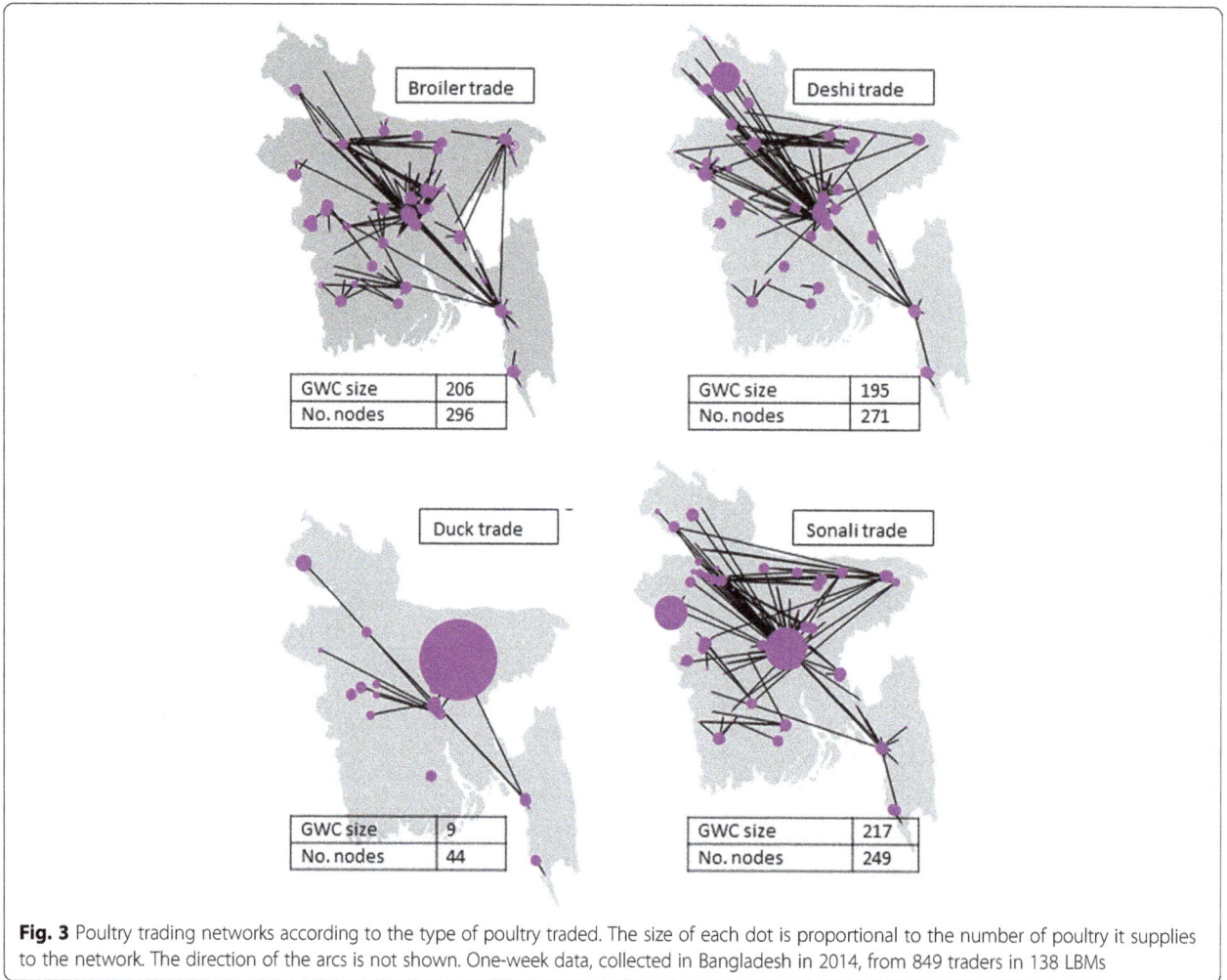

Fig. 3 Poultry trading networks according to the type of poultry traded. The size of each dot is proportional to the number of poultry it supplies to the network. The direction of the arcs is not shown. One-week data, collected in Bangladesh in 2014, from 849 traders in 138 LBMs

network) lay between 2 other nodes. Their betweenness was positively correlated with the size of the output domain (Pearson's correlation coefficient: 0.90, p-value $< 10^{-15}$). The LBM with the greatest betweenness score was a large wholesale LBM in DCC that supplied 6.1% of all the poultry traded through the network. The LBM with the 2nd greatest betweenness score was outside DCC and was supplied by the farm upazila with the largest output domain.

Comparing the impacts of node removal on the maximum output and input domains and the connectedness revealed that the removal of the nodes with the greatest betweenness scores would have the greatest impact on the network's connectedness, and maximum output and input domains (Fig. 5). Removing 25 LBMs (5.6% of the entire network) decreased the maximum size of the input domain by 66%, the maximum size of the output domain by 73% and the connectedness by 58%. These nodes were LBMs located throughout the country (including in DCC and CCC). They all sold broilers, sonalis and deshis and 3 of them also sold ducks.

Discussion

To the best of our knowledge, this study represents the first assessment of poultry trading practices and networks on a nationwide scale in Bangladesh. Trading patterns varied according to the type of poultry being traded. Broilers were the main type of poultry sold by interviewed traders and they were sourced from both farms and other LBMs located in the vicinity of the surveyed LBMs. In contrast, sonalis, deshis and ducks were mainly sourced from other LBMs and were mainly bought further away. Most deshis came from the North-west of Bangladesh, ducks from the North of the country, and sonalis from the central and North-Western districts. The overall poultry trading network was highly connected and disassortative. Removing a small fraction of nodes with the highest betweenness scores substantially reduced the networks' connectedness and maximum sizes of output and input domains.

This study had some limitations. Firstly, we used the centroid of the upazilas as a proxy for the exact coordinates of farm upazilas and non-surveyed LBMs. As the

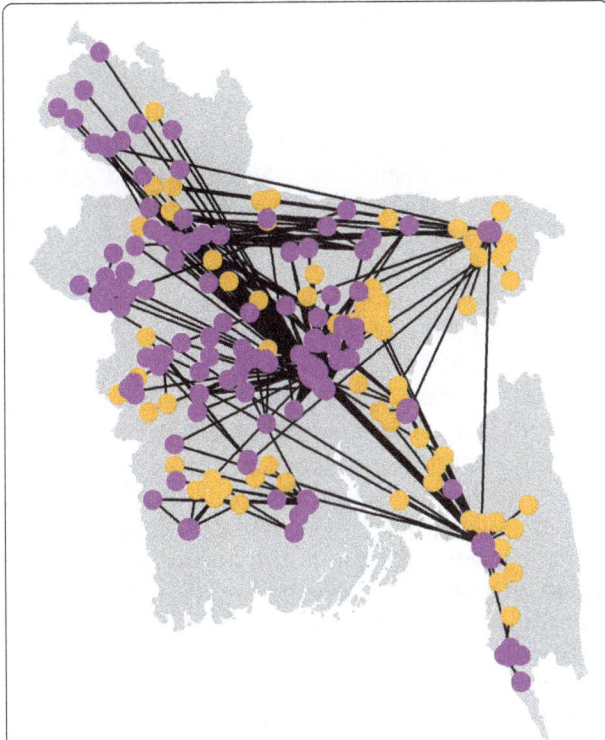

Fig. 4 Poultry trading network. Nodes are LBMs (*purple*), or farm upazilas (*orange*). When the GPS coordinates of the nodes were not available (for all the nodes that are not the surveyed LBMs) the GPS coordinates of the centroid of the upazila were used. The direction of the arcs is not shown. One-week data, collected in Bangladesh in 2014, from 849 traders in 138 LBMs

in this dataset. Secondly, the sampling strategy impacted the estimation of the network's metrics and their interpretation. As the proportion of traders interviewed in each LBM was unknown, some LBM-level metrics, such as the total number of poultry traded per LBM (weighted indegree), and the total number of arcs sent to and from a LBM (unweighted indegree and outdegree), could not be calculated. In addition, LBMs cited by interviewed traders as poultry sources were not surveyed, leading to the construction of an incomplete network, the total size of which was unknown. The proportion of connections identified in large LBMs was likely to be lower than in small LBMs as the proportion of interviewed traders was generally lower in large than in small LBMs. Therefore, the centrality measures of the largest LBMs might have been underestimated. Likewise, the impact of the removal of these nodes on network connectedness could have been underestimated. Nevertheless, the number of surveyed LBMs was large compared to other similar studies [26, 41, 42], and although network metrics might have been underestimated, the network properties highlighted here are likely to reflect relevant features of the real, complete network as well as poultry type-specific trade patterns.

Within-LBM variance in the number of poultry sold by traders was higher in the largest LBMs. This could be due to greater heterogeneity in traders' practices in large LBMs compared to small LBMs. Homogeneity of trading practices seemed indeed more likely in small LBMs, where only a limited number of middlemen come to sell small numbers of poultry. This pattern should be taken into account in the design of future surveys. A large proportion of traders operating in the largest LBMs would need to be surveyed in order to identify most poultry trading routes, and to appropriately assess the diversity in trading practices. On the other hand, homogeneity in the practices of traders operating in small LBMs would

exact GPS coordinates were unavailable for farms and non-surveyed LBMs, using upazila centroids as coordinates provided an estimate of the range of transaction distances according to poultry types traded. However, since upazilas can be as big as 300 to 400 km²; the resulting measurement error varies between observations

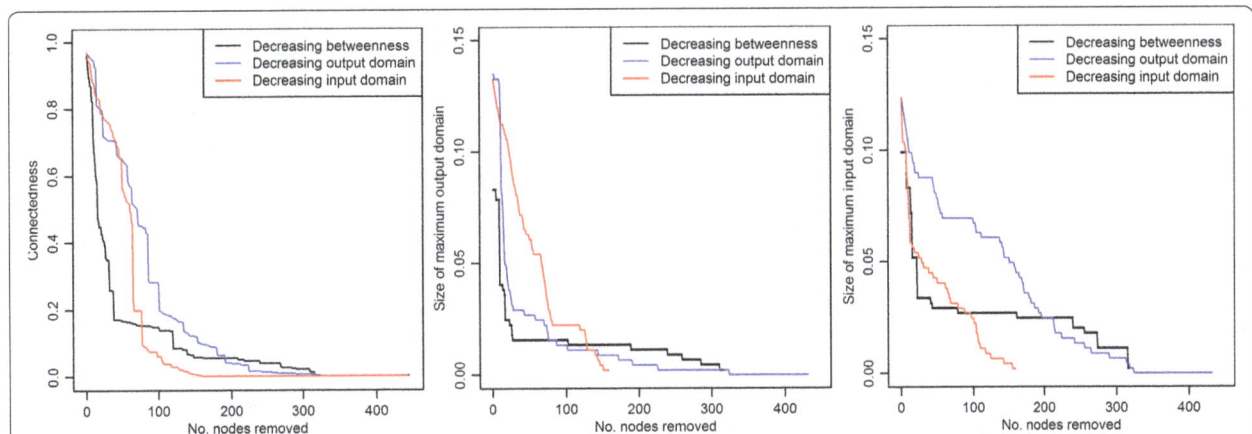

Fig. 5 Comparison of the impact of LBM removal on network metrics (maximum output and input domains and connectedness). Nodes were removed from the network one after the other, in decreasing order of their betweenness, output or input domains. One-week data, collected in Bangladesh in 2014, from 849 traders in 138 LBMs

mean that characteristics of non-surveyed small LBMs can be extrapolated based on a small survey sample.

This survey was conducted over the period of a month, it was therefore impossible to identify temporal variation in trading practices or network structure. However, in multiple settings, the number of traded poultry was reported to increase during seasonal and religious festivals, such as Chinese New Year celebrations and Ramadan [25, 26, 42, 43]. A study in China [42] also reported that during Chinese New Year celebrations, not only did the number of poultry traded increase, but so did the number of arcs and distances over which poultry was traded. Such trading patterns are likely to occur in Bangladesh as well, during religious festivals such as Ramadan and Eid-ul-Adha for example. They may particularly impact the trade of deshis, which are considered a delicacy. While increases in trading activities would be expected to promote AIV spread, as observed in Viet Nam [43], alteration of network structures may promote the mixing of poultry from a variety of geographical origins and farming systems and the reassortment of viral strains which otherwise might have remained isolated. Other factors that are likely to alter the structure of the poultry trading network include temporal variation in the market price of finished poultry or production inputs, e.g. day-old-chicks, and the occurrence of natural disasters, e.g. floods. Moreover, since duck production is linked to rice harvesting, duck trade is also likely to be seasonal [10]. Future investigations should aim at quantifying temporal variation in trading patterns for each poultry type in order to explore their impact on AIV epidemiology.

The distribution of the distances travelled by poultry between their sources and the LBMs surveyed in this study, were similar to those reported in other studies conducted in South-East Asia [11, 26, 44] and New Zealand [45]. Only in China were the maximum distances travelled by poultry greater than those described in our study [42]. In the first two HPAI H5N1 epidemic waves that occurred in Bangladesh in 2007–2008, clustering was seen within distances of 250-300 km [16]. In addition, genetically identical viruses caused outbreaks over an area of more than 200 km in 2007, indicating long-distance transmission events within Bangladesh [46]. Long-distance viral spread was also reported in 2010–11 [47], with movement of infected poultry or contaminated materiel being suggested as possible transmission routes. Transaction distances estimated in this study support the aforementioned results, and the possible role played by poultry trade in long distance spread of AIVs. In our study, cross-border trade was not reported. Nevertheless, such activities cannot be ruled out, as illegal importation of poultry across the porous Indian border has been mentioned in a previous study [48]. Cross-border trade would be worth

investigating further in order to address the role of poultry trade in regional spread of AIVs.

The high level of contamination of Bangladeshi LBMs with a variety of AIV strains [49–53] and the association between LBM density and the risk of HPAI H5N1 outbreaks in Bangladeshi farms [18] suggests that AIV surveillance and control programmes implemented in LBMs could be effective for reducing disease risk for the production sector, as well as for humans. Furthermore, the position of nodes, either live animal markets or farms, in networks of potentially infectious contacts has been associated with their roles in the spread of pathogens [14, 54]. Consequently, the knowledge of the network structure can be used to help identifying the most suitable targets for control and surveillance programmes. Poultry trading networks described in multiple settings were all heterogeneous, with a small number of LBMs having a major influence on the potential of viruses to spread through these networks [8, 33, 55–57]. Targeting surveillance and control programmes at this specific set of LBMs is likely to be the most cost-effective and realistic control option, especially in resource-poor settings such as Bangladesh. In this study, the removal of 25 LBMs (5.6% of all the nodes of the network) substantially reduced the network's connectedness, and the maximum size of input and output domains. We would therefore expect that the implementation of control measures, such as daily cleaning, disinfection and regular rest days [33, 58] in these LBMs to reduce the potential of pathogens to spread through the network. Indeed, reducing the maximum size of the output domain would decrease the potential of any contaminated LBM to spread viruses through poultry trade movements, and a reduced connectedness would reduce the maximum possible size of an epidemic. LBMs with the largest number of poultry sources and input domain received poultry from the most geographically diverse locations throughout Bangladesh. As a limited number of LBMs exhibited these features, targeting surveillance activities to these LBMs could allow the monitoring of the diversity of AIV strains circulating in the country. In addition, if control measures were implemented throughout the network, a reduction in viral diversity in LBMs could be an indicator of the effectiveness of control measures. However, even if a small number of LBMs is to be targeted, the effective implementation of such control strategies may prove challenging. Although poor hygiene in Bangladeshi LBMs was associated with a higher likelihood to detect AIVs in one study [52], studies have shown that only a limited number of biosecurity measures have been implemented in LBMs so far [49, 59]. Furthermore, one study revealed that those limited changes had not been sufficient to reduce the level of AIV contamination in LBMs with better biosecurity measures compared to

LBMs with less or no biosecurity measures in place [49]. These results suggest a suboptimal implementation of biosecurity measures. In order to ensure that control interventions are effective in the future, further studies should aim to assess their feasibility, their economic impact and the compliance of LBM traders. These interventions may need to be complemented by a reinforcement of biosecurity measures at farm-level and during poultry transportation from farms to LBMs, with the aim to reduce the viral load introduced into LBMs. Indeed, poultry management and infrastructure of small commercial chicken farms did not meet basic biosecurity requirements. In particular, vehicles were regularly allowed on farm premises without prior disinfection [60].

In Bangladesh, HPAI H5N1 was more frequently detected in ducks than in chickens [48, 53], and in deshi than in broiler chickens [48]. In a recent study, Bangladeshi LBM stalls selling ducks alongside other poultry types were more than twice as likely to test positive for AIVs as stalls that didn't sell any ducks [52]. While ducks are known to play an important role in HPAI H5N1 spread and maintenance, in particular due to their ability to remain asymptomatic [61–63], the structure of the Bangladeshi poultry trading network may further foster their impact on the epidemiology of the disease. Most deshis and ducks sold in surveyed LBMs were supplied by other LBMs, whereas broilers were equally sourced from other LBMs and farms, suggesting that more actors were involved in the trade of deshis and ducks than in the trade of broilers. Deshi and duck traders also travelled greater distances than broiler traders. This meant that deshi and duck movements could have been directly involved in the aforementioned long-distance HPAI H5N1 transmission events, and that these poultry types were likely to spend more time in traders' hands than broiler chickens. They might thus greatly contribute to the amplification and maintenance of the viral circulation along the live poultry trading network [33]. Deshi and duck trade networks were less connected than other poultry type-specific networks. However, interactions between those networks – through the mixing of multiple poultry types in LBMs – meant that the connectedness of the overall network was very high. Almost all farming systems and production areas across the country were thus potentially epidemiologically connected through the network. Although ducks, and to a lesser extent deshis, represented a small proportion of the overall poultry trade in Bangladesh, the network characteristics may create conditions for viral strains circulating in an otherwise isolated deshi or duck population to spread to geographically distant poultry populations. Further research should aim to quantify the time spent by different poultry types within the trade network and, therefore, their potential to amplify the level and increase the diversity of viral circulation. Also, when multiple traders, and LBMs, are involved in the transport of poultry from the farm gate to the end-user, the actual production areas from which poultry originate should be identified. This would allow a better assessment of the way in which poultry type-specific trading practices may shape the overall network structure and contribute to its potential to spread AIVs.

Conclusion

In conclusion, poultry trade practices varied according to the poultry type considered. It was the interaction between poultry type-specific networks that resulted in an overall live poultry trading network within which almost all poultry production areas across the country and LBMs identified during this survey were connected. While it appeared that control interventions targeted at a small number of key LBMs could be efficient and effective in controlling virus circulation, the feasibility of this strategy, taking into account the likelihood of behaviour change amongst all actors involved, would require further investigation.

Abbreviations
AI: Avian influenza; AIV: Avian influenza virus; CCC: Chittagong corporation; DCC: Dhaka city corporation; GPS: Global positioning system; GSC: Giant strong component; GWC: Giant weak component; HPAI H5N1: Highly pathogenic avian influenza; IQR: Inter-quantil range; LBM: Live bird market

Acknowledgements
The authors would like to thank all the participants involved in the field study, and Chang Yu-Mei for her assistance in designing the mixed effect model.

Funding
FAO ECTAD (Emergency Centre for Transboundary Animal Diseases, Food and Agriculture Organisation) Bangladesh provided support to collect data under the USAID funded project Strengthening national Capacity to prevent and control emerging and re-emerging pandemic threats including Influenza A in Bangladesh (OSOR/BGD/403/USA).
Natalie Moyen, Guillaume Fournié and Dirk U Pfeiffer currently receive funding from the BALZAC research programme 'Behavioural adaptations in live poultry trading and farming systems and zoonoses control in Bangladesh'. This is one of 11 programmes funded under ZELS, a joint research initiative between Biotechnology and Biological Sciences Research Council (BBSRC), Defence Science and Technology Laboratory (DSTL), Department for International Development (DFID), Economic and Social Research Council (ESRC), Medical Research Council (MRC) and Natural Environment Research Council (NERC).

Authors' contributions
GA, SG, RK, IK, ND and MY contributed to the design of the study, questionnaire design, training of the interviewers, coordination of the field work and data entry. TT contributed to the study design and preliminary analysis of the data. NM carried out the analysis and wrote the manuscript with considerable input from GF and DP. All authors approved and read the final manuscript.

Consent for publication
Not applicable

Competing interests
The authors declare that they have no competing interests.

Author details
[1]Department of Pathobiology and Population Sciences, Royal Veterinary College, University of London, Hatfield, Hertfordshire AL9 7TA, UK. [2]School of Veterinary Science, The University of Queensland, Gatton 4343, Qld, Australia. [3]Emergency Centre for Transboundary Animal Diseases, Food and Agriculture Organisation of the United Nations, Dhaka, Bangladesh. [4]National Centre for Animal Health, Thimphu, Bhutan. [5]College of Veterinary Medicine and Life Sciences, City University of Hong Kong, Tat Chee Avenue, Kowloon, Hong Kong.

References

1. WorldBank. Implementation completion and results report (IDA-43400 TF-90662) on a credit in the amount of SDR 10.5 million (US$16.0 million equivalent) to the People's republic of Bangladesh for an avian influenza preparedness and response project under the global program for avian influenza and human pandemic preparedness and response. Washington DC; 2013. Available at http://www-wds.worldbank.org/external/default/WDSContentServer/WDSP/IB/2013/07/04/000442464_20130704100805/Rendered/INDEX/ICR21770ICR0Av0Box0377341B00PUBLIC0.txt. Accessed 10 Jan 2018.
2. BBS. Statistical year book Bangladesh 2014. Dhaka. p. 2014. Available at http://www.bbs.gov.bd/site/page/29855dc1-f2b4-4dc0-9073-f692361112da/Statistical-Yearbook. Accessed 10 Jan 2018.
3. FAOstat http://www.fao.org/. Accessed 05 Feb 2017.
4. OIE. Update on highly pathogenic avian influenza in animals (type H5 and H7), 2013, Bangladesh follow-up report no. 42. Paris; 2013. Available at http://www.oie.int/wahis_2/public%5C..%5Ctemp%5Creports/en_fup_0000013335_20130430_165153.pdf. Accessed 10 Jan 2017.
5. Biswas PK, Christensen JP, Ahmed SS, Barua H, Das A, Rahman MH, Giasuddin M, Hannan AS, Habib MA, Ahad A, et al. Avian influenza outbreaks in chickens, Bangladesh. Emerg Infect Dis. 2008;14(12):1909–12.
6. FAO, OIE, WHO. FAO-OIE-WHO technical update: current evolution of avian influenza H5N1 viruses. 2011. Available at http://www.who.int/influenza/human_animal_interface/tripartite_notes_H5N1.pdf. Accessed 10 Jan 2017.
7. FAO. Approaches to controlling, preventing and eliminating H5N1 highly pathogenic avian influenza in endemic countries. Rome; 2011. Available at http://www.fao.org/docrep/014/i2150e/i2150e00.htm. Accessed 10 Jan 2017.
8. Webster RG. Wet markets–a continuing source of severe acute respiratory syndrome and influenza? Lancet. 2004;363(9404):234–6.
9. Sims LD. Lessons learned from Asian H5N1 outbreak control. Avian Dis. 2007;51(1 Suppl):174–81.
10. Dolberg F. Poultry sector country overview: Bangladesh. Dhaka; 2008. Available at http://www.fao.org/3/a-ai319e.pdf. Accessed 10 Jan 2017.
11. Fournie G, Tripodi A, Nguyen TT, Nguyen VT, Tran TT, Bisson A, Pfeiffer DU, Newman SH. Investigating poultry trade patterns to guide avian influenza surveillance and control: a case study in Vietnam. Sci Rep. 2016;6:29463.
12. Soares Magalhaes RJ, Ortiz-Pelaez A, Thi KL, Dinh QH, Otte J, Pfeiffer DU. Associations between attributes of live poultry trade and HPAI H5N1 outbreaks: a descriptive and network analysis study in northern Vietnam. BMC Vet Res. 2010;6:10.
13. Martin V, Zhou X, Marshall E, Jia B, Fusheng G, FrancoDixon MA, DeHaan N, Pfeiffer DU, Soares Magalhães RJ, Gilbert M. Risk- based surveillance for avian influenza control along poultry market chains in South China: the value of social network analysis. Prev Vet Med. 2011;102(3):196–205.
14. Rasamoelina-Andriamanivo H, Duboz R, Lancelot R, Maminiaina OF, Jourdan M, Rakotondramaro TM, Rakotonjanahary SN, de Almeida RS, Rakotondravao, Durand B, et al. Description and analysis of the poultry trading network in the Lake Alaotra region, Madagascar: implications for the surveillance and control of Newcastle disease. Acta Trop. 2014;135:10–8.
15. Fournié G, Guitian J, Desvaux S, Cuong VC, Dung dH, Pfeiffer DU, Mangtani P, Ghani AC. Interventions for avian influenza A (H5N1) risk management in live bird market networks. Proc Natl Acad Sci U S A. 2013;110(22):9177–82.
16. Ahmed SS, Ersbøll AK, Biswas PK, Christensen JP. The space-time clustering of highly pathogenic avian influenza (HPAI) H5N1 outbreaks in Bangladesh. Epidemiol Infect. 2010;138(6):843–52.
17. Ahmed S, Ersboll A, Biswas P, Christensen J, Toft N. Spatio-temporal magnitude and direction of highly pathogenic avian influenza (H5N1) outbreaks in Bangladesh. PLoS One. 2011:6(9).
18. Ahmed S, Ersboll A, Biswas P, Christensen J, Hannan A. Ecological determinants of highly pathogenic avian influenza (H5N1) outbreaks in Bangladesh. PLoS One. 2012:7(3).
19. Loth L, Gilbert M, Osmani MG, Kalam AM, Xiao X. Risk factors and clusters of highly pathogenic avian influenza H5N1 outbreaks in Bangladesh. Prev Vet Med. 2010;96(1–2):104–13.
20. FAO. Comparative performance of Sonali chickens, commercial broilers, layers and local non-descript (deshi) chickens in selected areas of Bangladesh. Rome; 2015. Available at www.fao.org/3/a-i4725e.pdf. Accessed 10 July 2017.
21. Bhuiyan AKFH, Bhuiyan MSA, Deb GK. Indigenous chicken genetic resources in Bangladesh: current status and future outlook. 2005. Available at www.fao.org/docrep/008/a0070t/a0070t0c.htm. Accessed 10 July 2017.
22. Biswas PK, Christensen JP, Ahmed SS, Barua H, Das A, Rahman MH, Giasuddin M, Hannan AS, Habib AM, Debnath NC. Risk factors for infection with highly pathogenic influenza A virus (H5N1) in commercial chickens in Bangladesh. Vet Rec. 2009;164(24):743–6.
23. Biswas PK, Christensen JP, Ahmed SS, Das A, Rahman MH, Barua H, Giasuddin M, Hannan AS, Habib MA, Debnath NC. Risk for infection with highly pathogenic avian influenza virus (H5N1) in backyard chickens, Bangladesh. Emerg Infect Dis. 2009;15(12):1931–6.
24. Raha S. Poultry industry in Bangladesh: present status and future potential. Mymensingh: Agricultural university of Mymensingh; 2000.
25. Molia S, Boly IA, Duboz R, Coulibaly B, Guitian J, Grosbois V, Fournie G, Pfeiffer DU. Live bird markets characterization and trading network analysis in Mali: implications for the surveillance and control of avian influenza and Newcastle disease. Acta Trop. 2016;155:77–88.
26. Van Kerkhove MD, Vong S, Guitian J, Holl D, Mangtani P, San S, Ghani AC. Poultry movement networks in Cambodia: implications for surveillance and control of highly pathogenic avian influenza (HPAI/H5N1). Vaccine. 2009;27(45):6345–52.
27. Vallee E, Waret-Szkuta A, Chaka H, Duboz R, Balcha M, Goutard F. Analysis of traditional poultry trader networks to improve risk-based surveillance. Vet J. 2013;195(1):59–65.
28. Poolkhet C, Chairatanayuth P, Thongratsakul S, Kasemsuwan S, Rukkwamsuk T. Social network analysis used to assess the relationship between the spread of avian influenza and movement patterns of backyard chickens in Ratchaburi, Thailand. Res Vet Sci. 2013;95(1):82–6.
29. R Core Team. R: A language and environment for statistical computing. Vienna: R Foundation for Statistical Computing; 2014.
30. Kao RR, Danon L, Green DM, Kiss IZ. Demographic structure and pathogen dynamics on the network of livestock movements in great Britain. Proc Biol Sci. 2006;273(1597):1999–2007.
31. Kiss IZ, Green DM, Kao RR. The network of sheep movements within great Britain: network properties and their implications for infectious disease spread. J R Soc Interface. 2006;3(10):669–77.
32. Newman ME. Mixing patterns in networks. Phys Rev E Stat Nonlinear Soft Matter Phys. 2003;67(2 Pt 2):026126.
33. Fournié G, Guitian FJ, Mangtani P, Ghani AC. Impact of the implementation of rest days in live bird markets on the dynamics of H5N1 highly pathogenic avian influenza. J R Soc Interface. 2011;8(61):1079–89.
34. Bates D, Machler M, Bolker B, Walker S. Fitting linear mixed-effects models using {lme4}. J Stat Softw. 2015;67(1):1–48.
35. Pebesma EJ, Bivand R. Classes and methods for spatial data in R. R news 5 (2). 2005.
36. Roger SB, Edzer P, Virgilio G-R. Applied spatial data analysis with R. 2nd ed. NY: Springer; 2013.
37. Bivand R, Lewin-Koh N. Maptools: tools for reading and handling spatial objects: R package; 2016. Accessible at: https://cran.r-project.org/web/packages/maptools/maptools.pdf. Accessed 24 July 2017.
38. Stabler B. Shapefiles: read and write ESRI Shapefiles: R package; 2013. Accessible at: https://cran.r-project.org/web/packages/shapefiles/shapefiles.pdf. Accessed 24 July 2017.
39. Carter TB. Social Network Analysis with sna. J Stat Softw. 2008;24(6):1–51.
40. Csardi G, Nepusz T. The igraph software package for complex network research. InterJournal. 2006; Complex Systems: 1695. http://igraph.org. Accessed 24 July 2017.
41. Fournie G, Guitian J, Desvaux S, Mangtani P, Ly S, Cong VC, San S, Dung DH, Holl D, Pfeiffer DU, et al. Identifying live bird markets with the potential to act as reservoirs of avian influenza A (H5N1) virus: a survey in northern Viet Nam and Cambodia. PLoS One. 2012;7(6):e37986.

42. Soares Magalhães RJ, Zhou X, Jia B, Guo F, Pfeiffer DU, Martin V. Live poultry trade in southern China provinces and HPAIV H5N1 infection in humans and poultry: the role of Chinese new year festivities. PLoS One. 2012;7(11):e49712.

43. Delabouglise A, Choisy M, Phan TD, Antoine-Moussiaux N, Peyre M, Vu TD, Pfeiffer DU, Fournie G. Economic factors influencing zoonotic disease dynamics: demand for poultry meat and seasonal transmission of avian influenza in Vietnam. Sci Rep. 2017;7(1):5905.

44. Paul M, Baritaux V, Wongnarkpet S, Poolkhet C, Thanapongtharm W, Roger F, Bonnet P, Ducrot C. Practices associated with highly pathogenic avian influenza spread in traditional poultry marketing chains: social and economic perspectives. Acta Trop. 2013;126(1):43–53.

45. Lockhart CY, Stevenson MA, Rawdon TG, Gerber N, French NP. Patterns of contact within the New Zealand poultry industry. Prev Vet Med. 2010;95(3–4):258–66.

46. Ahmed SS, Themudo GE, Christensen JP, Biswas PK, Giasuddin M, Samad MA, Toft N, Ersboll AK. Molecular epidemiology of circulating highly pathogenic avian influenza (H5N1) virus in chickens, in Bangladesh, 2007-2010. Vaccine. 2012;30(51):7381–90.

47. Osmani MG, Ward MP, Giasuddin M, Islam MR, Kalam A. The spread of highly pathogenic avian influenza (subtype H5N1) clades in Bangladesh, 2010 and 2011. Prev Vet Med. 2014;114(1):21–7.

48. Ansari WK, Parvej MS, El Zowalaty ME, Jackson S, Bustin SA, Ibrahim AK, El Zowalaty AE, Rahman MT, Zhang H, Khan MF, et al. Surveillance, epidemiological, and virological detection of highly pathogenic H5N1 avian influenza viruses in duck and poultry from Bangladesh. Vet Microbiol. 2016;193:49–59.

49. Biswas PK, Giasuddin M, Nath BK, Islam MZ, Debnath NC, Yamage M. Biosecurity and circulation of influenza A (H5N1) virus in live-bird Markets in Bangladesh, 2012. Transbound Emerg Dis. 2017;64(3):883–91. Epubdate: 2015/12/15. https://doi.org/10.1111/tbed.12454.

50. Negovetich NJ, Feeroz MM, Jones-Engel L, Walker D, Alam SM, Hasan K, Seiler P, Ferguson A, Friedman K, Barman S, et al. Live bird markets of Bangladesh: H9N2 viruses and the near absence of highly pathogenic H5N1 influenza. PLoS One. 2011;6(4):e19311.

51. Marinova-Petkova A, Shanmuganatham K, Feeroz MM, Jones-Engel L, Hasan MK, Akhtar S, Turner J, Walker D, Seiler P, Franks J, et al. The continuing evolution of H5N1 and H9N2 influenza viruses in Bangladesh between 2013 and 2014. Avian Dis. 2016;60(1 Suppl):108–17.

52. Sayeed MA, Smallwood C, Imam T, Mahmud R, Hasan RB, Hasan M, Anwer MS, Rashid MH, Hoque MA. Assessment of hygienic conditions of live bird markets on avian influenza in Chittagong metro, Bangladesh. Prev Vet Med. 2017;142:7–15.

53. Turner JC, Feeroz MM, Hasan MK, Akhtar S, Walker D, Seiler P, Barman S, Franks J, Jones-Engel L, McKenzie P, et al. Insight into live bird markets of Bangladesh: an overview of the dynamics of transmission of H5N1 and H9N2 avian influenza viruses. Emerg Microbes Infect. 2017;6(3):e12.

54. Ortiz-Pelaez A, Pfeiffer DU, Soares-Magalhaes RJ, Guitian FJ. Use of social network analysis to characterize the pattern of animal movements in the initial phases of the 2001 foot and mouth disease (FMD) epidemic in the UK. Prev Vet Med. 2006;76(1–2):40–55.

55. Guan Y, Smith GJ. The emergence and diversification of panzootic H5N1 influenza viruses. Virus Res. 2013;178(1):35–43.

56. Horm SV, Tarantola A, Rith S, Ly S, Gambaretti J, Duong V, Phalla Y, Sorn S, Holl D, Allal L, et al. Intense circulation of a/H5N1 and other avian influenza viruses in Cambodian live-bird markets with serological evidence of sub-clinical human infections. Emerg Microbes Infect. 2016;5(7):e70.

57. Chen J, Fang F, Yang Z, Liu X, Zhang H, Zhang Z, Zhang X, Chen Z. Characterization of highly pathogenic H5N1 avian influenza viruses isolated from poultry markets in central China. Virus Res. 2009;146(1–2):19–28.

58. Fournie G, Pfeiffer DU. Can closure of live poultry markets halt the spread of H7N9? Lancet. 2014;383(9916):496–7.

59. Sarker S, Talukder S, Chowdhury EH, Das PM. Knowledge, attitudes and practices on biosecurity of workers in live bird markets at Mymensingh, Bangladesh. ARPN J Agri Biol Sci. 2011;6(6):12–7.

60. Rimi NA, Sultana R, Muhsina M, Uddin B, Haider N, Nahar N, Zeidner N, Sturm-Ramirez K, Luby SP. Biosecurity conditions in small commercial chicken farms, Bangladesh 2011-2012. EcoHealth. 2017;14(2):244–58.

61. Hulse-Post DJ, Sturm-Ramirez KM, Humberd J, Seiler P, Govorkova EA, Krauss S, Scholtissek C, Puthavathana P, Buranathai C, Nguyen TD, et al. Role of domestic ducks in the propagation and biological evolution of highly pathogenic H5N1 influenza viruses in Asia. Proc Natl Acad Sci U S A. 2005;102(30):10682–7.

62. Webster RG, Bean WJ, Gorman OT, Chambers TM, Kawaoka Y. Evolution and ecology of influenza A viruses. Microbiol Rev. 1992;56(1):152–79.

63. Sturm-Ramirez KM, Hulse-Post DJ, Govorkova EA, Humberd J, Seiler P, Puthavathana P, Buranathai C, Nguyen TD, Chaisingh A, Long HT, et al. Are ducks contributing to the endemicity of highly pathogenic H5N1 influenza virus in Asia? J Virol. 2005;79(17):11269–79.

64. Dijkstra EW. A note on two problems in Connexion with graphs. Numer Math. 1959;1:269–71.

13

Eprinomectin pour-on (EPRINEX® Pour-on, Merial): efficacy against gastrointestinal and pulmonary nematodes and pharmacokinetics in sheep

Dietmar Hamel[1]*, Antonio Bosco[2], Laura Rinaldi[2], Giuseppe Cringoli[2], Karl-Heinz Kaulfuß[3], Michael Kellermann[1], James Fischer[4], Hailun Wang[5], Katrin Kley[1], Sandra Mayr[1], Renate Rauh[1], Martin Visser[1], Thea Wiefel[1], Becky Fankhauser[5] and Steffen Rehbein[1]

Abstract

Background: The anthelmintic efficacy of the 0.5% *w/v* topical formulation of eprinomectin (EPN), EPRINEX® Pour-on (Merial) when administered at 1 mg/kg body weight was evaluated in sheep in two dose confirmation laboratory studies and one multicenter field study. In addition, the pharmacokinetics of EPN when administered at that dosage to adult sheep was determined.

Results: In the two dose confirmation studies, which included 10 sheep each, sheep treated with topical EPN had significantly ($p < 0.05$) fewer of the following nematodes than the untreated sheep with overall reduction of nematode counts by >99%: adult *Dictyocaulus filaria, Haemonchus contortus, Teladorsagia circumcincta*(pinnata/ trifurcata), *Trichostrongylus axei, T. colubriformis, T. vitrinus, Cooperia curticei, Nematodirus battus, Strongyloides papillosus, Chabertia ovina* and *Oesophagostomum venulosum*, and inhibited fourth-stage *Teladorsagia* larvae.
A total of 196 sheep harboring naturally acquired gastrointestinal nematode infections were included in the field efficacy study at two sites each in Germany (48 Merino x Ile de France lambs, 52 adult Merino females) and in Italy (adult male and female Bagnolese, Lacaune, Lacaune x Bagnolese, Bagnolese x Sarda sheep; 48 animals per site). Animals were blocked on pre-treatment body weight and within each block, one animal was randomly assigned to the control (untreated) group and three animals were randomly assigned to be treated with topical EPN. Examination of feces 14 days after treatment demonstrated that, relative to the controls, topical EPN-treated sheep had significantly ($p < 0.0001$) lower strongylid egg counts. Reduction was ≥97% at each site and 98.6% across all sites.
Pharmacokinetics of EPN following single treatment with topical EPN were determined in eight ~4.5 year old female Merino cross sheep based on the analysis of plasma samples which were collected from two hours to 21 days following treatment. The main pharmacokinetic parameters were: C_{max} 6.20 ± 1.71 ng/mL, AUC_{last} 48.8 ± 19.2 day*ng/mL, T_{max} 3.13 ± 2.99 days and $T_{1/2}$ 6.40 ± 2.95 days.
No treatment-related health problems or adverse drug events were observed in any study.

Conclusion: These studies demonstrated 0.5% *w/v* EPN administered topically at 1 mg/kg body weight to be highly efficacious against a broad range of ovine gastrointestinal nematodes and *D. filaria* lungworms and well tolerated by sheep of different ages, breeds, gender and physiological status.

Keywords: Eprinomectin, Topical, Gastrointestinal nematodes, Lungworms, Pharmacokinetics, Sheep

* Correspondence: dietmar.hamel@merial.com
[1]Merial GmbH, Kathrinenhof Research Center, Walchenseestr. 8-12, 83101 Rohrdorf, Germany
Full list of author information is available at the end of the article

Background

Because of their ubiquitous occurrence, nematode endoparasites are a major concern to sheep farmers and are an important drain of resources worldwide. Nematode parasitism negatively impacts the production (meat, milk, wool) and reproduction of sheep and has the capability to seriously compromise the health and welfare of the animals. Even subclinical nematode infections cause losses of productivity as demonstrated repeatedly by treatment-induced improved performance e. g., [1–10]. Therefore, a prerequisite for economically sustainable sheep farming and efficient production is the effective control of ovine nematode parasites [8, 11].

Eprinomectin is a macrocyclic lactone registered as a broad spectrum endectocide as a 0.5% *w/v* topical formulation (EPRINEX® Pour-on, Merial) for use in cattle. In this formulation, eprinomectin dosed at 0.5 mg per kg body weight is characterized by a broad safety margin and a zero milk withholding in dairy cows due to a low milk partitioning coefficient, an exceptional pharmacokinetic property within the macrocyclic lactone class of anthelmintics [12, 13]. The excellent endoparasiticidal efficacy of eprinomectin in sheep has been known for more than 20 years because experimentally infected sheep dosed orally were used for screening avermectin/milbemycin analogs in the effort to identify a candidate compound allowing the use in all classes of cattle, including lactating animals [12]. However, reports on the topical treatment of sheep with eprinomectin have been published only quite recently [14–20].

While there are drugs from all anthelmintic classes available for effective treatment of ovine endoparasites, most products are not authorized for use in lactating dairy animals or require a period of withholding the milk because of the levels of residues excreted with milk. Products without disclaimer against use in lactating dairy sheep are of particular importance for the commercial sheep farming in the Mediterranean region where about two thirds of the world's sheep milk is produced [21].

Based on studies determining the excretion of eprinomectin in the milk of lactating sheep (Merial, unpublished data), 0.5% *w/v* eprinomectin (EPRINEX® Pour-on, Merial) administered at 1 mL per kg body weight (equivalent to 1 mg eprinomectin per kg body weight) topically to lactating sheep has been recently granted zero hours milk withdrawal by the European Medicines Agency.

Here we present the results of a series of four studies (two dose confirmation laboratory studies, one multicenter field efficacy study and one pharmacokinetic study) which were conducted between 2013 and 2015 in order to support the market authorization in sheep of 0.5% *w/v* topical formulation of eprinomectin (EPRINEX® Pour-on, Merial) when administered at 1 mg per kg body weight.

Methods

This series of studies consisted of two dose confirmation laboratory studies (Studies 1 and 2), one multicenter field efficacy study (Study 3), and one pharmacokinetic study (Study 4). The design of the Studies 1, 2 and 3 was in accordance with the International Cooperation on Harmonisation of Technical Requirements for Registration of Veterinary Medicinal Products (VICH) GL7, "Efficacy of Anthelmintics: General Requirements" and GL13, "Efficacy of Anthelmintics: Specific Recommendations for Ovine" [22] and the "World Association for the Advancement of Veterinary Parasitology (W.A.A.V.P.) second edition of guidelines for evaluating the efficacy of anthelmintics in ruminants (bovine, ovine, caprine)" [23]. The studies were conducted in compliance with VICH GL9, entitled Good Clinical Practice and were performed as blinded studies, i.e., all personnel involved in collecting efficacy data and making health observations were masked as to the treatment assignment of the animals.

Study 4 was conducted in accordance to Guidelines for the Conduct of Pharmacokinetic Studies in Target Animal Species, EMEA/CVMP/133/99-FINAL.

General study design

Studies 1, 2 and 3 were conducted as randomized block design studies with blocks of two (Studies 1 and 2) or four (Study 3) animals formed based on pre-treatment body weight. Within blocks, animals were allocated at random to treatment groups, Control (untreated) or to be treated with 0.5% *w/v* eprinomectin (EPRINEX® Pour-on, Merial) at 1 mL per 5 kg of body weight topically (1 mg eprinomectin per kg body weight). As per VICH GL 7, control (untreated) to 0.5% *w/v* eprinomectin (treated) ratio was 1:1 in the dose confirmation studies (Studies 1 and 2); however, the ration was 1:3 in the multicenter field efficacy study (Study 3) in order to gain further experience on the test product in a larger number of animals of different breeds, age, body weight, gender and physiological status. All eight sheep enrolled in Study 4 were treated with 0.5% *w/v* eprinomectin at 1 mL per 5 kg of body weight topically.

Pre-treatment body weight obtained with verified scales on Day −5 (Study 3/Sites 3 and 4), or Day −1 (Studies 1, 2, 3/Sites 1 and 2, and Study 4) was used for allocation and dose calculation, as appropriate. The calculated dose was rounded up to the next 0.5 mL (Study 4) or 1.0 mL (Studies 1, 2 and 3) increment, if it was not an exact 0.5 mL or 1.0 mL increment, respectively.

Treatment was administered once at Day 0 topically along the back line, from the withers to the tail head using appropriately sized syringes. For administration of formulation, the fleece was parted, and the formulation was administered directly onto the skin of the sheep.

In each study, general health observations were carried out daily. In addition, animals were observed hourly for the first four hours after treatment for reactions to treatment.

Study Animals: Studies 1, 2, 3 and 4

Sheep of different breeds, age, body weight, gender and physiological status were included in the four studies (Table 1). While the animals used in Study 3 were owned by private sheep farmers, sheep utilized in Studies 1, 2 and 4 were bought from commercial farms. None of the animals were treated with macrocyclic lactone products within six weeks of the start of the study.

Sheep included in Studies 1, 2 and 4 were kept indoors on straw, and following allocation to treatment groups, animals were housed in individual pens to prevent them from having physical contact with others. Animals were offered a roughage-based diet for ad libitum consumption. Sheep included at the four sites in the multicenter field Study 3 (Sites 1 and 2, Germany; Sites 3 and 4, Italy) were grazed on permanent pastures with the study animals (treated and untreated sheep) grazing together with sheep not enrolled in the study (remaining sheep at sites). Animals in all studies had continuous access to water.

Animals in Study 1 were tested negative for patent gastrointestinal and pulmonary nematode infections prior to first inoculation with gastrointestinal nematodes and *Dictyocaulus filaria* lungworms. At commencement of Studies 2 and 3, all sheep harbored naturally acquired gastrointestinal nematode infections as demonstrated through shedding strongylid (other than *Nematodirus*) eggs prior to treatment; in addition, *Nematodirus* eggs, *Trichuris* eggs, *Moniezia* eggs and/ or protostrongylid larvae were recovered from the feces of various animals.

Fecal Examination: Studies 1, 2 and 3

In Study 1, rectal fecal samples were collected from all animals and examined to confirm the absence of patent gastrointestinal and pulmonary nematode infections seven days prior to the initiation of experimental nematode infections. In Studies 2 and 3, rectal fecal samples were collected from all animals ten or five days prior to treatment, respectively, and examined to confirm the presence of natural infection of the animals with gastrointestinal nematodes and/or lungworms. In order to estimate the efficacy of the treatment in terms of the reduction of fecal nematode egg counts in Study 3, individual fecal samples were collected in addition 14 days after treatment and examined.

For fecal egg counting a modified McMaster method with one egg counted representing 10 eggs per gram of feces (EPG) was used with saturated sodium chloride solution for floatation [24] in Studies 1, 2 and 3/Sites 1 and 2. Samples collected in Study 3/Sites 3 and 4 were examined using the FLOTAC dual technique (sensitivity = 6 EPG) [25]. For lungworm larval recovery, 10-g (Studies 1, 2 and 3/Sites 1 and 2) or 5-g (Study 3/Sites 3 and 4) fecal samples were subjected to the Baermann technique [24] to establish lungworm larval counts per gram of feces. When present, eggs were referred to as 'strongylid' (nematode genera including *Bunostomum*, *Chabertia*, *Cooperia*, *Haemonchus*, *Oesophagostomum*, *Teladorsagia*, and *Trichostrongylus*), *Nematodirus* (a strongylid which was identified and counted independently), *Strongyloides* and/or *Trichuris*. Other findings in the fecal examination (*Moniezia* eggs and protostrongylid lungworm larvae) were recorded.

In addition, fecal culture procedures were employed for the identification of the larvae of strongylid nematodes developing from the eggs excreted by the sheep in the multicenter field Study 3. Composite fecal cultures were performed utilizing the fecal samples subjected to

Table 1 Description of study animals

Study	Number of animals	Breed	Sex	Age (range)	Pre-treatment (Days −5 to −1) body weight (kg), range
1	20	Merino	Male	~5–6 months	33.2–46.0
2	20	Merino Cross	Female[a]	~3–6 years	37.4–76.2
3, Site 1, Germany 1	48	Merino x Ile de France	Female	~6 months	25.2–44.2
3, Site 2, Germany 2	52	Merino	Female[a]	~2–7 years	37.8–81.6
3, Site 3, Italy 1	48	Bagnolese (44), Lacaune x Bagnolese (4)	Male (3), female (45)[b]	~1–6 years	55.5–104.7
3, Site 4, Italy 2	48	Bagnolese (26), Lacaune (16), Sarda x Bagnolese (5), Lacaune x Bagnolese (1)	Female[c]	~2–6 years	40.2–71.4
4	8	Merino Cross	Female[a]	~4.5 years	66.8–101.8

[a]Dry, not pregnant
[b]Dry, not pregnant (15); dry, pregnant (9); lactating, not pregnant (21)
[c]Lactating, not pregnant

lungworm larval recovery (fecal samples of all animals combined prior to treatment; fecal samples of animals combined by treatment group post-treatment) to determine composition by genera. For coproculture, samples of fecal composites were mixed with vermiculite and incubated for seven days after which the third-stage larvae were harvested. Per culture, 100 larvae were identified to genus using standard morphological identification keys [24, 26].

Inoculation of Sheep: Study 1

Sheep of Study 1 were inoculated with infective third-stage larvae (L3) of gastrointestinal and pulmonary nematode species by oral gavage. The inoculation schedule was designed so that nematodes were expected to be adults on Day 0 (= day of treatment): Day −56, Chabertia ovina, ~800 L3 per animal; Day −35, D. filaria, ~500 L3 per animal and Oesophagostomum venulosum, ~800 L3 per animal; Day −28, Teladorsagia circumcincta(pinnata/trifurcata), ~8000 L3 per animal; Day −25, Haemonchus contortus, ~2000 L3 per animal; Day −23, Trichostrongylus axei, ~5000 L3 and Nematodirus battus, ~2000 L3 per animal; Day −21, T. colubriformis and Cooperia curticei, ~5000 L3 per species and animal. The parasites used were recent field isolates from Germany as defined per VICH GL 7 [22]. The number of larvae given was generally in accord with the W.A.A.V.P. guidelines for testing of anthelmintics in ruminants [23].

Parasite counts: Studies 1, 2 and 3

In Studies 1 and 2, all animals were humanely euthanized and organs (the lungs, abomasum, small intestine and large intestine including cecum) were collected for parasite recovery and count 14 days after treatment administration. In Study 3, two sentinel animals with the same history as and thus representative of the study animals were randomly selected at each of the sites and necropsied prior to treatment of the study animals for parasite recovery and count.

Lungs were examined completely for lungworms by lengthwise opening of all accessible air passages. The contents of the abomasum, small and large intestines were collected separately and diluted with water. Abomasum and small intestine were incubated (saline soak) overnight to recover mucosal stages of the parasites for identification and counting. To facilitate isolation and counting of nematodes, organ contents and soaks were screened over sieves of appropriate mesh sizes (abomasum and small intestine contents: 150 µm; large intestine content: 250 µm; abomasal soak: 40 µm) to remove the debris. Gastrointestinal nematode counts were made on 10% aliquots (abomasum, abomasum soak and small intestine; Studies 1, 2 and 3), 20% aliquots (large intestine;

Study 1) or total content (large intestine; Studies 2 and 3); cestodes were collected directly from the small intestines during processing and counted totally. Counts of each nematode species for each animal were calculated by multiplying the number of worms actually counted from each organ by the aliquot factor and summing over all organs.

Teladorsagia male nematodes were identified to 'morphs' (T. circumcincta, pinnata and trifurcata), based on their distinct morphological characters. However, in accepting the concept of polymorphism [27, 28], total worm count was presented as 'T. circumcincta(pinnata/trifurcata)' by adding male T. circumcincta(pinnata/trifurcata) and female Teladorsagia spp. Female Trichostrongylus spp. nematodes were assigned based on location of recovery (i.e. abomasum or small intestine, respectively) to T. axei (abomasum) or T. capricola, T. colubriformis and T. vitrinus (small intestine). To estimate total counts per species for T. capricola, T. colubriformis and T. vitrinus, female Trichostrongylus spp. nematodes of the small intestine were proportioned according to the counts of males.

Analysis of parasite and fecal egg counts: Studies 1, 2 and 3

For Studies 1 and 2, nematode counts by species and stage, if applicable, were transformed to the natural logarithm (ln) of (count +1) for calculation of geometric means for each treatment group. Efficacy was determined by calculating the percent efficacy as $100 \times [(C-T)/C]$, where C is the geometric mean nematode count among the untreated controls and T is the geometric mean among the animals treated with 0.5% w/v eprinomectin. The log counts for each nematode species of the treated group were compared to the log-counts of the control group using an F-test adjusted for the allocation blocks used to randomize the animals to the treatment groups. The mixed procedure in SAS version 9.4 was used for the analysis, with the treatment groups listed as a fixed effect, and the allocation blocks listed as a random effect. All testing was two-sided at the significance level $\alpha = 0.05$.

For Study 3, fecal egg per gram (EPG) counts were transformed to the natural logarithm of (count + 1) for the calculation of geometric means by treatment group. Efficacy was determined based on post-treatment fecal egg counts by calculating the percent efficacy as $100 \times [(C-T)/C]$, where C is the geometric mean among the untreated controls and T is the geometric mean among the treated animals. The log-counts (EPG) of the treated group were compared to the log-counts of the untreated control group using analysis of variance for a generalized randomized block design. The mixed procedure in SAS version 9.4 was used for the analysis, with the treatment groups, sites and treatment-by-site inter-action term listed as fixed effects and blocks as random effects. Exclusion criterion for individual analysis for nematodes was based on a rate of <40% animals shedding nematode eggs or lungworm larvae in the untreated controls. All testing was two-sided at the significance level $\alpha = 0.05$.

Collection and analysis of plasma and pharmacokinetic analysis: Study 4

In Study 4, whole blood of all sheep was collected from the jugular vein into lithium heparinized tubes prior to treatment (Day −1), and approximately 2, 4, 6, 8, 12, 24 and 36 h after treatment. Additional samples were collected on Days 2, 3, 4, 5, 6, 10, 14, 17 and 21. Plasma was separated by centrifugation and stored at $\leq -20\,°C$ until assayed for eprinomectin (B1a component) concentration.

All plasma samples collected were analyzed for eprinomectin B1a using a fully validated high-performance liquid chromatography method with fluorescence detection which was described previously [29]. The lower limit of quantitation of the assays for eprinomectin was established as 0.75 ng/mL, and the lower limit of detection of the assays as 0.50 ng/mL.

The analytical method performed well during sample analyses. Individual quality control (QC) samples had eprinomectin B1a recoveries in plasma from 89.0% to 110% for three QC levels: 1.0, 10 and 40 ng/mL; %relative standard deviation was 4.57 for 27 QC samples.

Pharmacokinetic analysis was performed using WinNonlin® version 5.2.1 non-compartmental analysis (Pharsight Corporation, Mountain View, CA, USA) for each individual animal and parameters were then averaged for the group. Eprinomectin plasma concentrations below the limit of quantitation of the assay method (<0.75 ng/mL) were not used in the pharmacokinetic calculations. The maximum concentration (C_{max}) and time to maximum concentration (T_{max}), and last quantifiable concentration (C_{last}) and time to last quantifiable concentration (T_{last}) were determined directly from the plasma concentration data. The first order rate constant associated with the terminal log-linear portion of the curve (k_{el}) was estimated via linear regression of the log plasma concentration versus time curve and the terminal plasma half-life was calculated using $T_{1/2} = \ln(2)/k_{el}$. The area under the plasma concentration versus time curve (AUC) was determined using the linear trapezoidal rule for increasing plasma concentrations and the logarithmic trapezoidal rule for decreasing plasma concentrations (linear up/log down) from Day 0 to the last time the drug plasma concentration was above the lower limit of quantitation, AUC_{last}. AUCs were also extrapolated to infinity using the formula: $AUC_{inf} = AUC_{last} + C_{last}/k_{el}$. The calculations were assessed by examining the extent of extrapolation for the AUC_{inf} values, so the AUC percentage extrapolated (AUC_%Extrap) was also determined. Group means and standard deviations were calculated.

Results

No health problems or abnormal reactions to treatment were observed throughout the studies. In addition, all animals but one were reported to be healthy throughout the studies. This animal of Study 4 presented signs of respiratory disease at 18 and 19 days following treatment. It was thus medicated as appropriate and recovered within two days, and remained in the study until study end (Day 21).

Studies 1 and 2 – nematode counts and efficacy

The nematode counts of 0.5% w/v eprinomectin-treated animals and the untreated control animals and percentage efficacy are summarized in Table 2 for those parasites which were recovered from at least four control animals in one of the two studies. For the sheep included in Study 2, pre-treatment fecal strongylid egg counts did not differ ($p = 0.4725$) between sheep allocated to the untreated control group and sheep allocated to the topical 0.5% w/v eprinomectin-treated group (range, 120 to 3010 EPG vs. 170 to 3130 EPG, respectively).

Considering Studies 1 and 2 collectively, sheep treated with 0.5% w/v eprinomectin had significantly ($p < 0.05$) fewer of the following nematodes than the untreated control sheep with overall reduction of nematode counts by >99%: adult *D. filaria*, *H. contortus*, *T. circumcincta*(*pinnata/trifurcata*), *T. axei*, *T. colubriformis*, *Trichostrongylus vitrinus*, *C. curticei*, *N. battus*, *S. papillosus*, *Ch. ovina* and *O. venulosum*, and inhibited fourth-stage *Teladorsagia* larvae (Table 2).

Nematode parasites which were recovered from no more than three control animals per study and thus did not allow for a meaningful analysis were inhibited fourth-stage *Haemonchus* larvae (2/10 controls) in Study 1 and adult *Trichostrongylus capricola* (2/10 controls), *Capillaria musimon* (2/10 controls and 2/10 treated), *Trichuris ovis* (3/10 controls and 3/10 treated) and *Trichuris skrjabini* (3/10 controls and 1/10 treated) in Study 2. In addition, *Moniezia* cestodes were recovered from two controls and two 0.5% w/v eprinomectin-treated sheep in Study 2.

Multicenter Field Study 3 – parasite counts of sentinel animals, fecal nematode egg counts and efficacy

All 196 sheep enrolled in the study at four sites were naturally infected with gastrointestinal nematodes. By

Table 2 Nematode counts and therapeutic efficacy against pulmonary and gastrointestinal nematodes of topical 0.5% *w/v* eprinomectin (EPRINEX® Pour-on, Merial) administered once at 1 mg/kg body weight to experimentally infected sheep (Study 1) or sheep with naturally acquired nematode infections (Study 2)

Study	Nematode counts		EPRINEX® Pour-on		Probability[c]	Efficacy (%)[d]
	Control (untreated)					
	NI/NG[a]	GM[b] (Range)	NI/NG	GM (Range)		
Dictyocaulus filaria, adult						
1	10/10	30.1 (8–115)	0/10	0	<0.0001	100
Chabertia ovina, adult						
1	10/10	19.5 (5–70)	0/10	0	<0.0001	100
2	9/10	5.7 (0–38)	0/10	0	<0.0001	100
Cooperia curticei, adult						
1	10/10	99.4 (10–260)	0/10	0	<0.0001	100
2	9/10	674.9 (0–6820)	0/10	0	<0.0001	100
Haemonchus contortus, adult						
1	10/10	992.5 (450–1610)	0/10	0	<0.0001	100
2	9/10	477.2 (0–11,030)	0/10	0	<0.0001	100
Nematodirus battus, adult						
1	10/10	148.8 (10–450)	0/10	0	<0.0001	100
2	3/10	2.8 (0–130)	0/10	0	0.0465	100
Oesophagostomum venulosum, adult						
2	10/10	103.1 (3–893)	0/10	0	<0.0001	100
Strongyloides papillosus, adult[e]						
1	10/10	151.4 (30–700)	0/10	0	<0.0001	100
Teladorsagia circumcincta(pinnata/trifurcata), adult						
1	10/10	1700.6 (950–3770)	0/10	0.0	<0.0001	100
2	10/10	6826.2 (1040–23,430)	3/10	1.7 (0–40)	<0.0001	>99.9
Teladorsagia, inhibited fourth-stage larvae						
1	3/10	1.0 (0–10)	0/10	0	0.0465	100
2	5/10	6.2 (0–240)	0/10	0	0.0214	100
Trichostrongylus axei, adult						
1	10/10	1169.3 (180–2430)	0/10	0	<0.0001	100
2	8/10	99.7 (0–1280)	0/10	0	0.0003	100
Trichostrongylus colubriformis, adult						
1	10/10	1032.3 (560–1400)	0/10	0	<0.0001	100
2	9/10	424.0 (0–13,947)	0/10	0	<0.0001	100
Trichostrongylus vitrinus, adult						
2	4/10	7.8 (0–146)	0/10	0	0.0155	100

[a]NI/NG: Number of sheep Infected/Number of sheep in Group
[b]GM = geometric mean, computed by subtracting 1 from the anti-logarithm of the mean of ln(count + 1)
[c]Probability using the F-Test
[d]Efficacy (%) = 100×[(GM Control – GM EPRINEX® Pour-on)/GM Control]
[e]Naturally acquired infection

pre-treatment fecal examination, strongylid, *Nematodirus*, *Trichuris* and protostrongylid nematode infections were demonstrated in 196, 37, 25 and 52 sheep, respectively. In addition, pre-treatment fecal examination revealed *Moniezia* cestode eggs in 31 sheep.

Based on fecal examination, strongylid, *Nematodirus* and *Trichuris* nematode infections were present at all sites while evidence of protostrongylid lungworms and *Moniezia* cestodes was present only at Sites 2, 3 and 4 or Site 1, respectively.

Necropsy of two sentinel animals per site revealed a variety of gastrointestinal helminths (*H. contortus, T. circumcincta(pinnata/trifurcata), T. axei, T. capricola, T. colubriformis, T. vitrinus, N. battus, N. filicollis, Ch. ovina, O. venulosum, Tr. ovis, Tr. discolor* and/or *Moniezia* spp.) and/or *Protostrongylus rufescens* lungworms. The sentinel animals' parasite counts, which defined the parasite composition of the study animals and represented the natural nematode contamination, indicated the occurrence of at least 12 and 11, four and six species of gastrointestinal nematodes at Sites 1, 2, 3 and 4, respectively (Table 3).

Only strongylid egg counts were included in the analysis (Table 4). Analysis of strongylid egg counts did not reveal treatment-by-site interaction (pre-treatment, $p = 0.9263$; post-treatment, $p = 0.0621$); thus combined Sites 1 to 4 analysis of pre- and post-treatment strongylid egg counts comparing untreated control animals and topical 0.5% *w/v* eprinomectin-treated animals was performed. Pre-treatment fecal strongylid egg counts did not differ between the two groups ($p = 0.2528$). After treatment, topical 0.5% *w/v* eprinomectin-treated sheep had significantly ($p < 0.0001$) lower strongylid egg counts than the untreated control group across all sites. Reduction of strongylid egg counts was 98.6% across all sites and ≥97% at each site (Table 4). Pre-treatment coprocultures revealed larvae of the gastrointestinal nematode genera *Haemonchus, Teladorsagia* and *Trichostrongylus* for all sites while *Chabertia/Oesophagostomum* larvae were recovered from the coprocultures of Sites 1, 2 and 4 only. Identification of the larvae recovered from the post-treatment coprocultures

of both untreated control animals and topical 0.5% *w/v* eprinomectin-treated animals at each study site indicated no change in the spectrum of nematode genera composition.

Nematodirus eggs, *Trichuris* eggs and protostrongylid larvae were observed infrequently at fecal examinations with overall less than 40% of the animals in the control (untreated) group shedding eggs or larvae (Table 5) such that no meaningful analysis was possible.

Study 4 – pharmacokinetics of eprinomectin

The absence of eprinomectin (B1a component) was confirmed in the plasma samples of the animals prior to treatment with topical 0.5% *w/v* eprinomectin. The plasma concentration vs. time profile of eprinomectin following treatment is shown in Fig. 1, and the pharmacokinetic parameters are summarized in Table 6. Eprinomectin B1a was detected in the plasma of all sheep at quantifiable levels four hours after treatment and remained at quantifiable levels in all animals until Day 10 when the average concentration was 2.84 ± 1.48 ng/mL. The highest mean plasma eprinomectin (B1a component) level (5.46 ± 2.04 ng/mL) was observed 36 h post treatment followed by a continuous decline until Day 21 when three animals had quantifiable levels (0.804–1.03 ng/mL). Greater than 20% extrapolation (AUC_%Extrap) of the total AUC in four sheep indicates that the elimination phase was not adequately defined in these animals. Based on the four animals in which the elimination phase was adequately defined, AUC_{inf} was 69.8 ± 13.7 day*ng/mL.

Table 3 Parasite counts of sentinel animals at Sites 1 to 4 of Study 3

Parasite species/stage	Parasite count							
	Site 1, Germany 1		Site 2, Germany 2		Site 3, Italy 1		Site 4, Italy 2	
	Animal 1	Animal 2	Animal 1	Animal 2	Animal 1	Animal 2	Animal 1	Animal 2
Haemonchus contortus, adult	60	40	20	1030	2890	760	160	20
Teladorsagia circumcincta(pinnata/trifurcata), adult	4860	3360	1660	8370	380	140	2460	70
Trichostrongylus axei, adult	40	150	350	1360	30	20	150	20
Trichostrongylus capricola, adult	25	0	0	107	0	0	120	60
Trichostrongylus colubriformis, adult	172	68	1193	4169	0	0	0	0
Trichostrongylus vitrinus, adult	123	23	47	214	50	40	20	720
Nematodirus battus, adult	270	290	0	10	0	0	0	0
Nematodirus filicollis, adult	100	290	0	20	0	0	0	0
Moniezia spp.	2	18	0	0	0	0	0	0
Chabertia ovina, adult	19	18	54	21	0	0	63	74
Oesophagostomum venulosum, adult	4	22	25	166	0	0	0	0
Trichuris ovis, adult	5	2	2	2	0	0	0	0
Trichuris discolor, adult	0	5	0	0	0	0	0	0
Protostrongylus rufescens, adult	0	0	0	229	0	0	0	0

Table 4 Geometric mean fecal strongylid egg counts and percentage efficacy of topical 0.5% *w/v* eprinomectin (EPRINEX® Pour-on, Merial) administered once at 1 mg/kg body weight to naturally infected sheep under field conditions (Study 3)

Site(s)	Occasion	GM[a] (Range) strongylid eggs per gram counts		Efficacy (%)[b]
		Control (untreated)	EPRINEX® Pour-on	
Site 1, Germany 1	Pre-Treatment[c]	437.7 (60–2150)	454.5 (70–1790)	NC[d]
	Post-Treatment[e]	276.2 (10–1490)	8.4 (0–120)	97.0
Site 2, Germany 2	Pre-Treatment	290.9 (90 – 3920)	416.0 (100–6480)	NC
	Post-Treatment	322.0 (90–3740)	5.7 (0–180)	98.2
Site 3, Italy 1	Pre-Treatment	1042.7 (108–4590)	1219.0 (126–12,456)	NC
	Post-Treatment	601.1 (162–6594)	12.1 (0–126)	98.0
Site 4, Italy 2	Pre-Treatment	633.0 (144–2064)	807.3 (480–4668)	NC
	Post-Treatment	519.9 (144–1200)	1.5 (0–36)	99.7
Sites 1 to 4 combined	Pre-Treatment	531.8 (60–4590)	650.8[f] (70–6480)	NC
	Post-Treatment	406.4 (10–6594)	5.8[g] (0–180)	98.6

[a]GM = geometric mean, computed by subtracting 1 from the anti-logarithm of the mean of ln(count + 1)
[b]Efficacy (%) = 100×[(GM Control – GM EPRINEX® Pour-on)/GM Control]
[c]Pre-treatment fecal examination, Day −5
[d]NC = Not calculated
[e]Post-treatment fecal examination, Day 14
[f]Control vs. EPRINEX® Pour-on, $p = 0.2528$
[g]Control vs. EPRINEX® Pour-on, $p < 0.0001$

Discussion

The primary objective of the studies was to confirm the efficacy of the 0.5% *w/v* eprinomectin formulation (EPRINEX® Pour-on, Merial) against gastrointestinal and pulmonary nematode endoparasites in sheep when administered at 1 mg eprinomectin per kg body weight. Based on parasite burdens recovered from the sheep with induced and naturally acquired nematode infections and the reduction of fecal egg counts in the multicenter field study including the parasite counts of sentinel animals from all sites, results of this series of studies demonstrated consistently a very high efficacy against the major production-limiting gastrointestinal nematode parasites affecting sheep in temperate climates, i. e. *H. contortus, T. circumcincta(pinnata/trifurcata), T. axei, T. colubriformis, T. vitrinus, C. curticei, N. battus, Ch. ovina* and *O. venulosum* [8, 11], and *D. filaria* lungworms. These species of nematodes are representative of the spectrum of nematode parasites infecting sheep throughout Europe and are found to a greater or lesser extent in sheep in southern Europe, e. g. Spain, Italy and Greece [15, 30–32], and central and northern Europe, e. g. Austria, Czech Republic, Germany, the UK and Norway [33–38]. Parasitism of naturally infected sheep determined in the context of the studies reported here demonstrates that gastrointestinal nematode infections remain an important constraint to sheep in Europe such that appropriate control measures including anthelmintic use are needed to ensure appropriate levels of productivity as well as animal welfare [10, 11]. As shown with respect to dairy cattle, the availability of a broad spectrum anthelmintic for use in sheep (and goats) with a zero hours milk withholding period offers an unique advantage for the treatment of lactating animals which have been demonstrated to benefit substantially from efficacious nematode control [2, 4–7].

The results of the dose confirmation laboratory studies and the multicenter field study indicate some variability in efficacy in that, compared to untreated animals, sheep treated with topical 0.5% *w/v* eprinomectin demonstrated >99% efficacy with respect to nematode count reductions while efficacy in terms of reduction of fecal egg

Table 5 Fecal stages of intestinal and pulmonary nematodes in the naturally infected sheep of multicenter field Study 3 (Sites 1 to 4 combined) that were not analyzed because rate of detection was less than 40% in control (untreated) animals (*Nematodirus*, *Trichuris*, protostrongylid) and of *Moniezia* cestodes

Treatment group	Number of positive sheep/number of sheep in group							
	Nematodirus eggs		*Trichuris* eggs		Protostrongylid larvae		*Moniezia* eggs	
	PreT[a]	PostT[b]	PreT	PostT	PreT	PostT	PreT	PostT
Control (untreated)	11/49	8/49	4/49	8/49	12/49	14/49	6/49	4/49
EPRINEX® Pour-on	26/147	3/147	21/147	0/147	40/147	7/147	25/147	6/147

[a]PreT = pre-treatment fecal examination, Day −5
[b]PostT = post treatment fecal examination, Day 14

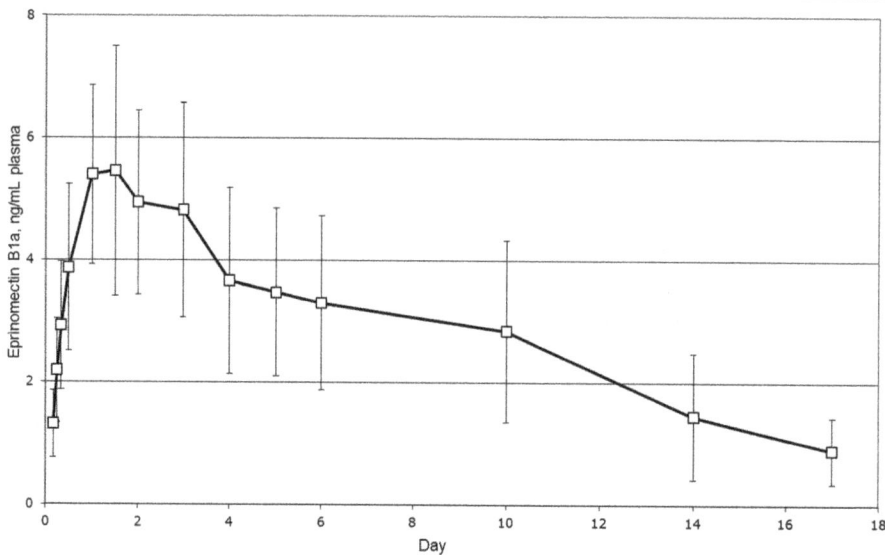

Fig. 1 Plasma profile of eprinomectin B1a in sheep following topical administration of 0.5% *w/v* eprinomectin (EPRINEX® Pour-on, Merial) at 1 mg eprinomectin per kg body weight (Study 4). Each point represents the mean of plasma concentrations of eight sheep. Error bars indicate standard deviations

counts varied from 97% to >99% at the field study sites. Considering that fecal cultures suggested no change in the spectrum of nematode population composition following treatment at the field study sites, this finding may, at least partly, reflect variability in the sensitivity of the respective nematode populations.

The high efficacy against all major gastrointestinal and pulmonary nematodes of sheep demonstrated in the present studies adds considerable knowledge regarding the spectrum of nematocidal activity of topical 0.5% *w/v* eprinomectin compared to observations reported previously which yielded three nematode species from one necropsy study [16] and gastrointestinal nematode egg and lungworm larval count reductions from field efficacy evaluations [15, 17, 20]. In addition, there is also

indication of efficacy of topical 0.5% *w/v* eprinomectin against *Oestrus ovis* nasal bot infestation [16, 17]. Overall, the therapeutic efficacy demonstrated in the present studies in sheep was very similar to the array of nematode parasites effectively treated by the administration of topical 0.5% *w/v* eprinomectin at 1 mg per kg body weight to goats [39, 40].

Any anthelmintic use raises concerns in terms of selection of resistant parasite populations. Recently published systematic reviews of peer-reviewed literature concluded that anthelmintic resistance in gastrointestinal nematodes of sheep is generally widespread in Europe but prevalence varies importantly by region and class of anthelmintic [41] and that high frequency of treatments is the major risk factor associated with anthelmintic resistance in

Table 6 Basic pharmacokinetic parameters describing the disposition of eprinomectin (B1a component) in plasma of sheep after administration of topical 0.5% *w/v* eprinomectin (EPRINEX® Pour-on, Merial), current Study 4 and data from other authors

Source	Topical eprinomectin (mg/kg body weight)	C_{max} (ng/mL)	T_{max} (day)	$T_{1/2}$ (days)	AUC_{last} (day*ng/mL)
Current Study 4	1.0 ($n = 8^a$)	6.20 (± 1.71)	3.13 (± 2.99)	6.40 (± 2.95)	48.8 (± 19.2)
[19]	0.5 ($n = 6^b$)	2.22 (± 0.88)	1.2 (± 0.4)	5.4 (± 0.7)	13.6 (± 4.8)
	1.0 ($n = 6^c$)	5.25 (± 2.7)	1.5 (± 0.5)	12.2 (± 5.8)	33.7 (± 22.5)
[16]	0.5 ($n = 6^d$)	NR^e	NR	NR	56.0 (± 26.2)
[18]	0.5 (n = 6^f)	2.28 (± 0.41)	3.17 (± 0.40)	2.20 (± 0.34)	16.2 (± 3.69)
	0.5 (n = 6^g)	2.30 (± 0.60)	3.00 (± 0.45)	1.85 (± 0.13)	15.5 (± 3.67)

[a]Female dry adult Merino Cross sheep, sampled up to 21 days post dose
[b]Female lactating adult Istrian Parmenka sheep, sampled up to 32 days post dose
[c]Female lactating adult Istrian Parmenka sheep, sampled up to 42 days post dose
[d]Five month old sheep, sampled up to 21 days post dose
[e]Not reported
[f]Female lactating (early-mid lactation) adult Pampina Cross sheep, sampled up to 35 days post dose
[g]Female lactating (mid-late lactation) adult Pampina Cross sheep, sampled up to 35 days post dose

sheep [42]. Therefore, monitoring the efficacy of treatments, appropriate grazing management, and exclusion of part of the nematode population from the exposure to the treatment (creation and/or maintaining of refugia) may be ways to reduce the selective advantage for resistant specimens. Overall, sustainable control requires responsible use of correctly administered anthelmintics providing a balance between maintaining acceptable levels of productivity and animal welfare and the inevitable evolution of anthelmintic resistance because the risk of losses from parasite infection may increase with further intensification of pastoral production systems [11, 42–44].

Regarding induced infection Study 1, inoculation produced adequate levels of infections as recommended by VICH GLs 7 and 13 [22] for all nematodes but *O. venulosum* which was not recovered from any animal. This finding is probably related to an antagonistic interaction between *Ch. ovina* and *O. venulosum* which is dominated by *Ch. ovina*. Both species of large intestinal nematodes under natural infection conditions frequently occur in co-infections [33, 34, 36, 45; this Study 2]. However, infection with *Ch. ovina* stimulates an immune response in the host and suppression of *O. venulosum*, when inoculated subsequently to challenge with *Ch. ovina*, has been observed previously in experimental studies [46, 47]. In addition to the nematodes inoculated, all untreated control animals harboured *Strongyloides papillosus* nematodes. As fecal samples of the lambs were negative for *Strongyloides* eggs before initiation of experimental infections, this inadvertent infection originated likely either from pre-patent infections present in at least some lambs at the time of the pre-inoculation fecal examination or lambs harbored very low level patent infections resulting in egg excretion below the detection limit of the McMaster method used for examination of the feces. *Strongyloides papillosus* is transmitted through the bedding (eggs can hatch in the bedding and third-stage larvae infect sheep by skin penetration) [48] such that infection may have spread among the study animals during the seven week indoor-housing period prior to treatment.

Plasma concentrations and basic pharmacokinetic parameters were comparable to those previously reported following the administration of topical 0.5% *w/v* eprinomectin at 1 mg per kg body weight to sheep [19]. Although some variability can be seen possibly due to different animal physiology (e.g., lactating vs. non-lactating) or breed, considering data of adult female sheep treated with topical 0.5% *w/v* eprinomectin at 0.5 mg per kg body weight [18, 19] indicates dose proportionality. However, one study indicated an exceptional high AUC_{last} of 56.0 ± 26.2 day*ng/mL following topical administration of 0.5% *w/v* eprinomectin at 0.5 mg/kg body weight to five months old lambs weighing 20 to 25 kg [16]. Results of this study are difficult to

interpret as only limited information on the pharmacokinetic profile and characteristics of study animals was reported. Compared to goats [cf. 40], T_{max} appears to occur later in sheep, indicating slower absorption possibly due to the difference of the structure of the skin/hair coat characteristics between the two species. Overall, similar pharmacokinetic profiles were demonstrated for topical 0.5% *w/v* eprinomectin in sheep and goats which translate to a similar spectrum of anthelmintic activity in sheep (these Studies 1, 2 and 3) and goats [39, 40, 49].

Conclusion

This series of studies demonstrated eprinomectin administered topically at 1 mg/kg body weight onto the skin of sheep to be highly efficacious against a broad range of ovine gastrointestinal nematodes and *D. filaria* lungworms and to be well tolerated by sheep of different ages, breeds, gender and physiological status.

Acknowledgements
Not applicable.

Funding
All studies reported herein were funded by Merial Inc., GA, USA. The funding company provided the conceptual aspect and design of the study and reviewed the final version of the manuscript.

Authors' contributions
DH participated in the design of the study, performed treatments in studies 1, 2, and 4, monitored field studies and drafted the paper. KK, SM, RR, MV and HW contributed in preparing the inoculum, blood sampling, nematode counts, data management and statistical analysis. AB, LR, GC and KHK collaborated in the multicenter field studies and collaborated in the manuscript preparation. MK, JF and TW analyzed plasma samples and performed PK analysis. BF and SR contributed to the design and supervision of the studies and helped drafting the manuscript. All the authors read and approved the final manuscript.

Author details
[1]Merial GmbH, Kathrinenhof Research Center, Walchenseestr. 8-12, 83101 Rohrdorf, Germany. [2]Department of Veterinary Medicine and Animal Production, University of Naples Federico II, Via della Veterinaria, 1, 80137 Naples, Italy. [3]Tierarztpraxis Hoffmann, Untere Schulstraße 8, 38875 Elbingerode, Germany. [4]Merial, Inc., North Brunswick Research Center, 631 Route 1 South, North Brunswick, NJ 08902, USA. [5]Merial, Inc., 3239 Satellite Blvd., Duluth, GA 30096-4640, USA.

References

1. Rehbein S, Corba J, Pitt SR, Várady M, Langholff WK. Evaluation of the anthelmintic efficacy of an ivermectin controlled-release capsule in lambs under field conditions in Europe. Small Ruminant Res. 1999;33: 123–9.
2. Fthenakis GC, Papadopoulos E, Himonas C. Effects of three anthelmintic regimes on milk yield of ewes and growth of lambs. J Vet Med A. 2005;52:78–82.
3. Fthenakis GC, Mavrogianni VS, Gallidis E, Papadopoulos E. Interactions between parasitic infections and reproductive efficiency in sheep. Vet Parasitol. 2015;208:56–66.
4. Cringoli G, Veneziano V, Pennacchio S, Mezzino L, Santaniello M, Schioppi M, Fedele V, Rinaldi L. Economic efficacy of anthelmintic treatments in dairy sheep naturally infected by gastrointestinal strongyles. Parassitologia. 2007; 49:201–9.
5. Cringoli G, Veneziano V, Jackson F, Vercruysse J, Greer AW, Fedele V, Mezzino L, Rinaldi L.. Effects of strategic anthelmintic treatments on the milk production of dairy sheep naturally infected with gastrointestinal strongyles. Vet Parasitol. 2008;156:340–5.
6. Cringoli G, Rinaldi L, Veneziano V, Mezzino L, Vercruysse J, Jackson F. Evaluation of targeted selective treatments in sheep in Italy: effects on faecal worm egg count and milk production in four case studies. Vet Parasitol. 2009;164:36–43.
7. Sechi S, Giobbe M, Sanna G, Casu S, Carta A, Scala A. Effects of anthelmintic treatment on milk production in Sarda dairy ewes naturally infected by gastrointestinal nematodes. Small Ruminant Res. 2010;88:145–50.
8. Sutherland I, Scott I. Gastrointestinal Nematodes of Sheep and Cattle. Biology and Control. Wiley-Blackwell, Chichester, West Sussex, UK. 2001.
9. Geurden T, Slootmans N, Glover M, Bartram DJ. Production benefit of treatment with a dual active oral formulation of derquantel-abamectin in slaughter lambs. Vet Parasitol. 2014;205:405–7.
10. Mavrot F, Hertzberg H, Torgerson P. Effect of gastro-intestinal nematode infection on sheep performance: a systematic review and meta-analysis. Parasites Vectors. 2015;8:557.
11. Sargison ND. Pharmaceutical control of endoparasitic helminth infections in sheep. Vet Clin Food Anim. 2011;27:139–56.
12. Shoop WL, DeMontigny P, Fink DW, Williams JB, Egerton JR, Mrozik H, Fisher MH, Skelly BJ, Turner MJ.. Efficacy in sheep and pharmacokinetics in cattle that led to the selection of eprinomectin as a topical endectocide for cattle. Int J Parasitol. 1996;26:1227–35.
13. Shoop W, Soll M. Ivermectin, abamectin and eprinomectin. In: Vercruysse J, Rew R. (Eds), Macrocyclic Lactones in Antiparasitic Therapy, CABI Publishing, Oxon, UK. 2002;pp. 1–29.
14. Panitz E, Godfrey RW, Dodson RE. Resistance to ivermectin and the effect of topical eprinomectin on faecal egg counts in St Croix white hair sheep. Vet Res Comm. 2002;26:443–6.
15. Cringoli G, Rinaldi L, Veneziano V, Capelli G. Efficacy of eprinomectin pour-on against gastrointestinal nematode infections in sheep. Vet Parasitol. 2003;112:203–9.
16. Hoste H, Lespine A, Lemercier P, Alvinerie M, Jacquiet P, Dorchies P. Efficacy of eprinomectin pour-on against gastrointestinal nematodes and the nasal bot fly (Oestrus ovis) in sheep. Vet Rec. 2004;154:782–5.
17. Habela M, Moreno A, Gragera-Slikker A, Gomez JM, Montes G, Rodriguez P, Alvinerie M. Efficacy of eprinomectin pour-on in naturally Oestrus ovis infested Merino sheep in Extremadura. South-West Spain Parasit Res. 2006; 99:275–80.
18. Imperiale F, Pis A, Sallovitz J, Lifschitz A, Busetti M, Suárez V, Lanusse C. Pattern of eprinomectin milk excretion in dairy sheep unaffected by lactation stage: comparative residual profiles in dairy products. J Food Prot. 2006;69:2424–9.
19. Hodošček L, Grabnar I, Milčinski L, Süssinger A, Eržen NK, Zadnik T, Pogačnik M, Cerkvenik-Flajs V. Linearity of eprinomectin pharmacokinetics in lactating dairy sheep following pour-on administration: excretion in milk and exposure of suckling lambs. Vet Parasitol. 2008;154:129–36.
20. Kırcalı Sevimli F, Kozan E, Doğan N. Efficacy of eprinomectin pour-on treatment in sheep naturally infected with Dictyocaulus filaria and Cystocaulus ocreatus. J Helminthol. 2001;85:472–5.
21. Pandya AJ, Ghodke KM. Goat and sheep milk products other than cheeses and yoghurt. Small Ruminant Res. 2007;68:193–206.
22. Vercruysse J, Holdsworth P, Letonja T, Barth D, Conder G, Hamamoto K, Okano K. International harmonisation of anthelmintic efficacy guidelines. Vet Parasitol. 2001;96:171–93.
23. Wood IB, Amaral NK, Bairden K, Duncan JL, Kassai T, Malone JB Jr, Pankavich JA, Reinecke RK, Slocombe O, Taylor SM, Vercruysse J. World Association for the Advancement of Veterinary Parasitology (W.A.A.V.P.) second edition of guidelines for evaluating the efficacy of anthelmintics in ruminants (bovine, ovine, caprine). Vet Parasitol. 1995;58:181–213.
24. MAFF. Manual of Veterinary Parasitological Laboratory Techniques, Reference Book 418. London: Her Majesty's Stationery Office; 1986.
25. Cringoli G, Rinaldi L, Maurelli MP, Utzinger J. FLOTAC: new multivalent techniques for qualitative and quantitative copromicroscopic diagnosis of parasites in animals and humans. Nat Protoc. 2010;5:503–16.
26. Van Wyk JA, Cabaret J, Michael LM. Morphological identification of nematode larvae of small ruminants and cattle simplified. Vet Parasitol. 2004;119:277–306.
27. Stevenson LA, Gasser RB, Chilton NB. The ITS-2 rDNA of Teladorsagia circumcincta, T. trifurcata and T. davtiani (Nematoda: Trichostrongylidae) indicates that these taxa are one species. Int J Parasitol. 1996;26:1123–6.
28. Leignel V, Cabaret J, Humbert JF. New molecular evidence that Teladorsagia circumcincta (Nematoda: Trichostrongylidea) is a species complex. J Parasitol. 2002;88:135–40.
29. Rehbein S, Visser M, Kellermann M. Letendre L (2012) Reevaluation of efficacy against nematode parasites and pharmacokinetics of topical eprinomectin in cattle. Parasitol Res. 2012;111:1343–7.
30. Papadopoulos E, Arsenos G, Sotiraki S, Deligiannis C, Lainas T, Zygoyiannis D. The epizootiology of gastrointestinal nematode parasite in Greek dairy breeds of sheep and goats. Small Ruminant Res. 2003;47:193–202.
31. Torina A, Dara S, Marino AMF, Sparagano OAE, Vitale F, Reale S, Caracappa S. Study on gastrointestinal nematodes of Sicilian sheep and goats. Ann N Y Acad Sci. 2004;1026:187–94.
32. Uriarte J, Llorente MM, Valderrábano J. Seasonal changes of gastrointestinal nematode burden in sheep under an intensive grazing system. Vet Parasitol. 2004;118:79–92.
33. Rehbein S, Kollmannsberger M, Visser M, Winter R. Untersuchungen zum Helminthenbefall von Schlachtschafen in Oberbayern. 1. Artenspektrum, Befallsextensität und Befallsintensität. Berl Münch Tierärztl Wochenschr. 1996;109:161–7.
34. Rehbein S, Visser M, Winter R. Ein Beitrag zur Kenntnis des Endoparasitenbefalls der Schafe auf der Schwäbischen Alb. Dtsch Tierärztl Wochenschr. 1998;105:419–24.
35. Rehbein S, Visser M, Winter R. Ein Beitrag zur Kenntnis des Parasitenbefallsvon Bergschafen aus dem Oberpinzgau (Salzburg). Mitt Österr Ges Tropenmed Parasitol. 1999;21:99–106.
36. Makovcová K, Langrová I, Vadljech J, Jankovská I, Lytvynets A, Borkovcová M. Linear distribution of nematodes in the gastrointestinal tract of tracer lambs. Parasitol Res. 2008;104:123–6.
37. Burgess CGS, Bartley Y, Redman E, Skuce PJ, Nath M, Whitelaw F, Tait A, Gilleard JS, Jackson F. A survey of the trichostrongylid nematode species present on UK sheep farms and associated anthelmintic control practices. Vet Parasitol. 2012;189:299–307.
38. Domke AV, Chartier C, Gjerde B, Leine N, Vatn S, Stuen S. Prevalence of gastrointestinal helminths, lungworms and liver fluke in sheep and goats in Norway. Vet Parasitol. 2012;194:40–8.
39. Rehbein S, Kellermann M, Wehner TA. Pharmacokinetics and anthelmintic efficacy of topical eprinomectin in goats prevented from grooming. Parasitol Res. 2014;113:4039–44.
40. Hamel D, Visser M, Kellermann M, Kvaternick V, Rehbein S. Anthelmintic efficacy and pharmacokinetics of pour-on eprinomectin (1 mg/kg body weight) against gastrointestinal and pulmonary nematode infections in goats. Small Ruminant Res. 2015;127:74–9.
41. Rose H, Rinaldi L, Bosco A, Mavrot F, de Waal T, Skuce P, Charlier J, Torgerson PR, Hertzberg H, Hendrickx G, Vercruysse J, Morgan ER. Widespread anthelmintic resistance in European farmed ruminants: a systematic review. Vet Rec. 2015;176:546.
42. Falzon LC, O'neill TJ, Menzies PI, Peregrine AS. Jones Bitton A, vanLeeuwen J, Mederos A. A systematic review and meta-analysis of factors associated with anthelmintic resistance in sheep. Prev Vet Med. 2014;117:388–402.
43. Sargison N. Responsible use of anthelmintics for nematode control in sheep and cattle. In Practice. 2011;33:318–27.
44. Rinaldi L, Morgan ER, Bosco A, Coles GC, Cringoli G. The maintenance of anthelmintic efficacy in sheep in a Mediterranean climate. Vet Parasitol. 2014;203:139–43.
45. Rehbein S, Lindner T, Kollmannsberger M, Winter R, Visser M. Untersuchungen

Eprinomectin pour-on (EPRINEX® Pour-on, Merial): efficacy against gastrointestinal and pulmonary...

129

zum Helminthenbefall von Schlachtschafen in Oberbayern. 3. Verteilung der Siedlungsorte der Dickdarmnematoden beim Schaf. Berl Münch Tierärztl Wschr. 1997;110:223–8.

46. Hörchner F. Versuche zur Immunisierung von Schafen gegen Chabertia ovina und Untersuchungen über die Spezifität der Antigene dieser Wurmart. Habilitation thesis, Faculty of Veterinary Medicine, Free University Berlin; 1967.

47. Hörchner F. Immunologische Untersuchungen an *Chabertia ovina*- bzw. *Oesophagostomum venulosum*-befallenen Schafen. Zschr Parasitenk. 1968;31:6–7.

48. Bürger HJ. Parasitosen der Wiederkäuer. Helminthen. In: Eckert J, Kutzer E, Rommel M, Bürger HJ, Körting W (Eds), Veterinärmedizinische Parasitologie, Verlag Paul Parey, Berlin and Hamburg. 1992;pp. 174–323.

49. Cringoli G, Rinaldi L, Veneziano V, Capelli G, Rubino R. Effectiveness of eprinomectin pour-on against gastrointestinal nematodes of naturally infected goats. Small Ruminant Res. 2004;55:209–13.

14

Characteristics of extended-spectrum β-lactamase–producing *Escherichia coli* isolated from fecal samples of piglets with diarrhea in central and southern Taiwan in 2015

Wan-Chen Lee[1] and Kuang-Sheng Yeh[1,2*]

Abstracts

Background: The production of extended-spectrum β-lactamases (ESBLs) confer resistance to the commonly used beta-lactam antimicrobials and ESBL–producing bacteria render treatment difficulty in human and veterinary medicine. ESBL–producing bacteria have emerged in livestock in recent years, which may raise concerns regarding possible transfer of such bacteria through the food chain. The swine industry is important in Taiwan, but investigations regarding the status of ESBL in swine are limited.

Results: We collected 275 fecal swab samples from piglets with diarrhea in 16 swine farms located in central and southern Taiwan from January to December 2015 and screened them for ESBL–producing *Escherichia coli*. ESBL producers were confirmed phenotypically by combination disc test and genotypically by polymerase chain reaction and DNA sequencing. The occurrence rate of ESBL–producing *E. coli* was 19.7% (54 of 275), and all were obtained in swine farms located in southern Taiwan. $bla_{CTX-M-1-group}$ and $bla_{CTX-M-9-group}$ were the two bla_{CTX-M} groups found. $bla_{CTX-M-55}$ (34 of 54; 63.0%) and $bla_{CTX-M-15}$ (16 of 54; 29.6%), which belong to the $bla_{CTX-M-1-group}$, were the two major *bla* gene types, whereas $bla_{CTX-M-65}$ was the only type found in the $bla_{CTX-M-9\ group}$. Twenty-seven strains contained bla_{TEM-1}, and the other 27 strains contained $bla_{TEM-116}$. One strain found in Pingtung harbored three *bla* genes: $bla_{TEM-116}$, $bla_{CTX-M-55}$, and $bla_{CTX-M-65}$. ESBL–producing *E. coli* exhibited a multidrug-resistant phenotype, and multilocus sequence typing revealed that the ST10 clonal complexes, including ST10, 167, 44, and 617 accounted for 35% (19 of 54) of these strains.

Conclusions: ESBL-producing *E. coli* from piglets with diarrhea were isolated from swine farms located in southern Taiwan. The most commonly detected *bla* were $bla_{CTX-M-15}$ and $bla_{CTX-M-55}$. The ST10 clonal complexes comprised most of our ESBL-producing *E. coli* strains. Fecal shedding from swine may contaminate the environment, resulting in public health concerns; thus, continued surveillance of ESBL is essential in swine and in other food animals.

Keywords: Extended spectrum β-lactamase, *Escherichia coli*, Multilocus sequence typing

* Correspondence: ksyeh@ntu.edu.tw
[1]Department of Veterinary Medicine, School of Veterinary Medicine, College of Bio-Resources and Agriculture, National Taiwan University, Taipei 106, Taiwan
[2]National Taiwan University Veterinary Hospital, Taipei 106, Taiwan

Background

Diarrhea is a common clinical syndrome in the swine industry and may be classified into three entities. Sucking piglets usually exhibit neonatal diarrhea a few days after birth, and young piglet diarrhea occurs from the first week after birth to weaning [1]. Older piglets, commonly 2 weeks after weaning, may contract post-weaning diarrhea. Neonatal and post-weaning diarrhea are caused by pathogenic *Escherichia coli*, and causative agents of young piglet diarrhea may include transmissible gastroenteritis virus, rotavirus, coccidia, and *E. coli* [1]. Occurrence of diarrheal disease can be reduced by the vaccination of sows to let piglets obtain maternal antibodies. However, measures such as the use of antibiotic supplements in feed are also frequently practiced along with vaccination to reduce the incidence of diarrhea. If prudent usage of antibiotics is not taken into consideration, the massive, indiscriminate, and long-term use of antibiotics in veterinary practice may contribute to the selection and spread of drug-resistant bacteria [2, 3].

Production of extended-spectrum β-lactamases (ESBLs) confers resistance to the frequently used beta-lactam antimicrobial agents, including the third-generation cephalosporins such as ceftriaxone, ceftazidime, and ceftiofur. However, ESBLs are inhibited by the β-lactamases inhibitors clavulanic acid, sulbactam, and tazobactam [4]. TEM, SHV, and CTX-M-types are the three major families of ESBL [4]. All CTX-M-types enzymes are ESBLs, whereas the TEM- and SHV- types of ESBL arise by point mutation at specific residues from the natural TEM-1/TEM-2 and SHV-1 β-lactamase [5]. The production of ESBLs is mainly plasmid mediated, and such plasmids often carry genes that encode resistance to other classes of antimicrobials, such as fluoroquinolones and aminoglycosides [6]. ESBLs are widely distributed in *Enterobacteriaceae*, particularly in *E. coli*, and the rapid emergence and spread of ESBL-producing *E. coli* have been reported in food animals globally [7]. Such findings raise concerns about the possible transfer of ESBL producers through the food chain, thus presenting a hazard to public health [8].

The status of ESBL-producing *E. coli* in food animals in Taiwan has only been reported in cows [9]. Although a foot and mouth disease outbreak in 1997 had a great impact on the swine industry [10], swine are still among the most important agricultural products in Taiwan. The objective of this study is to analyze the fecal carriage of ESBL-producing *E. coli* isolated from piglets with diarrhea in 16 pig farms located in central and southern Taiwan. It is important to screen for the ESBL producers from food animals such as swine from a public health perspective.

Methods

Sample collection

A total of 275 fecal swab samples were collected from the piglets with diarrhea before weaning from 16 swine farms in Taiwan (one in Taichung, one in Nantou, one in Chunghua, two in Yunlin, four in Chiayi, one in Tainan, and six in Pingtung) from January to December 2015. These 16 farms belong to the same swine industry corporation. Isolating *E. coli* from feces in piglets with diarrhea and preparing "tailored vaccine" has been routinely practiced in these farms. These *E. coli* were cultured, inactivated and used as a vaccine component to feed pregnant sows. Neonatal piglets will presumably obtain maternal antibodies when sucking colostrum. Occurrence of neonatal diarrhea due to *E. coli* infection may be reduced as long as piglets have enough maternal antibody. We shared these fecal swab samples and inoculated on CHROMagar ESBL (CHROMagar, Paris, France) to screen for ESBL-producing *E. coli*. Any pink colony that appeared on the agar after incubation at 37 °C for 16–18 h was initially designated as ESBL-producing *E. coli* since they were resistant to cefotaxime and/or ceftazidime, and its identity as *E. coli* was confirmed with the RapID™ ONE System (RapID™, Lenexa, KS, USA). Confirmed *E. coli* strains were stored at –80 °C for further study.

ESBL testing

E. coli isolates were tested phenotypically for ESBL production by combination disc tests with cefotaxime and ceftazidime (30 μg), with and without clavulanic acid (10 μg), as stated by the guidelines of the Clinical and Laboratory Standards Institute [11]. The tested *E. coli* strains were plated on Muller-Hinton agar at a concentration of 0.5 McFarland standards and grown at 35 °C for 16–18 h. A difference of 5 mm or more in the inhibition zones for at least one cefotaxime or ceftazidime/clavulanic acid combination versus the corresponding cefotaxime or ceftazidime alone was used to define an ESBL producer. *Klebsiella pneumoniae* ATCC 700603 and *E. coli* ATCC 25922 were used as the positive and negative controls, respectively [11].

Detection of *bla* genes

The *E. coli* strains that were phenotypically confirmed to be ESBL producers were examined with polymerase chain reaction (PCR) to detect their *bla* genes. The tested strains were cultured for 16–18 h at 37 °C on tryptic soy agar plates (Difco/Becton Dickinson, Franklin Lakes, NJ, USA), and a loopful of bacterial cells was resuspended in 200 μL ddH$_2$O and boiled for 10 min [12]. After centrifugation at 12,000 *g* for 10 min, the supernatant was saved as the source of template DNA for PCR. The primer sequences used to amplify *bla*$_{TEM}$,

bla_{SHV}, $bla_{CTX-M-1-group}$, $bla_{CTX-M-2-group}$, $bla_{CTX-M-8-group}$, $bla_{CTX-M-9-group}$, and $bla_{CTX-M-25-group}$, the annealing temperature, and the expected PCR product sizes are specified in Table 1. The PCR cycling program was set as follows using a LifeEco thermocycler (Bioer Technology, Hangzhou, China): initial denaturation at 95 °C for 5 min, followed by 35 cycles at 95 °C for 30 s, then the annealing temperature specified in Table 1 for 40 s, and 72 °C extension for 1 min. The reaction was then maintained at 72 °C for 10 min. Ten microliters of each PCR sample was loaded onto a 1.2% agarose gel and electrophoresed at 100 volts for 40 min. The gels were then stained with a fluorescent nucleic acid dye (Biotium, Hayward, CA, USA) and examined under a blue light LED illuminator (Smobio, Hsinchu City, Taiwan). The PCR products were then sliced from the agarose gel and subjected to further purification and sequenced by ABI 3130 x1 Genetic Analyzer (Applied Biosystems, Foster, CA, USA) in Center for Genomic Medicine, National Cheng Kung University, Tainan, Taiwan. The results were analyzed with MEGA 6.0 and examined with the NCBI BLAST program (http://www.ncbi.nlm.nih.gov/blast/) and β-lactamase database (http://www.ncbi.nlm.nih.gov/pathogens/submit-beta-lactamase).

Antimicrobial susceptibility testing

The ESBL-producing *E. coli* strains were tested for susceptibility to antimicrobial agents using the disc agar diffusion method [11]. The antimicrobial agents tested included amikacin 30 µg, ampicillin 10 µg, amoxyclav 30 µg, ceftiofur 30 µg, cephalothin 30 µg, ciprofloxacin 10 µg, doxycycline 30 µg, enrofloxacin 5 µg, florfenicol 30 µg, gentamicin 30 µg, nalidixic acid 30 µg, streptomycin 10 µg, co-trimoxazole 25 µg, tetracycline 10 µg; all of the discs were purchased from Oxoid (Oxoid, Hampshire, UK).

E. coli genotyping

Our *E. coli* strains were analyzed genotypically by multilocus sequence typing. DNA fragments derived from *adk*, *fumC*, *gyrB*, *icd*, *mdh*, *purA*, and *recA* were amplified by PCR, sequenced, and then uploaded to the MLST website (http://enterobase.warwick.ac.uk/) for comparison [13]. Phylogenetic analysis was performed using BioNumerics Software version 7.0 (Applied Maths, Sint-Martens-Latem, Belgium).

Results

Fifty-four samples exhibited pink colonies on CHROMagar ESBL, initially indicating an identity of ESBL−producing *E. coli*. All of these strains were then confirmed biochemically with RapID™ ONE System as *E. coli*, and they were not hemolytic when grown on blood agar. These 54 strains exhibited the ESBL phenotype when assayed by combination disc tests. From our results, we did not detect any ESBL−producing *E. coli* in diseased piglets from any of the five swine farms in Taichung, Nantou, Chunghua, and Yunlin, which were located in central Taiwan. ESBL−producing *E. coli* were all obtained in swine farms in southern Taiwan, including Chiayi, Tainan, and Pingtung, with the exception of one farm in Chiayi (farm ID CY-3) and one in Pingtung (farm ID PT-3). Geographic distribution of the swine farms and occurrence of ESBL were indicated in Fig. 1. Overall, the occurrence rate of ESBL-producing *E. coli* was 19.7% (54 of 275). Table 2 lists the occurrence of ESBL-producing *E. coli* in 16 swine farms.

Table 1 Sequences of primers used for ESBL gene detection

PCR target	primer	Sequences (5′–3′)	Annealing Tm (°C)	Predicted PCR size (bp)	Reference
bla_{TEM}	TEM-F	TCGGGGAAATGTGCGCG	55	972	[37]
	TEM-R	TGCTTAATCAGTGAGGCACC			
bla_{SHV}	SHV-F	GCCTTTATCGGCCCTCACTCAA	54	819	[38]
	SHV-R	TCCCGCAGATAAATCACCACAATG			
$bla_{CTX-M-1-group}$	CTX-M-1-F	CCCATGGTTAAAAAATCACTGC	54	942	[39]
	CTX-M-1-R	CAGCGCTTTTGCCGTCTAAG			
$bla_{CTX-M-2-group}$	CTX-M-2-F	CGACGCTACCCCTGCTATT	52	552	[40]
	CTX-M-2-R	CCAGCGTCAGATTTTTCAGG			
$bla_{CTX-M-8-group}$	CTX-M-8-F	CAAAGAGAGTGCAACGGATG	52	205	[40]
	CTX-M-8-R	ATTGGAAAGCGTTCATCACC			
$bla_{CTX-M-9-group}$	CTX-M-9-F	ATGGTGACAAAGAGAGTGCAAC	55	876	[26]
	CTX-M-9-R	TTACAGCCCTTCGGCGATGATT			
$bla_{CTX-M-25-group}$	CTX-M-25-F	GCACGATGACATTCGGG	52	327	[40]
	CTX-M-25-R	AACCCACGATGTGGGTAGC			

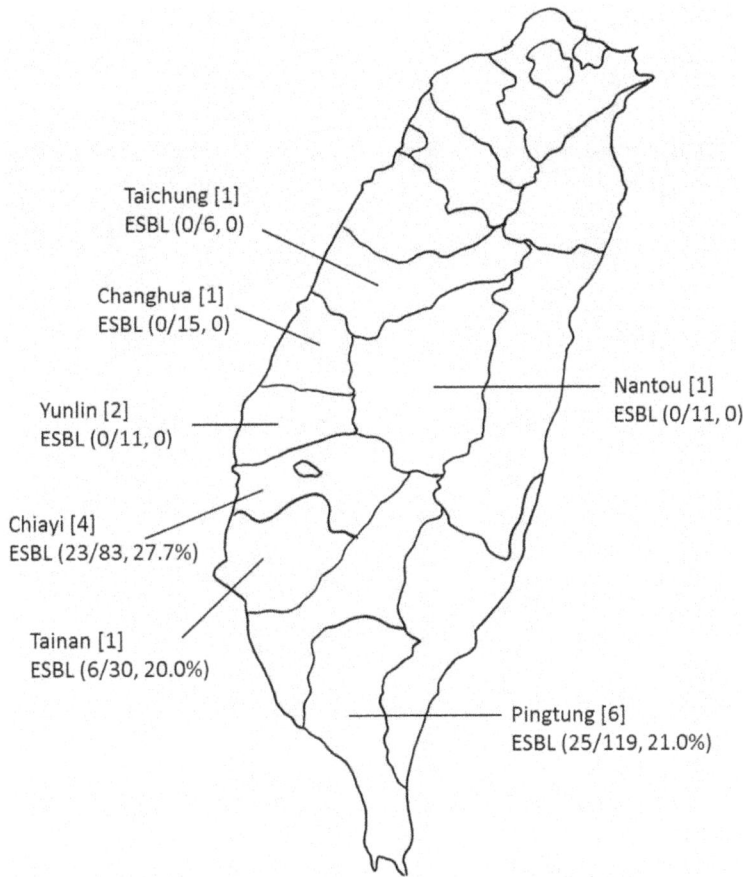

Fig. 1 Geographic distribution of the swine farms in Taiwan included in this study. The number of square brackets indicates the number of swine farms included from each region. The number and occurrence rate of ESBL-producing *E. coli* are denoted in parentheses

Table 2 Occurrence of ESBL–producing *E. coli* in 16 farms

Farm location	Farm ID	No. of fecal samples	No. of ESBL-producing E. coli	Occurrence (%)
Taichung	TC-1	6	0	0
Changhua	CH-1	15	0	0
Nantou	NT-1	11	0	0
Yunlin	YL-1	5	0	0
	YL-2	6	0	0
Chiayi	CY-1	40	19	47.5
	CY-2	8	2	25.0
	CY-3	29	0	0
	CY-4	6	2	33.3
Tainan	TN-1	30	6	20.0
Pingtung	PT-1	10	2	20.0
	PT-2	36	16	44.4
	PT-3	37	0	0
	PT-4	17	2	11.8
	PT-5	11	3	27.3
	PT-6	8	2	25.0

Table 3 lists the *bla* genes and sequence type of ESBL-producing *E. coli*. $bla_{\text{CTX-M-1-group}}$ and $bla_{\text{CTX-M-9-group}}$ were the two $bla_{\text{CTX-M}}$ groups found in ESBL-producing *E. coli*. The $bla_{\text{CTX-M-1-group}}$ contained $bla_{\text{CTX-M-55}}$ (34 of 54, 63.0%) and $bla_{\text{CTX-M-15}}$ (16 of 54, 29.6%), whereas $bla_{\text{CTX-M-65}}$ was the only type found from the $bla_{\text{CTX-M-9 group}}$. All 54 strains contained bla_{TEM}; 27 strains had $bla_{\text{TEM-1}}$ and the other 27 strains contained $bla_{\text{TEM-116}}$. One strain found in Pingtung harbored $bla_{\text{TEM-116}}$, $bla_{\text{CTX-M-55}}$, and $bla_{\text{CTX-M-65}}$. The $bla_{\text{CTX-M-2-group}}$, $bla_{\text{CTX-M-8-group}}$, $bla_{\text{CTX-M-25-group}}$, and bla_{SHV} types of ESBL were not detected in this study.

The results of the antibiotic susceptibility testing of the ESBL-producing *E. coli* isolated from Chiayi, Tainan, and Pingtung are shown in Table 4. The susceptibility testing showed that all 54 ESBL positive isolates were resistant to five antibiotics: ampicillin, cephalothin, ceftiofur, tetracycline, and enrofloxacin. Amikacin and gentamicin were active against 31 strains (57.4%) and 17 strains (31.5%) of ESBL producers, respectively. Overall, ESBL-producing *E. coli* exhibited a multi-drug-resistant phenotype.

Table 3 The *bla* genes and sequence type of ESBL-producing *E. coli*

Farm location	*bla* gene	Sequence Type
Chiayi	bla_{TEM-1} + $bla_{CTX-M-15}$ ($n = 9$)	4981 ($n = 5$), 1638 ($n = 1$), 3268 ($n = 1$), NT[a] ($n = 2$)
	$bla_{TEM-116}$ + $bla_{CTX-M-15}$ ($n = 3$)	1638 ($n = 1$), NT ($n = 2$)
	bla_{TEM-1} + $bla_{CTX-M-55}$ ($n = 5$)	167 ($n = 4$), NT ($n = 1$)
	bla_{TEM-6} + $bla_{CTX-M-55}$ ($n = 6$)	4981 ($n = 1$), 349 ($n = 2$), 44 ($n = 2$), NT ($n = 1$)
Tainan	bla_{TEM-1} + $bla_{CTX-M-55}$ ($n = 2$)	NT ($n = 2$)
	$bla_{TEM-116}$ + $bla_{CTX-M-55}$ ($n = 3$)	457 ($n = 2$), 617 ($n = 1$)
	$bla_{TEM-116}$ ($n = 1$)	NT ($n = 1$)
Pingtung	bla_{TEM-1} + $bla_{CTX-M-15}$ ($n = 2$)	617 ($n = 2$)
	bla_{TEM-1} + $bla_{CTX-M-55}$ ($n = 9$)	10 ($n = 3$), 648 ($n = 1$), 69 ($n = 3$), 617 ($n = 1$), NT ($n = 1$)
	$bla_{TEM-116}$ + $bla_{CTX-M-15}$ ($n = 2$)	10 ($n = 1$), 349 ($n = 1$)
	$bla_{TEM-116}$ + $bla_{CTX-M-55}$ ($n = 8$)	NT ($n = 1$), 10 ($n = 1$), 457 ($n = 2$), 617 ($n = 1$), 167 ($n = 1$), 69 ($n = 1$), 38 ($n = 1$)
	$bla_{TEM-116}$ + $bla_{CTX-M-55}$ + $bla_{CTX-M-65}$ ($n = 1$)	NT ($n = 1$)
	$bla_{TEM-116}$ ($n = 3$)	44 ($n = 1$), 167 ($n = 1$), NT ($n = 1$)

[a]: *NT* new type, there was no comparison standard in the databank

The most frequently seen sequence type of ESBL-producing *E. coli* was ST167 (ST10 clonal complex; 7 of 54; 13.0%), followed by ST4981 (6 of 54; 11.1%) and ST10 (ST10 clonal complex; 5 of 54; 9.3%). There were four strains of ST617 (ST10 clonal complex, 4/54; 7.4%), ST457, and ST69 (ST69 clonal complex) and three strains of ST44 (ST10 clonal complex; 3 of 54; 5.6%) and ST349 (ST349 clonal complex). ST1638 had two strains (2 of 54; 3.7%). ST38 (ST38 clonal complex), ST3268, and ST648 (ST648 clonal complex) had only one strain each. Nonetheless, we still had 13 strains whose sequence types were not matched to any type in the current databank. Figure 2 indicates the minimal spanning tree of the ESBL-producing *E. coli* STs based on the degree of allele sharing.

Discussion

The ESBL-producing *E. coli* were all obtained from the swine farms located in southern Taiwan. There were five swine farms in central Taiwan (Taichung, Changhua, Yunlin, and Nantou) that participated in our study, and only 43 fecal samples (43 of 275; 15.6%) were collected from piglets with diarrhea and screened for ESBL. Although the scale of these farms was similar to that of those located in southern Taiwan, the hygienic procedures or disease control management of individual farms may contribute to such differences in diarrheal cases. The specificity of this chromogenic agar was 100% because all of the pink colonies, indicative of ESBL-producing *E. coli*, were phenotypically and genotypically positive for ESBL. A previous report also suggested the high sensitivity and specificity of CHROMagar ESBL in the detection of clinical ESBL-producing *Enterobacteriaceae* [14]. However, our strategy may also lose some ESBL producers that could grow on blood agar or MacConkey agar but not on CHROMagar ESBL.

The occurrence of ESBL-producing *E. coli* in food animals has been increasing around the world [2]. For example, more than 40% of the ESBL-producing *E. coli* were detected from piglets with post-weaning diarrhea in Heilongjiang Province, China [15]. The authors also compared their findings with those of a similar study in healthy pigs in China and concluded that ESBL-producing *E. coli* were more commonly found in sick animals [16]. Because we did not investigate the prevalence of ESBL in a healthy swine population, there was no basis of comparison for healthy and diseased swine in Taiwan. Although diseased pigs are not likely to enter slaughter or market, fecal shedding from such pigs can contaminate the piggery environment and provide a reservoir for the exchange of drug-resistance genes [17].

TEM-116–producing *E. coli* was first identified in Korean hospitals in a nationwide survey in 2002 [18]. Consequently, a high prevalence of TEM-116 was also reported in Spain [19]. In animals, TEM-116–producing *E. coli* has been detected in dogs [20, 21]. Our results, to the best of our knowledge, demonstrate for the first time the presence of $bla_{TEM-116}$ genes in the ESBL–producing *E. coli* from porcine origin. Although most of our TEM-116–containing strains also had CTX-M, we did find four *E. coli* strains that harbored only TEM-116 that exhibited an ESBL phenotype. ESBL producers within the CTX-M group are becoming more common [22, 23]. In Europe, CTX-M-1 is broadly disseminated in animals, whereas CTX-M-14 is most prevalent in animals in Asian countries [8]. The most frequently found CTX-M types in our study were CTX-M-15 and CTX-M-55, whereas CTX-M-14 was reported in healthy and

Table 4 Antimicrobial susceptibility test of ESBL–producing E. coli

Antibiotic discs used	Chiayi, n = 23 (%)			Tainan, n = 6 (%)			Pingtung, n = 25 (%)			Total, N = 54 (%)		
	S[a]	I[b]	R[c]	S	I	R	S	I	R	S	I	R
Ampicillin	0	0	23 (100.0)	0	0	6 (100.0)	0	0	25 (100.0)	0	0	54 (100.0)
Amoxicillin/clavulanic acid	3 (13.0)	8 (34.8)	12 (52.2)	2 (33.3)	1 (16.7)	3 (50.0)	6 (24.0)	11 (44.0)	8 (32.0)	11 (20.4)	20 (37.0)	23 (42.6)
Cephalothin	0	0	23 (100.0)	0	0	6 (100.0)	0	0	25 (100.0)	0	0	54 (100.0)
Ceftiofur	0	0	23 (100.0)	0	0	6 (100.0)	0	0	25 (100.0)	0	0	54 (100.0)
Amikacin	15 (65.2)	8 (34.8)	0	5 (83.3)	1 (16.7)	0	11 (44.0)	3 (12.0)	11 (44.0)	31 (57.4)	12 (22.2)	11 (20.4)
Gentamicin	12 (52.2)	3 (13.0)	8 (34.8)	1 (16.7)	0	5 (83.3)	4 (16.0)	0	21 (84.0)	17 (31.5)	3 (5.6)	34 (62.9)
Streptomycin	0	1 (4.4)	22 (95.6)	0	0	6 (100.0)	0	1 (4.0)	24 (96.0)	0	2 (3.7)	52 (96.3)
Doxycycline	3 (13.0)	10 (43.5)	10 (43.5)	2 (33.3)	2 (33.3)	2 (33.3)	2 (8.0)	9 (36.0)	14 (56.0)	7 (13.0)	21 (38.9)	26 (48.1)
Tetracycline	0	0	23 (100.0)	0	0	6 (100.0)	0	0	25 (100.0)	0	0	54 (100.0)
Nalidixic acid	1 (4.4)	0	22 (95.6)	0	0	6 (100.0)	2 (8.0)	0	23 (92.0)	3 (5.6)	0	51 (94.4)
Ciprofloxacin	1 (4.4)	0	22 (95.6)	0	0	6 (100.0)	3 (12.0)	0	22 (88.0)	4 (7.4)	0	50 (92.6)
Enrofloxacin	0	0	23 (100.0)	0	0	6 (100.0)	0	0	25 (100.0)	0	0	54 (100.0)
Florfenicol	0	1 (4.4)	22 (95.6)	0	0	6 (100.0)	0	4 (16.0)	21 (84.0)	0	5 (9.3)	49 (90.7)
Co-trimoxazole	6 (26.1)	0	17 (73.9)	0	0	6 (100.0)	2 (8.0)	0	23 (92.0)	8 (14.8)	0	46 (85.2)

[a]: susceptible; [b]: intermediate resistant; [c]: resistant

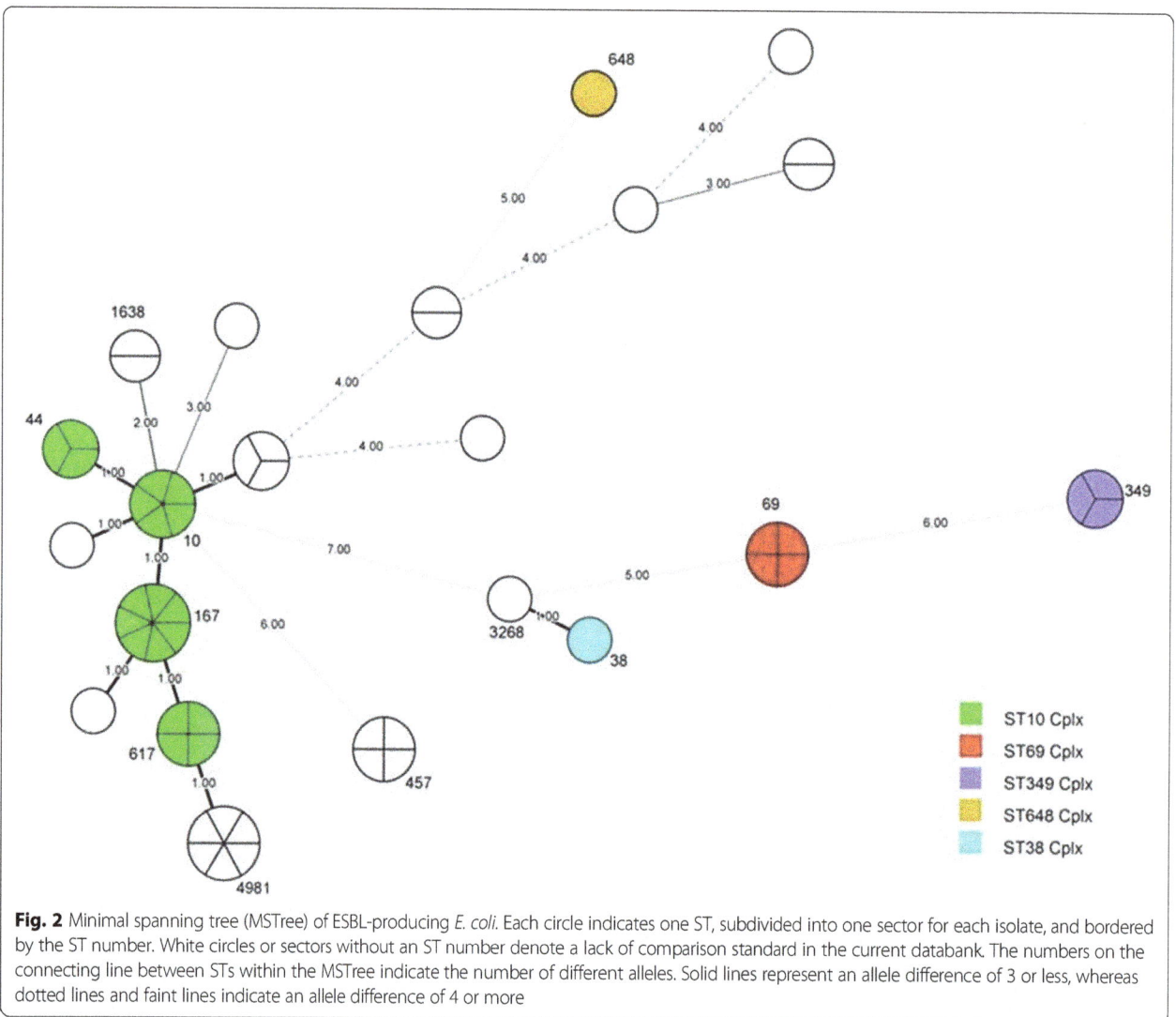

Fig. 2 Minimal spanning tree (MSTree) of ESBL-producing *E. coli*. Each circle indicates one ST, subdivided into one sector for each isolate, and bordered by the ST number. White circles or sectors without an ST number denote a lack of comparison standard in the current databank. The numbers on the connecting line between STs within the MSTree indicate the number of different alleles. Solid lines represent an allele difference of 3 or less, whereas dotted lines and faint lines indicate an allele difference of 4 or more

diseased swine in Korea and China [15, 16, 24]. CTX-M-55 was first isolated from patients in a hospital in Thailand; this novel CTX-M type was derived from CTX-M-15, with only a single substitution of valine instead of alanine at residue 77 [25]. The incidence of CTX-M-55 has been reported to exceed that of CTX-M-15 in outpatient infection cases in Chinese county hospitals [26]. Thus, the authors of that study hypothesized an animal-human transfer of CTX-M-55 because most of these outpatients in county hospitals live in rural areas and thus have more chances to come into contact with infected food animals and farm sewage [26]. In our study, CTX-M-55 was the predominant ESBL type isolated from swine with diarrhea in southern Taiwan. CTX-M-55–producing *E. coli* has also been detected in the milk of cows with clinical mastitis in the same region [9]. It is conceivable that ESBL-producing *E. coli* that possess CTX-M-55 have spread to the environment. One strain obtained in Pingtung possessed CTX-M-65

in addition to CTX-M-55 and TEM-116. Although the detection rate of CTX-M-65 was low compared to those of CTX-M-55 and CTX-M-15, the presence of CTX-M-65 has been reported in humans, animals, and vegetables [15, 24, 27]. These findings underscore the importance of screening and investigation of the genotypes of ESBL producers in food animals on a regular basis. Our investigation did not detect any CTX-M-8 or CTX-M-25, which were also not detected in previous studies [15, 28].

Antimicrobial susceptibility test revealed a high frequency of the resistance of ESBL-producing *E. coli* to most antimicrobial agents. Inappropriate use or overuse of antimicrobial agents, including third-generation cephalosporins, may be associated with the emergence of ESBL-producing *E. coli* in swine [29]. The selection of CTX-M-producing *E. coli* in swine by treatment with ceftiofur has been documented [30]. It is worthwhile to consider banning the use of third-generation of cephalosporins such as ceftiofur in food animals to decrease the occurrence of ESBL producers. For example, the occurrence of ESBL-producing *E. coli* was reduced when third-generation cephalosporins were banned in the Danish pig industry [31]. Forty-two percent of ESBL-producing *E. coli* were resistant to amoxicillin/clavulanic acid. Possible reasons that account for this phenotype may include hyper production of chromosomal class C β-lactamase, possession of plasmid-mediated TEM enzymes, production of oxacillinases, or production of inhibitor-resistant TEM by these isolates [32]. In addition, plasmid mediated AmpC may also cause resistance to amoxicillin/clavulanic acid [8].

The ST10 clonal complexes (ST10, 167, 44, 617) comprised most of our ESBL-producing *E. coli* strains. There were 13 strains that did not match any ST in the current database; however, six of these had only a one to three-allele difference from ST10 clonal complexes. It is fair to say that ST10 was the dominant clonal complex in our study. A recent investigation indicated that ESBL-producing *E. coli* were commonly isolated from river waters in southern Taiwan and that ST10 and ST58 was the most frequently found clonal complexes [33]. The authors of that study also observed a substantial association of these ESBL-producing *E. coli* with the presence of chicken farms at that region. Geographically, food animal farms, including swine, chicken, and cattle, are primarily situated in southern Taiwan. It is conceivable that livestock may spread ESBL-producing *E. coli* from feces, thus contaminating the environment. We did not detect ESBL-producing *E. coli* ST131 (O25:H4) that possessed CTX-M-15, a leading cause of urinary tract infections and bacteremia in human medicine globally, in our study. However, swine and other food animals may play a role as vectors in the transmission of bacteria to humans [34], so continued surveillance of food animals for ESBL-producing *E. coli* is essential.

Our study has some limitations. Fecal samples from healthy piglets were not collected and there was no comparison for the occurrence of ESBL-producing *E. coli* between the healthy and diseased populations. The virulence factors like K88, K99 or 987 P fimbriae genes in our ESBL-producing *E. coli* isolates were not screened and they were not hemolytic when grown on blood agar. It is possible that the ESBL-producing *E. coli* in the present study was not the causative agent for the diarrhea of these piglets. We did not detect if these *E. coli* isolates produced AmpC-β-lactamases, which also hydrolyze third-generation cephalosporins. AmpC-producing *E. coli* were also found in increasing numbers in food-producing animals [8]. In addition, profiles of resistant plasmids were not characterized in our study. Plasmid analysis methods like PCR-based replicon typing could assign the incompatibility (Inc) groups [35], whereas replicon sequence typing could discriminate IncF plasmid variants [36]. Inclusion of plasmid characterization could have provided insights into the epidemiology of the ESBL plasmid in our study.

Conclusions

ESBL-producing *E. coli* from piglets with diarrhea were isolated from swine farms located in Chiayi, Tainan, and Pingtung. $bla_{\text{CTX-M-15}}$ and $bla_{\text{CTX-M-55}}$ were the most commonly detected *bla* genes. The ST10 clonal complexes comprised most of our ESBL-producing *E. coli* strains. Fecal shedding from swine may contaminate the environment, from a public health perspective, continued surveillance of ESBL is essential in swine and in other food animals.

Abbreviations
ATCC: American type culture collection; BLAST: Basic local alignment search tool; CTX-M: Cefotaximase-Munich; ESBL: Extended-spectrum β-lactamase; h: hour; MEGA 6.0: Molecular evolutionary genetics analysis software version 6.0; min: minute; MLST: Multilocus sequence typing; NCBI: National center for biotechnology information; PCR: Polymerase chain reaction; s: second; SHV: Sulfhydryl variable; ST: Sequence type; TEM: Temoneira

Acknowledgements
The authors would like to thank Hsiuo-Tung Yeh for Fig. 1 graphic drawing and Dr. Lee-Jene Teng, Department of Clinical Laboratory Sciences and Medical Biotechnology, National Taiwan University, for providing *Klebsiella pneumoniae* ATCC 700603.

Funding
This research was supported by National Taiwan University, No. G049919.

Authors' contributions
W-CL performed all the experiments and was a major contributor in writing up this manuscript. K-SY coordinated this study and also helped draft this manuscript. Both authors read and approved the final manuscript.

Competing interests
The authors declare that they have no competing interests.

Consent for publication
Not applicable.

References
1. Biehl LG, Hoefling DC. Diagnosis, treatment, and prevention of diarrhea in 7-to 14-day-old pigs. J Am Vet Med Assoc. 1986;188:1144–6.
2. Seiffert SN, Hilty M, Perreten V, Endimiani A. Extended-spectrum cephalosporin-resistant gram-negative organisms in livestock; an emerging problem for human health. Drug Resist Updat. 2013;16:22–45.
3. Marshall BM, Levy SB. Food animals and antimicrobials: impacts on human health. Clin Microbiol Rev. 2011;24:718–33.
4. Bush K, Jacoby GA. Updated functional classification of β-lactamases. Antimicrob Agents Chemother. 2010;54:969–76.

5. Gniadkowski M. Evolution of extended-spectrum β-lactamase by mutation. Clin Microbiol Infect. 2008;14 Suppl 1:11–32.

6. Pitout JD, Laupland KB. Extended-spectrum β-lactamase-producing Enterobacteriaceae: an emerging public-health concern. Lancet Infect Dis. 2008;8:159–66.

7. Smet A, Martel A, Persoons D, Dewulf J, Heyndrickx M, Herman L, Haesebrouck F, Butaye P. Broad-spectrum β-lactamases among Enterobacteriaceae of animal origin: molecular aspects, mobility and impact on public health. FEMS Microbiol Rev. 2010;34:295–316.

8. Ewers C, Bethe A, Semmler T, Guenther S, Wieler LH. Extended-spectrum β-lactamase-producing and AmpC-producing Escherichia coli from livestock and companion animals, and their putative impact on public health: a global perspective. Clin Microbiol Infect. 2012;18:646–55.

9. Su Y, Yu CY, Tsai Y, Wang SH, Lee C, Chu C. Fluoroquinolone-resistant and extended-spectrum beta-lactamase-producing Escherichia coli from the milk of cows with clinical mastitis in Southern Taiwan. J Microbiol Immunol Infect. 2014. doi: 1016/j.jmii.2014.1010.1003.

10. Shieh HK. The FMD, situiation in Taiwan. J Chin Soc Vet Sci. 1997;23(5):395–402.

11. CLSI. Performance standards for antimicrobial susceptibility testing: twenty-first informational supplement M100-21. Wayne, PA, USA: CLSI; 2011.

12. Shaheen BW, Oyarzabal OA, Boothe DM. The role of class 1 and 2 integrons in mediating antimicrobial resistance among canine and feline clinical E. coli isolates from the US. Vet Microbiol. 2010;144:363–70.

13. Wirth T, Falush D, Lan R, Colles F, Mensa P, Wieler LH, Karch H, Reeves PR, Maiden MCJ, Ochman H, et al. Sex and virulence in Escherichia coli: an evolutionary perspective. Mol Microbiol. 2006;60:1136–51.

14. Saito R, Koyano S, Nagai R, Okamura N, Moriya K, JKoike K. Evaluation of a chromogenic agar medium for the detection of extended-spectrum β-lactamase-producing Enterobacteriaceae. Lett Appl Microbiol. 2010;51: 704–6.

15. Xu G, An W, Wang H, Zhang X. Prevalence and characteristics of extended-spectrum β-lactamase genes in Escherichia coli isolated from piglets with post-weaning diarrhea in Heilongjiang province, China. Front Microbiol. 2015;6:1103.

16. Zheng H, Zeng Z, Chen S, Liu Y, Yao Q, Deng Y, Chen X, Lv L, Zhuo C, Chen Z, et al. Prevalence and characterisation of CTX-M β-lactamases amongst Escherichia coli isolates from healthy food animals in China. Int J Antimicrob Agents. 2012;39:305–10.

17. Whitehead TR, Cotta MA. Stored swine manure and swine faeces as reservoirs of antibiotic resistance genes. Lett Appl Microbiol. 2013;56:264–7.

18. Jeong SH, Bae IK, Lee JH, Sohn SG, Kang GH, Jeon GJ, Kim YH, Jeong BC, Lee SH. Molecular characterization of extended-spectrum beta-lactamases produced by clinical isolates of Klebsiella pneumoniae and Escherichia coli from a korean nationwide survey. J Clin Microbiol. 2004;42:2902–6.

19. Romero EDV, Padilla TP, Hernández AH, Grande RP, Vázquez MF, García IG, García-Rodríguez JA, Bellido JLM. Prevalence of clinical isolates of Escherichia coli and Klebsiella spp. producing multiple extended-spectrum β-lactamases. Diagn Microbiol Infect Dis. 2007;59:433–7.

20. Ewers C, Grobbel M, Stamm I, Kopp PA, Diehl I, Semmler T, Fruth A, Beutlich J, Guerra B, Wieler LH, et al. Emergence of human pandemic O25: H4-ST131 CTX-M-15 extended-spectrum- β-lactamase-producing Escherichia coli among companion animals. J Antimicrob Chemother. 2010;65:651–60.

21. Rzewuska M, Stefanska I, Kizerwetter-Swida M, Chrobak-Chmiel D, Szczygielska P, Lesniak M, Binek M. Characterization of extended-spectrum-β-lactamases produced by Escherichia coli strains isolated from dogs in Poland. Pol J Microbiol. 2015;64(3):285–8.

22. Zhao WH, Hu ZQ. Epidemiology and genetics of CTX-M extended-spectrum β-lactamases in Gram-negative bacteria. Crit Rev Microbiol. 2013;39(1): 79–101.

23. Bonnet R. Growing group of extended-spectrum betya-lactamases: the CTX-M enzymes. Antimicrob Agents Chemother. 2004;48:11–4.

24. Tamang MD, Nam HM, Kim SR, Chae MH, Jang GC, Jung SC, Lim SK. Prevalence and molecular characterization of CTX-M β-lactamase-producing Escherichia coli isolated from healthy swine and cattle. Foodborne Pathog Dis. 2013;10:13–20.

25. Kiratisin P, Apisarnthanarak A, Saifon P, Laesripa C, Kitphati R, Mundy LM. The emergence of a novel ceftazidime-resistant CTX-M extended-spectrum beta-lactamase, CTX-M-55, in both community-onset and hospital-acquired infections in Thailand. Diagn Microbiol Infect Dis. 2007;58:349–55.

26. Zhang J, Zheng B, Zhao L, Wei Z, Ji J, Li L, Xiao Y. Nationwide high prevalence of CTX-M and an increase of CTX-M-55 in Escherichia coli isolated from patients with community-onset infections in Chinese county hospitals. BMC Infect Dis. 2014;14:659–69.

27. Riccobono E, Di Pilato V, Di Maggio T, Revollo C, Bartoloni A, Pallecchi L, Rossolini GM. Characterization of IncI1 sequence type 71 epidemic plasmid lineage responsible for the recent dissemination of CTX-M-65 extended-spectrum beta-lactamase in the Bolivian Chaco region. Antimicrob Agents Chemother. 2007;59:5340–7.

28. Geser N, Stephan R, Hächler H. Occurrence and characteristics of extended-spectrum β-lactamase (ESBL) producing Enterobacteriaceae in food producing animals, minced meat and raw milk. BMC Vet Res. 2012;8:21.

29. Barton MD. Impact of antibiotic use in the swine industry. Curr Opin Microbiol. 2014;19:9–15.

30. Cavaco LM, Abatih E, Aarestrup FM, Guardabassi L. Selection and persistence of CTX-M-producing Escherichia coli in the intestinal flora of pigs treated with amoxicillin, ceftiofur, or cefquinome. Antimicrob Agents Chemother. 2008;52:3612–6.

31. Agersø Y, Aarestrup FM. Voluntary ban on cephalosporin use in Danish pig production has effectively reduced extended-spectrum cephalosporinase-producing Escherichia coli in slaughter pigs. J Antimicrob Chemother. 2013; 68:569–72.

32. Leflon-Guibout V, Speldooren V, Heym B, Nicolas-Chanoine M-H. Epidemiological survey of amoxicillin-clavulanate resistance and corresponding molecular mechanisms in Escherichia coli isolates in France: new genetic features of bla$_{TEM}$ genes. Antimicrob Agents Chemother. 2000;44:2709–14.

33. Chen PA, Hung CH, Huang PC, Chen JR, Huang IF, Chen WL, Chiou YH, Hung WY, Wang JL, Cheng MF. Characteristics of CTX-M extended-spectrum β-lactamase-producing Escherichia coli strains isolated from multiple rivers in Southern Taiwan. Appl Environ Microbiol. 2016;82(6):1889–97.

34. Dahms C, Hübner N, Kossow A, Mellmann A, Dittmann K, Kramer A. Occurrence of ESBL-producing Escherichia coli in livestock and farm workers in Mecklenburg-Western Pomerania, Germany. PLoS One. 2015;10:e0143326.

35. Carattoli A, Bertini A, Villa L, Falbo V, Hopkins KL, Threlfall EJ. Identification of plasmids by PCR-based replicon typing. J Microbiol Methods. 2005;63:219–28.

36. Villa L, García-Fernández A, Fortini D, Carattoli A. Replicon sequence typing of IncF plasmids carrying virulence and resistance determinants. J Antimicrob Chemother. 2010;65:2518–29.

37. Stucliffe JG. Nucleotide-sequence of ampicillin resistance gene of Escherichia coli plasmid pBR322. Proc Natl Acad Sci U S A. 1978;75:3737–41.

38. Chia JH, Chu C, Su LH, Chiu CH, Kuo AJ, Sun CF, Wu TL. Development of a multiplex PCR and SHV melting-curve mutation detection system for detection of some SHV and CTX-M-lactamases of Escherichia coli, Klebsiella pneumoniae, and Enterobacter cloacae in Taiwan. J Clin Microbiol. 2005;43:4486–91.

39. Yu Y, Ji S, Chen Y, Zhou W, Wei Z, Li L, Ma Y. Resistance of strains producing extended-spectrum β-lactamases and genotype distribution in China. J Infect. 2007;54:53–7.

40. Woodford N, fagan EJ, Ellington MJ. Multiplex PCR for rapid detection of genes encoding CTX-M extended-spectrum β-lactamases. J Antimicrob Chemother. 2006;57:154–5.

15

Birth delivery method affects expression of immune genes in lung and jejunum tissue of neonatal beef calves

Carla Surlis[1]* ⓘ, Keelan McNamara[1], Eoin O'Hara[1], Sinead Waters[1], Marijke Beltman[2], Joseph Cassidy[2] and David Kenny[1]*

Abstract

Background: Caesarean section is a routine veterinary obstetrical procedure employed to alleviate dystocia in cattle. However, CS, particularly before the onset of labour, is known to negatively affect neonatal respiration and metabolic adaptation in humans, though there is little published information for cattle. The aim of this study was to investigate the effect of elective caesarean section (ECS) or normal trans-vaginal (TV) delivery, on lung and jejunal gene expression profiles of neonatal calves.

Results: Paternal half-sib Angus calves (gestation length 278 ± 1.8 d) were delivered either transvaginally (TV; $n = 8$) or by elective caesarean section (ECS; $n = 9$) and immediately euthanized. Lung and jejunum epithelial tissue was isolated and snap frozen. Total RNA was extracted using Trizol reagent and reverse transcribed to generate cDNA. For lung tissue, primers were designed to target genes involved in immunity, surfactant production, cellular detoxification, membrane transport and mucin production. Primers for jejunum tissue were chosen to target mucin production, immunoglobulin uptake, cortisol reaction and membrane trafficking. Quantitative real-time PCR reactions were performed and data were statistically analysed using mixed models ANOVA. In lung tissue the expression of five genes were affected ($p < 0.05$) by delivery method. Four of these genes were present at lower (*LAP, CYP1A1, SCN11α* and *SCN11β*) and one (*MUC5AC*) at higher abundance in ECS compared with TV calves. In jejunal tissue, expression of *TNFα, Il-1β* and *1 l-6* was higher in ECS compared with TV calves.

Conclusions: This novel study shows that ECS delivery affects the expression of key genes involved in the efficiency of the pulmonary liquid to air transition at birth, and may lead to an increased inflammatory response in jejunal tissue, which could compromise colostral immunoglobulin absorption. These findings are important to our understanding of the viability and management of neonatal calves born through ECS.

Keywords: Mode of delivery, Elective caesarean, Calf, Bovine, Lung, Jejunum, Birth

Background

Presently, caesarean section (CS) is the most common surgical procedure carried out in cattle [1, 2]. Since its introduction at the start of the twentieth century, CS is commonly employed as a final remedy to save the calf and/or the cow in complicated obstetrical cases, such as dystocia as a consequence of malpresentation or foetal oversize of the calf [2–4]. Dystocia represents the most common cause of bovine perinatal mortality, and is a significant issue for subsequent dam reproductive efficiency, with CS providing an essential role in saving both dam and calf during a difficult birth [4, 5]. The use of elective caesarean section (ECS) without an initial attempt at a natural birth, however, has become common practice in a number of breeds [5]. For example, double-muscled breeds of cattle such as the Belgian blue are typically prone to dystocia due to feto-maternal disproportion, with a high proportion of calves born through ECS. Indeed, ECS is commonly carried out in the early stages of parturition before labour has progressed [3, 5–7] In

* Correspondence: carla.surlis@teagasc.ie; david.kenny@teagasc.ie
[1]Animal and Grassland Research and Innovation Centre, Teagasc, Grange, Dunsany, Co. Meath, Ireland

human obstetrics, ECS before the onset of labour is known to negatively affect neonatal health in early life [8–10]. Transition from intrauterine to extrauterine life is a complex process that requires a number of systemic physiological changes to take place in the neonate. Firstly, a rapid and coordinated clearance of foetal lung fluid, and the secretion of surfactants required for successful inspiration are key steps in establishing the transition from placental to gaseous exchange at birth [10, 11]. As gestational full term approaches, fluid production within the foetal lung slows, and there is also a marked increase in the pulmonary expression of sodium membrane channels, which appear to play an important role in liquid absorption [8, 10, 12]. The act of compression of the thorax during vaginal birth also appears to have a role in the expulsion of fluid during natural parturition, but to a lesser extent than the physical reabsorption of fluid at birth [9, 10]. Preterm labour and operative delivery prior to even the earliest stages of labour, have been shown to cause excessive retention of lung fluid in some mammals including preterm rabbits and foetal lambs [13].

Calves are born agammaglobulinemic, and must acquire immunoglobulin (Ig) through passive transfer from dam to neonate via colostrum consumption, which is a critical step in protection against disease [1, 14]. Passive transfer occurs in the small intestine, with reports suggesting the jejunum plays a primary role in this absorption [15–17]. Failure of passive transfer is relatively common, and is a prominent causative agent of neonatal mortality [1]. Colostrum contains additional bioactive factors, including growth factors, which the calf is able to absorb directly for the initial 20–24 h following parturition before the ability for macromolecule absorption ceases [1, 18, 19]. Previous research by Sangild 2003 has suggested that method of delivery may have an effect on the uptake of Ig in the neonatal calves and pigs, however prematurity was an additional variable alongside mode of delivery.

The aim of the research presented in this study was to establish the effect of birth delivery method on the expression of a number of key genes involved in two critical tissues- the lung and the jejunum, of neonatal Angus calves. To the author's knowledge, this is the first study to examine transcript abundance for key genes involved in the immediate post-natal function of these two critically important tissues.

Methods

All procedures involving animals were approved for the use of live animals in experiments by the Teagasc Animal Ethics Committee and were licensed by the Health Products Regulatory Authority in accordance with the Cruelty to Animals Act (Ireland 1897) and European Community Directive 86/609/EC.

Animal model

Commercially purchased 18 month old oestrous synchronised Aberdeen Angus heifers ($n = 21$) were inseminated by AI with frozen thawed semen from one Aberdeen Angus sire. Foetal sex was determined in all calves at day 100 of gestation. All animals were housed during the last 2 months of gestation and allowed *ab libitum* access to a high-energy low forage diet. One week prior to predicted calving date, heifers were blocked on foetal sex to one of 2 groups; TV (calves to be delivered trans-vaginally ($n = 10$) and ECS (calves to be delivered via elective caesarean section ($n = 11$). Heifers in the TV group were induced to calve by administering 2 ml of a prostaglandin F2$_\alpha$ analogue (Estrumate, Merck), to facilitate a staggered calving schedule. Caesarean sections were carried out by an experienced veterinary surgeon using standard protocols. Local anaesthetic (Lidocaine) was used in animals undergoing caesarean section in accordance with the protocol of normal veterinary practice. All calves were euthanized within 5 min of birth by lethal injection (Dolethal (Vetoquinol), Euthatal (Merial)) administered through the jugular vein.

Tissue sample collection

Tissue was harvested within 10 mins of slaughter using sterilized and RNase zap treated instruments. A transverse sample from the midsection of the jejunum was harvested from the small intestine from all calves, washed in DPBS and snap frozen in liquid nitrogen. Lung tissue was sampled from the centre of the right lung from all animals in a consistent manner. Tissue samples were washed in DPBS, and subsequently snap frozen in liquid nitrogen until analysis.

Histology

In order to ensure consistency in cellular content for all samples, histological staining and imaging was conducted prior to transcriptomic analysis. Tissue was sampled from an area approx. 2 cm from the center of the right lung in each animal and placed immediately in 10% buffered formalin (pH 7.4). After 24 h, the samples were processed using an automated processor (Tissue-TEK VIP, Sakura Finetek) and embedded in paraffin wax. Sections of 5 μm were cut using a microtome (Leitz 1512), and mounted onto glass slides. Samples were stained using a haematoxylin and eosin stain and were visualized using Image-Pro Plus software (Version 5, Media Cybernetics) [20].

RNA extraction

Total RNA was isolated from jejunum and lung tissue samples (TV $n = 8$, ECS $n = 9$) using the Qiagen RNeasy plus universal mini kit (Qiagen, UK), according to manufacturer's instructions. We failed to recover samples, due to logistical reasons, from three calves delivered by TV and one calf delivered by ECS. Approximately 75 mg of frozen tissue was used for RNA extraction. RNA quality was determined by measuring the absorbance at 260 nm using a Nanodrop spectrophotometer ND-100, to ensure all samples had an absorbance of between 1.8 and 2.0 (Nanodrop Technologies, Wilmington, DE, USA). RNA quality was assessed on the Agilent Bioanalyser 2100 (Agilent Technologies Ireland Ltd., Dublin, Ireland) using the RNA 6000 Nano lab chip kit, with RIN value of between 8 and 10 were deemed to be of sufficiently high quality. RIN values of samples that did not meet these requirements were further purified using the RNA clean and concentrate kit (Zymo Research, UK), according to manufacturer's instructions.

Complementary DNA synthesis

cDNA was synthesized using a High Capacity cDNA Reverse Transcription kit (Applied Biosystems, Foster City, CA, USA) according to the manufacturer's instructions. 2 μg of total RNA from each sample was reverse transcribed into cDNA using MultiScribe reverse transcriptase. The converted cDNA was quantified by absorbance at 260 nm and stored at –20 °C for subsequent analyses.

Primers

Specific genes involved in immune response and pulmonary liquid-gas transition in neonates were targeted in this study and are outlined in Table 2. All primers for real-time PCR were designed using Primer3web (http://primer3.ut.ee/). Primers were then subjected to BLAST analysis (http://www.ncbi.nlm.nih.gov/) to confirm their specificity and ensure that they were homologous to the bovine sequence. Details of primer sets used in this study are listed in Tables 1 and 2. Primers for reference and target genes were commercially synthesized (Sigma-Aldrich Ireland, Dublin, Ireland). A total of 13 genes were chosen for target in the lung tissue, with the majority of the genes selected to examine alterations in the ability of ECS delivered calves to produce the necessary surfactants and mucin proteins for normal lung function. A number of immune and stress related genes such as LAP and CYP1A1 were also chosen for target, to

Table 1 Sequences of oligonucleotide primers used for qPCR analysis of lung tissue

Gene ID[a]	Primers (5′-3′)	Accession No.	Amplicon length
MUC5AC	Forward: CAGTACAGAGTGCATGGGGA Reverse: TTCACAAACACCTCCCCACT	XM_015470102.1	185
MUC5B	Forward: AAAACGCCCTTCACCTTCAC Reverse: TGCCTCAGGTTCTCGAATGT	XM_015470101.1	176
LAP	Forward: CCCTGGAAGCATGAGACAGA Reverse: TTTCTGACTCCGCATCCAGT	S76279.1	109
SP-A	Forward: TGGGGAGGCATCTTGTTAGG Reverse: TGTTCATCAGCAGGCAGGTA	NM_001077838.2	194
SP-B	Forward: GACATGTGGAAGCCGATGAC Reverse: TGAGTCCTGGAAAATGGCCT	NM_001075311.2	92
SP-C	Forward: GAGATCCAGGAGCAAAGGGT Reverse: CCTCCCACAGTCCCATTTCT	NM_174462.4	95
SP-D	Forward: CCTGTACCCTGGTCATGTGT Reverse: AGCAGAGCCATTGTCTCCTT	NM_181026.2	179
CYP1A1	Forward: GTCACAACTGCCCTTTCCTG Reverse: AAAGGAGGAGTGTCGGAAGG	XM_005192890.3	180
CYP1A2	Forward: TCCTCTTCCTGGCCATCTTG Reverse: CAGAACGCCAGCAACTTCTT	XM_015468527.1	165
ABCA3	Forward: GAGCACACCTTCAACCACAG Reverse: AAAGGAGCCTGTCTGAGTCC	NM_001113746.1	108
SCN11A	Forward: ATGCTGGCTTTAATCTGCGG Reverse: CCTGGAAGCACGAATGGATG	NM_174598.3	177
SCN11B	Forward: CTGAAGGACCTGGACGAACT Reverse: TTGATAAAGACCAGGGGCGT	NM_001098075.1	128
SGK-1	Forward: GGCTCGATTCTATGCTGCTG Reverse: ACGTTGTGCCATTGTGTTCA	NM_001102033.1	121

[a]MUC5AC mucin 5 ac, MUC5B mucin 5B, LAP lingual antimicrobial peptide, SP-A Surfactant protein A, SP-B Surfactant protein B, SP-C surfactant protein C, SP-D surfactant protein D, CYP1A1 cytochrome P45 family 1 subfamily A member 1, CYP1A2 cytochrome P450 family 1 subfamily A member 2, ABCA3 ATP binding cassette subfamily A member 3, SCN11A Sodium channel voltage-gated type 2 alpha subunit, SCN11B Sodium channel voltage-gated type 2 beta subunit, SGK-1 Serum/glucocorticoid regulated kinase 1

Table 2 Sequences of oligonucleotide primers used for qPCR analysis of jejunum tissue

Gene ID[a]	Primers (5'-3')	Accession No.	bp[2]
NR3C1	Forward: TACAGGCAGCAATGGTCTCA Reverse: AAGAGGGTGGTCATTCTGGG	NM_001206634.1	154
MUC1	Forward: CCCAACTCTGTTCTGGGCTA Reverse: TTCCAGCCAGTATTCCAGCA	AJ400824.1	163
MUC2	Forward: TGCTACTACGTGCTGACCAA Reverse: ACGTTCTTCTTGTTGTCGGC	XM_010806231.2	131
TNF- α	Forward: CGTGGACTTCAACTCTCCCT Reverse: GGACACCTTGACCTCCTGAA	NM_173966.3	179
Il-1 β	Forward: CCAGCTGCAGATTTCTCACC Reverse: TCACACAGAAACTCGTCGGA	NM_174093.1	195
il-4	Forward: TGCCCCAAAGAACACAACTG Reverse: GAACAGGTCTTGCTTGCCAA	NM_173921.2	144
Il-6	Forward: ACTTCTGCTTTCCCTACCCC Reverse: TGTCGACCATGCGCTTAATG	NM_173923.2	121
VIP	Forward: TCGACTCCCAGGACTTCAAC Reverse: GAGAAGAGCACGCTGAACAG	NM_173970.3	148
FCGRT	Forward: ACTATCGCTCGCTCCAGTAC Reverse: CCTGCGCCCGTAGATTATTG	NM_176657.1	126
B2M	Forward: GTTCACTCCCAACAGCAAGG Reverse: TCTCGATGGTGCTGCTTACA	NM_173893.3	109
RAB11A	Forward: GCAACAAGAAGCATCCAGGT Reverse: TAAGGCACCTACAGCTCCAC	NM_001038162.2	120
RAB25	Forward: AGCTGAGAGTTGAGGGCATT Reverse: TCGGCTCTGTTTCCCATCTT	NM_001017936.1	102
STX3	Forward: GAGCAGCATCAAGGAGCTTC Reverse: TACCAATTTCTTCCGGGCCT	NM_001101971.1	181
plg	Forward: GGTGTGCTGGTTCCTTCTTG Reverse: ATAAGCTGCAGGTGGGAACT	NM_174143.1	93
MYO5B	Forward: GAAGCAATACCGCATGCAGA Reverse: TTCTGGATGATGGTGGCCTT	XM_015469176.1	147

[a]NR3C1 Nuclear receptor subfamily 3 group C member 1, MUC1 Mucin 1, MUC2 Mucin 2, TNFα Tumor necrosis factor alpha, IL-1β Interleukin 1 beta, IL-4 Interleukin 4, IL-6 Interleukin 6, VIP Vasoactive intestinal polypeptide, FCGRT fc fragment of Ig receptor and transporter, B2M Beta −2- microglobulin, RAB11A Ras related protein Rab11a, RAB25 Ras related protein RAB25, STX3 Syntaxin 3, plg Polymeric immunoglobulin receptor, MYO5B Myosin VB

investigate a potential effect of both by ECS on the stress response and subsequent immune response in neonatal calves. In addition to these, two sodium channel genes were also chosen, to examine whether or not there could be a potential molecular basis for fluid retention in the ECS delivered neonate. A total of 15 genes were chosen for target in jejunal tissue including a number of interleukins to examine alterations to the immune response, and a number of receptors including a component of the FC receptor complex, which plays an important role in Ig absorption from colostrum. A number of genes involved in extracellular matrix structure were also examined, to identify any structural conformation changes in the jejunum of ECS delivered calves that may affect Ig absorption. Five housekeeping genes were selected from commonly used reference genes for lung and jejunum tissue: β-actin (ACTB), glyceraldehyde-3-phosphate-dehydrogenase (GAPD), 40s Ribosomal Protein S9 (RSP9), Ubiquitously Expressed (FAU) and Hypoxanthine Phosphoribosyltransferase 1 (HPRT1). The gene expression levels were measured by real-time qPCR, and the expression stabilities were evaluated by the M value of geNorm.

Real-time quantitative PCR

Real-time quantitative PCR was carried out using the ABI 7500 Fast real-Time PCR System with SYBR green Master Mix (Applied Biosystems, Warrington, UK). Reactions were carried out in a 96-well plate (Applied Biosystems, Warrington, UK) and prepared in a total volume of 20 μl, with 2 μl cDNA, 10 μl SYBR green master mix, 7 μl of nuclease free H20 and 1 μl of 5 μM forward and reverse primer mix. Optimal cDNA concentration, primer efficiencies and concentrations were determined. Thermal cycling conditions applied to each assay consisted of an initial denaturation step at 95 °C for 15 min followed by 40 cycles of denaturation at 95 °C for 5 s and annealing and extension at 60 °C for 40 s. At the end of each cycle, SYBR green fluorescence was detected to monitor the quantity of PCR product. The efficiency of the qPCR reaction was calculated for each gene by creating a standard curve from two-fold serial dilutions of cDNA. Only primers with PCR efficiencies between 90 and 110% were used. The software package GenEx 5.2.1.3 (MultiD Analyses AB, Gothenburg, Sweden) was used for efficiency correction of the raw Ct values, normalisation to the reference genes, and calculation of quantities relative to the average Ct value for each gene.

Statistical analyses

Data were checked for normality using the UNIVARIATE procedure of Statistical Analysis Software (SAS, version 9.3). Data were transformed by raising coefficients to the appropriate power of λ using the TransReg procedure when appropriate, and subsequently analysed using mixed model methodology within the MIXED procedure of SAS. The Tukey critical difference test was performed to determine the presence of statistical differences, with $P < 0.05$ was considered significant.

Results

Histology of lung tissue

Tissue was analysed to ensure samples taken were of sufficient similarity prior to transcriptomic analysis. A representative image from the lung tissue sampled from trans-vaginally delivered and elective caesarean delivered calves can be seen in Figs. 1 and 2 respectively. Following staining and visualisation, it was determined that samples were of a consistent cellular similarity for transcriptional analysis.

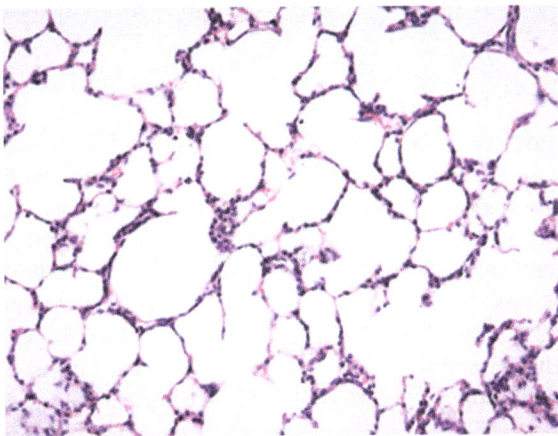

Fig. 2 Histology findings for lung tissue taken from the centre right lobe of the right lung in calves delivered by elective caesarean section at ×20 magnification

Differential expression of genes in lung and jejunum tissue

Lung tissue

A total of 13 genes were chosen for analysis in the lung tissue of the TV and ECS delivered calves (See Table 1 for primer list) and the effect of delivery method on transcript abundance is outlined in Table 3. There was a statistically significant effect of delivery method on the relative transcriptional abundance of 5 genes from the 13 chosen. Transcript abundance for *LAP*, an antimicrobial peptide, was 2.06 fold lower in ECS compared with TV calves ($p = 00.75$; Table 4). Similarly *CYP1A1*, a member of the detoxifying cytochrome P450 superfamily, was 3.95 fold lower in ECS calves ($p = 0.0016$). *MUC5AC*, which encodes a secreted mucin protein, was present at higher levels of abundance (2.55 fold difference; $p = 0.0021$; Table 4) in the ECS delivered calves. Two genes which encode for cellular sodium voltage gated channel components were lower in abundance in calves delivered via ECS (*SCN11a*; 2.73 fold; $p = 0.0002$ and *SCN11b*: 1.72 fold; $p = 0.0051$).

Jejunum tissue

Fifteen genes were chosen for comparative quantitative analysis of jejunal function between calves delivered by caesarean section compared with natural transvaginal birth. Differences in transcript abundance between the treatments for three out of the 15 genes analysed reached statistical significance Table 4. Two of the genes were members of the Interleukin superfamily; *Il-6* (1.23 fold increase; $p = 0.0018$; Table 4) and *IL-1β* (1.77 fold increase; $p = 0.0013$, Table 4) were higher in calves delivered by ECS compared with TV calves. An additional gene from the interleukin superfamily, *IL-4* was present at an increased transcript abundance in CS calves, with a tendency towards significance (1.13 fold increase; $p = 0.057$; Table 4).

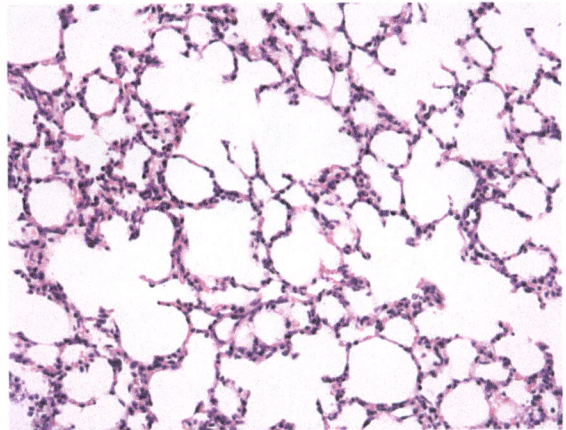

Fig. 1 Histology findings for lung tissue taken from the centre right lobe of the right lung in calves delivered natural by vaginal birth at ×20 magnification

Table 3 Effect delivery method on the expression of genes in calf lung and jejunum tissue. Table highlights expression levels of genes in ECS delivered calves relative to the TV delivered controls

Symbol	Gene annotation	Fold change[a]	P value[2]
Lung Tissue			
LAP	Lingual antimicrobial peptide	−2.06	**0.0075**
SP-A	Surfactant protein A	−1.33	0.2890
SP-B	Surfactant protein B	−1.20	0.5915
SP-C	Surfactant protein C	−1.29	0.2616
SP-D	Surfactant protein D	1.47	0.2065
SCN11A	Sodium voltage-gated channel alpha subunit 11 alpha	−2.73	**0.0002**
SCN11B	Sodium voltage-gated channel beta subunit 11 beta	−1.72	**0.0051**
SGK1	Serine/threonine-protein kinase 1	1.29	0.8690
MUC5AC	Mucin 5 subtype AC	2.55	**0.0021**
MUC5B	Mucin 5B	−2.07	0.6330
ABCA3	Bovine ATP-binding cassette sub-family A member 3	−1.31	0.4254
CYP1A1	Cytochrome P450 family 1 subfamily A member 1	−3.95	**0.0016**
CYP1A2	Cytochrome p450 family 1 subfamily 1 member 2	−1.42	0.2811
Jejunum Tissue			
IL-6	Interleukin 6	1.23	**0.0018**
IL-4	Interleukin 4	1.77	0.162
IL1-β	Interleukin 1 beta	1.92	**0.0013**
FCGRT	FC fragment of Ig receptor and transporter	1.24	0.1643
MYO5B	Myosin 5B	−1.41	0.2863
VIP	Vasoactive intestinal polypeptide	1.08	0.6519
MUC1	Mucin 1	−1.76	0.6519
MUC2	Mucin 2	−1.80	0.1456
TNFα	Tumour necrosis factor alpha	1.92	**0.001**
B2M	Beta-2-microglobulin	1.28	0.3555
NR3C1	Nuclear receptor subfamily 3 group C member 1	−1.05	0.1001
Rab11a	Ras related protein Rab11a	1.38	0.4954
Rab25	Ras related protein Rab25	1.02	0.3995
STX3	Syntaxin 3	−1.08	0.3794
plg	Polymeric Immunoglobulin receptor	1.13	0.6741

P values in bold indicate significant change to the transcriptional levels of the gene
[2]P values were corrected (Bonferroni correction)
[a]Fold change represents observed reduction or increase in abundance of gene in CS delivered calves

There was a tendency towards significance ($p = 0.10$) on the effect of birthing method on *NR3C1*, present at lower abundance in the calves delivered by ECS. This gene encodes for a receptor for glucocorticoid, a corticosteroid hormone which is heavily involved in neonatal adaptation during birth and early life.

Discussion
Lung function in calves delivered by elective caesarean section
Fluid absorption
The transition of the lungs from an aqueous environment to one of oxygen is arguably the most vulnerable and

critical adaptation phase in neonatal development immediately following birth. The neonate must rapidly assume responsibility for oxygenating its own blood from atmospheric air, by clearance of fluid from the lung cavity and by the production of sufficient amounts of glycerophospholipid-rich surfactant lipoproteins [21, 22]. The lung, rather than the amniotic sac itself, is the source of liquid in the foetal lung [10, 23]. As parturition approaches, the rate of liquid formation and the volume of liquid within the lung decreases [10]. The first challenge for the neonate to overcome upon delivery is to rapidly clear air spaces of the remaining fluid [24]. The passage of the neonate through the birth canal is thought

Table 4 Sequences of oligonucleotide primers selected as reference genes for qPCR analysis

Gene ID[a]	Primers (5'-3')	Accession No.	Amplicon length
Act- β	Forward: CGGCATCGAGGACAGGAT Reverse: CATCGTACTCCTGCTTGCTGAT	NM_173979.3	169
GAPDH	Forward: CCTGCCCGTTCGACAGATA Reverse: GGCGACGATGTCCACTTTG	NM_001034034.1	150
FAU	Forward: CCGCATGCTTGGAGGTAAAG Reverse: CACAACATTGACAAAGCGCC	NM_174731.3	154
RPS9	Forward: CCGCATGCTTGGAGGTAAAG Reverse: CACAACATTGACAAAGCGCC	NM_001101152.2	188
HPRT1	Forward:GGATTACATCAAAGCACTGAACA Reverse: CATTGTCTTCCCAGTGTCAATT	NM_001034035	194

[a] Act- β β-actin, GAPDH Glyceraldehyde-3-Phosphate Dehydrogenase, FAU Finkel-Biskis-Reilly Murine Sarcoma Virus (FBR-MuSV) Ubiquitously Expressed, RPS9 Ribosomal Protein S9, HPRT1 Hypoxanthine Phosphoribosyltransferase 1

to be important for aiding the expulsion of liquid from the lungs, due to the "Vaginal squeeze", and calves born by caesarean before the onset of labour are completely deprived of this [10]. Hormonal changes immediately prior to labour, which continue throughout are thought to play an important role in preparing the lungs for liquid removal and for aerobic respiration, in particular, the release of epinephrine from the adrenal medulla. Amiloride sensitive sodium transport is a key event in the trans-epithelial movement of alveolar fluid in the lungs, which is enabled by sodium channels lining the epithelial layer [10, 25]. At late stage gestation, an increased expression of epithelial sodium channels has been demonstrated in mammals, with many authors reporting much lower expression of sodium channel subunits in preterm human neonates suffering from respiratory distress [10, 26]. These sodium channel subunits were also shown to be present at significantly reduced levels of abundance in neonates that had been delivered via elective caesarean section [26]. This reduction is most likely due to the lack of the surge in hormones normally induced by labour. Cortisol, one of the main labour induced hormones is known to activate these pumps following parturition [11] The clearance of fluid by the action of these sodium channels involves a two-step process: firstly, the passive movement of sodium (Na$^+$) from the lumen, across the apical membrane and into the cell through the sodium voltage gated channels, followed by the active extrusion of Na$^+$ from the cell, across the basolateral membrane and into the serosal space [25]. The remaining fluid is then rapidly reabsorbed as water and the lungs cleared [27] The sodium channel ENaC is comprised of three subunits, and we demonstrated a significant decrease in the abundance of the alpha and beta subunits in the calves delivered by elective caesarean section. The decreased relative abundance of these subunits indicates a reduced availability of sodium channels in the lung epithelial layer of calves delivered by ECS, and such channels may significantly affect transition of the calf at birth. The reduction in the transcriptional abundance of these genes may be due to the lack of the hormonal surge

in the prepartum period, and this, combined with the lack of the mechanical fluid clearance mechanisms during the uterine contractions of a vaginal delivery, may lead to a much slower clearance of fluid from the lungs of ECS delivered calves and could affect potentially affect future pulmonary function [10, 12, 28].

Secretion of normal lung proteins

As the end of gestation approaches, foetal lung fluid declines in preparation for delivery. At this time, the production of surfactant proteins initiates, with labour prompting the secretion of these proteins into the remaining foetal lung fluid, which increases the overall surfactant concentration of the lung [29]. Without the onset of labour, the initiation of this surfactant secretion does not occur, meaning that the overall concentration does not reach the necessary levels needed for rapid transition at birth [11]. ECS deliveries before the prepartum increase of hormones also results in a reduction in the expression of additional proteins such as antioxidants and does not occur at its peak and necessary level [30]. This was highlighted in our study, as we showed an almost four fold reduction to the relative abundance of CYP1A1. The lungs are a major target for all inhaled toxins, and also of endogenously derived cellular toxins such as ROS. This gene is part of the cytochrome p450 super family of isoenzymes, which are capable of detoxifying both endogenous and exogenous toxins and has been shown to play a critical role in protection against hyperoxic lung injury by reducing lipid peroxidation and oxidative stress in both human and murine studies [31, 32]. Hyperoxia is commonly seen in both preterm infants and those delivered by ECS suffering from respiratory difficulty [31]. The reduced abundance in transcript levels of this gene may indicate that calves delivered by ECS may be more susceptible to toxins and hyperoxic stress, which could affect transition at birth and in the subsequent days that follow. The expression levels of two Mucin genes were also examined for alterations in their abundance in the ECS delivered calves.

Lung immunity

Transcript abundance for *LAP*, an antimicrobial peptide (AMP) found in the lung, was more than two-fold lower in the calves delivered by ECS compared with TV. As a member of the β-defensin group of AMPs, LAP plays a critical role in protection against opportunistic invading pathogens. Defensins have been shown to exhibit a developmentally regulated expression from foetus to neonate, with much lower expression found in foetal sheep and humans in comparison to levels expressed in their neonatal counterparts [33, 34]. Natural trans-vaginal delivery is associated with a steady increase in stress in the neonate, important for the expression of genes critical to adaptation. For example, an increased expression of proinflammatory cytokines during labour of both term and preterm neonates causes a correlated increase in the levels of SP-A produced [22]. Here, we suggest that the significantly lower levels of *LAP* could be due to the absence of a rise in the neonatal stress levels associated with natural TV delivery. The lower level of transcriptional abundance demonstrated here could have serious implications leading to an inactivated and impaired lung immunity in the ECS delivered calves [35]. An additional gene, *MUC5AC*, was found to be significantly higher in the calves delivered by ECS. *MUC5AC* encodes the protein backbone of the MUC5AC glycoprotein. Its presence here at higher levels of expression in the calves delivered by section indicates a possible excess production of mucous, which is symptomatic of many lung disease states, including bronchial hyperactivity in asthma [36] Mucin proteins also serve as binding points for bacteria, potentially problematic for the already impaired immunity of ECS delivered calves [27, 37].

Jejunal gene expression and implications for immunoglobulin absorption

Immediately following birth, the new-born must respond to a huge influx of potentially dangerous pathogens [1, 38]. While passive immunity exists in humans at birth, calves, like other ruminant neonates and some other species like pigs, are born agammaglobulinemic [18]. Transfer of Ig to the foetus in cows is not possible due to epitheliochorial placentation which interposes large numbers of epithelial layers between the maternal and foetal blood supplies [39, 40]. Immune transfer of immunoglobulins (Ig) and other bioactive factors such as lysozyme and growth hormones must occur rapidly after birth through passive transfer from colostrum to the neonate [1, 41]. Failure of sufficient passive transfer is thought to be a major contributing factor in perinatal mortality in calves, and the aim of the expression analysis in this study was to ascertain whether there was evidence that delivery by ECS prior to the normal foetal

cortisol surge may affect the immune response of the jejunum or on the potential uptake of Ig within the intestinal tract [18].

To examine a possible effect of mode of delivery on the absorption of Ig through passive transfer in the neonatal calves, a number of genes with proposed function in gastrointestinal absorption were selected. *FCGRT*, encodes for the alpha chain of the bovine receptor FcRn. This receptor plays a critical role in the absorption of Ig through passive transfer in calves [42] There was no observed difference in the expression of this gene between the ECS and TV delivered calves. The expression of polymeric immunoglobulin receptor (pIg) was also analysed, with no significant difference to its relative abundance observed between the ECS and TV delivered calves. pIg is a receptor present in epithelial cells along the intestinal tract which is known to play a vital role in Ig passage through the epithelial layer [43, 44]. The absence of a significant change to the abundance of these Ig receptors suggests that mode of delivery does not affect the absorption of Ig, at least at this early time point. Upon colostrum intake, a number of genes involved in cell growth and the immune response are known to become highly expressed, and possibly colostrum ingestion also influences the expression of Ig absorption associated genes, which may have been apparent had the calves been permitted to ingest colostrum [45, 46]. A number of genes involved in intracellular trafficking with possible roles in the transport of Ig following uptake by pinocytosis were also examined for alterations to their relative transcriptional abundance [14, 18]. *Rab11a* and *Rab25*, which both encode Ras-related proteins involved in the absorption of molecules such as calcium through endocytosis in the intestinal epithelial membrane were also examined for changes in their expression levels [47, 48]. Neither gene was changed at a level that was deemed statistically significant, but possibly would have presented at altered levels in calves following colostrum absorption.

Immune gene expression in the neonatal jejunum

Three genes, out of the 15 in total that were analysed for quantitative variations between the calves delivered either naturally by TV or by ECS, were present at higher abundance in the latter; *Il-6*, *Il-1β* and *TNFα*. All three genes are immune related genes, and members of the proinflammatory cytokine group. Proinflammatory cytokines are known to play an important role during labour and delivery, such as ripening of the cervix, and are critical to immune protection of the neonate in response to influx of gastrointestinal pathogens [49, 50]. The birth process itself has recently received attention as a possible stimulus for both immediate and long term disease susceptibility in both humans and in animals, with the altered immune response in jejunal tissue observed in our study highlighting this as a possibility [51]. There have been conflicting results with regard to the effect of mode of delivery on the abundance of cytokines, with similar expressional analysis demonstrating opposite results in both cord and blood samples. For example, Ly et al. (2006) demonstrated increased abundance of Il-1β, Il-6 and IFNγ in cord blood samples of human neonates that had been delivered before labour by ECS. In contrast, a similar study using cord blood samples of neonatal humans born by ECS found these genes plus Il-4 and TNFα, were found at a lower level of abundance [52]. An additional study found further conflicting results, with quantitative analysis of plasma samples showing an increase in the abundance of Il-6 but a reduction to the abundance of Il- 2 in neonates delivered by ECS [53]. The increased abundance of certain proinflammatory cytokines including the three that were increased in the calves delivered before the onset of labour in our study, have been linked with certain allergenic diseases in later life, including asthma [52]. Cytokines are also known to play an interactive role between the immune and the neuroendocrine system, with the expression of certain types (e.g. Il-1β and TNFα) known to activate the hypothalamic-pituitary axis, also acting on the central nervous system [53]. During normal delivery, the neonate is subjected to increased levels of stress over time, compared to the stress caused by ECS, which is more immediate and short-term, and this may be the cause of the heightened expression of these pro-inflammatory cytokines [35]. A previous study observing expression of cytokines in human neonates suggested that altered abundance in the transcript levels of these cytokines could be due to the use of operative anaesthetic, both regional and general, during the procedure, which influences expression within the foetus before birth [54]. However, conflicting results were demonstrated in an additional comparative mode of delivery study using neonatal piglets, where an increased abundance of TNFα was seen in piglets delivered by caesarean section, but delivery was carried out immediately after stunning of the dams in the trial, with no anaesthesia or analgesia used [55]. The increase in abundance of these genes could potentially be due to an overall stress response in the calves when delivered so abruptly before endogenous hormonal preparation, but is unlikely as the process of natural labour and birth is thought to cause the greatest stress to both neonate and dam [53]. The functional integrity of the intestine is critical to the immunocompetence of the bovine neonate. Immediately after birth, the neonatal animal must, alongside other adaptations rapidly required for successful transition, adjust from an aqueous environment sterile environment to that of an atmospheric one laden with opportunistic pathogens.

The lack of passage through the birth canal poses an immediate threat to neonates born by caesarean section, particularly exaggerated in those born before the pre-labour hormonal surge where a compromised immune response as observed here may mitigate against the animals' immunocompetence [56].

Two genes from the mucin family, MUC1 and MUC2 were also examined for changes to their relative abundance in the TV compared with ECS calves. A healthy epithelial mucin layer in the gastrointestinal tract is vital for many functions including lubrication of food, maintenance of a physical barrier protection against pathogens, and for provision of a permeable gel layer through which gaseous and nutrient exchange can take place [57]. Despite their putative roles in the absorption of bioactive factors in colostrum, we failed to observe an effect of mode of delivery on the relative expression of these genes [57–59].

Conclusion

In the current study, we provide some evidence that lung tissue in calves delivered by ECS may be compromised for efficient transition to neonatal life. Specifically, we observed reduced abundance of two sodium channel subunits, critical for rapid clearance of fluid from the lungs. Comparison of gene expression for the jejunal tissue between the two birthing processes employed highlighted a number of alterations to the ECS delivered calves that could have serious consequences for subsequent post-natal health. While we did not observe direct evidence that the potential for immunoglobulin absorption may be negatively affected by ECS in calves, it is possible that such effects may be manifested at a later stage following colostrum ingestion.

The work here highlights the potential risks to neonatal calf health following elective caesarean section before the onset of labour. Further work is required to ascertain whether or not colostrum immunoglobulin uptake is compromised, and to examine whether latent effects of birth process on the functionality of the immune system exist. These data also offer value to understanding potential negative effects from the use of elective caesarean section procedures in humans.

Abbreviations

AI: Artificial insemination; Ct value: Cycle threshold value; DPBS: Dulbecco's phosphate-buffered saline; ECS: Elective caesarean section; Mins: Minutes; PCR: Polymerase chain reaction; qPCR: Quantitative polymerase chain reaction; RIN: RNA integrity value; TV: Transvaginal delivery

Acknowledgments

Not applicable.

Funding

Not applicable.

Authors' contributions

KMN and EOH coordinated tissue collection, and KMN completed RNA extraction. CS and KMN completed qPCR of all genes, and writing of manuscript. EOH and SMW contributed to biological interpretation of the data and editing of the manuscript. MB conducted Caesarean sections, advised on animal health and contributed to manuscript preparation. JC carried out histological analysis of tissues. DAK designed and conceptualised the study, contributed to the biological interpretation of the data and editing of the manuscript. All authors have read and approved the final manuscript.

Consent for publication

Not applicable.

Competing interests

The authors declare that they have no competing interests.

Author details

[1]Animal and Grassland Research and Innovation Centre, Teagasc, Grange, Dunsany, Co. Meath, Ireland. [2]School of Veterinary Medicine, University College Dublin, Belfield, Dublin 4, Ireland.

References

1. Sangild P. Uptake of Colostral Immunoglobulins by the com-promised newborn farm animal. Acta Vet Scand. 2003;98:105–22.
2. Kolkman I, et al. Pre-operative and operative difficulties during bovine caesarean section in Belgium and associated risk factors. Reprod Domest Anim. 2010;45(6):1020–7.
3. Kolkman I, et al. Protocol of the caesarean section as performed in daily bovine practice in Belgium. Reprod Domest Anim. 2007;42(6):583–9.
4. Nix J, et al. A retrospective analysis of factors contributing to calf mortality and dystocia in beef cattle. Theriogenology. 1998;49(8):1515–23.
5. Uystepruyst C, et al. Optimal timing of elective caesarean section in Belgian white and blue breed of cattle: the calf's point of view. Vet J. 2002;163(3):267–82.
6. Hanset, R., B.B. Herd-Book, and R. des champs Elysées, Emergence and selection of the Belgian Blue breed. Belgian Blue Herd-Book. Ciney, University of Liege, 1998.
7. Greger M. Trait selection and welfare of genetically engineered animals in agriculture. J Anim Sci. 2010;88(2):811–4.
8. Estorgato GR, et al. Surfactant deficiency in full-term newborns with transient tachypnea delivered by elective C-section. Pediatr Pulmonol. 2015;
9. Ramachandrappa A, Jain L. Elective cesarean section: its impact on neonatal respiratory outcome. Clin Perinatol. 2008;35(2):373–93.
10. Bland RD. Loss of liquid from the lung lumen in labor: more than a simple "squeeze". Am J Phys Lung Cell Mol Phys. 2001;280(4):L602–5.
11. Hillman NH, Kallapur SG, Jobe AH. Physiology of transition from intrauterine to extrauterine life. Clin Perinatol. 2012;39(4):769–83.
12. Hooper SB, te Pas AB, Kitchen MJ. Respiratory transition in the newborn: a three-phase process. Arch Dis Child Fetal Neonatal Ed. 2016;101(3):F266–71.
13. Brown M, et al. Effects of adrenaline and of spontaneous labour on the secretion and absorption of lung liquid in the fetal lamb. J Physiol. 1983;344:137.
14. Weaver DM, et al. Passive transfer of colostral immunoglobulins in calves. J Vet Intern Med. 2000;14(6):569–77.
15. Partridge I. Studies on digestion and absorption in the intestines of growing pigs. Br J Nutr. 1978;39(03):527–37.
16. Gravitt, T.A.D.A.R., Porcine Colostrum and milk stimulate visceral organ and skeletal muscle protein synthesis in neonatal Piglets1. 1992.
17. Cabrera R, et al. Early postnatal kinetics of colostral immunoglobulin G absorption in fed and fasted piglets and developmental expression of the intestinal immunoglobulin G receptor. J Anim Sci. 2013;91(1):211–8.
18. Quigley J. The role of oral immunoglobulins in systemic and intestinal immunity of neonatal calves. Cedar Rapid, Iowa, USA: Diamond V Mills; 2004.
19. Blum J. Nutritional physiology of neonatal calves*. J Anim Physiol Anim Nutr. 2006;90(1–2):1–11.
20. Wheadon N, et al. Plasma nitrogen isotopic fractionation and feed efficiency in growing beef heifers. Br J Nutr. 2014;111(9):1705–11.
21. Poulsen KP, McGuirk SM. Respiratory disease of the bovine neonate. Vet Clin N Am Food Anim Pract. 2009;25(1):121–37.
22. Mendelson CR. Minireview: fetal-maternal hormonal signaling in pregnancy and labor. Mol Endocrinol. 2009;23(7):947–54.
23. Hooper SB, Polglase GR, Roehr CC. Cardiopulmonary changes with aeration of the newborn lung. Paediatr Respir Rev. 2015;16(3):147–50.
24. Bleul U. Respiratory distress syndrome in calves. Vet Clin N Am Food Anim Pract. 2009;25(1):179–93.
25. Jain L, Eaton DC. Physiology of fetal lung fluid clearance and the effect of labor. In Seminars in perinatology. (Vol. 30, No. 1). WB Saunders; 2006. p. 34–43.
26. Katz C, Bentur L, Elias N. Clinical implication of lung fluid balance in the perinatal period. J Perinatol. 2011;31(4):230–5.
27. Ackermann MR, Derscheid R, Roth JA. Innate immunology of bovine respiratory disease. Vet Clin N Am Food Anim Pract. 2010;26(2):215–28.
28. Bland R, et al. Clearance of liquid from lungs of newborn rabbits. J Appl Physiol. 1980;49(2):171–7.
29. Gallacher DJ, Hart K, Kotecha S. Common respiratory conditions of the newborn. Breathe. 2016;12(1):30.
30. McCartney J, et al. Mineralocorticoid effects in the late gestation ovine fetal lung. Physiol Reports. 2014;2(7):e12066.
31. Lingappan K, et al. Mice deficient in the gene for cytochrome P450 (CYP) 1A1 are more susceptible than wild-type to hyperoxic lung injury: evidence for protective role of CYP1A1 against oxidative stress. Toxicol Sci. 2014;141(1):68–77.
32. Wang L, et al. Disruption of Cytochrome P4501A2 in mice leads to increased susceptibility to hyperoxic lung injury. Free Radic Biol Med. 2015;82:147–59.
33. Starner TD, et al. Expression and activity of β-defensins and LL-37 in the developing human lung. J Immunol. 2005;174(3):1608–15.
34. Mitchell GB, et al. Effect of corticosteroids and neuropeptides on the expression of defensins in bovine tracheal epithelial cells. Infect Immun. 2007;75(3):1325–34.
35. Cho CE, Norman M. Cesarean section and development of the immune system in the offspring. Am J Obstet Gynecol. 2013;208(4):249–54.
36. Roy MG, et al. Mucin production during prenatal and postnatal murine lung development. Am J Respir Cell Mol Biol. 2011;44(6):755–60.
37. Chen Y, et al. Dexamethasone-mediated repression of MUC5AC gene expression in human lung epithelial cells. Am J Respir Cell Mol Biol. 2006;34(3):338–47.
38. Gensollen T, et al. How colonization by microbiota in early life shapes the immune system. Science. 2016;352(6285):539–44.
39. Williams PP. Immunomodulating effects of intestinal absorbed maternal colostral leukocytes by neonatal pigs. Can J Vet Res. 1993;57(1):1.
40. Rooke J, Bland I. The acquisition of passive immunity in the new-born piglet. Livest Prod Sci. 2002;78(1):13–23.
41. Boudry C, et al. Effects of oral supplementation with bovine colostrum on the immune system of weaned piglets. Res Vet Sci. 2007;83(1):91–101.
42. Laegreid WW, et al. Association of bovine neonatal Fc receptor a-chain gene (FCGRT) haplotypes with serum IgG concentration in newborn calves. Mamm Genome. 2002;13(12):704–10.
43. Hurley WL, Theil PK. Perspectives on immunoglobulins in colostrum and milk. Nutrients. 2011;3(4):442–74.
44. Kacskovics I. Fc receptors in livestock species. Vet Immunol Immunopathol. 2004;102(4):351–62.
45. Xu R-J. Development of the newborn GI tract and its relation to colostrum/milk intake: a review. Reprod Fertil Dev. 1996;8(1):35–48.
46. Blais M, et al. A gene expression programme induced by bovine colostrum whey promotes growth and wound-healing processes in intestinal epithelial cells. J Nutr Sci. 2014;3:e57.
47. Cayouette S, et al. Involvement of Rab9 and Rab11 in the intracellular trafficking of TRPC6. Biochimica et Biophysica Acta (BBA)-molecular. Cell Res. 2010;1803(7):805–12.
48. Bastin G, Heximer SP. Rab family proteins regulate the endosomal trafficking and function of RGS4. J Biol Chem. 2013;288(30):21836–49.
49. Malamitsi-Puchner A, et al. The influence of the mode of delivery on circulating cytokine concentrations in the perinatal period. Early Hum Dev. 2005;81(4):387–92.
50. Nanthakumar NN, et al. Inflammation in the developing human intestine: a possible pathophysiologic contribution to necrotizing enterocolitis. Proc Natl Acad Sci. 2000;97(11):6043–8.

51. Hyde MJ, et al. The health implications of birth by caesarean section. Biol Rev. 2012;87(1):229–43.
52. Khafipour E, Ghia J-E. Mode of delivery and inflammatory disorders. J Immunol. 2013;1(1004)
53. Bessler H, et al. Labor affects cytokine production in newborns. Am J Reprod Immunol. 1998;39(1):27–32.
54. Rizzo A, et al. Update on anesthesia and the immune response in newborns delivered by cesarian section. Immunopharmacol Immunotoxicol. 2011; 33(4):581–5.
55. Daniel J, et al. Evaluation of immune system function in neonatal pigs born vaginally or by cesarean section. Domest Anim Endocrinol. 2008;35(1):81–7.
56. Zanardo V, et al. Cytokines in human colostrum and neonatal jaundice. Pediatr Res. 2007;62(2):191–4.
57. Forstner J. Intestinal mucins in health and disease. Digestion. 1978;17(3):234–63.
58. Bansil R, Turner BS. Mucin structure, aggregation, physiological functions and biomedical applications. Curr Opin Colloid Interface Sci. 2006;11(2):164–70.
59. Deplancke B, Gaskins HR. Microbial modulation of innate defense: goblet cells and the intestinal mucus layer. Am J Clin Nutr. 2001;73(6):1131S–41S.

Quantification of resistant alleles in the β-tubulin gene of field strains of gastrointestinal nematodes and their relation with the faecal egg count reduction test

Myriam Esteban-Ballesteros[1,2], Francisco A. Rojo-Vázquez[1,2], Philip J. Skuce[3], Lynsey Melville[3], Camino González-Lanza[2] and María Martínez-Valladares[1,2*]

Abstract

Background: Benzimidazole (BZ) resistance in gastrointestinal nematodes is associated with a single nucleotide polymorphism (SNP) at codons 167, 198 and 200 in the isotype 1 of beta-tubulin gene although in some species these SNPs have also been associated with resistance to macrocyclic lactones. In the present study we compared the levels of resistance in *Teladorsagia circumcincta* and *Trichostrongylus colubriformis* by means of the faecal egg reduction test (FECRT) and the percentage of resistant alleles obtained after pyrosequencing. The study was conducted in 10 naturally infected sheep flocks. Each flock was divided into three groups: i) group treated with albendazole (ABZ); ii) group treated with ivermectin (IVM); iii) untreated group. The number of eggs excreted per gram of faeces was estimated at day 0 and 14 post-treatment.

Results: Resistance to ABZ was observed in 12.5% (1/8) of the flocks and to IVM in 44.4% (4/9) of them. One flock was resistant to both drugs according to FECRT. Coprocultures were performed at the same dates to collect L3 for DNA extraction from pooled larvae and to determine the resistant allele frequencies by pyrosequencing analysis. In *T. circumcincta*, SNPs were not found at any of the three codons before treatment; after the administration of ABZ, SNPs were present only in two different flocks, one of them with a frequency of 23.8% at SNP 167, and the other 13.2% % at SNP 198. In relation to *T. colubriformis*, we found the SNP200 before treatment in 33.3% (3/9) of the flocks with values between 48.5 and 87.8%. After treatment with ABZ and IVM, the prevalence of this SNP increased to 75 and 100% of the flocks, with a mean frequency of 95.1% and 82.6%, respectively.

Conclusion: The frequencies observed for SNP200 in *T. colubriformis* indicate that the presence of resistance is more common than revealed by the FECRT.

Keywords: Sheep, *Teladorsagia circumcincta*, *Trichostrongylus colubriformis*, Anthelmintic resistance, FECRT, Single nucleotide polymorphism, Beta-tubulin, Pyrosequencing

* Correspondence: mmarva@unileon.es
[1]Departamento de Sanidad Animal, Facultad de Veterinaria, Universidad de León, Campus de Vegazana s/n, 24071 León, Spain
[2]Instituto de Ganadería de Montaña (CSIC-Universidad de León), Finca Marzanas, Grulleros, 24346 León, Spain
Full list of author information is available at the end of the article

Background

Infections by gastrointestinal nematodes (GIN) are a serious problem for extensive systems of sheep farming worldwide. GIN can cause serious losses in animal production since they affect the production of milk, wool and meat, and also interfere with reproduction [1]. The most prevalent GIN species infecting sheep in temperate areas of the world are *Teladorsagia circumcincta*, *Trichostrongylus* spp., *Haemonchus contortus*, *Chabertia ovina* and *Cooperia* spp.

The usual mode of controlling GIN infections in ruminants is by chemotherapy and the most commonly used anthelmintics are grouped into 3 families: benzimidazoles (BZs), imidazothiazoles and macrocyclic lactones (MLs). Recently, two new chemical groups have been introduced to the market, namely, the aminoacetonitrile derivative (monepantel) [2] and the spiroindole (derquantel), in combination with abamectin [3]. Anthelmintics have been used with great success in the past. However, their frequent use and the underdosing of animals has favored the emergence of anthelmintic resistance (AR), among other factors. There are some reports describing AR against all anthelmintic groups worldwide, even against the most recent anthelmintic drug, monepantel, in New Zealand, The Netherlands and Uruguay in *T. circumcincta*, *Trichostrongylus colubriformis* and *H. contortus* [4–7]. Therefore, the high prevalence of AR to several drugs against GIN in small ruminants continues to threaten the viability of small ruminant farms [8].

The BZs and MLs are the major groups of anthelmintics used to control GIN infections in some countries like Spain [9]. The mode of action of BZ involves binding to β-tubulin and disrupting microtubule polymerization of tubulin [10–13]. Genetic studies have shown that BZ resistance is associated with a single nucleotide polymorphism (SNP) in the gene encoding isotype-1 β-tubulin. The substitution of a phenylalanine (Phe, TTC) for a tyrosine (Tyr, TAC) at codon 200 (F200Y) has been linked to BZ resistance in *H. contortus*, *T. colubriformis*, *T. circumcincta*, *C. oncophora* and *O. ostertagi* [14]. Less frequently, the same SNP was found at codon 167 (F167Y) in resistant strains of *H. contortus*, *T circumcincta* and *O. ostertagi* [15]. Furthermore, a point mutation of alanine (Ala, GCA) to glutamine (Glu, GAA) at codon 198 has been described in resistant strains of *H. contortus*, *C. oncophora* and *O. ostertagi* [16].

Previous assays have shown that resistant strains of *H. contortus* also carried the resistant allele at codons 200 and 167 after treatment with an ML, specifically, ivermectin (IVM). These results suggest a possible association between resistance mechanisms in ML and BZ [17].

Since AR is increasing recently around the world [8, 18, 19], its control is required with the aim to use anthelmintic drugs more effectively and sustainably and to avoid the development of new resistant strains. The faecal egg count reduction test (FECRT) is the most widely used diagnostic method for the detection of AR in vivo [20, 21]. However, this technique lacks sensitivity and is not able to detect resistance when the level of genetically resistant individuals in the population is below ~25% [22]. Therefore, the development of new in vitro techniques is necessary for an early diagnosis of AR.

In this context, the aim of the present study was the measurement of the frequency of the resistant allele at codons 200, 198 and 167 of isotype-1 β-tubulin in GIN field isolates of sheep flocks from the Northwest of Spain before and after administration of a BZ and/or ML anthelmintic.

Methods

Faecal egg count reduction test (FECRT)

The study was conducted on 10 sheep flocks located in the province of León, Northwest Spain. In each flock, two groups of 10 sheep, naturally infected by GIN, were selected. Each group was treated with a BZ, albendazole (ABZ) (7.5 mg/kg bw), or IV (0.2 mg/kg bw). Faecal samples were collected on day 0 and day 14 post-treatment (pt). The number of eggs per gram of faeces (EPG) was determined by the modified McMaster method [23]. The faecal egg count reduction was calculated according to the recommendations of WAAVP (World Association for the Advancement of Veterinary Parasitology) [24] and using the following formula:

$$
\begin{aligned}
\text{FECRT \%} = \ & (\text{Arithmetic mean epg day 0} \\
& - \text{Arithmetic mean epg days} \\
& + 10\text{–}14)/\text{Arithmetic mean epg day 0} \\
& \times 100
\end{aligned}
$$

When the percentage reduction in egg count was <90%, the flock was considered resistant to the anthelmintic; if the faecal reduction was between 90 and 95%, the flock was classified as borderline or suspicious of resistance, and when values were higher than 95%, the flock was considered susceptible to the anthelmintic.

From each group, pooled faeces were cultured on days 0 and 14 pt to recover third stage larvae (L3). After collecting the cultures, a minimum of 100 L3 per culture were identified using the morphological keys in MAFF [23]; in flocks with a reduction of 100% we recovered a few L3, between 20 and 225. The remainder was stored at −20 °C until the DNA extraction was carried out.

DNA extraction

DNA was extracted from each L3 pool, before and after treatment, using the Speed Tools Tissue DNA Extraction kit (Biotools), according to the manufacturer's instructions. The DNA samples were stored at −20 °C until used.

Determination of allele frequencies

Two pairs of PCR primers were designed to amplify two fragments of the gene encoding the β-tubulin of *T. circumcincta* including the SNPs 167 and 198/200, respectively (Table 1). For *Trichostrongylus colubriformis*, we only designed one pair of primers to amplify a fragment encompassing codons 198 and 200 jointly (Table 1).

With the aim to amplify these three fragments, we firstly carried out a PCR with the primers described in Table 1, but without biotinylation. Cycling conditions were 95 °C for 10 min followed by 40 cycles of 95 °C for 30 s, Tm for 30 s and 72 °C for 45 s followed by 10 min at 72 °C and 4 °C to finish using *Taq* DNA polymerase MasterMix 2x, $MgCl_2$ 2,0 mM (Biotools, Madrid, Spain). PCR products were run on a 1.5% agarose gel, and the corresponding band was cut out and purified using PCR Clean-Up SpeedTools kit (Biotools, Madrid, Spain). The final elution volume was 50 µl. Then, this same PCR was carried out again in a 50 µl reaction volume but, in this case, one of the primers of each pair was labeled with biotin at the 5′, as shown in Table 1. Prior to pyrosequencing, a 7 µl aliquot of each PCR product was tested by agarose gel electrophoresis.

The specificity of the primers used to amplify these regions was confirmed initially after the amplification of the two regions in *T. circumcincta* adult worms and of one region in *T. colubriformis* adult worms. These PCRs were run in the same way as previously described but

using 1 µl of DNA template. The resulting bands were excised from an agarose gel 1,5% and purified using PCR Clean-Up SpeedTools kit (Biotools) and then were sequenced in the "Laboratorio de Técnicas Instrumentales" (University of León, Spain).

Pyrosequencing of the three resulting PCR fragments, targeting SNPs 167 and 198/200 in *T. circumcincta* and SNPs 198/200 in *T. colubriformis*, was carried out using the sequencing primers (Seq) (Table 1). The sequencing primer for the SNP167 of *T. circumcincta* was previously described by Skuce et al. [25]. The pyrosequencing assay was carried out using a PyroMark ID Pyrosequencer (Biotage, Sweden) according to the manufacturer's recommendations. After the initial PCR amplification, 40 µl of PCR product was added to 37 µl 2x Binding buffer (Biotage, Sweden), 3 µl streptavidin sepharose beads (Roche) in a 96 well plate and then agitated for 5 min at room temperature to allow binding of biotin-labelled DNA to the beads. The beads were processed using the sample preparation tool and reagents (Biotage, Sweden) dispensed into the assay plate with 40 µl of 0.4 µM sequencing primer per well. Positive controls representing gDNA extracted from susceptible and resistant individual adults from each species were included in each assay.

The determination of the allele frequencies was carried out at least 3 times for each sample and the arithmetic mean was calculated. The frequencies of the resistant allele with values equal to or lower than <5% were considered as technical background and, therefore, not classed as resistant.

Results

Faecal egg count reduction test (FECRT)

The results of the FECRT carried out in the 10 sheep flocks are shown in Tables 3 and 4.

Table 1 Primer sequences for *T. circumcincta* SNPs 167 and 198/200, and *T. colubriformis* SNP 198/200, and sequence primers for both species

Primer name	Sequence 5'-3'	Modifications	Expected product size (pb)	Tm (°C)
T. circumcincta				
Tc SNP200 F	CACTCTTTCTGTACACCAATTG	[Btn]5	128	60
Tc SNP200 R	AGTGATTGAGATCGCCATAA	-		
Tc SNP167 F	CAAAATTCGCGAGGAGTAT	-	276	60
Tc SNP167 R	TTCTACCAATTGGTGTACAGAAAG	[Btn]5		
T. colubriformis				
Tri SNP200 F	TACTTTATCAGTCCATGAGCTGG	[Btn]5	128	62
Tri SNP200 R	ATGGTTGAGATCTCCATAGGTTG	-		
Sequencing				
Tc 200 Seq	AGAGCTTCATTATCGATG	-	-	60
Tc 167 Seq	CGGATAGAATCATGGCT	-	-	60
Tri 200 Seq	AGAGCTTCGTTATCGATGCA	-	-	62

According to the results of FECRT, and taking into account the resistant and borderline flocks together, 40% of flocks were resistant to one drug. Resistance to ABZ was observed in 12.5% (1/8) of the flocks and to IVM in 44.4% (4/9). One flock (10%) was resistant to both drugs.

Before treatment, the most frequent species were *T. circumcincta* (43–55% of larvae) and *Trichostrongylus* sp (38–47% of larvae), however, other GIN species such as *H. contortus*, *Bunostomun* sp, *Nematodirus* sp, *Chabertia ovina* and *Cooperia oncophora* were also observed at a lower percentage (1-13% of larvae) (Table 2). After treatment, only *T. circumcincta* and *Trichostrongylus* sp were identified, with the exception of farm 7 where *Bunostomun* sp was also identified (6% of larvae) (Tables 3 and 4).

β-Tubulin allele frequencies
The assays were capable of detecting BZ resistance-associated SNPs because, with gDNA from resistant adult samples, included as a positive control, the mean frequencies of the resistant allele at codon 200 were 89% in *T. circumcincta* and 92% in *T. colubriformis*.

Allele frequencies before treatment
In relation to *T. circumcincta*, the resistant allele was not found in any of the flocks tested at codons 167, 198 or 200. In *T. colubriformis*, the resistant allele was found in 11.1% (1/9) of the flocks at codon 198, with a very low percentage (6.8%). At codon 200, the resistant allele was found in 33.3% (3/9) of the flocks, with high frequencies ranging from 48.5 to 87.8% (Table 2).

Allele frequencies after treatment with ABZ
For *T. circumcincta*, after the administration of ABZ, the percentage of flocks with the resistant allele at codon 167 was 16.7% (1/6), with a frequency of 23.8%. The resistant allele at codon 198 was found in one out of five

flocks (20%) with a frequency of 13.2%. On the other hand, the resistant allele carrying SNP200 was not found in any flock.

In relation to *T. colubriformis*, the resistant allele at codon 198 was found in 25% (2/8) of flocks, with values of 5.5 and 19%, and at position 200 in 75% (6/8) of flocks with very high frequencies, between 89.8 and 99.3% (Table 3).

Allele frequencies after treatment with IVM
After treatment with IVM, for *T. circumcincta*, the resistant allele carrying SNP167 was found in 11.1% of the flocks (1/9) but with a very low frequency (6.1%). At codons 198 and 200, the resistant allele was not found in any sample.

For *T. colubriformis*, the SNP198 was found in 3 of the 8 flocks (37.5%) at low frequency, between 6.1 and 14%. At codon 200, all flocks in which the determination was done carried the resistant allele, with frequencies ranging from 23.3 to 100% (Table 4).

Discussion
This study describes the frequencies of the resistant alleles present in the gene encoding isotype-1 β-tubulin of field populations of GIN, collected before and after treatment with ABZ or IVM. The resistance status of these flocks situated in the Northwest of Spain was firstly determined in vivo by means of the FECRT.

Since the FECRT cannot detect low resistance levels, especially under field conditions when the infections by GIN are typically of mixed species composition [26], determining the frequency of resistant alleles in pools of L3 could be an alternative to or proxy for the FECRT, with the added benefit, if sufficiently robust and repeatable, of not needing to treat animals to determine resistance status. Since the resistant phenotype is only

Table 2 Allele frequencies before treatment for *T. circumcincta* SNPs 167 and 198/200, and *T. colubriformis* SNP 198/200, and morphological identification

Farm	T. colubriformis		T. circumcincta			% L3 species						
	% SNP200	% SNP198	% SNP200	% SNP198	% SNP167	Tc	Tri	Hc	Bu	Ne	Ch	Co
1	0	2.5	0	0	0	51	42	4	1	2	0	0
2	51.6	6.8	0	0	-	49	47	3	0	1	0	0
3	-	-	3.4	0	0	52	45	2	0	0	0	1
4	0	3.3	0	0	2.6	43	41	2	12	1	2	0
5	0	0	0	0	2.8	55	42	2	0	0	0	0
6	3.1	0	0	0	2.8	48	38	0	1	0	13	0
7	2	2	0	2	3.1	46	43	0	3	0	8	0
8	87.8	3.9	-	-	0	-	-	-	-	-	-	-
9	0	3.1	0	0	-	54	46	0	0	0	0	0
10	48.5	0	0	0	-	53	46	0	1	0	0	0

Tc: *T. circumcincta*, Tri: *Trichostrongylus* spp, Hc: *H. contortus*, Bu: *Bunostomun* spp, Ne: *Nematodirus* spp, Ch: *Chabertia ovina* and Co: *Cooperia oncophora*. -: Failed samples

Table 3 Allele frequencies after ABZ treatment for *T. circumcincta* SNPs 167 and 198/200, and *T. colubriformis* SNP 198/200, and morphological identification

Farm	% Egg reduction	R/S classification	*T. colubriformis*		*T. circumcincta*			% L3 species		
			% SNP200	% SNP198	% SNP200	% SNP198	% SNP167	Tc	Tri	Bu
1	91.4	Borderline	1.7	4.8	2.9	0	3	50	50	0
2	98.9	S	97.6	3	0	0	23.8	55	45	0
3	100	S	96	1.5	0	13.2	4.4	56	44	0
4	98.5	S	99.3	5.5	0	0	-	58	42	0
5	100	S	89.8	0	0	0	4.8	52	48	0
6	99.3	S	1.3	3	-	-	-	62	38	0
7	100	S	90.6	19	-	-	0	47	47	6
10	100	S	97.2	4.8	-	-	0	38	62	0

Tc: *T. circumcincta*, Tri: *Trichostrongylus* spp and Bu: *Bunostomun* spp. -: Failed samples

detected by the FECRT when the frequency of resistant alleles in the population is over 25% [22], the FECRT only represents an estimate of the resistance in a flock naturally infected by GIN.

In this study, we used pyrosequencing to determine the resistant allele frequency in the gene encoding isotype-1 β-tubulin at codons 167, 198 and 200 of *T. circumcincta* and codons 198 and 200 of *T. colubriformis*, in DNA samples from pools of L3 collected before and after treatment. These species were the most frequent detected in all sampled flocks and represent the 84–100% of the GIN burden on all of them.

According to Kwa et al. [14] and Elard and Humbert [27], the mutation of Phe to Tyr at position 200 of β-tubulin isotype 1 is the most important mechanism responsible for conferring resistance to BZ. According to Silvestre and Cabaret [15], this same mutation at codon 167 is also involved in the development of resistance in the absence of the mutation at position 200. The SNP167 is rare in field populations, but for *T. circumcincta* may account for the survival of parasites which

do not carry the mutation at codon 200 [15]. In the present study, this SNP was present before treatment; after treatment with ABZ, it was only shown in one susceptible flock with a low frequency of 23.8% and, therefore, we do not correlate it with AR. However, Demeler et al. [28] found resistant alleles at codons 167 and 200, but not at 198 in an *Ostertagia ostertagi* BZ resistant isolate, indicating that BZ treatment does not necessarily select for the SNP198. In the present study, the SNP at codon 198 was not found before treatment in *T. circumcincta* and, after administration of ABZ, only one susceptible flock, with a low frequency (13.2%) was detected in the five flocks tested. Our results regarding the SNP198 in *T. circumcincta* are not conclusive, due to the low number of analyses. However, Ghisi et al. [16] found the SNP198 in 90% of resistant isolates of *H. contortus* which did not carry the mutation in codon 200. Therefore, SNP198 has been related to BZ resistance. In *T. colubriformis*, before treatment, SNP198 was only found in one flock with a very low frequency (6.8%), and after treatment with

Table 4 Allele frequencies after IVM treatment for *T. circumcincta* SNPs 167 and 198/200, and *T. colubriformis* SNP 198/200, and morphological identification

Farm	% Egg reduction	R/S classification	*T. colubriformis*		*T. circumcincta*			% L3 species	
			% SNP200	% SNP198	% SNP200	% SNP198	% SNP167	Tc	Tri
1	37.3	R	23.3	6.4	0	0	0	63	37
2	100	S	97.3	0	0	0	0	54	46
3	86.0	R	79.2	14	0	0	2.4	41	59
4	−11.9	R	94.7	0	0	0	0	60	40
5	99.3	S	93.7	1	-	-	0	50	50
6	100	S	-	-	0	0	3	54	46
7	95.9	S	100	14	0	0	0	70	30
8	100	S	89.9	6.1	-	-	0	-	-
9	93.5	Borderline	95	2.4	0	0	6.1	44	56

Tc: *T. circumcincta* and Tri: *Trichostrongylus* spp. -: Failed samples

ABZ in two different flocks with frequencies of 5.5 and 19%.

Therefore, in the present study, the frequencies of resistant alleles carrying SNPs 167 and 198 in *T. circumcincta*, and SNP 198 in *T. colubriformis*, are either zero or very low and consequently we did not find any relation between them and AR status/phenotype.

In relation to the SNP200 in *T. circumcincta*, this was not found before or after treatment with ABZ. These results are in agreement with the previous study of Martinez-Valladares et al. [29], who did not describe any resistant allele at codons 167, 198 and 200 in individual *T. circumcincta* L3 collected from flocks in the same study area (León, Spain). Due to the absence of resistant alleles, but also because we only found one flock with borderline resistance to ABZ, we cannot conclude that there is an association between resistance and resistant allele frequency at codon 200 in *T. circumcincta*. On the other hand, Skuce et al. [25], after studying different strains of *T. circumcincta*, reported a resistant allele frequency of 64.9% in a multidrug-resistant strain (MTci5) at codon 200 and 0% in a susceptible strain (MTci1).

However, the results shown in the current study for the frequencies at SNP 200 in *T. colubriformis* are totally different. Before treatment with ABZ, the resistant allele was present in 33.3% of the farms tested, with values between 48.5 and 87.8% (Mean = 62.6%) and, after ABZ treatment, in 75% of the farms, with very high allele frequencies, ranging from 89.8 to 99.3% (Mean = 95.1%). However, these data are not consistent with the results of the FECRT, since we found high levels of the resistant allele in susceptible farms classified according to the FECRT. This result could be due to the resistant allele(s) being diluted in the population because of the presence of other (susceptible) species, before treatment. The FECRT can only detect resistance when ~25% of the total population is resistant. The resistant allele at codon 200 has been previously reported in different BZ resistant strains of *T. colubriformis* [30, 31] and also in *Trichostrongylus axei* adult worms, with a frequency of 63% recovered from lambs after treatment with BZ [32].

In the current study, we also determined the frequency of the resistant alleles for these species before and after the administration of IVM. Freeman et al. [33] were the first authors to describe a relation between resistance to IVM and the β-tubulin gene in *H. contortus*. These authors found a marked alteration in the amphid neurons, which are formed by bundles of microtubules, heterodimers of α-tubulin and β-tubulin in IVM resistant strains of *H. contortus*. Moreover, Mottier and Prichard [17] reported that repetitive use of IVM and moxidectin in *H. contortus* strains produced changes in the frequency of the alleles at codons 167, 198 and 200 of β-tubulin.

After the administration of IVM, in *T. circumcincta* the resistant allele was not found at codons 198 and 200. At position 167, the resistant allele was found in only one flock but with a very low frequency of 6.1%. These results are in agreement with a previous study in which none of these SNPs (167, 198 and 200) were detected in *T. circumcincta* when sheep were treated with IVM; the authors concluded that other molecular mechanisms than beta-tubulin could be implicated in the development of resistance against IVM in this case [29]. The present study would add further evidence to support this hypothesis, since 4 out of 9 studied flocks were resistant or borderline to IVM in the absence of the acknowledged resistant alleles. Therefore, we suggest that *T. circumcincta* is not responsible for the presence of resistance in these flocks and/or there could be different mechanisms or genes implicated in resistance development, especially to IVM. Previous studies suggest that the F200Y mutation could be related to IVM resistance in different species as observed by Njue et al. [34] in IVM-resistant strains of *C. oncophora* and by Eng and Prichard [35] in *Onchocerca volvulus*.

In contrast, for *T. colubriformis*, the frequency of the resistant allele at codon 200 after the administration of IVM was between 23.3 and 100% (Mean = 82.6%). These data suggest that there could be an association between the SNP 200 in *T. colubriformis* and IVM resistance. However, according to FECRT, only 4 flocks were resistant or borderline resistant to this drug. The effect of the resistant allele could be diluted by the presence of other susceptible species, like it happened after the treatment with ABZ. Recently, Ashraf et al. [36] reported that that IVM binds to *H. contortus* α and β tubulins at low micromolar affinities and stabilizes the microtubules. However, in a different study, Ashraf et al. [37] concluded that the SNPs 167 and 200 cause no difference in the polymerization of wild and mutant tubulins, and therefore, neither of the SNPs reduced IVM binding. The hypothesis is that the SNPs 167 and 200 could be part of a signaling mechanism that results in over expression of P-glycoproteins, which ultimately leads to IVM resistance [37].

In relation to these results, we suggest that the presence of resistance is more common than the FECRT indicates. Therefore using only this in vivo test we could obtain false negatives. The percentage of resistant alleles show a more detailed perspective on the presence of resistance at the molecular level; in consequence the level of resistance of a strain will depend on the percentage of resistant alleles in each species. Therefore, it would be worthy monitoring the percentage of resistant alleles after the FECRT, in susceptible flocks to avoid the development of resistance to BZs, and in resistant ones to dilute the percentage of resistant alleles, in both cases applying different management practices at the same time.

Conclusions

In conclusion, comparing the results of FECRT and pyrosequencing, we suggest that the presence of resistance is more common than expected and that would have been declared using FECRT alone. The SNP200 in *T. circumcincta* is not present in the Spanish nematode populations tested although its association with resistance has been described in others field isolates from other countries, most notably, UK [25] and France [31]. In contrast, in *T. colubriformis*, the SNP200 is present on multiple farms and at high frequency before (between 48.5 and 87.8%) and after treatment with ABZ (mean in positive flocks = 95.1%) and IVM (mean = 82.6%). In relation to the other SNPs, 167 and 198, these were not detected in most of the analyses, so were either not present or present at very low frequency i.e. below the sensitivity of detection of the pyrosequencing assays used. Therefore, we cannot conclude that these SNPs are related with ABZ and/or IVM resistance in any of the Spanish GIN populations tested.

Abbreviations

ABZ: Albendazole; AR: Anthelmintic resistance; BZ: Benzimidazole; BZs: Benzimidazoles; EPG: Eggs per gram of faeces; FECRT: Faecal egg count reduction test; GIN: Gastrointestinal nematodes; IVM: Ivermectin; L3: Third stage larvae; MLs: Macrocyclic lactones; SNP: Single nucleotide polymorphism; WAAVP: World association for the advancement of veterinary parasitology

Acknowledgements

This study has been funded by the national project INIA (Instituto Nacional de Investigaciones Agrarias: RTA2013-00064-C02-02) of the Ministry of Economy and Competitiveness (Ministerio de Economía y Competitividad), the European Regional Development Fund (Fondos Feder), the Spanish "Ramón y Cajal" Programme of the Ministry of Economy and Competitiveness (MMV, RYC-2015-18368), and the Cooperativa Bajo Duero, COBADU. The authors would like to thank all animal owners for their willingness to collaborate in the study.

Funding

Not applicable.

Authors' contributions

MEB performed DNA extractions, PCR amplification, data analysis, and drafted the manuscript. FARV and CGL collected samples and revised the manuscript. PJS and LN performed the pyrosequencing and revised the manuscript. MMV designed and supervised the experiment, collected samples and drafted manuscript. All authors read and approved the final manuscript.

Competing interests

The authors declare that they have no competing interests.

Consent for publication

Not applicable.

Author details

[1]Departamento de Sanidad Animal, Facultad de Veterinaria, Universidad de León, Campus de Vegazana s/n, 24071 León, Spain. [2]Instituto de Ganadería de Montaña (CSIC-Universidad de León), Finca Marzanas, Grulleros, 24346 León, Spain. [3]Moredun Research Institute, Pentland Science Park, Bush Loan, Edinburgh, UK.

References

1. Kloosterman A, Parmentier HK, Ploeger HW. Breeding cattle and sheep for resistance to gastrointestinal nematodes. Parasitol Today. 1992;8:330–5.
2. Kaminsky R, Ducray P, Jung M, Clover R, Rufener L, Bouvier J, Weber SS, Wenger A, Wieland-Berghausen S, Goebel T, Gauvry N, Pautrat F, Skripsky T, Froelich O, Komoin-Oka C, Westlund B, Sluder A, Maser P. A new class of anthelmintics effective against drug-resistant nematodes. Nature. 2008;452:176–80.
3. Little PR, Hodges A, Watson TG, Seed JA, Maeder SJ. Field efficacy and safety of an oral formulation of the novel combination anthelmintic, derquantel-abamectin, in sheep in New Zealand. N Z Vet J. 2010;58:121–9.
4. Scott I, Pomroy WE, Kenyon PR, Smith G, Adlington B, Moss A. Lack of efficacy of monepantel against *Teladorsagia circumcincta* and *Trichostrongylus colubriformis*. Vet Parasitol. 2013;198:166–71.
5. Dobson RJ, Hosking BC, Jacobson CL, Cotter JL, Besier RB, Stein PA, Reid SA. Monepantel resistance reported on Dutch sheep farms. Vet Rec. 2014;175:418.
6. Mederos AE, Banchero GE, Ramos Z. First report of monepantel *Haemonchus contortus* resistance on sheep farms in Uruguay. Parasit Vectors. 2014;7:598.
7. Van den Brom R, Moll L, Kappert C, Vellema P. *Haemonchus contortus* resistance to monepantel in sheep. Vet Parasitol. 2015;209:278–80.
8. Kaplan RM, Klei TR, Lyons ET, Lester G, Courtney CH, French DD, Tolliver SC, Vidyashankar AN, Zhao Y. Prevalence of anthelmintic resistant cyathostomes on horse farms. J Am Vet Med Assoc. 2004;225:903–10.
9. Rojo-Vázquez FA, Hosking BC. A telephone survey of internal parasite control practices on sheep farms in Spain. Vet Parasitol. 2013;192:166–72.
10. Lubega GW, Prichard RK. Specific interaction of benzimidazole anthelmintics with tubulin: high-affinity binding and benzimidazole resistance in *Haemonchus contortus*. Mol Biochem Parasitol. 1990;38:221–32.
11. Lubega GW, Prichard RK. Beta-tubulin and benzimidazole resistance in the sheep nematode *Haemonchus contortus*. Mol Biochem Parasitol. 1991;47:129–37.
12. Lubega GW, Prichard RK. Interaction of benzimidazole anthelmintics with *Haemonchus contortus* tubulin: binding affinity and anthelmintic efficacy. Exp Parasitol. 1991;73:203–13.
13. Coles GC, Klei TR. Animal parasites, politics and agricultural research. Parasitol Today. 1995;11:276–8.
14. Kwa MS, Veenstra JG, Roos MH. Benzimidazole resistance in *Haemonchus contortus* is correlated with a conserved mutation at amino acid 200 in beta-tubulin isotype 1. Mol Biochem Parasitol. 1994;63:299–303.
15. Silvestre A, Cabaret J. Mutation in position 167 of isotype 1 β-tubulin gene of Trichostrongylid nematodes: role in benzimidazole resistance? Mol Biochem Parasitol. 2002;120:297–300.
16. Ghisi M, Kaminsky R, Mäser P. Phenotyping and genotyping of *Haemonchus contortus* isolates reveals a new putative candidate mutation for benzimidazole resistance in nematodes. Vet Parasitol. 2007;144:313–20.
17. Mottier ML, Prichard RK. Genetic analysis of a relationship between macrocyclic lactone and benzimidazole anthelmintic selection on *Haemonchus contortus*. Pharmacogenet Genomics. 2008;18:129–40.
18. Sutherland IA, Leathwick DM. Anthelmintic resistance in nematode parasites of cattle: a global issue? Trends Parasitol. 2011;27:176–81.
19. Martínez-Valladares M, Geurden T, Bartram DJ, Martínez-Pérez JM, Robles-Pérez D, Bohórquez A, Florez E, Meana A, Rojo-Vázquez FA. Resistance of gastrointestinal nematodes to the most commonly used anthelmintics in sheep, cattle and horses in Spain. Vet Parasitol. 2015;211:228–33.
20. Waller PJ. Anthelmintic in nematodes parasites of sheep. In: Russel GE, editor. Agricultural zoology reviews, Vol. 1. Poteland, UK: Intercept; 1986. p. 33–7.
21. Jackson F, Coop RL. The development of anthelmintic resistance in sheep nematodes. Parasitology. 2010;120(Suppl):95–107.
22. Martin PJ, Anderson N, Jarrett RG. Detecting benzimidazole resistance with faecal egg count reduction tests and in vitro assays. Aust Vet J. 1989;66:236–40.

Quantification of resistant alleles in the β-tubulin gene of field strains of gastrointestinal nematodes...

157

23. MAFF (Ministry of Agriculture, Fisheries and Food). Manual of veterinary Parasitology laboratory techniques. 3rd ed. London: GB; 1986.

24. Coles GC, Bauer C, Borgsteede FH, Geerts S, Klei TR, Taylor MA, Waller PJ. World Association for the Advancement of Veterinary Parasitology (W.A.A.V.P.) methods for the detection of anthelmintic resistance in nematodes of veterinary importance. Vet Parasitol. 1992;44:35–44.

25. Skuce P, Stenhouse L, Jackson F, Hypša V, Gilleard J. Benzimidazole resistance allele haplotype diversity in United Kingdom isolates of *Teladorsagia circumcincta* supports a hypothesis of multiple origins of resistance by recurrent mutation. Int J Parasitol. 2010;40:1247–55.

26. McKenna PB. Further potential limitations of the undifferentiated faecal egg count reduction test for the detection of anthelmintic resistance in sheep. N Z Vet J. 1997;45:244–6.

27. Elard L, Humbert JF. Importance of the mutation of amino acid 200 of the isotype 1 beta-tubulin gene in the benzimidazole resistance of the small-ruminant parasite *Teladorsagia circumcincta*. Parasitol Res. 1999;85:452–6.

28. Demeler J, Krüger N, Krücken J, von der Heyden VC, Ramünke S, Küttler U, Miltsch S, López Cepeda M, Knox M, Vercruysse J, Geldhof P, Harder A, von Samson-Himmelstjerna G. Phylogenetic characterization of β-tubulins and development of pyrosequencing assays for benzimidazole resistance in cattle nematodes. PLoS One. 2013;8:e70212.

29. Martínez-Valladares M, Donnan A, Geldhof P, Jackson F, Rojo-Vázquez FA, Skuce P. Pyrosequencing analysis of the beta-tubulin gene in Spanish *Teladorsagia circumcincta* field isolates. Vet Parasitol. 2012;184:371–6.

30. Grant WN, Mascord LJ. Beta-tubulin gene polymorphism and benzimidazole resistance in *Trichostrongylus colubriformis*. Int J Parasitol. 1996;26:71–7.

31. Silvestre A, Humbert JF. Diversity of benzimidazole-resistance alleles in populations of small ruminant parasites. Int J Parasitol. 2002;32:921–8.

32. Palcy C, Silvestre A, Sauve C, Cortet J, Cabaret J. Benzimidazole resistance in *Trichostrongylus axei* in sheep: long-term monitoring of affected sheep and genotypic evaluation of the parasite. Vet J Jan. 2010;183:68–74.

33. Freeman AS, Nghiem C, Li J, Ashton FT, Guerrero J, Shoop WL, Schad GA. Amphidial structure of ivermectin-resistant and susceptible laboratory and field strains of *Haemonchus contortus*. Vet Parasitol. 2003;110:217–26.

34. Njue AI, Prichard RK. Cloning two full-length beta-tubulin isotype cDNAs from *Cooperia oncophora*, and screening for benzimidazole resistance-associated mutations in two isolates. Parasitology. 2003;127:579–88.

35. Eng JK, Prichard RK. A comparison of genetic polymorphism in populations of *Onchocerca volvulus* from untreated- and ivermectin-treated patients. Mol Biochem Parasitol. 2005;142:193–202.

36. Ashraf S, Beech RN, Hancock MA, Prichard RK. Ivermectin binds to *Haemonchus contortus* tubulins and promotes stability of microtubules. Int J Parasitol. 2015;45:647–54.

37. Ashraf S, Mani T, Beech RN, Prichard RK. Macrocyclic lactones and their relationship to the SNPs related to benzimidazole resistance. Mol Biochem Parasitol. 2015;201:128–34.

17

Sarcoptes infestation in two miniature pigs with zoonotic transmission

Alexander Grahofer[1]* iD, Jeanette Bannoehr[2], Heiko Nathues[1] and Petra Roosje[3]

Abstract

Background: Scabies is a contagious skin disease rarely described in miniature pigs. To the best of the authors' knowledge, a zoonotic transfer from infected pet pigs to humans has not been reported previously.

Case presentation: This case report describes the infestation with *Sarcoptes scabiei* mites in two miniature pigs presenting with unusual clinical signs, and disease transmission to a child.
Two 7-month-old male castrated miniature pig siblings were examined. Both had developed skin lesions, one animal was presented for neurological signs and emaciation. They were housed together in an indoor- and outdoor enclosure. Dermatological examination revealed a dull, greasy coat with generalized hypotrichosis and multifocal erythema. Microscopic examination of skin scrapings, impression smears of affected skin and ear swabs revealed high numbers of Sarcoptes mites in both animals as well as bacterial overgrowth. A subcutaneous injection of ivermectin 0.3 mg/kg was administered to both animals and repeated after 2 weeks. Both miniature pigs received subcutaneous injections with butafosfan and cyanocobalamin, were washed with a 3% chlorhexidine shampoo and were fed on a well-balanced diet. Pig enclosures were cleaned. The infested child was examined by a physician and an antipruritic cream was prescribed. Both miniature pigs and the child went into clinical remission after treatment.

Conclusion: Sarcoptic mange is rare or even eradicated in commercial pig farming in many countries but miniature pigs may represent a niche for *Sarcoptes scabiei* infections. This case report indicates that miniature pigs kept as pets can efficiently transmit zoonotic disease to humans. In addition, these animals may represent a niche for *Sarcoptes scabiei* infestation in countries where sarcoptic mange in commercial pig farms has been eradicated and could therefore pose, a hazard for specific pathogen free farms.

Keywords: Pet pig, Zoonosis, Sarcoptic mange, Pruritus, Mite, Hypotrichosis

Background

This case report describes a severe infestation of scabies in two miniature pigs kept as pets and subsequent zoonotic infection of a child.

Sarcoptic mange is regarded as one of the most important ectoparasitic diseases of swine worldwide, caused by the burrowing mite *Sarcoptes (S.) scabiei var. suis*, which shows a certain degree of host specificity [1–3]. However, it is a highly contagious skin disease with the potential to affect a variety of different animal host species [4]. All life stages of the parasite are found in either burrows of the epidermis or on the surface of the skin [1, 5]. Transmission mainly occurs directly, by prolonged skin to skin contact, but also indirectly, as mites may survive in humid and cold environment for several weeks [1, 6]. Mites can be found distributed over the entire body of infested hosts, but are generally concentrated in the ears and on the outer pinnae [2]. In pigs, two distinct clinical presentations of scabies are recognized and the occurrence of one or the other depends on the age of the animal. In growing pigs, a pruritic hypersensitive form is commonly seen, whereas a chronic, hyperkeratotic form with the presence of aural crusts and a large number of mites on the animal is recognized in sows [7]. A high prevalence of up to 95% within infected herds causes tremendous economic losses in swine reproduction and compromises animal welfare [2, 8]. A conclusive diagnosis of swine scabies is difficult, because a single, reliable and sensitive diagnostic tool is not available [9, 10]. The diagnosis in pigs with crusted scabies can usually be made

---

* Correspondence: alexander.grahofer@vetsuisse.unibe.ch
[1]Clinic for Swine, Department of Clinical Veterinary Medicine, Vetsuisse Faculty, University of Bern, Bremgartenstrasse 109a, 3012 Bern, CH, Switzerland

easily, as thousands of mites are harbored in the skin and can be detected by skin scrapings. However, a timely diagnosis and successful elimination of mange can be challenging in pigs with only mild symptoms, due to the presence of only very few mites in the early stages of disease, leading to negative skin scrapings and asymptomatic or only mildly symptomatic animals ([1, 2, 7]. Furthermore, sarcoptic mange mimics other skin diseases such as atopic dermatitis, insect bites and skin conditions caused by irritating agents [1]. Thus, diagnostic blood tests for scabies have been developed to improve specificity and sensitivity [1]. Several ELISA systems are available to detect antibodies in infected pigs 5 to 6 weeks post infection with scabies mites [11]. In addition, examination of pig carcasses in the slaughterhouse provides further information on the scabies status of particular pig herds. In conclusion, clinical signs consistent with sarcoptic mange, microscopic analysis of skin scrapings and blood sampling for detection of specific antibodies by ELISA should ideally be combined to reach an accurate diagnosis.

Nowadays, miniature pigs have become popular companion animals and are more frequently seen by veterinarians [12]. While information on swine sarcoptic mange caused by S. scabiei var. suis is available, scabies in miniature pigs is rarely described in the scientific literature. "Dippity" syndrome is an important differential diagnosis in miniature pigs with skin lesions and behavioral problems. It is a poorly understood disease and once affected the animals recover within a few days without any treatment [13, 14].

Zoonotic transfer from infected pigs kept as pets has not been reported previously and little is known about zoonotic transmission of Sarcoptes mites from pigs to humans [15]. Transmission of the mite from infected domestic animals to humans occurs during close contact and causes intense pruritus and irritation in affected humans due to a hypersensitivity reaction against the mites and their products. Young children and immunocompromised adults are more susceptible to the disease [10]. Infestations are typically self-limiting but cases of persistent infection requiring several months to resolve have been described in the literature [16, 17].

Case presentation

A 7- month-old, castrated male miniature pig with a body weight of 3.5 kg was referred for abnormal behavior consisting of continuous screaming, increased periods of lateral recumbency with uncontrolled pedaling motions, decreased activity and ataxia. The first symptoms appeared approximately 1 week prior to referral to the swine clinic. Previous therapy consisted of danofloxacin and corticosteroid injections with unknown dosage.

According to the owner, the other miniature pig, a castrated male littermate, did not show any abnormal

clinical signs. Both animals shared an indoor pen in the house and an outdoor enclosure with a hut. The animals were fed a commercial horse feed combined with fresh vegetables and fruit, and offered water ad libitum. The miniature pigs had not been vaccinated and had never been treated for endo- or ectoparasites. Both animals had been bought 1 month before from a specialized miniature pig breeder. At presentation, the referred miniature pig was lethargic, and had a cachectic body condition. General examination revealed a slightly decreased body temperature of 36.5 °C (physiological range: 37–38 °C), a mildly increased heart rate of 100 beats per minute (physiological range: 68–98 per minute), and an increased respiratory rate of 36 (physiological range: 11–29 per minute).

Dermatological examination of the miniature pig revealed a dull, greasy coat with generalized hypotrichosis and multifocal erythema (Fig. 1). Extensive areas of hyperkeratosis were observed on the head, bilateral on the shoulders, extremities, the abdomen and perineal area. The entrance of both ear canals was obstructed with cerumen and squames, bordered by hyperkeratotic crusts. Microscopic examination of skin scrapings and cerumen revealed numerous Sarcoptes mites, instars, fecal pellets and eggs (Fig. 2). Cytological examination of the cerumen and crusts from the entrance to both ear canals revealed heavy colonization with rod-shaped bacteria, and a low number of coccoid bacteria and yeasts. Impression smears of the skin showed some neutrophils, extracellular cocci and corneocytes.

Upon diagnosis of sarcoptic mange in this patient, the other miniature pig was also presented and examined. This animal showed a better body condition and comparable but milder skin lesions, consisting of mild hypotrichosis over the head and trunk and greasy skin with

Fig. 1 Clinical photograph of the severely affected miniature pig with generalized hypotrichosis, multifocal hyperkeratotic crusts and erythema

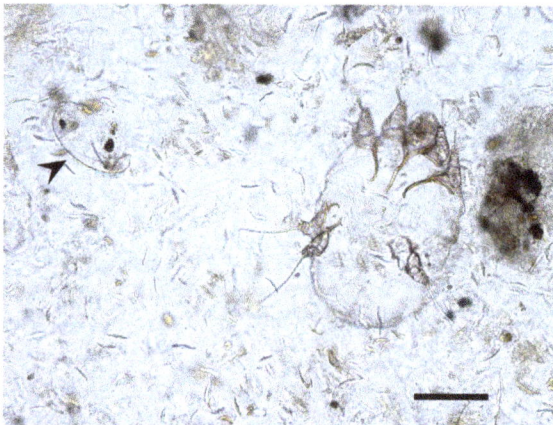

Fig. 2 Photomicrograh of a Sarcoptes scabiei mite and an egg (arrow) in skin scrapings. Bar = 0.15 mm

large scales. Pale hyperkeratotic crusts were absent. Microscopic examination revealed sarcoptes mites, but in smaller numbers. Further, the owner reported that her 7- year old daughter - often in contact with the miniature pigs - had recently developed pruritic papular skin lesions on the upper legs (Fig. 3) and was referred to a physician who confirmed a scabies infestation.

Both miniature pigs remained hospitalized for 14 days. During this time, an adequate feeding regime was started and a heating source was installed to control body temperature. Both pigs were treated with a subcutaneous injection of ivermectin (0.3 mg/kg; Ivomec®; Merial Ltd, Duluth, GA, USA), which was repeated after 2 weeks, combined with an intramuscular injection of butafosfan and cyanocobalamin (vitamin B12) (0.3 ml/kg; Catosal® 10%; Bayer, Shawnee Mission, KS, USA). Furthermore, the severely affected miniature pig was washed with a 3% chlorhexidine shampoo (Pyoderm; Virbac®, SA, Carros Cedex 06511, France) three times a week until scales and crusts resolved. The other miniature pig was washed only twice. No clinical adverse effects were observed throughout the treatment period. The owner

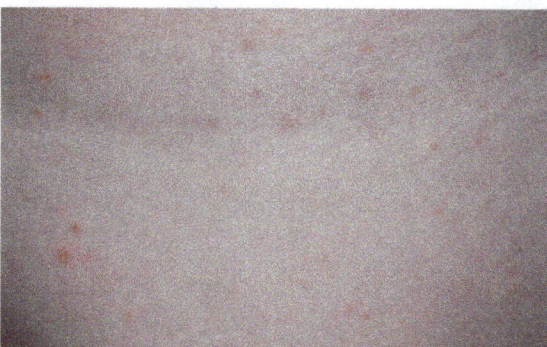

Fig. 3 Papules and erythema on the leg of the infested child

was advised to remove the straw from the hut and to thoroughly clean both pens with hot water. After cleaning the outdoor area, the pen and hut were sprayed with a 1% cypermethrin solution (Intermitox®, SaVet Pharma GmbH, Vechta, Germany). The bedding of the indoor sleeping place was washed at 60 °C. The owner's child was treated with an antipruritic ointment prescribed by the physician.

Skin scrapings were repeated after 2 weeks of hospitalization and were negative in both animals. Subsequently, both animals were discharged from the clinic. The owner was advised on how to feed the animals properly. Both miniature pigs were examined at their home at two and 4 weeks after their return. During this time, the animals further improved and displayed normal behavior and weight gain with complete remission of skin lesions. The owner was contacted by phone 4 months later and declared that both animals and the child remained without skin lesions.

Discussion

Nowadays, miniature pigs are exotic but relatively popular pets [13]. In the last decades, the number of miniature pigs kept as pets increased significantly. However, finding veterinarians with expertise in this particular animal species may prove challenging for animal owners [14]. Hence, many miniature pigs do not receive the recommended preventive health care. Swine specialists have the best knowledge in treating pet pigs, but are not always within easy reach for the owners. Furthermore, owners can be easily offended if routine techniques to handle swine are used on their pets [14].

A survey of clinical problems identified in 102 miniature pigs revealed that the locomotor system and the integument were the most common locations for problems in pet pigs [18]. In our case, a *Sarcoptes* infestation was diagnosed in both miniature pigs. Depending on the country, sarcoptic mange can be either absent or endemic with variable prevalence in conventionally held pigs [19]. Although sarcoptic mange is eradicated in farmed pigs in Switzerland, infestations with *Sarcoptes* mites were recently recognised in wild boars [20]. However, the presented miniature pigs, not having any evidence for wild boar contact, had most likely been infected at the breeding farm. Based on history and on-site examination, any transmission by fomites or direct contact with wild boars or other pigs in their home environment was excluded.

Pruritus in artificially infested pigs may develop at the earliest 2–3 weeks post infestation but often starts later [21, 22]. Although intense pruritus is considered the hallmark sign of sarcoptes infestations in dogs and other species, it was not obvious as such in either presented miniature pigs. Neither of the animals was observed to scratch or rub themselves intensely during hospitalisation, but the pedalling motions in the recumbent animal may have been a sign of pruritus combined with general weakness. However, this animal ate with great appetite when offered food, and was much more active 1 day after the first treatment. Malnutrition caused by inappropriate diet may have contributed to the development of crusted scabies and atypical clinical signs in this miniature pig. Development of crusted scabies with a high mite burden is generally associated with an inadequate immune response and is often associated with immunosuppression in humans and dogs [23]. This observation is supported by the fact that *Sarcoptes*-infected pigs develop crusted scabies when treated with dexamethasone [24]. It remains a moot point whether the previous glucocorticoid injection contributed to the hyperkeratotic lesions in one of the miniature pigs.

Although the health status of adult pigs with crusted scabies has not been described in detail, it has not been reported either that these animals suffer from poor health. However, these animals typically do not show pruritus and therefore do not remove mites by pruritic behavior. This lack of pruritic behavior may have resulted in the large mite numbers found in the miniature pig described here.

Whereas the distribution of clinical lesions in pigs affected by sarcoptic mange usually includes the head and lateral side of the extremities, the proximal parts of the pinnae around the entrance to the ear canals were severely affected in one of the miniature pigs described here. Of note, this area is reported to be a predilection site with high mite burden [21, 25].

Thorough cleaning of the pigs' enclosures is mandatory, as transmission of sarcoptes mites by fomites has been demonstrated in pigs [22]. Transmission of sarcoptes mites from dogs or other species to people and induction of pruritus in affected humans is well known. Close contact with infected animals and lack of host specificity are contributing factors [23]. More recently, DNA-based studies on sarcoptes mites from different animal hosts including humans indicate that there is only limited genetic diversity, and one clade included both human and pig genotypes [26]. This finding supports the likelihood of cross-species infestations. In the case described here, the clinical signs of pruritus and erythematous papules on the legs of the child, the time point of their appearance and later recovery were suggestive of infestation with *Sarcoptes scabiei var. suis*.

Sarcoptic mange in pigs is successfully managed by two subcutaneous injections of ivermectin or a single intramuscular injection of doramectin [27, 28]. Because of malnutrition and poor to moderate body condition, both miniature pigs described here received an injection with a metabolic stimulant containing butafosfan and

cyanocobalamin [29, 30]. The additional chlorhexidine shampoo therapy was administered to counteract the changes in the skin microbiota as observed by cytology. Alterations in the skin microbiome were demonstrated in a porcine scabies model where an increase of *Staphylococcus* species and a shift within the staphylococcal population in pigs with scabies infestation were shown [31]. In principle, this could have implications for the human skin microbiota as well, especially when pigs are kept as companion animals in close proximity with their owners sharing the same environment [32].

Conclusions

This report describes the first case of a severe infestation of scabies in two miniature pigs kept as pets causing zoonotic infection of a child.

Pruritus is a clinical hallmark of sarcoptes infestations in most species, but can be absent or aberrant in miniature pigs. The inadequate nutrition may have influenced the course of disease in the case presented. Keeping miniature pigs as pets facilitates zoonotic transfer of *Sarcoptes* mites and microbiota. In addition, these animals may represent a niche for *Sarcoptes scabiei* infestation in countries where sarcoptic mange in commercial pig farms has been eradicated and can therefore pose a hazard for specific pathogen free farms.

Acknowledgments
The authors gratefully acknowledge Regula Cao, Division of Clinical Dermatology, Vetsuisse Faculty University of Bern for assistance with the clinical workup. The authors also thank the owner for the collaboration.

Funding
Funding information is not applicable.

Authors' contributions
AG performed the clinical examination, developed the diagnosis, designed the treatment of the miniature pigs, summarized the results of the cases and drafted the manuscript. JB performed the dermatological skin examination, skin sampling and microscopic analysis and drafted parts of the manuscript. HN and PR supervised and coordinated the project. All authors contributed to the development and the revisions of the manuscript and approved the final version.

Consent for publication
The mother of the child gave a written and verbal consent for all or any part of this material to appear in this case report.

Competing interests
There are no competing interests of any of the authors that could inappropriately influence or bias the content of the paper.

Author details
[1]Clinic for Swine, Department of Clinical Veterinary Medicine, Vetsuisse Faculty, University of Bern, Bremgartenstrasse 109a, 3012 Bern, CH, Switzerland. [2]Dermatology Department, Animal Health Trust, Lanwades Park, Kentford, Newmarket, Suffolk, Cardiff CB8 7UU, UK. [3]Division of Clinical Dermatology, Department of Clinical Veterinary Medicine, Vetsuisse Faculty, University of Bern, Bern, Länggassstrasse 128, 3012 Berne, CH, Switzerland.

References

1. Arlian LG, Morgan MS. A review of sarcoptes scabiei: past, present and future. Parasit Vectors. 2017;10:297. https://doi.org/10.1186/s13071-017-2234-1.
2. Davis DP, Moon RD. Dynamics of swine mange: a critical review of the literature. J Med Entomol. 1990;27:727–37. https://doi.org/10.1093/jmedent/27.5.727.
3. Sokolova TV, Lange AB. The parasite-host specificity of the itch mite sarcoptes scabiei (acariformes: sarcoptidae) in man and animals (a review of the literature). Parazitologiia. 1992;26:97–104.
4. Currier RW, Walton SF, Currie BJ. Scabies in animals and humans: history, evolutionary perspectives, and modern clinical management. Ann N Y Acad Sci. 2011;1230:E50–60. https://doi.org/10.1111/j.1749-6632.2011.06364.x.
5. Burgess I. Sarcoptes scabiei and scabies. Adv Parasitol. 1994;33:235–92.
6. Fang F, Bernigaud C, Candy K, Melloul E, Izri A, Durand R, et al. Efficacy assessment of biocides or repellents for the control of sarcoptes scabiei in the environment. Parasit Vectors. 2015;8:416. https://doi.org/10.1186/s13071-015-1027-7.
7. Laha R. Sarcoptic mange infestation in pigs: an overview. J Parasit Dis. 2015;39:596–603. https://doi.org/10.1007/s12639-014-0419-5.
8. Davies PR. Sarcoptic mange and production performance of swine: a review of the literature and studies of associations between mite infestation, growth rate and measures of mange severity in growing pigs. Vet Parasitol. 1995;60:249–64.
9. Rueda-López M. Elimination of sarcoptic mange due to sarcoptes scabiei var suis from a 1800-sow farrow-to-finish farm. Vet Rec. 2006;159:595–7.
10. Walton SF, Currie BJ. Problems in diagnosing scabies, a global disease in human and animal populations. Clin Microbiol Rev. 2007;20:268–79.
11. Kessler E, Matthes H-F, Schein E, Wendt M. Detection of antibodies in sera of weaned pigs after contact infection with sarcoptes scabiei var. suis and after treatment with an antiparasitic agent by three different indirect ELISAs. Vet Parasitol. 2003;114:63–73.
12. Sipos W, Schmoll F, Stumpf I. Minipigs and potbellied pigs as pets in the veterinary practice – a retrospective study. J Vet Med Ser A. 2007;54:504–11. https://doi.org/10.1111/j.1439-0442.2007.00968.x.
13. Carr J, Wilbers A. Pet pig medicine: 1. The normal pig. In Pract. 2008;30:160–6. https://doi.org/10.1136/inpract.30.3.160.
14. Tynes VV. Emergency care for potbellied pigs. Vet Clin North Am Exot Anim Pract. 1998;1:177–89.
15. Chakrabarti A. Pig handler's itch. Int J Dermatol. 1990;29:205–6. https://doi.org/10.1111/j.1365-4362.1990.tb03801.x.
16. Gallegos JL, Budnik I, Pena A, Canales M, Concha M, López J. Sarna sarcóptica: comunicación de un brote en un grupo familiar y su mascota. Rev Chil Infectol. 2014;31:47–52. https://doi.org/10.4067/S0716-10182014000100007.
17. Taplin D, Meinking TL, Porcelain SL, Athey RL, Chen JA, Castillero PM, et al. Community control of scabies: a model based on use of permethrin cream. Lancet. 1991;337:1016–8. https://doi.org/10.1016/0140-6736(91)92669-S.
18. Carr J. Survey of clinical problems identified in pet pigs in the UK. Vet Rec. 2004;155:269–71. https://doi.org/10.1136/VR.155.9.269.
19. Goyena E, Ruiz de Ybáñez R, Martínez-Carrasco C, Balseiro A, Alonso de Vega F, Casais R, et al. On the aggregated nature of chronic sarcoptes scabiei infection in adult pigs. Vet Parasitol. 2013;192:301–6. https://doi.org/10.1016/j.vetpar.2012.10.007.
20. Haas C, Origgi FC, Akdesir E, Batista Linhares M, Giovannini S, Mavrot F, et al. First detection of sarcoptic mange in free-ranging wild boar (sus scrofa) in Switzerland. Schweiz Arch Tierheilkd. 2015;157:269–75. https://doi.org/10.17236/sat00020.
21. Sheahan BJ. Experimental sarcoptes scabiei infection in pigs: clinical signs and significance of infection. Vet Rec. 1974;94:202–9.
22. Smith HJ. Transmission of sarcoptes scabiei in swine by fomites. Can Vet J. 1986;27:252–4.
23. Currier RW, Walton SF, Currie BJ. Scabies in animals and humans : history, evolutionary perspectives, and modern clinical management. Ann N Y Acad Sci. 2012;1230:50–60.
24. Mounsey KE, Murray HC, Bielefeldt-Ohmann H, Pasay C, Holt DC, Currie BJ, et al. Prospective study in a porcine model of sarcoptes scabiei indicates the association of Th2 and Th17 pathways with the clinical severity of scabies. PLoS Negl Trop Dis. 2015;9:e0003498. https://doi.org/10.1371/journal.pntd.0003498.
25. Davis DP, Moon RD. Density, location, and sampling of sarcoptes scabiei (Acari: Sarcoptidae) on experimentally infested pigs. J Med Entomol. 1990;27:391–8.
26. Mofiz E, Seemann T, Bahlo M, Holt D, Currie BJ, Fischer K, et al. Mitochondrial genome sequence of the scabies mite provides insight into the genetic diversity of individual scabies infections. PLoS Negl Trop Dis. 2016;10:e0004384. https://doi.org/10.1371/journal.pntd.0004384.
27. Jacobson M, Bornstein S, Wallgren P. The efficacy of simplified eradication strategies against sarcoptic mange mite infections in swine herds monitored by an ELISA. Vet Parasitol. 1999;81:249–58.
28. Seaman JT, Thompson DR, Barrick RA. Treatment with ivermectin of sarcoptic mange in pigs. Aust Vet J. 1993;70:307–8.
29. Nuber U, van Dorland HA, Bruckmaier RM. Effects of butafosfan with or without cyanocobalamin on the metabolism of early lactating cows with subclinical ketosis. J Anim Physiol Anim Nutr (Berl). 2016;100:146–55. https://doi.org/10.1111/jpn.12332.
30. Van Der Staay FJ, De Groot J, Van Reenen CG, Hoving-Bolink AH, Schuurman T, Schmidt BH. Effects of butafosfan on salivary cortisol and behavioral response to social stress in piglets. J Vet Pharmacol Ther. 2007;30:410–6. https://doi.org/10.1111/j.1365-2885.2007.00884.x.
31. Swe PM, Zakrzewski M, Kelly A, Krause L, Fischer K. Scabies mites alter the skin microbiome and promote growth of opportunistic pathogens in a porcine model. PLoS Negl Trop Dis. 2014;8:e2897. https://doi.org/10.1371/journal.pntd.0002897.
32. Song SJ, Lauber C, Costello EK, Lozupone CA, Humphrey G, Berg-Lyons D, et al. Cohabiting family members share microbiota with one another and with their dogs. elife. 2013;2:e00458. https://doi.org/10.7554/eLife.00458.

18

Development and application of a monoclonal antibody-based blocking ELISA for detection of antibodies to Tembusu virus in multiple poultry species

Lijiao Zhang[1] , Zhanhong Li[1], Huan Jin[1], Xueying Hu[2] and Jingliang Su[1*]

Abstract

Background: Tembusu virus (TMUV) is a member of the genus *Flavivirus*. Outbreak of this virus infection in duck flocks was first observed in China in April 2010, causing severe egg drop and neurological signs in laying ducks. Recently reported duck infections in southeastern Asia highlighted the need for well-validated diagnostic methods of TMUV surveillance to understand its epidemiological characteristics and maintenance in nature. Several enzyme-linked immunosorbent assays (ELISAs) for the detection of TMUV infection have been reported, but none have been applied to high-throughput diagnostics.

Results: In this study, a monoclonal antibody (MAb) against TMUV was generated and characterized. MAb 9E4 was shown to bind specifically to a disulfide bond-dependent epitope on the domain I/II of TMUV E protein, and a blocking ELISA was established based on this MAb. The cut-off percentage inhibition value for negative sera was set at 30%. By comparison with the virus neutralization test, the specificity and sensitivity of the blocking ELISA were 96.37% and 100%, respectively, and the kappa value was 0.966, based on 416 serum samples collected from both experimentally and clinically infected ducks, geese and chickens. A good correlation ($r^2 = 07998$, $P < 0.001$) was observed between the blocking ELISA and plaque reduction neutralization test (PRNT) titers. Using archived duck serum samples collected between 2009 and 2015, the seroprevalence in duck flocks raised in Northern China was estimated by blocking ELISA.

Conclusions: Our MAb-based blocking ELISA provides a reliable and rapid diagnostic tool for serological monitoring of TMUV infection and evaluation of immune status following TMUV vaccination in multiple poultry species.

Keywords: Duck, Tembusu virus, Flavivirus, Monoclonal antibody, Blocking ELISA

Background

Tembusu virus (TMUV) is a mosquito-borne virus that belongs to the genus *Flavivirus* within the family *Flaviviridae*. The virus was originally isolated from mosquitoes in Malaysia in 1953 and the first animal infection case was reported in a 4-week-old broiler chicken flock in Perak State, characterized by encephalitis, growth retardation and increased blood glucose levels [1]. In April 2010, a severe outbreak of duck TMUV infection causing egg drop and signs of central nerve system involvement was reported in China [2]. Thereafter, the disease spread diffusely throughout duck-producing regions, involving ducks, geese and even laying chickens [3–5]. The recent emergence of TMUV infections in duck flocks in Malaysia and Thailand provided warnings of the increasing impact on animal health [6, 7]. In addition to domestic birds, TMUV has been occasionally isolated from mosquitos, pigeons and house sparrows in the vicinity of duck farms in China [8–10], suggesting that wild birds may serve as a natural reservoir, carrying and

* Correspondence: suzhang@cau.edu.cn
[1]Key Laboratory of Animal Epidemiology of the Ministry of Agriculture, College of Veterinary Medicine, China Agricultural University, Beijing 100193, China

disseminating the virus. However, the epidemiological characteristics and maintenance in nature of TMUV remain unclear, largely owing to the absence of convenient, sensitive and specific diagnostic tests.

Currently, identification of TMUV infection in ducks is based on isolation of the virus or detection of the viral nucleic acid by reverse transcriptase PCR. The diagnostic efficiency is largely dependent on the longevity of the virus in samples, which is closely related to the time when the samples are collected after infection. Since serologic evidence of infection may present rapidly and last long after infection, detection of TMUV-specific antibodies in serum samples could provide a convenient way to determine virus infection in animal populations. In this respect, enzyme-linked immunosorbent assays (ELISAs) are much more suitable for vaccination assessment and epidemiological analyses involving a large number of samples. Several ELISAs for the detection of TMUV infection have been reported recently [11–13], but none of them have been applied to large-scale clinical serum sample testing. We have recently described the uses of a *Flavivirus* group-specific monoclonal antibody-based blocking ELISA for detection of antibody responses in ducks immunized with an inactivated TMUV vaccine under laboratory conditions [14]. The drawback of this ELISA is the cross-reactivity with antibodies to other flaviviruses that infect animals in the field. In this paper, we describe the development of a blocking ELISA based on a TMUV-specific MAb and evaluate its potential application for high throughput of clinical serum samples.

Methods

Preparation of virus antigen
Duck TMUV strain JXSP was isolated from an infected duck flock as described previously [9]. After the initial two passages in duck embryos, the virus was propagated in baby hamster kidney cells (BHK-21) cells and used as stock virus (designated as $JXSP_{2-4}$) for antigen preparation or virus neutralization test (VNT). For antigen preparation, BHK-21 cells were grown and infected with $JXSP_{2-4}$ at a multiplicity of infection (MOI) of 0.001. When the cytopathic effect (CPE) reached approximately 75%, the infected supernatant was harvested by three freeze-thaw cycles, followed by centrifugation at 10,000×g for 45 min at 4 °C. To inactivate the virus, beta-propiolactone (BPL) (FERAK Berlin Gmbh, Berlin, Germany; NMR ≥ 98.5%) was added to the clarified virus suspensions to the final concentration of 1: 4000 and incubated at 4 °C for 24 h [14]. Virus particles were pelleted by ultracentrifugation at 160,000×g for 2.5 h at 4 °C, then resuspended in PBS and stored at − 80 °C until use.

Production of monoclonal antibody
Five female six-week-old BALB/c mice (Vitalriver, China) were injected subcutaneously with 100 μg of BPL-inactivated virus antigen emulsified with complete Freund's adjuvant (Sigma-Aldrich, St Louis, MO), followed by two subcutaneous boosters of the same antigen with incomplete Freund's adjuvant and one intraperitoneal inoculation of the antigen without adjuvant at ten days intervals. After the fourth inoculation, mouse spleen cells were harvested to prepare hybridomas using the standard method. Hybridomas secreting antibody against TMUV were screened by indirect ELISA, and sub-cloned three times by limiting dilution. The supernatant of the hybridoma culture was collected for immunoglobulin isotyping using the Mouse Monoclonal Antibody Isotyping Kit (Sigma-Alrich) according to the manufacturer's instructions. The selected hybridoma was inoculated into BALB/c mice and ascitic fluid was purified by saturated ammonium sulfate (SAS) precipitation as described [15].

Western blot analysis
To investigate the antigen binding of the generated MAbs, virus concentrated by ultracentrifugation was resuspended in reducing or non-reducing lane marker sample buffer (Thermo scientific, USA) and boiled for 6 min before SDS-PAGE separation. The separated proteins were transferred onto a PVDF (Polyvinylidene Fluoride) membrane, followed by incubation in blocking buffer (5% skim milk in PBS with 0.05% Tween-20) overnight at 4 °C. After washing, the protein was probed with the MAb and horseradish peroxidase (HRP)-conjugated goat anti-mouse IgG at a dilution of 1:5000. The signal was developed with chemiluminescence substrate (ECL reagent, Cwbiotech, Beijing, China). To further analyze the MAb binding domain, full length E protein of TMUV, domain I/II and domain III of E protein were individually expressed in *E. coli* using the pET32α vector (see Additional file 1). Purified and renatured recombinant protein was separated by SDS-PAGE under non-reducing condition and analyzed using the generated MAbs by Western blot as described above.

Immunofluorescence assay and immunochemistry
For the immunofluorescence assay (IFA), BHK-21 cells were cultured in 96-well-plates. Cells were infected with $JXSP_{2-4}$, Japanese encephalitis virus or duck-origin Batai virus at an MOI of 0.001 for 1 h and maintained in DMEM with 2% FBS for 36 h in a CO_2 incubator. The cells were then fixed with an ice-cold acetone/methanol (1:1) mixture for 20 min at room temperature. After washing three times with PBS, 200 μL of the blocking buffer was added and incubated at 37 °C for 30 min. Wells were then gently washed with PBS, the hybridoma culture supernatant or diluted murine ascitic fluid was

added and incubated at 37 °C for 45 min. Wells were washed and FITC-conjugated goat anti-mouse IgG (Eathox, USA) was added at a dilution of 1:800, followed by 30 min incubation at 37 °C. After three times washes, nuclei of the cells were stained with DAPI (Solarbio, China) for 10 min at room temperature. Wells were washed again and observed under fluorescence microscopy. For immunochemistry, BHK-21 cells were cultured on coverslips in a 24-well-plate, infected as described above and fixed with 4% paraformaldehyde for 30 min. Paraformaldehyde was removed by washing with PBS and cells were stained with the MAb as previously described [16].

Virus neutralization test

The plaque reduction neutralization test (PRNT) was performed in 12-well plates as previously described with slight modification [17] to verify the presence of TMUV-specific antibodies in serum samples and to quantitate antibody titers. Briefly, sera were inactivated at 56 °C for 30 min and serially diluted with DMEM. The stock virus $JXSP_{2-4}$ was diluted, mixed with an equal volume of diluted serum and incubated at 37 °C for 1 h. The mixture was transferred to BHK-21 cell monolayers in a 12-well plate in duplicate to a concentration of 100 plaque forming units (PFU) of infectious virus per well. After incubation at 37 °C for 1 h, the supernatant was removed and overlaid with 2.5 mL DMEM containing 1.0% (w/v) LMP agarose and 2% FBS. Following 3 days of incubation at 37 °C, infected cells were stained with 0.03% (w/v) neutral red and plaques were counted. Compared with the negative serum control, tested samples which showed more than 50% plaque reduction ($PRNT_{50}$) at a five-fold dilution were considered positive. The $PRNT_{50}$ titer for each sample tested was determined by identifying the well containing the highest dilution of serum with a plaque count < 50% of the average negative serum controls.

Development of MAb-based blocking ELISA

Ninety-six-well microtiter plates (Costar, USA) were coated with 1 μg inactivated TMUV antigen (corresponding to 1.7×10^5 PFU of the infectious virus) per well in 100 μL of 0.05 M carbonate-bicarbonate buffer (pH 9.6) and incubated at 4 °C overnight. The plate was washed three times with PBST (0.05% Tween in PBS, v/v) and 200 μL of blocking buffer was added for 2 h at 37 °C. The plate was washed three times and 100 μL of duck serum diluted with blocking buffer was added. After incubation at 37 °C for 1 h, the plate was washed with PBST, then 100 μL of SAS-purified ascitic MAb (0.85 μg/mL) was added and incubated at 37 °C for 1 h. After three washes, HRP conjugated goat anti-mouse IgG (1:5000 dilution) was added and incubated for 45 min at 37 °C. The plate was thoroughly washed and 100 μL of TMB substrate solution (3,3′,5,5′-tetramethyl benzidine 0.24 mg/mL and 0.003% H_2O_2) was added. Following incubation at room temperature in the dark for 15 min, the chromogenic reaction was stopped with 50 μL of 0.5 M sulfuric acid and optical density (OD) was measured at 450 nm in a microplate reader (Thermo Scientific, USA). Control wells without primary serum or without MAb were also prepared. The percentage inhibition (PI) of each test sample was calculated by the following formula: PI (%) = [(OD$_0$-OD$_{sample}$)/(OD$_0$-OD$_{100}$)] × 100% as described, where OD$_0$ was the mean optical density of the negative control serum (0% inhibition), OD$_{100}$ was the background optical density (100% inhibition) [17].

To determine the optimal dilution of serum samples to be tested, five TMUV antibody-negative sera and five positive sera were serially diluted and analyzed by blocking ELISA (bELISA). Subsequently, 400 duck serum samples collected from TMUV-free farms were tested for determination of the cut-off value. These sera were confirmed to be TMUV-specific antibody-negative by PRNT.

Validation of the blocking ELISA

To validate the bELISA, the specificity and sensitivity were determined. Serum samples with antibodies against duck enteritis virus (DEV), duck hepatitis virus (DHV), Newcastle disease virus (NDV), duck reovirus (DRV), egg drop syndrome virus (EDSV) and avian influenza virus (AIV) subtypes H5 and H9 were analyzed with bELISA to evaluate cross-reactivity.

Two experiments were further performed to evaluate the bELISA. In the first experiment, one-day-old ducklings purchased from a TMUV-free farm were kept in positive pressure specific-pathogen-free (SPF) chicken isolators with ad libitum access to feed and water. Ten ducklings were infected subcutaneously with 3×10^5 PFU of a moderately attenuated TMUV in 0.5 mL at the age of 7 days. Negative control ducklings were kept in a separate isolator. Serum samples were collected on days 2, 4, 7, 14 and 21 post-vaccination for detection of antibody titers by both PRNT and bELISA. Each sample was initially diluted five-fold and then subjected to doubling dilution. In the second experiment, 340 duck and 46 goose serum samples collected from flocks on farms with a history of TMUV infection in the previous two years were tested by bELISA and PRNT in parallel. Meanwhile, 10 chicken sera collected from SPF chickens immunized twice with inactivated TMUV were also tested. Compared with the PRNT, the sensitivity and specificity of the bELISA were calculated according to the following formulae: sensitivity = true positives × 100/ (true positives + false negatives), specificity = true negatives × 100/ (true negatives + false positives).

Field samples

A total of 2349 serum samples belonging to 17 flocks in Northern China were tested by bELISA for antibodies to TMUV. These sera were submitted to our laboratory for assessment of AIV vaccination from 2009 to 2015 by the duck farm owners (Table 1).

Results

Characterization of the monoclonal antibody

After three cycles of subcloning/screening with indirect ELISA and IFA, two hybridomas secreting antibody against duck TMUV were isolated and designated as 9E4 and 4C10. The MAb 9E4 was selected for development of the bELISA as it displayed high affinity to the coating virus antigen in the preliminary indirect ELISA test. The heavy chain subclass of the 9E4 MAb was determined as IgG1 and the light chain was kappa type. Immunofluorescence and immunochemistry detection of TMUV-infected BHK-21 cells exhibited strong staining in the cytoplasm of infected cells (Fig. 1a). Cells infected with Japanese encephalitis virus or duck-origin Batai virus were stained in parallel and no positive signal was observed, suggesting that the 9E4 MAb did not react with the two arboviruses reported to infect domestic ducks in China. Using the standard PRNT in BHK-21 cell, the murine ascitic 9E4 MAb showed 90% plaque reduction ($PRNT_{90}$) at a dilution of 1:10 and $PRNT_{50}$ at a dilution of 1:50, demonstrating that this MAb possessed weak neutralizing activity against TMUV (see Additional file 2). Western

blot analysis with the viral antigen revealed that the 9E4 MAb recognized a band with a molecular weight of 52Kd, corresponding to the E protein of TMUV (Fig. 1b). Of note, the MAb reacted with E protein only when non-reducing sample buffer without 2-mercaptoethanol was used in sample preparation for SDS-PAGE separation, indicating its epitope binding was related to the presence of disulfide bonds in the protein. Further analysis with recombinant fragments covering different domains of the E protein under non-reducing condition demonstrated that the MAb bound an epitope within domain I/II (Fig. 1c).

Optimization of the blocking ELISA protocol

To determine the optimal dilution of the test serum sample in the bELISA, five positive duck sera with $PRNT_{50}$ titers ranging from 10 to 1280 were assessed in serial dilution. As shown in Fig. 2, the PIs of antibody-negative sera did not show significant variation in serial dilution, while the PI of the positive sera decreased with dilution. Sera with lower neutralizing titer ($PRNT_{50}$ = 10 and 20) declined rapidly for each doubling dilution, with 1:20 diluted sera producing < 60% inhibition. To ensure effective and sufficient blocking of the epitope recognized by MAb 9E4, the working dilution of the test serum samples was fixed at 1:10 in this study.

To determine the cut-off value for the bELISA, a panel of 400 duck sera lacking antibodies to TMUV was tested for non-specific inhibition of MAb binding to the coating antigen. These sera showed less than 20% plaque

Table 1 Tembusu virus (TMUV) seroprevalence in domestic ducks in Northern China, 2009–2015

Sampling date	Sampling site	Species	No. sera	b-ELISA positive	Positive rate
2009	Hebei	laying duck	120	0	0%
2009	Beijing	laying duck	80	0	0%
2011.03	Hebei	layingduck[a]	289	289	100%
2011.03	Beijing	breeding duck[a]	80	80	100%
2011.03	Shandong	breeding duck[a]	60	60	100%
2012. 11	Beijing	laying duck[b]	152	18	11.84%
2012. 12	Beijing	laying duck[b]	96	9	9.38%
2013. 01	Beijing	laying duck[b]	122	12	9.83%
2013.03	Beijing	table duckling[b]	90	0	0%
2013. 04	Beijing	table duckling[b]	176	0	0%
2013.06	Beijing	laying duck[b]	180	6	3.33%
2013.08	Beijing	laying duck[b]	190	25	13.16%
2013.12	Beijing	table duckling[b]	106	4	3.77%
2015.05	Beijing	table duckling[b]	50	0	0%
2015.09	Hebei	breeding duck[b]	80	0	0%
2015.12	Hebei	breeding duck[c]	241	79	32.78%
2015.12	Beijing	breeding duck[c]	237	226	95.36%

[a]The sera were from the animals suffered from DTMUV infection in December 2010
[b]The sera were from the animals with no history of DTMUV infection and immunization
[c]The sera were from the animals immunized with autogenous vaccine of DTMUV within one year

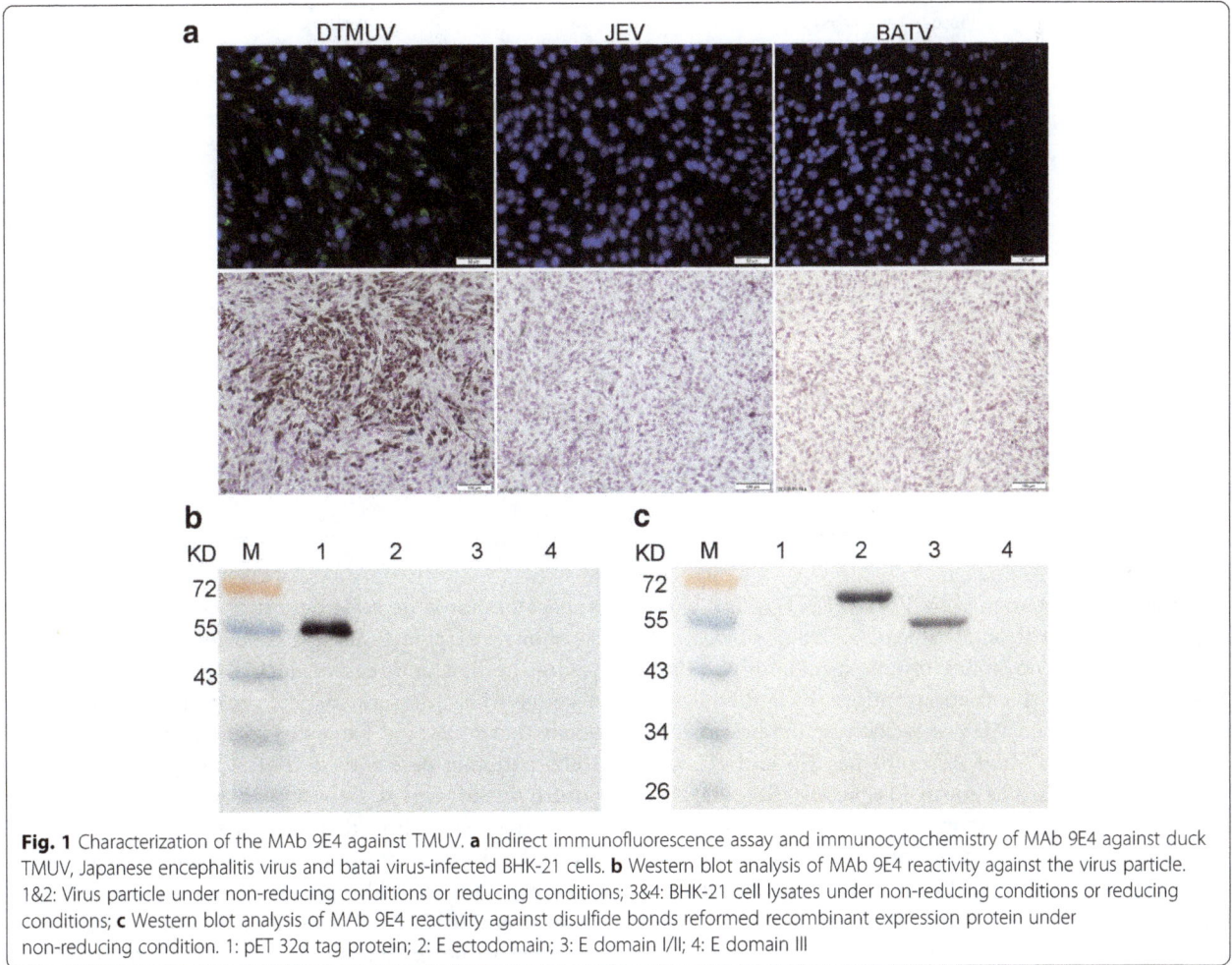

Fig. 1 Characterization of the MAb 9E4 against TMUV. **a** Indirect immunofluorescence assay and immunocytochemistry of MAb 9E4 against duck TMUV, Japanese encephalitis virus and batai virus-infected BHK-21 cells. **b** Western blot analysis of MAb 9E4 reactivity against the virus particle. 1&2: Virus particle under non-reducing conditions or reducing conditions; 3&4: BHK-21 cell lysates under non-reducing conditions or reducing conditions; **c** Western blot analysis of MAb 9E4 reactivity against disulfide bonds reformed recombinant expression protein under non-reducing condition. 1: pET 32α tag protein; 2: E ectodomain; 3: E domain I/II; 4: E domain III

reduction at a dilution of 1:5 by VNT. The mean PI was -0.65% for these sera with a standard deviation (SD) of 10.22%. The cut-off value of the test serum was set at 30% based on the criteria of "mean PI of the negative sera plus $3 \times$ SD", with 99% confidence of the PI$< 30\%$. Positive serum was set as 30%. Confirmation of

serum samples exhibiting a PI between 20% (X + 2SD) and 30% was required by repeating the test and it was considered to be negative if the PI was still $\leq 30\%$. Accordingly, the antibody titer was calculated as the highest dilution with a PI$> 30\%$.

Comparison of the blocking ELISA with virus neutralization assay

To test the performance of the bELISA, sera collected from ducks experimentally infected with TMUV were tested. As shown in Fig. 3, the antibody response to TMUV was detectable by bELISA on day 4 after infection, with the titers ranging from 1:40 to 1:320. In parallel, these sera were found to have weak neutralizing antibodies to TMUV as detected by VNT with PRNT$_{50}$ titers ranging from 5 to 20 (Fig. 3). Thereafter, the antibody levels determined by both bELISA and VNT increased significantly. For the sera of the five non-infected ducks, no TMUV-specific antibodies were detected by bELISA or VNT throughout this experiment. Linear regression analysis using GraphPad Prism showed a strong correlation between the serum

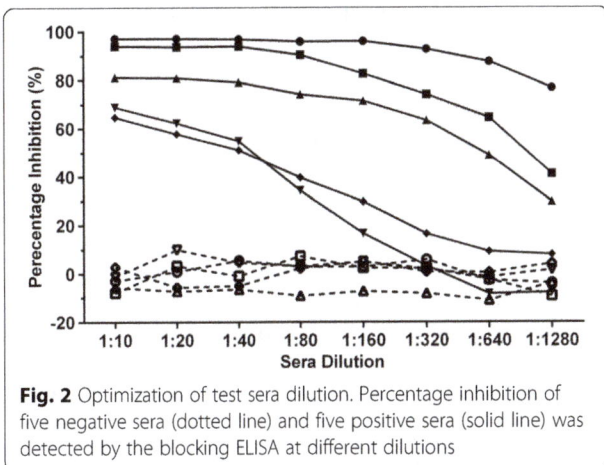

Fig. 2 Optimization of test sera dilution. Percentage inhibition of five negative sera (dotted line) and five positive sera (solid line) was detected by the blocking ELISA at different dilutions

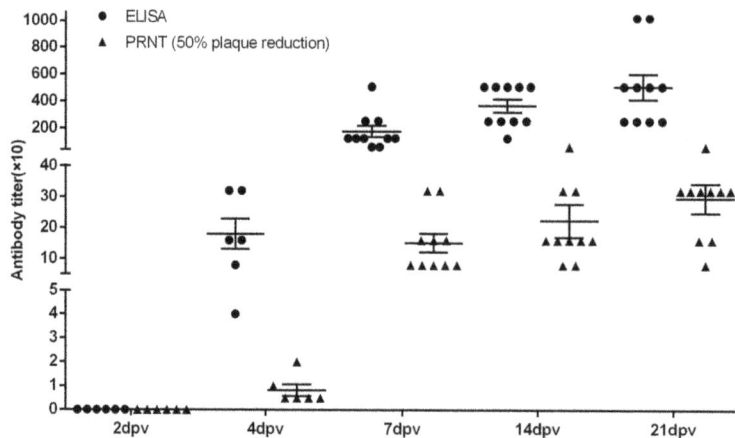

Fig. 3 Antibody responses in ducks inoculated with attenuated duck TMUV JXSP. The serum samples were collected for antibody titer testing both by bELISA and VNT at 2, 4, 7, 14 and 21 days post vaccination. Symbols represent results from individual samples, bars indicate the mean titer values ± SE

antibody levels determined by VNT ($PRNT_{50}$) and blocking ELISA titers, with an r^2 value of 0.7998 ($P < 0.001$).

To exclude the possibility of non-specific inhibition of MAb binding to the coating antigen by antibodies to viruses other than TMUV, sera with antibodies to DEV, DHAV, DRV, EDSV and AIV subtypes H5 and H9 were individually tested. The mean PIs plus $3 \times SD$ of immunized sera were lower than 20%, therefore non-specific binding to the epitope did not occur.

We next sought to test the specificity and sensitivity of the bELISA for detection in clinical samples. A total of 360 duck sera collected from different duck farms were tested by the bELISA and VNT. Among the duck serum samples, 189 sera were determined to be positive by bELISA with a PI higher than 30%, and 182 samples were considered positive by VNT with a $PRNT_{50}$ titer ≥1:5. When the results for individual samples were compared, seven samples were detected by bELISA but not VNT, resulting in 98.05% agreement between the two techniques. The 182 VNT-positive serum samples were all detected by the bELISA. It is reasonable that the discrepant seven positive sera were detected by the bELISA as ELISA is more sensitive for antibody detection. To extend the experiment, 46 serum samples collected from a goose flock with a history of TMUV infection were tested in the same manner. Highly consistent results were obtained in that 31 sera were positive by both bELISA and VNT. Moreover, when sera collected from SPF chickens immunized with the inactivated virus were evaluated, all samples exhibited titers ranging from 1:320 to 1:2560 in the bELISA ($n = 10$). Taken together, these results indicated that the specificity and sensitivity of the bELISA were 96.37% and 100%, respectively, in comparison with the VNT. The agreement rate between these two methods was 98.32% with the kappa value 0.966.

Application for large scale field sample detection

As shown in Table 1, all sera collected in 2009, the year before the first outbreak of duck TMUV infection, were negative while the sera collected from three laying duck flocks that recovered from the infection in 2011 showed 100% antibody positivity to TMUV, suggesting that exposure to the virus is extensive during an outbreak. This is in agreement with other studies showing that TMUV transmitted efficiently among ducks and caused severe egg drop [18, 19]. The positivity rate of the sera collected from 2012 to 2015 decreased significantly. It was reasoned to be related to the cessation of the epidemic and replacement of breeding flocks in the duck production industry. However, antibodies detected in a few samples suggest the existence of sporadic infection in duck flocks. Interestingly, duck flocks immunized with inactivated vaccine exhibited a high rate of antibody response but the positivity rate showed a significant difference.

Discussion

Since the outbreak of duck TMUV infection in China in 2010, several serological diagnostic techniques have been reported [11–13], but they vary by antigen used and the degree of validation. In this study, a blocking ELISA was developed based on a MAb specific to TMUV, which fulfils the need for detection of kinetic antibody responses in ducks infected experimentally with TMUV or immunized with vaccine. In addition, as the assay is based on the blockade of epitope binding, it would presumably detect all types of TMUV antibodies from any host species which can bind the epitope recognized by the MAb.

The E protein is the major surface protein of flaviviruses and it plays a critical role in virus infection, as well as being a principle target of neutralizing antibodies

[20, 21]. A number of anti-E protein MAbs have been successfully used in the development of ELISAs for the diagnosis of flavivirus infections [22–24]. The MAb 9E4 used in this study was mapped to recognize an epitope in the domain I/II of TMUV E protein. Results of IFA and IHC tests with virus-infected cells demonstrated that the MAb binding epitope is accessible on the virion surface (Fig. 1a). Western blot analysis revealed that the MAb 9E4 binds the epitope only when disulphide bonds in the fragment were intact without reduction by 2-mercaptoethanol (Fig. 1b and c). Since the existence of intact disulphide bonds in the E protein were shown to be necessary for induction of neutralizing antibodies in other flaviviruses [25, 26], the conformationally dependent binding of MAb 9E4 implied that the epitope might preferentially be blocked by specific neutralizing antibodies to the native E protein.

Analysis of the values of 400 negative duck serum samples with 2SD and 3SD statistical approaches yielded an optimal diagnostic cut-off value of 30% for the bELISA, which was highly consistent with the ROC analysis (see Additional file 2). In order to ascertain that the bELISA can effectively detect TMUV antibodies in birds, experimentally immunized chicken serum samples, and field-collected duck and goose serum samples were further tested. Because there is no government-approved serological diagnostic technique for TMUV infection, we chose the VNT as a reference, using the $PRNT_{50}$ titer as criteria for positive sera. The bELISA results displayed a high level of agreement with the VNT ($PRNT_{50}$). The discrepant seven positive duck samples detected by bELISA gave a PI in the borderline range, indicating that a low level of antibody was present in these samples. It is unsurprising that it was not detected by VNT since this technique is less sensitive than ELISA. Our blocking ELISA was able to detect TMUV-specific antibody response from 4 days post experimental infection, and the titers determined by bELISA correlated well with $PRNT_{50}$ titers throughout the experiment (Fig. 3). Comparison of antibody levels determined by bELISA and $PRNT_{50}$ also indicated that the former was superior in sensitivity.

Our bELISA was applied to 2359 domestic duck serum samples from semi-open duck farms in Northern China. As these samples were submitted for antibody evaluation after vaccination with inactivated avian influenza vaccines, the results are more likely to be a true reflection of the seroprevalence levels of TMUV in these farms. Antibodies to TMUV were detected in all samples in March 2011, being consistent with our earlier study [2], in which 100% antibody positivity was found in ducks after TMUV infection, supporting that TMUV is highly infectious and abundant in duck flocks during outbreaks of the disease. Several factors could contribute to the decline of positivity rates for the samples between 2012 and 2014. First, the infected duck flocks were culled or replaced as part of the duck production system. Second, owners paid more attention to biosecurity on duck farms; they did not introduce breeding eggs or ducklings from endemic areas. Importantly, high rates of antibody responses were detected in the samples collected from duck flocks experimentally immunized with autogenous inactivated TMUV vaccines. This result indicates potential use of vaccine for the control of TMUV infection in ducks, but it also presents a challenge for the differential diagnosis between ducks naturally infected by TMUV and those immunized with vaccine.

Conclusions
The well-validated epitope-blocking ELISA is useful for a range of serological investigations of TMUV infection in multiple poultry species. It can also be used to measure the antibody responses and assess vaccine efficacy in birds after immunization.

Abbreviations
AIV: Avian influenza virus; bELISA: Blocking ELISA; BPL: Beta-propiolactone; CPE: Cytopathic effect; DAPI: 4, 6-diamidino-2-phenylindole; DEV: Duck enteritis virus; DHV: Duck hepatitis virus; DRV: Duck reovirus; EDSV: Egg drop syndrome virus; ELISA: Enzyme-linked immunosorbent assays; HRP: Horseradish peroxidase; IFA: Immunofluorescence assay; MAb: Monoclonal antibody; NDV: Newcastle disease virus; PFU: Plaque forming units; PI: Percentage inhibition; PRNT: Plaque reduction neutralization test; PRNT50: 50% plaque reduction; PVDF: Polyvinylidene fluoride; SAS: Saturated ammonium sulfate; SD: Standard deviation; TMUV: Tembusu virus

Acknowledgements
The authors would like to thank Dr. Shuang Li at North China University of Science and Technology for technical assistance with the Western blot analysis.

Funding
This work was supported by grants from China Ministry of Science and Technology National Key R&D Program (No. 2016YFD0500106) and National Natural Science Foundation of China (No. 31372461 & 31672567).

Authors' contributions
JS and LZ conceived the study and participated in its design. LZ, ZL, HJ and XH performed experiments. LZ drafted the manuscript. JS helped interpret results and write the manuscript. All authors read and approved the final manuscript.

Consent for publication
Not applicable

Competing interests
The authors declare that they have no competing interests.

Author details
[1]Key Laboratory of Animal Epidemiology of the Ministry of Agriculture, College of Veterinary Medicine, China Agricultural University, Beijing 100193, China. [2]College of Veterinary Medicine, Huazhong Agricultural University, Wuhan 430070, China.

References

1. Kono Y, Tsukamoto K, Abd Hamid M, Darus A, Lian TC, Sam LS, Yok CN, Di KB, Lim KT, Yamaguchi S, Narita M. Encephalitis and retarded growth of chicks caused by sitiawan virus, a new isolate belonging to the genus flavivirus. Am J Trop Med Hyg. 2000;63(1,2):94–101.

2. Su J, Li S, Hu X, Yu X, Wang Y, Liu P, Lu X, Zhang G, Hu X, Liu D, Li X, Su W, Lu H, Mok NS, Wang P, Wang M, Tian K, Gao GF. Duck egg-drop syndrome caused by BYD virus, a newTembusu-related flavivirus. PLoS One. 2011;6(3):e18106.

3. Yan P, Zhao Y, Zhang X, Xu D, Dai X, Teng Q, Yan L, Zhou J, Ji X, Zhang S, Liu G, Zhou Y, Kawaoka Y, Tong G, Li Z. An infectious disease of ducks caused by a newly emerged Tembusu virus strain in mainland China. Virology. 2011;417(1):1–8.

4. Huang X, Han K, Zhao D, Liu Y, Zhang J, Niu H, Zhang K, Zhu J, Wu D, Gao L, Li Y. Identification and molecular characterization of a novel flavivirus isolated from geese in China. Res Vet Sci. 2013;94(3):774–80.

5. Chen S, Wang S, Li Z, Lin F, Cheng X, Zhu X, Wang J, Chen S, Huang M, Zheng M. Isolation and characterization of a Chinese strain of Tembusu virus from Hy-line Brown layers with acute egg-drop syndrome in Fujian, China. Arch Virol. 2014;159(5):1099–107.

6. Homonnay ZG, Kovács EW, Bányai K, Albert M, Fehér E, Mató T, Tatár-Kis T, Palya V. Tembusu-like flavivirus (Perak virus) as the cause of neurological disease outbreaks in young Pekin ducks. Avian Pathol. 2014;43(6):552–60.

7. Thontiravong A, Ninvilai P, Tunterak W, Nonthabenjawan N, Chaiyavong S, Angkabkingkaew K, Mungkundar C, Phuengpho W, Oraveerakul K, Amonsin A. Tembusu-related Flavivirus in ducks, Thailand. Emerg Infect Dis. 2015; 21(12):2164–7.

8. Tang Y, Diao Y, Chen H, Ou Q, Liu X, Gao X, Yu C, Wang L. Isolation and genetic characterization of a tembusu virus strain isolated from mosquitoes in Shandong, China. Transbound Emerg Dis. 2015;62(2):209–16.

9. Liu P, Lu H, Li S, Moureau G, Deng YQ, Wang Y, Zhang L, Jiang T, de Lamballerie X, Qin CF, Gould EA, Su J, Gao GF. Genomic and antigenic characterization of the newly emerging Chinese duck egg-drop syndrome flavivirus: genomic comparison with Tembusu and Sitiawan viruses. J Gen Virol. 2012;93(10):2158–70.

10. Y T, Diao Y, Yu C, Gao X, Ju X, Xue C, Liu X, Ge P, Qu J, Zhang D. Characterization of a Tembusu virus isolated from naturally infected house sparrows (Passer domesticus) in northern China. Transb Emerg Dis. 2013; 60(2):152–8.

11. Li X, Li G, Teng Q, Yu L, Wu X, Li Z. Development of a blocking ELISA for detection of serum neutralizing antibodies against newly emerged duck Tembusu virus. PLoS One. 2012;7(12):e53026.

12. Yin X, Lv R, Chen X, Liu M, Hua R, Zhang Y. Detection of specific antibodies against tembusu virus in ducks by use of an E protein-based enzyme-linked immunosorbent assay. J ClinMicrobiol. 2013;51(7):2400–2.

13. Fu Y, Ji Y, Liu B, Dafallah RM, Zhu Q. Development of a solid-phase competition ELISA to detect antibodies against newly emerged duck Tembusu virus. J Virol Methods. 2015;224:73–6.

14. Zhang L, Z L, Zhang Q, Sun M, Li S, Su W, Hu X, He W, Su J. Efficacy assessment of an inactivated Tembusu virus vaccine candidate in ducks. Res Vet Sci. 2017;110:72–8.

15. Darcy E, Leonard P, Fitzgerald J, Danaher M, O'Kennedy R. Purification of antibodies using affinity chromatography. Methods Mol Biol. 2011;681:369–82.

16. Li S, Li X, Zhang L, Wang Y, Yu X, Tian K, Su W, Han B, Su J. Duck Tembusu virus exhibits neurovirulence in BALB/c mice. Virol J. 2013;10:260.

17. Blitvich BJ, Marlenee NL, Hall RA, Calisher CH, Bowen RA, Roehrig JT, Komar N, Langevin SA, Beaty BJ. Epitope-blocking enzyme-linked immunosorbent assays for the detection of serum antibodies to west nile virus in multiple avian species. J Clin Microbiol. 2003;41(3):1041–7.

18. Liu P, Lu H, Li S, Wu Y, Gao GF, Su J. Duck egg drop syndrome virus: an emerging Tembusu-related flavivirus in China. Sci China Life Sci. 2013;56(8):701–10.

19. Li X, Shi Y, Liu Q, Wang Y, Li G, Teng Q, Zhang Y, Liu S, Li Z. Airborne transmission of a novel Tembusu virus in ducks. J Clin Microbiol. 2015;53(8): 2734–6.

20. Mukhopadhyay S, Kuhn RJ. RossmannMG. A structural perspective of the flavivirus life cycle. Nat Rev Microbiol. 2005;3(1):13–22.

21. Yu K, Sheng ZZ, Huang B, Ma X, Li Y, Yuan X, Qin Z, Wang D, Chakravarty S, Li F, Song M, Sun H. Structural, antigenic, and evolutionary characterizations of the envelope protein of newly emerging duck Tembusu virus. PLoS One. 2013;8(8):e71319.

22. Sotelo E, Llorente F, Rebollo B, Camuñas A, Venteo A, Gallardo C, Lubisi A, Rodríguez MJ, Sanz AJ, Figuerola J, Jiménez-Clavero MÁ. Development and evaluation of a new epitope-blocking ELISA for universal detection of antibodies to West Nile virus. J Virol Methods. 2011;174(1-2):35–41.

23. Zanluca C, Mazzarotto GA, Bordignon J. Duarte dos Santos CN. Development, characterization and application of monoclonal antibodies against Brazilian dengue virus isolates. 2014. PLoS One. 2014;9(11):e110620.

24. Adungo F, Yu F, Kamau D, Inoue S, Hayasaka D, Posadas-Herrera G, Sang R, Mwau M, Morita K. Development and characterization of monoclonal antibodies to yellow fever virus and application in antigen detection and IgM capture enzyme-linked immunosorbent assay. Clin Vaccine Immunol. 2016;23(8):689–97.

25. Roehrig JT, Volpe KE, Squires J, Hunt AR, Davis BS, Chang GJ. Contribution of disulfide bridging to epitope expression ofthe dengue type 2 virus envelope glycoprotein. J Virol. 2004;78(5):2648–52.

26. Wengler G, Wengler G. An analysis of the antibody response against West Nile virus E protein purified by SDS-PAGE indicates that this protein does not contain sequential epitopes for efficient induction of neutralizing antibodies. J Gen Virol. 1989;70(4):987–92.

19

Sub-chronic toxicopathological study of lantadenes of *Lantana camara* weed in Guinea pigs

Rakesh Kumar[1], Rinku Sharma[1*], Rajendra D. Patil[2], Gorakh Mal[1], Adarsh Kumar[2], Vikram Patial[3], Pawan Kumar[3] and Bikram Singh[3]

Abstract

Background: In the field conditions, animals regularly consume small quantities of lantana leaves either while grazing or due to mixing with regular fodder.

The hypothesis of this study was that consumption of lantana toxins over a long period of time leads to progression of sub-clinical disease.

Toxicopathological effects of sub-chronic (90 days) administration of lantadenes of *L. camara* were investigated in guinea pigs. For this, a total of 40 animals were divided into 5 groups whereby groups I, II, III and IV were orally administered lantadenes, daily at the dose of 24, 18, 12, and 6 mg/kg bw, respectively while group V was control. The animals were evaluated by weekly body weight changes, haematology, serum liver and kidney markers, tissue oxidative markers and histopathology.

Results: The results of significant decrease in weekly body weights, haematology, liver and kidney marker enzymes (alanine aminotransaminase, aspartate aminotransaminase, acid phosphatase and creatinine), oxidation stress markers (lipid peroxidation, reduced glutathione, superoxide dismutase and catalase) in liver and kidneys, histopathology, and confirmation of fibrous collagenous tissue proliferation by Masson's Trichome stain showed that lantadenes led to a dose-dependent toxicity in decreasing order with the highest dose (24 mg/kg bw) producing maximum lesions and the lowest dose (6 mg/kg bw) producing minimum alterations.

Conclusions: The study revealed that lantadenes which are considered to be classical hepatotoxicants in acute toxicity produced pronounced nephrotoxicity during sub-chronic exposure. Further studies are needed to quantify the levels of lantadenes in blood or serum of animals exposed to lantana in field conditions which would help to assess the extent of damage to the vital organs.

Keywords: Lantadenes, Sub-chronic toxicity, Haematology, Serum markers, Pathology, Oxidation stress, Guinea pigs

Background

Lantana camara is a noxious weed growing in tropical and subtropical parts of the world [1]. *L. camara* toxicity caused by lantadenes is characterized by intrahepatic cholestasis, liver damage and photosensitization. Both ruminants including cattle, sheep, buffalo, goats and non-ruminants like horses, guinea pigs, rabbits, female rats are susceptible to lantana toxicity. Guinea pigs exhibit most typical signs comparable to experimental or field cases of ruminants affected with lantana toxicity. So, guinea pigs are often used as a preferred animal model for lantadene toxicity studies. The leaves of red flower variety (*L. camara* var. *aculeata*) of lantana are mainly toxic to animals and contain pentacyclic triterpenoids [2, 3]. Lantadene A (LA) and lantadene B (LB), lantadene C (LC) and lantadene D (LD) are present in major quantity while reduced lantadene A (RLA) and reduced lantadene B (RLB) are the minor components having lesser importance as compared to other constituents present [4, 5]. Among these compounds, LA is the

* Correspondence: rinku.sharma@icar.gov.in; rinkusharma99@gmail.com
[1]Disease Investigation Laboratory, ICAR-Indian Veterinary Research Institute, Regional Station, Palampur, Himachal Pradesh, India

most hepatotoxic component while others are of little importance [2].

Generally this plant is not used as a fodder, but in field conditions, the outbreaks of lantana poisoning are seen during fodder scarcity, drought and flood where animals may consume small quantities of lantana leaves either while grazing or due to mixing with regular fodder. The consumption of lantana toxins over a long period of time may lead to progression of sub-clinical disease followed by death of the animal. However, the cause of death in such cases remains obscured due to complete lack of information about the toxicopathological effects which could be induced by sub-chronic ingestion of lantana leaves in animals. Any specific treatment for lantana toxicity is not available however some conventional therapies like fluid therapy, activated charcoal (5 g/kg), tefroli powder, liv-15 etc. can be used. The present study was conducted to investigate whether sub-chronic ingestion of lantadenes would lead to any adverse effects in guinea pig laboratory animal model.

Methods
Procurement, maintenance and housing of experimental animals
Forty guinea pigs weighing approximately 200–300 g and of either sex were procured from Laboratory Animal Resource Section, ICAR-IVRI, Izatnagar for the experimental study. All the animals were examined for any ailment or abnormality and their body weights were recorded. The animals were maintained in the Laboratory Animal Housing Facility of ICAR-IVRI Regional Station, Palampur. All the guinea pigs were kept in polypropylene cages and provided with 12 h light/dark cycle, temperature (23 ± 2 °C) and humidity ($55 \pm 10\%$ RH). The experimental protocols were reviewed and approved by the Institutional Animal Ethics Committee (No. PLP-IAEC 8). All sanitary and hygienic measures were observed as per the CPCSEA guidelines. The animals were provided ad libitum access to standard laboratory animal diet supplemented with vitamin C (Limcee, 1000 mg/kg feed) and clean water during the experimental trial.

Collection, processing and isolation of lantadenes from L. camara leaves
Leaves of red flower variety of L. camara var. aculeate were collected during the month of August–September from an area adjoining Palampur town located at an altitude of 1200 m above mean sea level. The samples were oven dried at 55 °C and ground to a fine powder of 1 mm particle size. The extraction of lantadenes was carried out by protocol described by Parimoo and co-workers [6]. The isolated lantadenes were stored in sealed vials at room temperature until further use. The purification was monitored by thin layer chromatography

(TLC). The quantification and characterization of lantadenes was done with the help of reversed-phase HPLC. The TLC and reversed-phase HPLC analysis protocols were as described earlier [6].

Experimental study
The animals were provided with an acclimatization period of 7 days. All the animals were weighed and divided into 5 groups with 8 animals (4 males and 4 females) in each group. The experimental animals (group I, II, III and IV) were administered graded doses of lantadenes, orally, in gelatin capsules, once daily for 90 days. The dose selection was made on the basis of previous study on sub-acute toxicity of L. camara by Parimoo and co-workers [7]. An interval of one hour was kept between administration of lantadenes and feed. Group V was control group and it received normal feed and water. All the animals were euthanized using chloroform at day 90 and different samples for laboratory analysis were collected. During euthanasia, sufficient ventilation and a means of exhausting waste gases in the closed container was made.

Collection of blood and separation of serum
Approximately 4 ml of blood was drawn from posterior vena cava in glass tubes and kept at room temperature for 2–4 h in slanting position, followed by centrifugation at 2000 rpm for 15 min to collect maximum quantity of serum. The collected serum was then stored in screw cap vials and kept at -70 °C for further estimation of various serum biochemical parameters.

Serum biochemical analysis
Serum biochemicals such as alanine aminotransferase (ALT), aspartate aminotransferase (AST), alkaline phosphatase (ALP), bilirubin, creatinine and total protein were estimated using commercial kits (Span Diagnostic Ltd., India) by employing automatic biochemistry analyzer (Bayer RA 50). The manufacturer's protocol was followed for all the estimations. However, acid phosphatase (ACP) was determined by the method described by Bergmeyer [8] and was expressed in IU/L.

Oxidation stress determination
Viable tissues samples of liver and kidneys were collected aseptically in sterile, screw capped polypropylene vials using sterile scissors and forceps, transported on ice and stored at -70 °C after proper labeling till further processing.

Preparation of homogenate
Tissue samples of liver and kidneys were homogenized (Remi, 12 U-56) in the ratio of 1:10 (200 mg tissue in 2 ml PBS) in ice cold 0.1 M PBS (pH 7.4). The homogenate was

centrifuged for 10 min at 10,000 rpm. The supernatant was used for the estimation of superoxide dismutase, catalase, lipid peroxidation, reduced glutathione and total protein.

Lipid peroxidation
Lipid peroxidation (LPO) was estimated by the method described by Dawra and co-workers [9]. The absorbance was read at 535 nm. Results were calculated from ΔE using molar extinction coefficient of $1.56 \times 10^5/\text{M/cm}$. The results were expressed as μmoles of MDA (malondialdehyde) production per g of wet tissue.

Reduced glutathione
Reduced glutathione (GSH) was estimated by the method described by Sedlak and Lindsay [10]. The absorbance was measured at 412 nm. Calculation was done by using extinction co-efficient (EC = 13,100/M/cm) and results were expressed in nM/g of wet tissue.

Catalase
Catalase enzyme was estimated by the method of Aebi [11]. The decrease in absorbance was monitored at 240 nm. The results were expressed as k/mg protein where k stands for nano moles of H_2O_2 utilized/min.

Superoxide dismutase
Superoxide dismutase (SOD) enzyme was estimated by the method described by Nishikimi and co-workers [12]. The increase in absorbance was measured at 560 nm. The unit of superoxide dismutase was defined as the amount of enzyme causing 50% inhibition of the rate of reduction of nitro blue tetrazolium. The results were expressed as units/mg protein.

Protein estimation
The protein content in the homogenate was estimated by Lowry and co-workers [13]. Bovine serum albumin (BSA) (1 mg/ml, Lobachemie) was used as standard. The absorbance of blue colour developed was recorded at 578 nm. The protein content in the sample was calculated from standard curve prepared using different concentrations of BSA.

Gross and histopathological examination
A detailed necropsy examination was conducted on animals and the gross findings were recorded. Neutral buffered formalin fixed tissues (liver, kidneys, heart, brain, spleen, lymph nodes, stomach, intestine etc.) were subjected to histopathological processing and staining as per standard procedures [14]. Haematoxylin and Eosin (HE) stained individual sections were microscopically examined and the histopathological alterations were recorded and digitally photomicrographed (Olympus

BX53). The histopathological lesions in liver and kidneys were graded with ordinal score of 0, no change; 1, mild (< 25% organ affected); 2, moderate (26–50% organ affected); and 3, severe (51–75% organ affected); which reflects the changes of low, medium and high grades, respectively [15]. The degree of fibrosis in liver and kidneys were assessed as per Scheuer/Batts, Ludwig/Tsui fibrosis scoring system, where the stages 0, 1, 2, 3, 4 indicates no fibrosis, portal/peri-portal fibrosis, septal fibrosis and cirrhosis, respectively [16].

Masson's Trichome stain (MST)
Parallel sections of tissues showing fibrous tissue proliferations in HE were stained with special stain, Masson's trichome stain [14].

Statistical analysis
One-way analysis of variance (ANOVA) was used to detect differences among groups and the means were compared by Dunnett's Multiple Comparison test using critical difference at 5% level of significance ($P \leq 0.05$). All analyses were performed with Graph Pad InStat software (San Diego, USA). Kruskal-Wallis Test (Non-parametric ANOVA) was used for analysis of microscopic scoring in liver and kidneys.

Results
Approximately 650 g of green lantana leaves yielded 100 g of dried leaf powder from which 490.20 mg lantadenes were obtained by the protocol described earlier [6]. In the present study, the LA and LB content in the leaves was estimated to be 49.18% and 13.09%, respectively by reversed phase-HPLC (Figs. 1 and 2a and b).

Weekly body weight gain
A significant decline in the body weights of animals of group I was observed as compared with the animals of control group (Table 1).

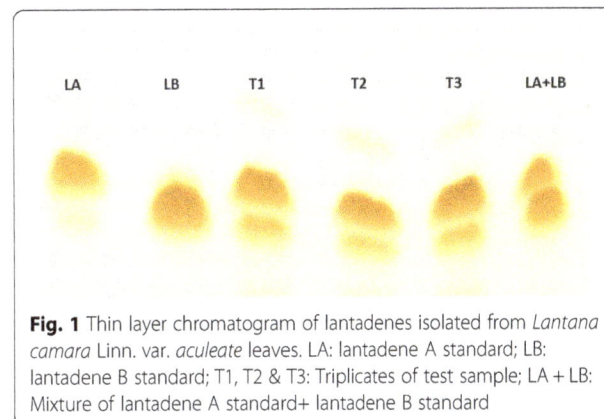

Fig. 1 Thin layer chromatogram of lantadenes isolated from *Lantana camara* Linn. var. *aculeate* leaves. LA: lantadene A standard; LB: lantadene B standard; T1, T2 & T3: Triplicates of test sample; LA + LB: Mixture of lantadene A standard+ lantadene B standard

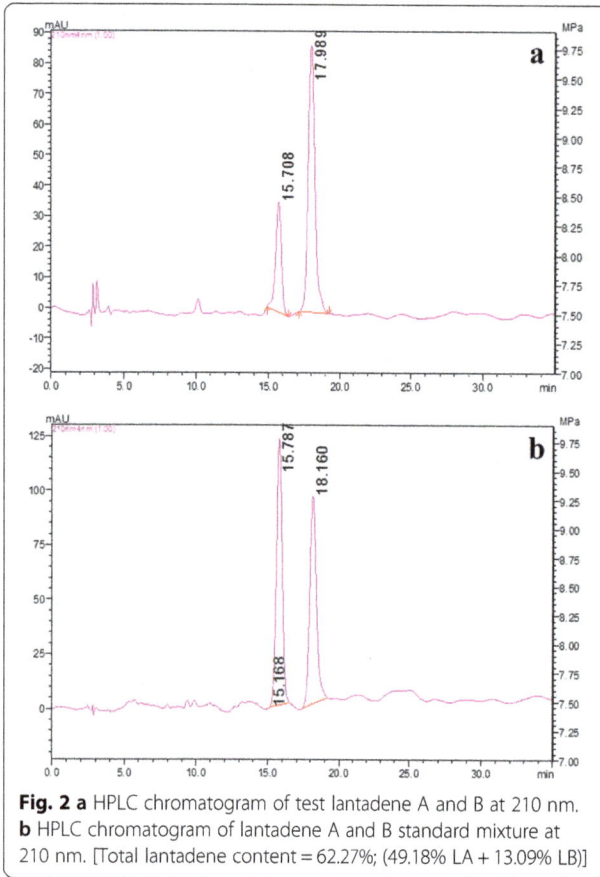

Fig. 2 a HPLC chromatogram of test lantadene A and B at 210 nm. **b** HPLC chromatogram of lantadene A and B standard mixture at 210 nm. [Total lantadene content = 62.27%; (49.18% LA + 13.09% LB)]

Haematology

There was a significant decrease in Hb and PCV levels of animals of group I (12.18 ± 0.55 g/dl and 36.28 ± 2.07%, respectively) as compared to group V (control). Total platelet counts of groups I (451.83 ± 55.03 × 10^3/µl) and II (464.5 ± 60.24 × 10^3/µl) were significantly decreased as compared to control (677 ± 58.26 × 10^3/µl). The results of haematological changes have been presented in Table 2.

Serum biochemical estimation

AST (107.07 ± 8.64 IU/L) levels of animals of group I were significantly ($P < 0.05$) elevated as compared to control group. However, there was no significant difference in AST values amongst groups II (100.03 ± 10.77 IU/L), III (91.08 ± 19.03 IU/L), IV (80.15 ± 15.14 IU/L) and V (58.10 ± 7.58 IU/L). ALT (77.82 ± 8.65 IU/L) levels of animals of group I were significantly elevated as compared to control group. But there was no significant difference in ALT values amongst groups II (63.88 ± 4.28 IU/L), III (62.32 ± 6.79 IU/L), IV (55.72 ± 7.30 IU/L) and V (52.14 ± 6.60 IU/L). There was no significant difference in ALP levels amongst different groups as compared to control. ACP (111.20 ± 15.30 IU/L) levels of animals of group I showed significant elevation as compared to control group. However, there was no significant difference in ACP values amongst groups II (90.74 ± 14.74 IU/L), III (75.96 ± 20.77 IU/L), IV (67.05 ± 14.13 IU/L) and V (43.83 ± 14.4 IU/L). The values of bilirubin were unchanged in all treatment groups as compared to control. Creatinine levels of animals of groups I (1.21 ± 0.22 mg/dl) and II (1.14 ± 0.17 mg/dl) were significantly increased as compared to control. However, creatinine values of groups III

Table 1 Weekly body weight (g) (Mean ±SE) of different treatment groups during sub-chronic *L. camara* toxicity study in guinea pigs

	GI	GII	GIII	GIV	GV
Day −0	241.17 ± 7.93[a]	250.17 ± 12.52[a]	260.33 ± 8.09[a]	256.17 ± 12.24[a]	268.50 ± 10.97[a]
Day −7	253.50 ± 11.87[a]	258.67 ± 14.21[a]	271.17 ± 8.56[a]	272.50 ± 12.59[a]	288.00 ± 15.91[a]
Day −14	262.83 ± 13.53[a]	277.17 ± 15.78[a]	277.33 ± 14.93[a]	275.67 ± 19.79[a]	307.17 ± 17.92[a]
Day −21	268.17 ± 10.34[a]	279.83 ± 8.02[a]	281.33 ± 22.86[a]	283.67 ± 26.31[a]	320.00 ± 24[a]
Day −28	256.33 ± 26.56[a]	284.50 ± 21.37[a]	296.83 ± 41.27[a]	296.00 ± 30.20[a]	337.83 ± 24.38[a]
Day −35	258.33 ± 18.38[a]	279.33 ± 24.12[a]	291.00 ± 32.75[a]	294.17 ± 31.73[a]	346.67 ± 22.23[a]
Day −42	263.50 ± 25.69[a]	278.17 ± 36.08[a]	287.00 ± 40.51[a]	292.50 ± 37.17[a]	368.17 ± 22.30[a]
Day −49	267.17 ± 17.27[a]	285.50 ± 42.69[a]	288.00 ± 42.40[a]	304.17 ± 36.12[a]	376.17 ± 22.75[a]
Day −56	276.00 ± 23.94[a]	283.00 ± 40.17[a]	290.17 ± 39.73[a]	341.00 ± 45.68[a]	385.17 ± 25.11[a]
Day −63	277.00 ± 13.46[a]	297.83 ± 32.11[ab]	327.83 ± 32.52[ab]	337.17 ± 37.41[ab]	395.00 ± 16.25[b]
Day −70	288.83 ± 13.97[a]	299.67 ± 32.36[ab]	342.33 ± 39.98[ab]	369.33 ± 41.95[ab]	403.83 ± 16.43[b]
Day −77	298.00 ± 26.31[a]	313.67 ± 39.98[ab]	354.33 ± 29.03[ab]	374.33 ± 25.09[ab]	409.50 ± 16.70[b]
Day −84	313.50 ± 20.86[a]	327.17 ± 19.73[ab]	356.17 ± 31.91[ab]	383.33 ± 27.46[ab]	415.17 ± 24.37[b]
Day −90	312.50 ± 15.94[a]	333.33 ± 35.11[ab]	347.17 ± 34.80[ab]	389.33 ± 25.04[ab]	425.33 ± 26.72[b]

[a-b]Values within rows with different superscripts differ significantly by ANOVA ($P \leq 0.05$). GI = Lantadenes @ 24 mg/kg bw; GII = Lantadenes @ 18 mg/kg bw; GIII = Lantadenes @ 12 mg/kg bw; GIV = Lantadenes @ 6 mg/kg bw; GV = control group, $n = 6$

Table 2 Hematological values (Mean ± SE) of different groups during sub-chronic *L. camara* toxicity in guinea pigs

	GI	GII	GIII	GIV	GV
TLC (×10³/μl)	8.82 ± 0.61	8.52 μ ± 1.18	7.67 ± 1.85	7.55 ± 2.03	6.73 ± 0.51
TEC (× 10⁶/μl)	5.583 ± 0.34	5.59 ± 0.18	5.87 ± 0.18	5.93 ± 0.21	6.03 ± 0.12
Hb (g/dl)	12.18 ± 0.55[a]	12.52 ± 0.65[ab]	12.68 ± 0.54[ab]	13.16 ± 0.47[ab]	14.31 ± 0.33[b]
PCV (%)	36.28 ± 2.07[a]	39.3 ± 2.51[ab]	43.07 ± 2.28[ab]	44.87 ± 1.90[ab]	45.59 ± 1.06[b]
MCV (fl)	69.22 ± 2.89	71.26 ± 2.08	72.22 ± 0.62	74.83 ± 1.55	75.63 ± 0.74
MCH (pg)	20.92 ± 1.30	21.913 ± 0.85	23.23 ± 0.58	23.22 ± 0.51	23.74 ± 0.13
MCHC (g/dl)	29.62 ± 1.02	29.72 ± 1.18	31.13 ± 0.78	31.33 ± 0.30	31.59 ± 0.26
Platelet Count (×10³/μl)	451.83 ± 55.3[a]	464.5 ± 60.24[a]	596.5 ± 50.30[ab]	623.17 ± 44.29[ab]	677 ± 58.26[b]
Heterophil Count (×10³/μl)	2.19 ± 0.15	3.65 ± 0.56	3.43 ± 0.36	3.39 ± 0.56	2.28 ± 0.36
Lymphocyte Count (×10³/μl)	5.48 ± 0.44	4.32 ± 0.50	4.09 ± 0.69	3.98 ± 0.63	4.03 ± 0.34
Monocyte Count (×10³/μl)	0.48 ± 0.11	0.42 ± 0.05	0.37 ± 0.07	0.3 ± 0.12	0.34 ± 0.04
Eosinophil Count (×10³/μl)	0.32 ± 0.16	0.45 ± 0.23	0.32 ± 0.17	0.38 ± 0.20	0.48 ± 0.22
Basophil Count (×10³/μl)	0.03 ± 0.01	0.02 ± 0.00	0.02 ± 0.01	0.01 ± 0.00	0.02 ± 0.00
Heterophil (%)	27.75 ± 2.18	37.2 ± 5.40	36.13 ± 4.27	37.08 ± 4.67	31.39 ± 5.06
Lymphocyte (%)	63.27 ± 2.89	53.98 ± 6.25	55.63 ± 3.45	54.51 ± 4.49	58.75 ± 5.36
Monocyte (%)	5.75 ± 0.94	5.55 ± 0.76	5.07 ± 0.53	4.95 ± 0.84	5.25 ± 0.40
Eosinophil (%)	2.93 ± 1.52	3 ± 0.99	2.93 ± 1.49	3.33 ± 1.12	4.39 ± 1.16
Basophil (%)	0.3 ± 0.07	0.27 ± 0.03	0.23 ± 0.06	0.13 ± 0.03	0.23 ± 0.06

[a-b]Values within rows with different superscripts differ significantly by ANOVA ($P \le 0.05$). GI = Lantadenes @ 24 mg/kg bw; GII = Lantadenes @ 18 mg/kg bw; GIII = Lantadenes @ 12 mg/kg bw; GIV = Lantadenes @ 6 mg/kg bw; GV = Control group, $n = 6$

(0.99 ± 0.10 mg/dl), IV (0.75 ± 0.10 mg/dl), and V (0.54 ± 0.07 mg/dl), did not exhibit any significant difference. Serum protein levels of group IV and V did not exhibit any significant difference. While serum protein levels of animals of groups I (4.05 ± 0.27 g/dl), II (4.42 ± 0.45 g/dl) and III (4.52 ± 0.49 g/dl) were significantly decreased as compared to control (6.60 ± 0.42 g/dl). The results of various biochemical parameters have been shown in Table 3.

Oxidation stress estimation

Malondialdehyde is considered as a marker of free radical-mediated lipid peroxidation injury. So, increased level of MDA lead to the failure of free-radical scavenging mechanisms. In the present study, LPO levels in liver of animals of almost all groups (II, 1.16 ± 0.16; III, 0.88 ± 0.260; IV, 0.74 ± 0.17; V, 0.73 ± 0.11) were

increased but the increase was significant ($P < 0.05$) only in animals of group I (1.50 ± 0.23 μM of MDA/g wet tissue). MDA levels in kidneys of lantadene administrated animals were significantly increased in animals of group I (1.78 ± 0.34 μM of MDA/g wet tissue) as compared to the control (0.64 ± 0.09 μM of MDA/g wet tissue), while this elevation was not significant in other groups. Superoxide dismutase (SOD) has been reported as one of the most important enzyme in enzymatic antioxidant defence system. It scavenges superoxide anion to form H_2O_2, hence diminishes the toxic effects induced by this free radical. In the present study, there was a significant decrease in SOD values in the liver of animals of groups I (4.82 ± 0.38 U/mg protein) (24 mg/kg bw) and II (6.04 ± 0.50 U/mg protein) (18 mg/kg bw) as compared to the control group (7.58 ± 0.27 U/mg protein). Similar pattern of decrease in SOD levels was

Table 3 Biochemical values (Mean ± SE) in different groups during sub-chronic *L. camara* toxicity in guinea pigs

GROUP	AST (IU/L)	ALT (IU/L)	ALP (IU/L)	ACP (IU/L)	Protein (g/dl)	Bilirubin (mg/dl)	Creatinine (mg/dl)
GI	107.07 ± 8.64[a]	77.82 ± 8.65[a]	116.60 ± 16.96[a]	111.20 ± 15.30[a]	4.05 ± 0.27[a]	1.01 ± 0.15[a]	1.21 ± 0.22[a]
GII	100.03 ± 10.77[ab]	63.88 ± 4.28[ab]	105.90 ± 17.71[a]	90.74 ± 14.74[ab]	4.42 ± 0.45[a]	0.84 ± 0.07[a]	1.14 ± 0.17[a]
GIII	91.08 ± 19.03[ab]	62.32 ± 6.79[ab]	100.48 ± 15.8[a]	75.96 ± 20.71[ab]	4.52 ± 0.49[a]	0.76 ± 0.16[a]	0.99 ± 0.10[ab]
GIV	80.15 ± 15.14[ab]	55.72 ± 7.30[ab]	98.69 ± 13.48[a]	67.05 ± 14.13[ab]	5.21 ± 0.41[ab]	0.73 ± 0.21[a]	0.75 ± 0.10[ab]
GV	58.10 ± 7.58[b]	52.14 ± 6.60[b]	78.03 ± 14.8[a]	43.83 ± 14.4[b]	6.60 ± 0.42[b]	0.49 ± 0.11[a]	0.54 ± 0.07[b]

[a-b]Values within columns with different superscripts differ significantly by ANOVA ($P \le 0.05$). GI = Lantadenes @ 24 mg/kg bw; GII = Lantadenes @ 18 mg/kg bw; GIII = Lantadenes @ 12 mg/kg bw; GIV = Lantadenes @ 6 mg/kg bw; GV = control group, $n = 6$

Table 4 Lipid peroxidation, catalase, superoxide dismutase, reduced glutathione, and protein values (Mean ± SE) in tissue homogenate of liver during sub-chronic *L. camara* toxicity in guinea pigs

Group	LPO (µM of MDA/g wet tissue)	Catalase (K/mg protein)	Superoxide dismutase (U /mg protein).	Reduced glutathione (nM/ g of wet tissue)	Protein (mg/dl)
GI	1.50 ± 0.23[a]	14.27 ± 1.62[a]	4.82 ± 0.38[a]	19.12 ± 1.29[a]	50.81 ± 5.70[a]
GII	1.16 ± 0.16[ab]	20.01 ± 3.15[ab]	6.04 ± 0.50[ab]	24.04 ± 1.90[ab]	66.49 ± 5.70[abc]
GIII	0.88 ± 0.260[b]	19.35 ± 1.73[ab]	7.22 ± 0.26[bc]	28.43 ± 1.75[bc]	106.2 ± 19.02[b]
GIV	0.74 ± 0.17[b]	22.91 ± 1.39[b]	7.57 ± 0.55[c]	32.59 ± 1.38	101.52 ± 11.22[b]
GV	0.73 ± 0.11[b]	22.39 ± 0.67[b]	7.58 ± 0.27[c]	32.51 ± 1.03[c]	112.72 ± 15.90[bc]

[a-c]Values within columns with different superscripts differ significantly by ANOVA ($P \leq 0.05$). GI = Lantadenes @ 24 mg/kg bw; GII = Lantadenes @ 18 mg/kg bw; GIII = Lantadenes @ 12 mg/kg bw; GIV = Lantadenes @ 6 mg/kg bw; GV = control group, $n = 6$

seen in the kidneys of the animals of groups I (3.94 ± 0.16 U/mg protein) and II (4.46 ± 0.31 U/mg protein) as compared to the control group (6.76 ± 0.66 U/mg protein). Reduced glutathione (GSH) is one of the most abundant non-biological antioxidant present in liver. GSH levels in liver homogenate of animals of groups I (19.12 ± 1.29 nM/g of wet tissue) and II (24.04 ± 1.90 nM/g of wet tissue) were significantly decreased as compared to control group. Reduction of this enzyme in kidney homogenate was only seen in the animals of group I.

Catalase decomposes H_2O_2 and protects the tissues from highly reactive hydroxyl radicals. Reduction in the anti-oxidative activities of this enzyme may result in a number of deleterious effects due to accumulation of superoxide radicals and H_2O_2. In the present study, the catalase levels were significantly lower in the liver (14.27 ± 1.62 K/mg protein) as well as kidneys (12.83 ± 0.46 K/mg protein) of animals of group I. The protein levels in liver homogenate were significantly decreased in animals of group I as compared to control, while the levels in other groups did not differ significantly from control. The protein levels in kidney homogenate were also decreased in animals of groups I and II as compared to control. The values of oxidation stress in liver and kidney homogenate have been presented in Tables 4 and 5, respectively.

Gross and histopathological evaluation

Grossly, the liver of animals of groups I (24 mg/kg bw) and II (18 mg/kg bw) showed presence of variable areas of

necrosis which were embedded deep and margins of liver lobes were friable. Group I histopathologically showed lymphatic distension, mild bile duct proliferation, collagen fibres deposition in peribiliary, periportal (Fig. 3a, b) and centrilobular (Fig. 4a, b) regions which was confirmed by MST staining. The results of MST staining showed the presence of mild to moderate collagen fibres deposition around central vein and portal triad. Binucleated hepatocytes, a few mitotic figures and increased Kupffer cell activity which could be assessed by their increased number and size were noticed (Fig. 5). Besides these, at certain places, ductular reaction, focal areas of karyomegaly, karyorrhexis, and degenerative changes along with infiltration of heterophils were evident in the hepatic parenchyma, which were indicative of coagulative necrosis (Fig. 6). While in certain areas, diffuse mononuclear cells (mainly lymphocytes and macrophages) infiltration was also seen. Almost similar changes were seen in the animals of group II. But the intensity of degenerative changes, necrosis and apoptosis was more pronounced in group I as compared with group II. Although scattered areas of necrosis were evident in the animals of groups I and II, no relevant level of significance was observed. However, they showed significant change in the lesion score values, namely, peribiliary and periportal fibrosis, bile duct proliferation, increased Kupffer cell activity, inflammatory changes, apoptotic changes, degenerative changes and regenerative changes shown by binucleated hepatocytes and mitotic figures as shown in Table 6. The animals of other groups also showed these changes but

Table 5 Lipid peroxidation, catalase, superoxide dismutase, reduced glutathione, and protein values (Mean ± SE) in tissue homogenate of kidneys during sub-chronic *L. camara* toxicity in guinea pigs

Group	LPO (µM of MDA/g wet tissue)	Catalase (K/mg protein)	Superoxide dismutase (U /mg protein)	Reduced glutathione (nM/ g of wet tissue)	Protein (mg/dl)
GI	1.78 ± 0.34[a]	12.83 ± 0.46[a]	3.94 ± 0.16[a]	20.12 ± 1.92[a]	68.46 ± 4.54[a]
GII	1.24 ± 0.20[ab]	15.22 ± 1.14[ab]	4.46 ± 0.31[a]	25.54 ± 1.78[ab]	74.64 ± 7.42[a]
GIII	0.92 ± 0.21[b]	15.78 ± 1.48[ab]	5.23 ± 0.50[ab]	25.71 ± 1.66[ab]	95.66 ± 16.52[ab]
GIV	0.68 ± 0.13[b]	18.52 ± 2.64[ab]	5.19 ± 0.33[ab]	30.57 ± 2.04[b]	116.92 ± 11.3[b]
GV	0.64 ± 0.09[b]	20.31 ± 2.21[b]	6.76 ± 0.66[b]	31.45 ± 0.98[b]	118.19 ± 12.65[b]

[a-b]Values within columns with different superscripts differ significantly by ANOVA ($P \leq 0.05$). GI = Lantadenes @ 24 mg/kg bw; GII = Lantadenes @ 18 mg/kg bw; GIII = Lantadenes @ 12 mg/kg bw; GIV = Lantadenes @ 6 mg/kg bw; GV = control group, $n = 6$

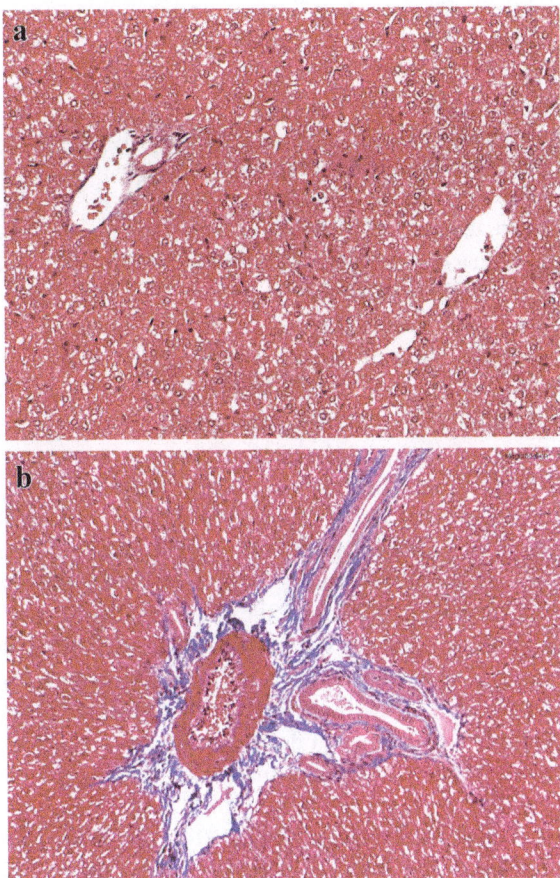

Fig. 3 a Liver (control, GV): Normal liver. MSTx100. **b** Liver (GI, lantadenes, 24 mg/kg bw): Bile duct proliferation and lymphatic distention along with mild peribiliary and periportal fibrosis. MSTx200

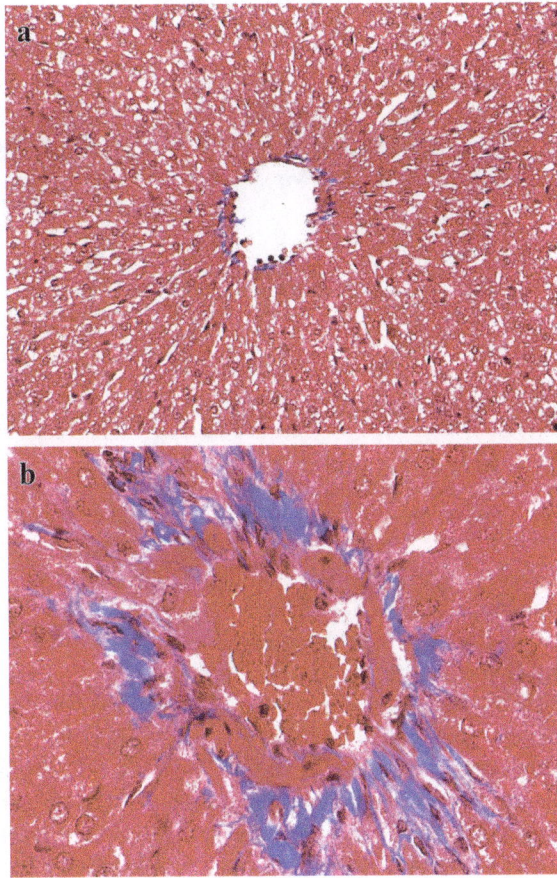

Fig. 4 a Liver (control, GV): Normal liver without centrilobular fibrosis.MSTx100. **b** Liver (GI, lantadenes, 24 mg/kg bw): Mild centrilobular fibrous tissue proliferation. MSTx200

were of lesser intensity as compared to groups I and II. The lesion score for histopathological changes in liver has been tabulated in Table 6.

In kidneys, the gross lesions were minimal and on histopathological examination, the animals of groups I and II showed severe degenerative and focal necrotic changes in tubules along with the infiltration of leucocytic cells (Fig. 7). In addition, there was marked deposition of hyaline and epithelial casts (Fig. 8) in the renal tubular lumens of animals of groups I and II. The periglomerular and peritubular fibrosis was also observed in sub-chronic lantadene toxicity mainly in the animals of groups I and II which was further confirmed by MST staining (Fig. 9a, b). The lesion score for histopathological changes in kidneys has been tabulated in Table 7. The gall bladder of animals of groups I and II was grossly distended. On histopathology, the gall bladder showed mild fibrosis along with inflammatory cells infiltration. The mesenteric lymph nodes (MLNs) of most of the treatment groups showed congestion of medullary region as compared to control group. Histopathologically, the MLNs of animals of group I showed

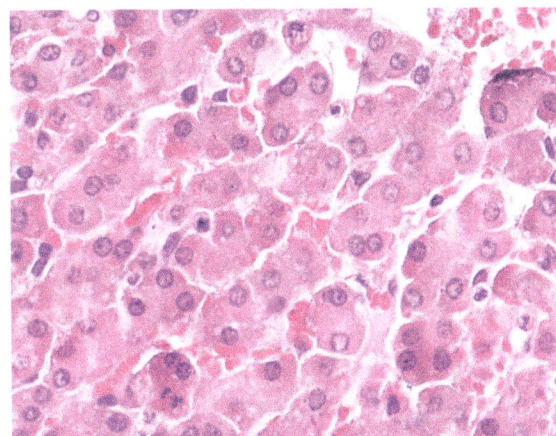

Fig. 5 Liver (GI, lantadenes, 24 mg/kg bw): Binucleated hepatocytes along with mitotic figures indicating regenerative changes. HEx200

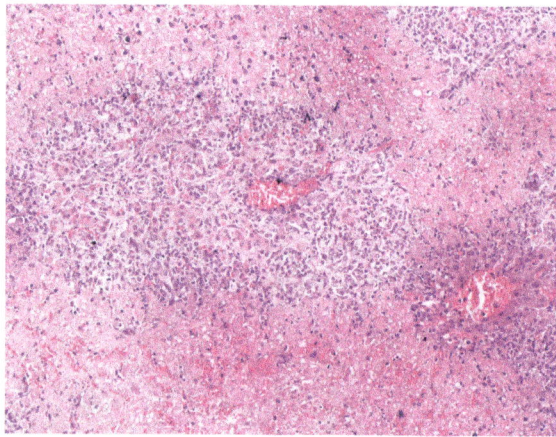

Fig. 6 Liver (GI, lantadenes, 24 mg/kg bw): Severe multifocal necrotic areas along with inflammatory cells and ducutular reaction in hepatic parenchyma. HEx100

lymphocytic depletion while other groups did not exhibit such changes. Grossly, the stomach of animals of groups I and II showed diffuse haemorrhages (Fig. 10) and similar lesions along with infiltration of MNCs (mainly lymphocytes and macrophages) on histopathological examination were seen. Brain of most of the treated group animals did not show any gross changes. However, on histopathology, cerebral cortex revealed mild increase in the perivascular space, neuronal shrinkage and focal areas of mild glial cell proliferation around degenerated neurons i.e. neuronal satellitosis (Fig. 11) was evident as compared to control group. This was suggestive of toxic effect even on the nervous system during sub-chronic exposure to lantadenes. The adrenal glands also showed scattered pyknotic nuclei in the cortical region during lantadene toxicity.

Discussion

Lantana camara introduced as an ornamental shrub has become a threat to our livestock and biodiversity [2]. This plant is capable to cause mortality in ruminant as well as non-ruminant species. Among the non-ruminants, guinea pigs are the most susceptible [17] species and therefore have been used as a model in the present study. This weed leads to hepatotoxicity and photosensitization in grazing animals and has allelopathic effect on other vegetation [2]. Major toxic components present in this plant are lantadenes [18, 19]. In the present study, LA and LB content in the leaves was estimated to be 49.18% and 13.09%, respectively by reverse phase-HPLC.

The orally administered lantadenes can be absorbed from stomach, small as well as large intestine. But small intestine is the most important route. The absorption is affected by presence of ingesta in the intestine [20]. Therefore, the animals in the present study were provided with feed and water one hour after the oral administration of lantadenes. After absorption, the transportation of toxin to the liver occurs via portal route and bile has no effect in this absorptive mechanism [2]. For the maintenance of effect of toxin, continuous absorption of toxin is required. This toxin can lead to ruminal stasis in cattle due to inhibitory neural impulses arising from damaged liver after 4–6 h of administration [21, 22]. In the present study, there was a significant decrease in weekly body weights of the animals of group I, while decrease in the body weights of animals of groups II, III and IV was not significant as compared to control. After absorption, lantadenes mainly acts on bile canalicular plasma membrane (CPM) which is the main target of its action in the hepatocytes [23, 24]. The biliary compounds from the blood are mainly taken to sinusoidal membrane and from there stored and metabolized by hepatocytes and ultimately secreted by CPM. The damaged CPM is not capable to secrete the bile thus leads to impaired hepatobiliary excretion and thereby intrahepatic cholestasis [23, 24]. The cholestasis is accompanied by dilated bile duct canaliculi, microvilli loss, alteration in enzymes and jaundice [25, 26]. Jaundice has been reported in acute and sub-acute lantadene toxicity and its development depends upon the dose of lantadenes administered and action of lantadenes on biliary secretions, which leads to cholestasis and thereby jaundice [2, 27, 28]. However, ictericity was not observed in sub-chronic toxicity as the biliary mechanism is least altered and can be compared with minimal effect of lantadenes on the values of ALP, which gives a clear

Table 6 Lesion score (Mean ± SE) for histopathological changes in liver during sub-chronic *L. camara* toxicity in guinea pigs

Group	Necrosis	Fibrosis	Bile duct proliferation	Increased Kupffer cell activity	Inflammatory change	Apoptosis	Degenerative change	Regenerative change
GI	2.00 ± 0.26[a]	1.67 ± 0.21[b]	1.50 ± 0.34[b]	2.50 ± 0.22[c]	1.83 ± 0.31[b]	1.33 ± 0.21[bc]	1.67 ± 0.21[b]	2.17 ± 0.31[b]
GII	1.33 ± 0.33[ab]	1.50 ± 0.22[b]	1.33 ± 0.21[b]	1.67 ± 0.21[bc]	1.67 ± 0.21[b]	1.83 ± 0.17[c]	1.33 ± 0.21[b]	1.67 ± 0.21[b]
GIII	1.17 ± 0.31[ab]	0.83 ± 0.17[ab]	1.00 ± 0.26[ab]	1.17 ± 0.17[abc]	1.17 ± 0.31[ab]	1.00 ± 0.26[abc]	1.00 ± 0.26[ab]	1.33 ± 0.21[ab]
G1V	0.67 ± 0.21[ab]	0.67 ± 0.21[ab]	0.50 ± 0.22[ab]	0.67 ± 0.21[ab]	0.67 ± 0.21[ab]	0.33 ± 0.21[ab]	0.83 ± 0.31[ab]	1.00 ± 0.26[ab]
GV	0.00 ± 0.00[a]	0.00 ± 0.00[a]	0.00 ± 0.00[a]	0.00 ± 0.00[a]	0.00 ± 0.00[a]	0.00 ± 0.00[a]	0.00 ± 0.00[a]	0.00 ± 0.00[a]

[a-c]Values within columns with different superscripts differ significantly by ANOVA ($P \leq 0.05$). GI = Lantadenes @ 24 mg/kg bw; GII = Lantadenes @ 18 mg/kgbw; GIII = Lantadenes @ 12 mg/kg bw; GIV = Lantadenes @ 6 mg/kg bw; GV = control group, n = 6

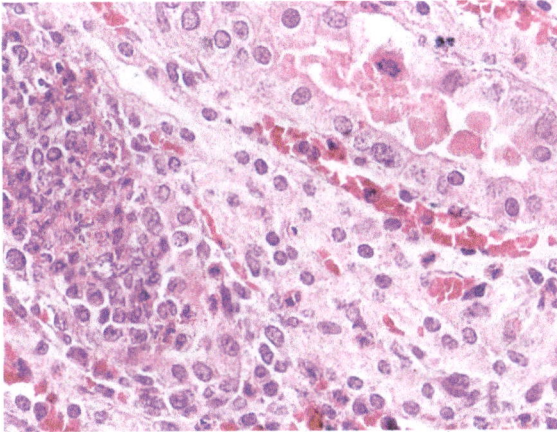

Fig. 7 Kidney (GI, lantadenes, 24 mg/kg bw): Formation of hyaline and granular casts in tubular lumens. HEx200

evidence for no development of cholestatic mechanism and ictericity in animals. But the presence of lower doses of this toxin in circulation might be capable to cause more sub-chronic changes and degenerations, which were minimal in previous studies on acute and sub-acute lantadenes exposure [2, 7].

The total platelet counts of groups I and II were significantly decreased as compared to control. The possible mechanism for this can be the haemorrhages caused by the action of lantadenes in stomach leading to anaemia. The significant decline in the values of Hb in the animals of group I might be because of defective haemopoietic mechanism, whereas decrease in the values of PCV indicated shrinkage in the size of erythrocytes as a result of lantadenes intoxication as studied earlier [29]. In one of the studies, transient decrease in PCV, TEC, Hb values and increase in neutrophil count has been reported in cattle and buffaloes [30, 31].

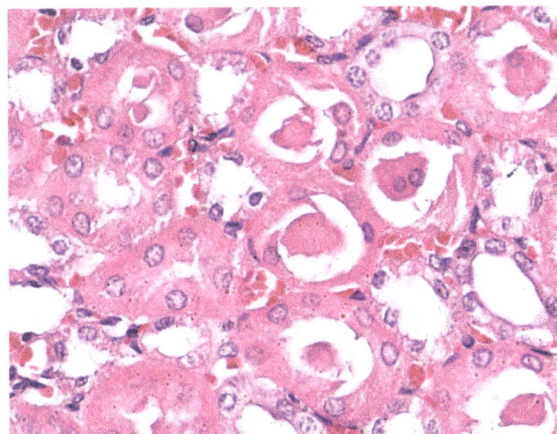

Fig. 8 Kidney (GII, lantadenes, 18 mg/kg bw): Tubules infiltrated with epithelial and hyaline casts. HEx400

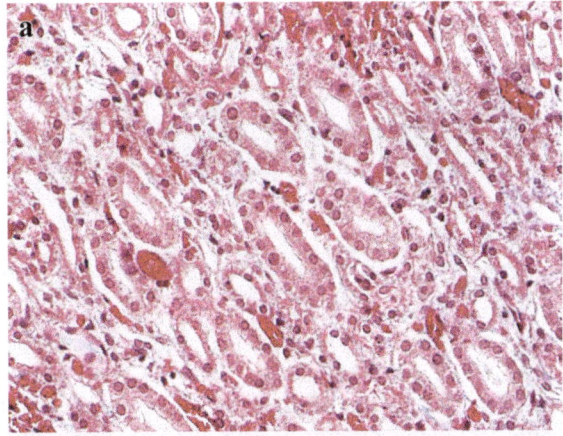

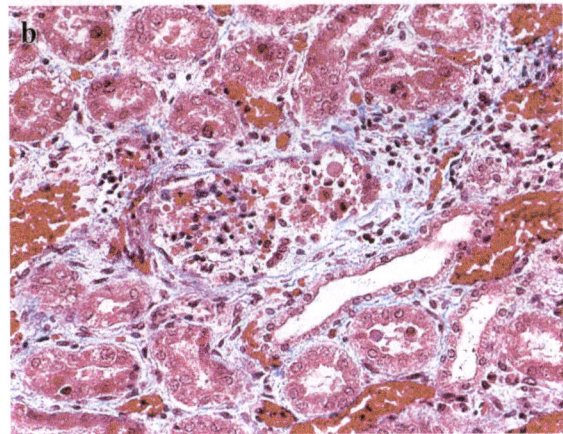

Fig. 9 a Kidney (control, GV): Normal, without any peritubular and periglomerular collagenous fibrous tissue proliferation. MSTx100. **b** Kidney (GI, lantadenes, 24 mg/kg bw): Mild to moderate peritubular and periglomerular collagenous fibrous tissue proliferation. MSTx200

Increased levels of ALT, AST and ACP of animals of groups I, II, III and IV indicated hepatic damage as these enzymes leaked out from the hepatocytes into blood due to tissue damage. However, this increase was significant only in group I, which was the highest dose group (24 mg/kg bw) and the elevation in this group could be correlated with marked histopathological changes. A slight increase in the values of ALP was observed in the animals of groups I and II, but this elevation was not statistically significant. The increase in the level of ALP is often associated with profound hepato-biliary injury and cholestasis. In present study, non-significant elevation in ALP levels indicates less significant involvement of biliary system in sub-chronic lantadenes toxicity. The changes in the liver specific serum enzymes indicated significant hepatic damage and were well supported by histopathological alterations in the liver. The previous study on sub-acute lantadene toxicity in guinea-pigs showed an increase in the values of ALT, AST, and ALP while no significant increase was seen in ACP levels [7]. Sharma and Sharma [32] reported that the oral

Table 7 Lesion score (Mean ± SE) for histopathological changes in kidneys during sub-chronic *L. camara* toxicity in guinea pigs

Group	Necrosis	Fibrosis	Degenerative change	Inflammatory change	Hyaline casts formation	Epithelial casts formation	Apoptosis
GI	2.33 ± 0.21[c]	2.00 ± 0.26[c]	2.50 ± 0.22[c]	2.17 ± 0.31[b]	2.83 ± 0.17[c]	2.67 ± 0.21[c]	1.67 ± 0.21[b]
GII	2.00 ± 0.26[bc]	1.50 ± 0.22[bc]	2.50 ± 0.22[c]	1.83 ± 0.31[b]	2.50 ± 0.22[bc]	2.33 ± 0.21[bc]	1.00 ± 0.37[ab]
GIII	1.17 ± 0.31[abc]	0.83 ± 0.17[abc]	1.33 ± 0.21[abc]	1.17 ± 0.31[ab]	1.50 ± 0.34[abc]	1.17 ± 0.31[abc]	0.83 ± 0.17[ab]
G1V	0.67 ± 0.21[ab]	0.33 ± 0.21[ab]	0.67 ± 0.21[ab]	0.83 ± 0.31[ab]	0.50 ± 0.22[ab]	0.50 ± 0.22[ab]	1.00 ± 0.26[ab]
GV	0.00 ± 0.00[a]	0.00 ± 0.00[a]	0.00 ± 0.00[a]	0.00 ± 0.00[a]	0.00 ± 0.00[a]	0.00 ± 0.00[a]	0.00 ± 0.00[a]

[a-c]Values within columns with different superscripts differ significantly by ANOVA ($P \leq 0.05$). GI = Lantadenes @ 24 mg/kg bw; GII = Lantadenes @ 18 mg/kg bw; GIII = Lantadenes @ 12 mg/kg bw; GIV = Lantadenes @ 6 mg/kg bw; GV = control group, $n = 6$

administration of LA to guinea pigs elicited an increase in activities of ALT, AST, ALP and ACP. There was no significant change in the total bilirubin levels in sub-chronic lantadene toxicity. Hence, the sub-chronic oral lantadene administration did not lead to the development of icteric subcutaneous tissues and mucus membranes in the experimental animals. However, in a previous study, the total bilirubin levels were significantly elevated at high doses (100 and 50 mg/kg bw) leading to icteric subcutaneous tissue and renal medulla. The creatinine levels were also found to increase in sub-acute lantadene toxicity (25 mg/kg bw). The lower doses of lantadenes (25 and 12.5 mg/kg bw) were supposedly the main cause for renal damage due to cumulative toxicity [7]. In the present study, the changes in the kidneys were more pronounced than in liver while in sub-acute lantadene toxicity study, hepatotoxic effects were more pronounced [7]. The possible reason for this could be the regeneration of hepatocytes at lower doses of lantadenes used in the present study. The concept of regeneration was supported by presence of mitotic figures and binucleated hepatocytes. A decrease in the values of protein, increase in values of cholesterol and cholesterol: phospholipid ratio was seen on oral administration of lantadenes [33]. In present study

also, the values of proteins were significantly decreased in animals of groups I, II and III as compared to group IV and control.

Malondialdehyde (MDA), the main oxidative degradation product of lipid-peroxidation, functions as a marker of oxidative injury of cellular membranes [34]. An increased MDA level in liver suggests enhanced lipid peroxidation leading to tissue damage and failure of antioxidant mechanism to prevent formation of excessive free radicals. LPO was exhibited by almost all the tissues in lantana poisoning in the order adrenals> liver> kidneys> heart> lungs> testes> brain [35]. The present study showed significant increase in the values of LPO in liver and kidney homogenates mainly in animals of group I. The SOD levels were declined in liver and kidneys of animals of group I and II at a significant level as compared to groups III and IV. Reduced glutathione (GSH) is one of the most abundant non-biological antioxidant present in liver [36]. GSH is a key component of overall antioxidant defense system that protects cell from deleterious effects of reactive oxygen species. In the present study, GSH levels of animals of groups I and II were significantly decreased as compared to control group. However, in sub-acute lantadene toxicity, there

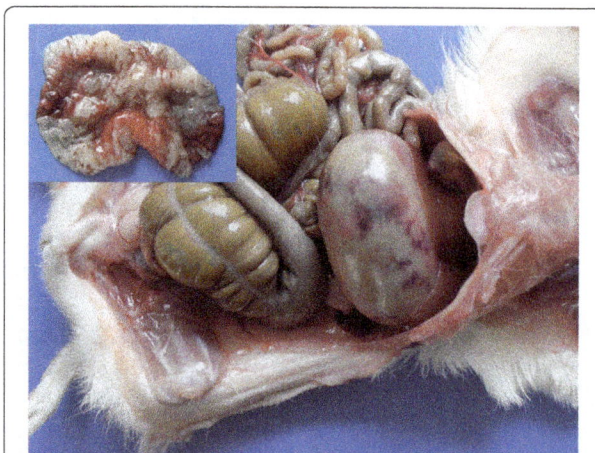

Fig. 10 Stomach (GI, lantadenes, 24 mg/kg bw): Areas of haemorrhages on the serosa and mucosal surface (inset)

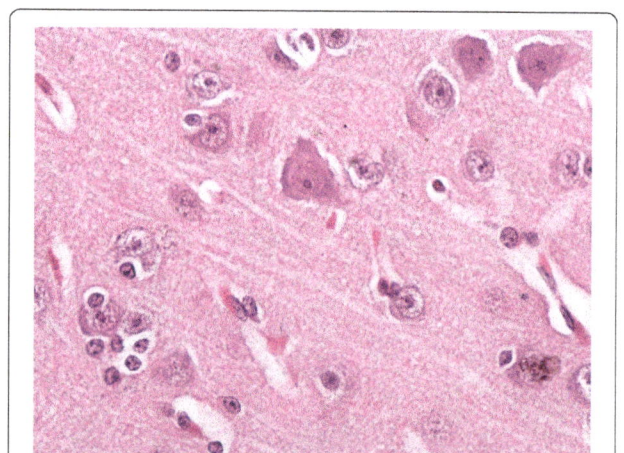

Fig. 11 Brain (GII, lantadenes, 18 mg/kg bw): Cerebral cortex showing neuronal satellitosis. HEx400

was no significant decrease in the values of GSH [7]. Reduced glutathione values were also found to decrease in kidneys of animals of group I as compared to control. There was a significant reduction in enzymatic antioxidants including SOD and catalase which are essential for endogenous antioxidative defense system to scavenge reactive oxygen species and maintain cellular redox defense [37]. In the present study, the catalase levels were significantly lower in liver as well as kidneys of animals of group I and this decrease in catalase levels was also observed in sub-acute lantadene toxicity [7]. The protein levels of liver homogenate of animals of group I was significantly decreased as compared to other groups. There was a significant reduction in the values of protein in kidney homogenate of animals of groups I and II while this decrease was not significant in other groups. In earlier study on lantadene toxicity, there was an increase in protein values in liver homogenate, while no significant difference was seen in kidneys as compared to control [7]. In the present study, it has been observed that the alteration in antioxidative enzymatic system was mainly recorded in groups I (24 mg/kg bw) and II (18 mg/kg bw) as compared to groups III, IV and V. These alterations were well supported by histopathology, special staining, oxidation stress determination and biochemical changes in liver as well as in kidneys.

Lantadenes mainly damages the peripheral parenchymal cells of liver while cells around central vein are normal [2]. Liver shows variable lesions in acute and sub-acute lantadene toxicity ranging from swelling, pale yellow and fragile appearance, with diffuse areas of necrosis and haemorrhagic streaks. The kidneys were yellowish in colour [2, 7, 38]. However, in the present study, the various doses of lantadenes did not produce severe characteristic gross lesions in liver and kidneys. The liver of animals of groups I and II exhibited a few necrotic foci on parenchyma and pale kidneys. The liver and kidneys of animals of groups III and IV did not show any significant gross changes as compared to control. However, the lesions in liver and kidneys were well evident on histopathological examination.

On histopathology, the liver of animals of groups I and II showed focal areas of hepatocytic necrosis. The nuclear changes included pyknotic nuclei, karyomegaly, karyorrhexis and dissolution of nuclei in the hepatocytes which also supports that the oxidative damage could be one of the mechanisms for hepatocellular injury [39]. In the present study, infiltration of MNCs admixed with heterophils, peribiliary and periportal fibrosis, bile duct proliferation, binucleated hepatocytes, mitotic figures, apoptotic changes and increased Kupffer cell activity was observed. Binucleated hepatocytes and mitotic figures (1–3 cells/high power fields) are important consequences of hepatocytic injury and chromosomal

hyperplasia. These changes are often appreciated in the cells which are undergoing regeneration [40]. Kupffer cells are often associated with liver injury and hepatocellular necrosis and often implicated as a cause of TGF-beta production required for the transformation of stellate cells into myofibroblasts and ultimately leads to fibrosis. The increased number and activity of Kupffer cells indicated the defence mechanism of detoxification associated with hepatic injury as studied earlier [41]. The remaining treatment groups also showed similar lesions but were of lesser severity or absent in some animals. The fibrotic changes around periportal and peribiliary regions were evident in the liver of animals of groups I and II which were further confirmed by MST as well as seen in previous study on sub-acute lantadene toxicity [7]. In sub-acute lantadene toxicity study, the liver showed large areas of coagulative necrosis, heterophilic infiltration, biliary fibrosis, haemorrhages and engorgement. The collagen tissue deposition in liver of these animals was confirmed by MST [7]. In the present study, the renal lesions were more characteristic which included marked degenerative changes, necrosis, hyaline and epithelial casts, apoptosis, periglomerular and peritubular fibrosis. The degree of fibrosis was further confirmed by MST. Similar fibrotic lesions in kidneys were also seen in sub-acute toxicity of lantadenes in guinea pigs where collagen tissue deposition was also confirmed by the same special stain [7]. The severity of collagen tissue deposition in the kidneys was more pronounced in sub-chronic toxicity. Our study on sub-chronic exposure to lantadenes is also supported by a similar kind of sub-chronic toxicity trial of 90 days, where animals were administrated with methanolic extract of Rhaphidophora decursiva [42]. On similar lines as of our study, no changes were evident grossly, while on histopathology the changes included increased Kupffer cell activity, karyomegaly and karyorrhexis in the liver and there was formation of cellular casts and pyknotic cells in kidneys [39, 43].

In the present study, the gall bladder was grossly distended. However, collagen tissue deposition was noticed on histopathological examination which was further confirmed by MST staining. In earlier acute and sub-acute lantadene toxicity studies, the contents of gall bladder were found to be inspissated, yellowish, opaque, thick and tarry and fibrous tissue formation was not evident on histopathology [2, 7]. Grossly the MLNs of all the treatment groups showed congestion in medulla as compared to control. Lantadenes are capable to cause immunosuppression by reducing both the cellular and humoral immune system as observed in sheep [44]. In the present study, MLNs of animals of group I showed lymphocytic depletion which indicated immunosuppressive nature of lantadenes. The brain of treatment groups on histopathology showed mild glial cells proliferation around a few degenerated neurons as compared to control group, which was suggestive of toxic effect even on the nervous system during sub-chronic exposure to lantadenes. The adrenal glands showed pyknotic nuclei indicative of physiological stress induced by lantadene toxicity at lower doses.

Lantadenes produced maximum damage in kidneys followed by liver and stomach during sub-chronic exposure. The more severe changes in kidneys revealed more nephrotoxic action of lantadenes as compared to hepatotoxic action in sub-chronic toxicity study (90 days). This could be because liver has power of regeneration so it might have developed some compensatory mechanism in liver parenchymal cells or in liver canaliculi to excrete bile not leading to the development of jaundice as studied in earlier experiments on lantadene toxicity at the dose of 50 mg/kg bw. However, at lower doses of lantadenes (25 mg/kg bw) in sub-acute exposure, the associated pathological alterations like cellular degenerations and hepatic necrosis were evident as also observed in the present study on sub-chronic administration of lantadenes [7]. While the kidneys showed more pronounced damage because of persistent effect of toxins and exhibited typical signs of sub-chronic toxicity.

Conclusions
In conclusion, the present study brought to light the finding that sub-chronic exposure to lantadenes produce a dose-dependent toxicity of lantadenes in decreasing order in the guinea pigs, with the highest dose (24 mg/kg bw) producing maximum and the lowest dose (6 mg/kg bw) producing minimum alterations in kidneys as well as in liver. The results of significant decline in the body weights, hematology, serum marker enzymes, oxidation stress levels, gross and histopathological examination and confirmation of collagen tissue by MST staining showed that the lantadenes, which are generally considered hepatotoxic, rather resulted in pronounced nephrotoxicity during sub-chronic exposure.

Abbreviations
GSH: Reduced glutathione; LA: Lantadene A; LB: Lantadene B; LPO: Lipid peroxidation; MLN: Mesenteric lymph node; MNC: Mononuclear cells; MST: Masson's trichome stain; SOD: Superoxide dismutase

Acknowledgements
The authors gratefully acknowledge the Director, ICAR-Indian Veterinary Research Institute, Izatnagar for providing necessary facilities for the study. The authors would also like to acknowledge the reviewers for improving the manuscript.

Funding
This work was supported by the research fund of ICAR-Indian Veterinary Research Institute, Izatnagar, Bareilly, India (Project Code: IVRI/PALAM/13–16/ 010). The funder had no role in study conception, design or analysis of results.

Authors' contributions
RS conceptualized, designed and supervised the study. RK conducted experimental trial, laboratory work and drafted the manuscript. RDP contributed in pathological interpretation. GM performed thin layer chromatography. AK contributed in clinical examination of animals. VP performed haematological analysis. PK and BS carried out HPLC analysis. RS, RK and RDP revised the manuscript. All authors read and approved the final revision.

Consent for publication
Not applicable.

Competing interests
The authors declare that they have no competing interests.

Author details
[1]Disease Investigation Laboratory, ICAR-Indian Veterinary Research Institute, Regional Station, Palampur, Himachal Pradesh, India. [2]DGCN COVAS, CSK HPKV, Palampur, Himachal Pradesh, India. [3]CSIR-IHBT, Palampur, Himachal Pradesh, India.

References
1. Sharma S, Sharma OP, Singh B, Bhat TK. Biotransformation of lantadenes, the pentacyclic triterpenoid hepatotox in of lantana plant, in Guinea pig. Toxicon. 2000;38:1191–202.
2. Sharma OP, Sharma S, Pattabhi V, Mahato SB, Sharma PD. A review of the hepatotoxic plant *Lantana camara*. Crit Rev Toxicol. 2007;37:313–52.
3. Pass MA. Poisoning of livestock by lantana plants. In: Handbook of natural Toxins,Toxicology of plant and fungal compounds R. Keeler, and T. Anthony, eds. Marcel Dekker, Inc. N Y 1991;6:297–311.
4. Sharma OP, Vaid J, Sharma PD. Comparison of lantadenes content and toxicity of different taxa of the lantana plant. J Chem Ecol. 1991;17:2283–91.
5. Sharma OP, Sharma S, Dawra RK. Reversed-phase high-performance liquid chromatographic separation and quantification of lantadenes using isocratic systems. J Chromatogr A. 1997;786:181–4.
6. Parimoo HA, Sharma R, Patil RD, Sharma OP, Kumar P, Kumar N. Hepatoprotective effect of *Ginkgo biloba* leaf extract on lantadenes induced hepatotoxicity in Guinea pigs. Toxicon. 2014;81:1–12.
7. Parimoo HA, Sharma R, Patil RD, Patial V. Sub-acute toxicity of lantadenes isolated from *Lantana camara* leaves in Guinea pig animal model. Comp Clinic Pathol. 2015;24:1541–52.
8. Bergmeyer HU. U.V: method of acid phosphatase assay. In: Methods of enzymatic analysis. 3rd ed. Weinheim: Deer field beach, Florida; 1974. p. 223.
9. Dawra RK, Sharma OP, Makkar HP. Evidence for a novel antioxidant in bovine seminal plasma. Biochem Int. 1984;8:655–9.
10. Sedlak J, Lindsay RH. Estimation of total, protein-bond and non-protein sulfhydryl groups in tissue with Ellman's reagent. Anal Biochem. 1968;25: 192–205.
11. Aebi HE. Catalase. In: Bergmeyer HU, editor. Methods of enzymatic analysis. 3rd ed. Weinheim, Florida: VerlagChemie; 1983. p. 273–86.
12. Nishikimi M, Rao NA, Yagi K. The occurrence of superoxide anion in the reaction of reduced Phenazinemethosulphate and molecular oxygen. Biochem Biophysic Res Commu. 1972;46:849–54.
13. Lowry OH, Rosebrough NJ, Farr AL, Randall RJ. Protein measurement with Folin phenol reagent. J Biol Chem. 1951;193:265–75.
14. Luna LG. Manual of histologic staining methods of the armed forces Institute of Pathology. 3rd ed. New York: McGraw-Hill; 1968.
15. Corley KNG, Olivier AK, Meyerholz DK. Principles for valid histopathologic scoring in research. Vet Pathol. 2013;50(6) https://doi.org/10.1177/0300985813485099.
16. Linda Ferrell, MD Distinguished Professor Vice Chair Director of Surgical Pathology Dept of Pathology. UCSF. http://labmed.ucsf.edu/uploads/472/ 227_Ferrell,%20LiverUpdateOnStagingOfFibrosisAndCirrhosis.pdf.
17. Gopinath C, Ford EJH. The effect of *Lantana camara* on the liver of sheep. J Pathol. 1969;99:75–85.
18. Hart NK, Lamberton JA, Sioumis AA, Saures H. New triterpenes of *Lantana camara*. A comparative study of the constituents of several taxa. Aust J Chem. 1976;29:655–71.
19. Hart NK, Lamberton JA, Sioumis AA, Saures H, Seawright AA. Triterpenes of toxic and nontoxic taxa of *Lantana camara*. Experientia. 1976;32:412.

20. Pass MA, McSweeney CS, Reynoldson JA. Absorption of the toxins of Lantana camara L. from the digestive system of sheep. J Appl Toxicol. 1981;1:38–42.

21. Mc Sweeney CS, Pass MA. The mechanism of ruminal stasis in lantana poisoned sheep. Quarterly J Expt Physiol Cog Med Sci. 1983;68:301–13.

22. Pass MA, Gemmell RT, Heath TJ. Effect of lantana on the ultrastructure of the liver of sheep. Toxicol Appl Pharmacol. 1978;43:589–96.

23. Pass MA, Pollitt S, Goosem MW, McSweeney CS. The pathogenesis of lantana poisoning. In Plant Toxicology, AA Seawright, MP Hegarty, LF James, RF Keeler (Eds.). Queensland poisonous plants committee, Yeerongpilly, Australia. Plant Toxicol. 1985:487–94.

24. Pass MA, Goosem MW. Observation of metabolises of reduced lantadene a in bile canalicular membranes of rats with triterpene-induced cholestasis. Chem Biol Interact. 1982;40:375–8.

25. Koopen NR, Muller M, Venk RJ, Zimniak P, Kuipers F. Molecular mechanisms of cholestasis: causes and consequences of impaired bile formation. Biochem Biophys Acta. 1998;1408:1–17.

26. Kellerman TS, Coetzer JAW. Hepatogenous photosensitivity diseases in South Africa. Onderstep J Vet Res. 1985;52:157–73.

27. Phillips MJ, Oshio C, Miyairi M, Smith CR. Intrahepatic cholestasis as a canalicular motility disorder. Evidence using cytochalasin. Lab Investig. 1983;48:205–11.

28. Trauner M, Meier PJ, Boyer JL. Mechanisms of disease: molecular pathogenesis of cholestasis. New Eng J Med. 1998;339:1217–27.

29. Kumar KV, Sharief SD, Rajkumar R, Ilango B, Sukumar E. Influence of Lantana aculeata stem extract on haematological parameters in rats. Adv Biores. 2011;2:79–81.

30. Alfonso HA, Figueredo JM, Merino N. Photodynamic dermatitis caused by Lantana camara in Cuba. A preliminary study. Revista de Salud Ani. 1982;4: 141–50.

31. Kalra DS, Dixit SN, Verma PC, Dwivedi P. Studies on experimental lantana poisoning in buffalo calves with special reference to its pathology and histochemistry. Haryana Vet. 1984;23:98–105.

32. Sharma S, Sharma OP. Effect of lantadene A-induced toxicity on enzymes associated with cholestasis in blood plasma of Guinea pigs. Med Sci Res. 1999;27:157–8.

33. Sharma OP, Dawra RK. Effect of lantana toxicity on canalicular plasma membrane of Guinea pig liver. Chemico Biologic Interactions. 1984;49: 369–74.

34. Yao P, Li K, Jin Y, Song F, Zhou S, Son X. Oxidative damage after chronic ethanol intake in rat tissues: prophylaxis of Ginkgo biloba extract. Food Chem. 2006;99:305–14.

35. Sharma DP, Dawra RK, Makkar HP. Effect of Lantana camara toxicity on lipid peroxidation in Guinea pig tissues. Res Commun Chem Pathol Pharmacol. 1982;38:153–6.

36. Meister A. New aspects of glutathione biochemistry and transport selective alterations glutathione metabolism. Nutr Rev. 1984;42:397–40.

37. Molina MF, Sanchez-Reus I, Iglesias I, Benedi J. Quercetin, a flavinoid antioxidant prevents and protects against ethanol induced oxidative stress in mouse liver. Biol Pharm Bull. 2003;26:1398–402.

38. Seawright AA, Allen JG. Pathology of the liver and kidney in lantana poisoning of cattle. Aust Vet J. 1972;48:323–31.

39. Jarrar BM, Taib NT. Histological and histochemical alterations in the liver induced by lead chronic toxicity. SJBS. 2012;19:203–10.

40. Gerlyng P, Abyholm A, Grotmol T, Erikstein B, Huitfeldt HS, Stokke T, Seglen PO. Binucleation and polyploidization patterns in developmental and regenerative rat liver growth. Cell Prolif. 2008;26:557–65.

41. Kolios G, Valatas V, Kouroumalis E. Role of Kupffer cells in the pathogenesis of liver disease. World J Gastroenterol. 2006;12:7413–20.

42. Arsad SS, Esa NM, Hamzah H. Histopathologic changes in liver and kidney tissues from male Sprague Dawley rats treated with Rhaphidophora decursiva (Roxb.) Schott extract. J Cytol Histol. 2014;S4:001. https://doi.org/10.4172/2157-7099.S4-001.

43. Neyrinck A. Modulation of Kupffer cell activity: physiopathological consequences on hepatic metabolism. Bull Mem Acad R Med Belg. 2004;159:358–66.

44. Ganai GN, Jha GJ. Immunosuppression due to chronic Lantana camara L. toxicity in sheep. Indian J Exp Biol. 1991;29:762–6.

20

Characterization of the DNA binding activity of structural protein VP1 from chicken anaemia virus

Guan-Hua Lai[1†], Ming-Kuem Lin[2†], Yi-Yang Lien[3], Jai-Hong Cheng[4], Fang-Chun Sun[5], Meng-Shiunn Lee[6], Hsi-Jien Chen[7] and Meng-Shiou Lee[2*] (ID)

Abstract

Background: Chicken anaemia virus (CAV) is commonly found in poultry. VP1 is the sole structural protein of CAV, which is the major component responsible for capsid assembly. The CAV virion consists of the VP1 protein and a viral genome. However, there is currently no information on the protein-nucleic acid interactions between VP1 and DNA molecules.

Results: In this study, the recombinant VP1 protein of CAV was expressed and purified to characterize its DNA binding activity. When VP1 protein was incubated with a DNA molecule, the DNA molecule exhibited retarded migration on an agarose gel. Regardless of whether the sequence of the viral genome was involved in the DNA molecule, DNA retardation was not significantly influenced. This outcome indicated VP1 is a DNA binding protein with no sequence specificity. Various DNA molecules with different conformations, such as circular dsDNA, linear dsDNA, linear ssDNA and circular ssDNA, interacted with VP1 proteins according to the results of a DNA retardation assay. Further quantification of the amount of VP1 protein required for DNA binding, the circular ssDNA demonstrated a high affinity for the VP1 protein. The preferences arranged in the order of affinity for the VP1 protein with DNA are circular ssDNA, linear ssDNA, supercoiled circular dsDNA, open circular DNA and linear dsDNA.

Conclusions: The results of this study demonstrated that the interaction between VP1 and DNA molecules exhibited various binding preferences that were dependent on the structural conformation of DNA. Taken together, the results of this report are the first to demonstrate that VP1 has no sequence-specific DNA binding activity. The particular binding preferences of VP1 might play multiple roles in DNA replication or encapsidation during the viral life cycle.

Keywords: Chicken anaemia virus, Capsid protein, VP1, DNA binding

Background

Chicken anaemia virus (CAV) is a common viral agent in chickens worldwide. CAV belongs to the genus *Gyrovirus* of the *Anelloviridae*, which have characteristics of circular single-stranded DNA viruses [1]. This virus frequently results in immunosuppression and anaemia in young chickens due to the destruction of T lymphoid tissue and aplasia of bone marrow, respectively, during virus infection [2–5]. Over 55% of the mortality rate and 80% of the morbidity rate were reported once the chicks

* Correspondence: leemengshiou@mail.cmu.edu.tw
†Equal contributors
²Department of Chinese Pharmaceutical Science and Chinese Medicine Resources, China Medical University, 91, Hsueh-Shih Road, Taichung, Taiwan

were infected with CAV [6]. Therefore, determining how to prevent CAV infection in the poultry industry has becomes an important challenge. The CAV virion lacks an envelope around its capsid coat, and it shows significantly high resistance to environmental stress or chemical agents. Currently, an attenuated live vaccine is available and effective for immunization of chickens for controlling CAV infection. However, young chicks less than 2 weeks-old are susceptible to CAV infection when the live vaccine was used [5]. This consequence has led to the development of a subunit vaccine, including DNA or protein based vaccine, over the past decade.

CAV is a relatively small virus approximately 23 nm in diameter. A total of three open reading frames (ORFs) were involved in the viral genome and have a length of 2.3 kb [2]. These ORFs respectively encode a 51 kDa VP1 protein, a 28 kDa VP2 protein and a 13 kDa VP3 protein. VP2 has dual-specificity phosphatase activity. VP3 is also referred to as apoptin with apoptosis-inducing activity. VP1 is the sole structural protein, which is the major component responsible for capsid assembly [5, 7, 8]. Currently, VP2 and VP3 proteins have been the focus of investigations of virus pathogenicity [9, 10]. In addition to its importance in the viral life cycle, VP3 has also demonstrated apoptosis-inducing activity as well as medical applications for anti-cancer treatments for humans in many previous studies [11–14]. VP1 can interact with VP2 and then significantly elicit the production of virus-neutralizing antibodies in the host in terms of immunogenicity studies [15]. Therefore, VP1 is thought to be a good candidate for an immunogen to develop a subunit vaccine [15].

DNA replication of DNA viruses usually occurs in the nucleus of infected cells. Thus, to establish a productive infection, viral DNA with a high molecular weight needs to cross the nuclear envelope through protein-mediated nuclear transportation after infecting the cells [16]. Approximately 90% of karyophilic proteins containing nuclear localization signals (NLSs) are directed to the nucleus. The NLS sequences usually overlap with the DNA binding domains. Therefore, proteins for nuclear transport possess both DNA binding and NLS activities [17, 18]. CAV is first Gyrovirus to be discovered and isolated [3]. By cloning and sequencing the viral genome, previous studies have reported an N-terminal 40 amino acid sequence within the predicted amino acid sequence of VP1 that demonstrated a significant (46%) degree of similarity to the protamine protein in Japanese quails. This specific region within the N-terminus of VP1 contains high arginine content and might confer an ability to VP1 to bind and protect DNA [19]. Using online software, including PSORT II (http://psort.hgc.jp) and DP-Bind (lcg.rit.albany.edu/dp-bind/), the VP1 protein was analysed in this study. A total of four putative DNA-binding motifs and two putative NLSs were found and predicted within the CAV VP1, as illustrated in Fig. 1. A previous researcher reported that transient expression of GFP-VP1 in the plant cells has been observed throughout the nucleoplasm [20]. This outcome demonstrated that VP1 protein might be a nuclear protein. Other circular single-stranded DNA virus, such as duck circovirus (DuCV) and beak and feather disease virus (BFDV), have exhibited a pattern of N-terminal amino acid residues within the capsid protein that are highly basic amino acid rich sequences with nuclear localization signals and DNA binding activity [21, 22]. Based on these findings, N-terminal amino acid residues within the capsid protein of circovirus are very similar to the CAV of Gyrovirus. However, there is still a lack of

direct evidence to prove and characterize the DNA binding ability or nuclear localization activity of VP1.

In this study, to gain insight into the role of the capsid protein VP1 in the life cycle of CAV, we have investigated the physical interactions of CAV VP1 with the viral DNA. A recombinant E. coli expression system was used to express the recombinant VP1 of CAV following our previous study [29]. The intracellular localization of the CAV VP1 was observed in MDCC-MSB1 cells or CHO-K1 cells using fluorescent green protein in the nucleoplasmic compartment. The DNA-binding activity of VP1 was also systemically examined. To the best of our knowledge, this is the first report to verify the DNA binding activity of the CAV capsid protein, VP1.

Results
Functional prediction of the CAV VP1 protein
Previous studies have shown that only the VP1 protein is located in the CAV virion. Thus, VP1 is also thought to be a DNA-binding protein that is responsible for the encapsidation of a viral genome during virus assembly. Presently, VP3 is the only one of three CAV viral proteins that has exhibited DNA binding activity in previous studies [23]. However, the function of VP1 on nucleic acid binding is still unknown. To gain insight into the role of the capsid protein VP1 in DNA binding, the bioinformatics software DP-Bind (http://lcg.rit.albany.edu/dp-bind/) was applied to analyse the features of DNA binding motifs within the amino acid sequence of the VP1 protein. Computational results of the DNA binding motif from the VP1 protein are shown in Fig. 1. Four potential DNA binding motifs were predicted by the DP-Bind program, and the putative motif position spanned from amino acids residues 3 to 22, 27 to 47, 62 to 67 and 333 to 349. According to these predicted results, VP1 might be having potential activity to bind DNA molecules. However, further investigation is still needed to verify the DNA binding activity of VP1.

Expression and purification of the recombinant CAV viral protein, VP1 and VP3
To examine the DNA-binding activity of VP1, E. coli was used to express recombinant VP1 protein. Recombinant VP3 protein was also expressed as a positive control for the evaluation of DNA-binding activity. As shown in Fig. 2a, after purification by a GST affinity column, the purity and antigenicity of purified GST-fused VP1 and VP3 were determined by SDS-PAGE and Western blotting, respectively. This result confirmed the integrity of the two recombinant proteins.

CAV VP1 is a nuclear protein that binds DNA with no sequence specificity
To elucidate whether the CAV VP1 protein is a DNA-binding protein, purified recombinant VP1 protein was

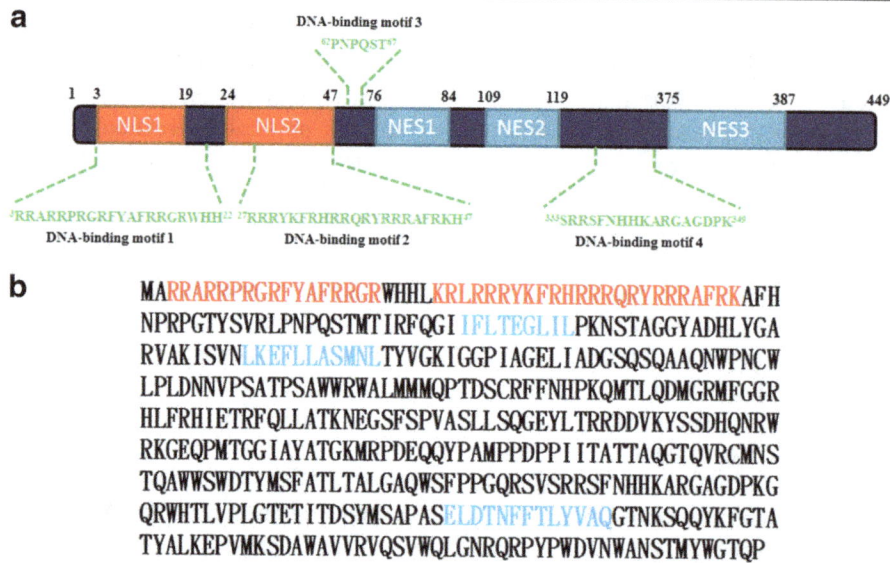

Fig. 1 Prediction results for putative NLS, NES and DNA-binding motifs on CAV VP1 proteins. **a** Schematic diagram representing the distribution regions of putative functional motifs: nuclear localization signals (NLS), nuclear export signals (NES) and DNA-binding motifs on CAV VP1. Two NLS that separate at the N-terminus of VP1 were predicted by PSORT II software (http://psort.hgc.jp/form2.html) and three NES, which are mainly at the C-terminus of VP1 were predicted by NetNES 1.1 Server software (http://www.cbs.dtu.dk/services/NetNES/), respectively. The software DP-Bind (lcg.rit.albany.edu/dp-bind) was used for putative DNA-binding motif prediction, and the predicted amino acid sequences were described on the diagram. **b** Predicted amino acid sequence results of NLS and NES for CAV VP1 were indicated by red (NLS)- and green (NES)-labels, respectively

added to circular dsDNA, pCAV, pcDNA3.1 and pGEM-T plasmids and incubated for 1 hour under 37 °C. After incubation, the occurrence of protein-DNA interaction was analysed by DNA movement on agarose gel. As illustrated in Fig. 2b, c and d, the migrations of pCAV, pcDNA3.1 and pGEM-T plasmid on the agarose gel were significantly reduced and shifted towards a pattern with a higher molecular weight. This result is very similar to the reduction in DNA migration that arose from binding VP3 to DNA (Fig. 2b, c and d). In contrast, no reduction in DNA migration occurred when the VP1 protein was absent or when GST protein was loaded with the addition of circular dsDNA plasmid. Moreover, when VP1 protein was pre-treated with 1% sodium dodecyl sulfate (SDS), the denatured VP1 no longer had DNA binding activity (Fig. 2b, c and d).

With respect to pCAV, which is a pcDNA3.1 plasmid carrying the entire CAV genome, the VP1 protein also displayed its DNA binding activity in terms of altered DNA migration pattern, as illustrated in Fig. 2b. In other words, regardless of whether the plasmid DNA used in the protein-DNA binding reaction was from pcDNA3.1 or pCAV, there was no significant effect on the resulting pattern of DNA migration (Fig. 2b, c and d). However, it is worth noting that the DNA migration of the circular dsDNA plasmid in the agarose gel displayed open circular dsDNA (form I, with a higher molecular weight pattern) and supercoiled dsDNA (form II, with a lower

molecular weight pattern), simultaneously (Fig. 2b). The VP1 protein was bound to supercoiled dsDNA. which demonstrated that DNA shifting was more obvious than open circular dsDNA.

Next, to confirm that VP1 not only has DNA binding activity but also has nuclear localization activity, we constructed a transit expression plasmid, pEGFP-VP1, which is a pcDNA3.1 vector carrying the VP1 gene fused to a GFP gene for cell transfection (Fig. 3a). When pEGFP-VP1 was respectively transfected into chicken lymphocytes, MDCC-MSB1 cells and Chinese Hamster Ovary (CHO) K1, the localization of GFP fluorescence was observed using confocal microscopy (Fig. 3b and c). As illustrated in Fig. 3b and c, GFP-VP1 and DAPI staining coincided significantly in the nuclei of MDCC-MSB1 cells (Fig. 3c). Additionally, GFP-VP1 was partially distributed and displayed in the cytoplasm of MDCC-MSB1 cells (Fig. 3c). A similar pattern of the distribution of GFP-VP1 was also presented in the CHO-K1 cells (Fig. 3b). These results clearly demonstrated VP1 is also a nuclear protein and distributed within the nucleocytoplasmic compartment. Taken together, these results indicated that CAV VP1 is a DNA-binding protein with nuclear localization activity, and its DNA binding is not specific to a particular sequence.

CAV VP1 binds DNA with a conformational preference

According to the results in Fig. 2b, the supercoiled dsDNA seems to interact with the VP1 protein more

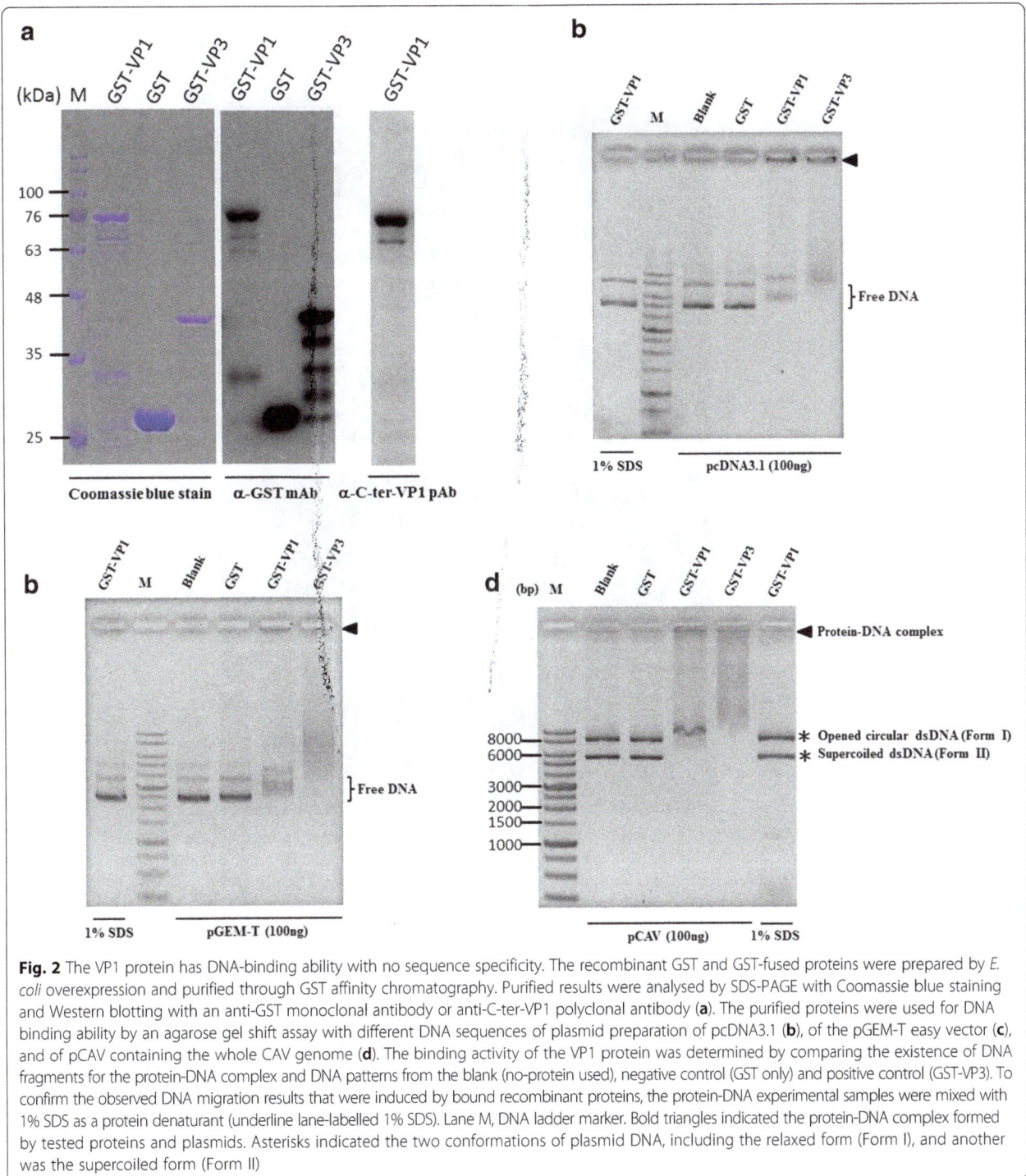

Fig. 2 The VP1 protein has DNA-binding ability with no sequence specificity. The recombinant GST and GST-fused proteins were prepared by *E. coli* overexpression and purified through GST affinity chromatography. Purified results were analysed by SDS-PAGE with Coomassie blue staining and Western blotting with an anti-GST monoclonal antibody or anti-C-ter-VP1 polyclonal antibody (**a**). The purified proteins were used for DNA binding ability by an agarose gel shift assay with different DNA sequences of plasmid preparation of pcDNA3.1 (**b**), of the pGEM-T easy vector (**c**), and of pCAV containing the whole CAV genome (**d**). The binding activity of the VP1 protein was determined by comparing the existence of DNA fragments for the protein-DNA complex and DNA patterns from the blank (no-protein used), negative control (GST only) and positive control (GST-VP3). To confirm the observed DNA migration results that were induced by bound recombinant proteins, the protein-DNA experimental samples were mixed with 1% SDS as a protein denaturant (underline lane-labelled 1% SDS). Lane M, DNA ladder marker. Bold triangles indicated the protein-DNA complex formed by tested proteins and plasmids. Asterisks indicated the two conformations of plasmid DNA, including the relaxed form (Form I), and another was the supercoiled form (Form II)

than opened dsDNA. Thus, to further address whether the DNA-binding activity of VP1 is affected by DNA conformation, various species of DNA molecules, such as linear dsDNA, circular ssDNA and linear ssDNA were used to confirm VP1 DNA binding activity. Because the DNA binding activity of VP1 has no sequence specificity as illustrated in Fig. 2b, c and d, the commercial circular single-stranded genome of the M13 phage was used as

sample DNA instead of the real circular ssDNA genome of CAV. As illustrated in Fig. 4 with respect to all DNA species, DNA retardation occurred when the recombinant VP1 proteins were added to DNA molecules, such as linear pcDNA3.1 (linear dsDNA, Fig. 4a), the linear single strand of the CAV genome (minus and strand, Fig. 4b) and the genome of the M13 phage (circular ssDNA, Fig. 4c). Similarly, with respect to the VP3

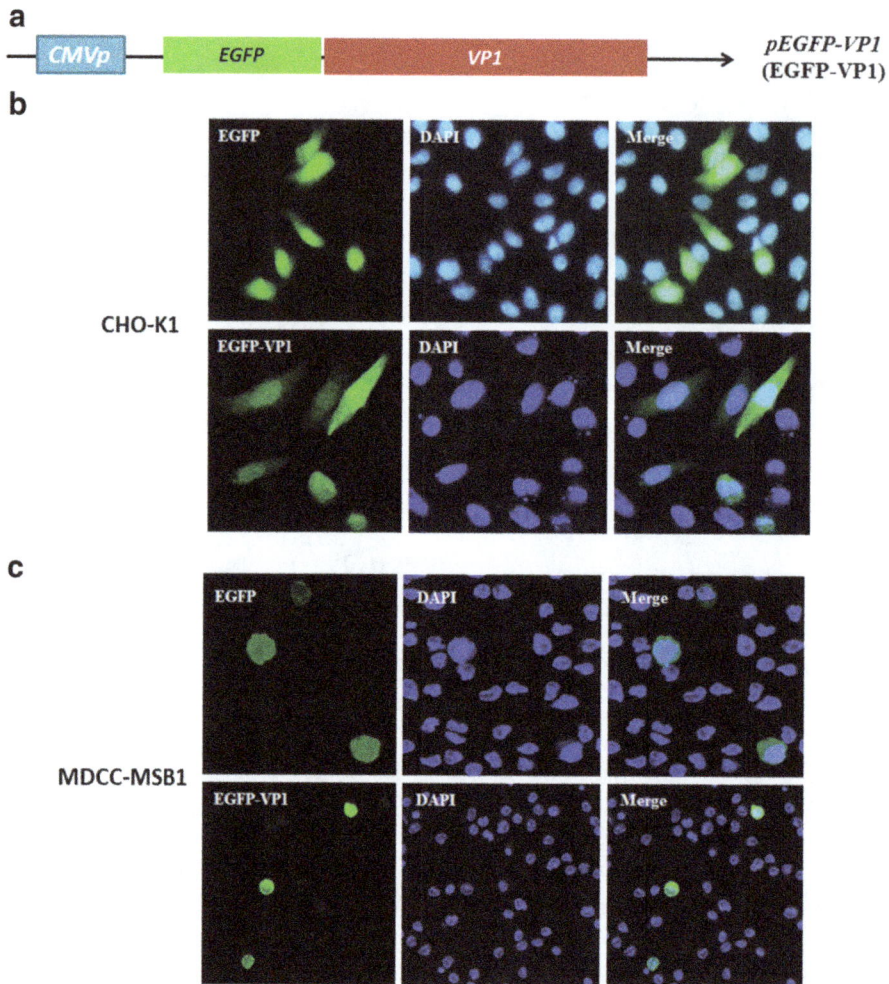

Fig. 3 The nucleocytoplasmic distribution characterization of VP1 protein in CHO-K1 and MDCC-MSB1 cells. To realize the subcellular distribution of VP1, the full-length VP1 gene included the fused whole EGFP gene at the 5′-terminus to generate the EGFP-VP1 expressing plasmid pEGFP-VP1 as illustrated in a schematic diagram (**a**). After 48 h post-transfection with the above plasmid in CHO-K1 cells (**b**) and MDCC-MSB1 cells (**c**), both cell types were fixed and stained with DAPI to reveal the nuclei. The subcellular localization of VP1 was determined by green fluorescence detection through confocal fluorescence microscopy

protein, all kinds of DNA molecules were also bound by VP3 and reduced the migration of DNA. However, comparing the significance of DNA patterns between various protein-DNA complexes, different DNA molecules bound by VP1 protein demonstrated there were distinct migration patterns of DNA (Fig. 4). Therefore, to address the binding preferences of the VP1 protein to DNA molecules, various amounts of VP1 protein were added to equal amounts of different DNA molecules for analysis of protein-DNA interactions. By quantifying the amount of VP1 protein required for DNA binding with respect to pCAV (circular dsDNA), especially for its supercoiled form dsDNA, the results showed 200 µg of VP1 were required to initiate VP1 binding to DNA molecules (Fig. 5a, supercoiled, form II). At least 300 µg of VP1 were required for this interaction to occur between VP1 and opened dsDNA (Fig. 5a, opened, form I). Higher amounts

of VP1 protein were used to bind circular dsDNA and reduced DNA migration patterns more significantly (Fig. 5a). Similarly, other DNA molecules, such as linear dsDNA (linearized pcDNA3.1, Fig. 5b), a linear single strand of the CAV genome (minus strand, Fig. 6a) and circular ssDNA (genome of the M13 phage, Fig. 6b), showed a similar pattern for protein-DNA interactions in the reaction mixture, with approximately 300 µg of VP1 required for linear dsDNA, 200 µg for the minus strand of linear ssDNA and 100 µg for circular ssDNA. In contrast, residual unbound DNA molecules representing the amount of DNA binding on the gel decreased if protein-DNA interaction occurred. Based on these results, the preferences in order of affinity to DNA with the VP1 protein in terms of the estimation of the percentage of unbound DNA molecules, which were sorted from low to high, were circular ssDNA (46.5% with respect to 300 µg of VP1), the linear minus strand of

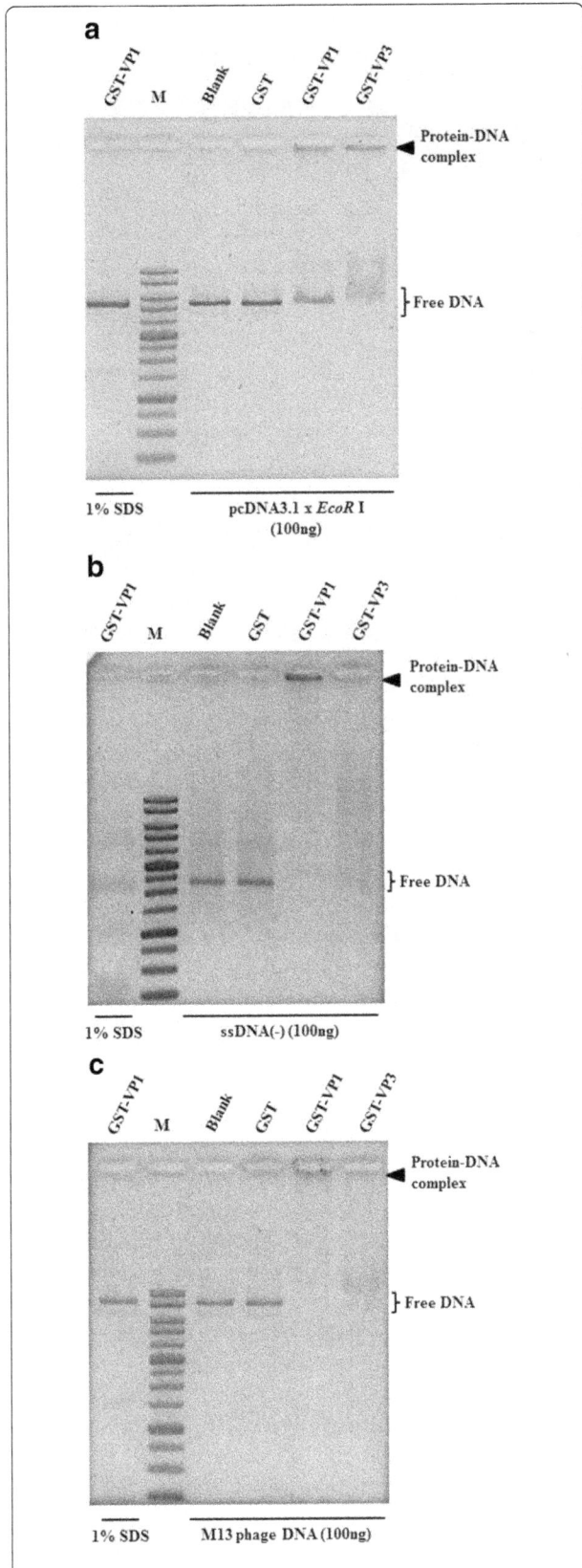

Fig. 4 VP1 protein binds to various DNA molecules. Purified GST and GST-fused proteins were used for analysing the interaction of recombinant proteins with various DNA samples, such as linear dsDNA (**a**), minus-strand ssDNA (**b**), and M13mp18 phage DNA (**c**). All DNA samples were generated by different preparations as described in the Materials and Methods. After the agarose gel shift assay, the DNA fragment signals were observed by EtBr staining. The 1% SDS (underline lane-labelled 1% SDS) was also used to confirm the retardation caused by tested proteins. Lane M, DNA ladder marker. Bold triangles indicate the protein-DNA complex formed by the tested protein and DNA molecules. The "pcDNA3.1 x EcoR I" indicated generation of the linear form of pcDNA3.1 DNA digested by EcoR I

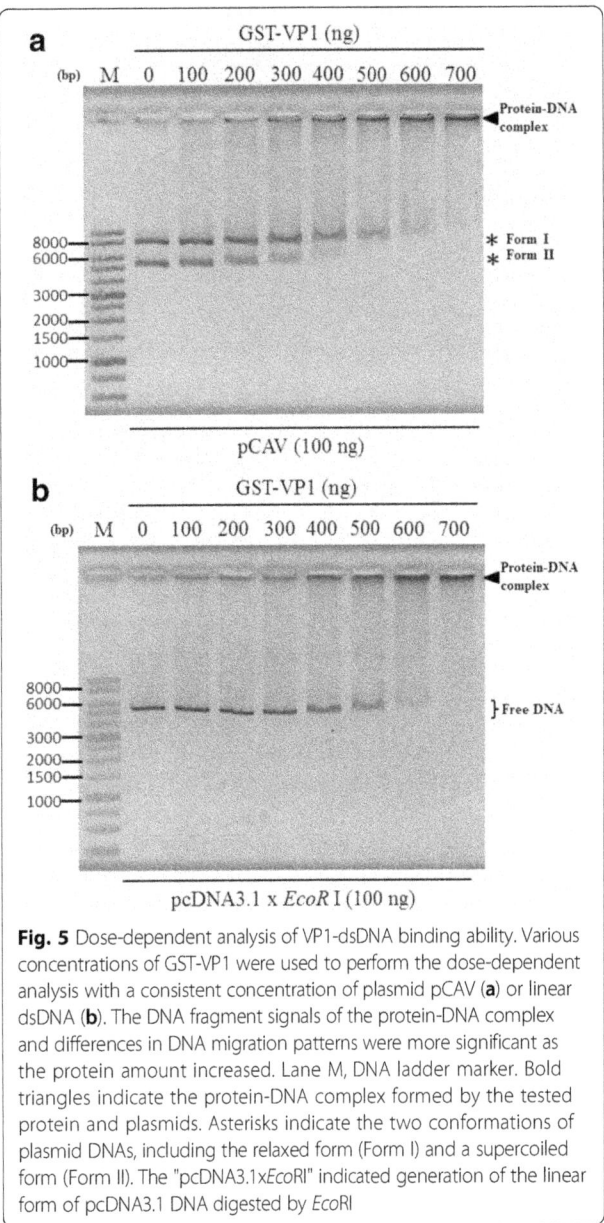

Fig. 5 Dose-dependent analysis of VP1-dsDNA binding ability. Various concentrations of GST-VP1 were used to perform the dose-dependent analysis with a consistent concentration of plasmid pCAV (**a**) or linear dsDNA (**b**). The DNA fragment signals of the protein-DNA complex and differences in DNA migration patterns were more significant as the protein amount increased. Lane M, DNA ladder marker. Bold triangles indicate the protein-DNA complex formed by the tested protein and plasmids. Asterisks indicate the two conformations of plasmid DNAs, including the relaxed form (Form I) and a supercoiled form (Form II). The "pcDNA3.1xEcoRI" indicated generation of the linear form of pcDNA3.1 DNA digested by EcoRI

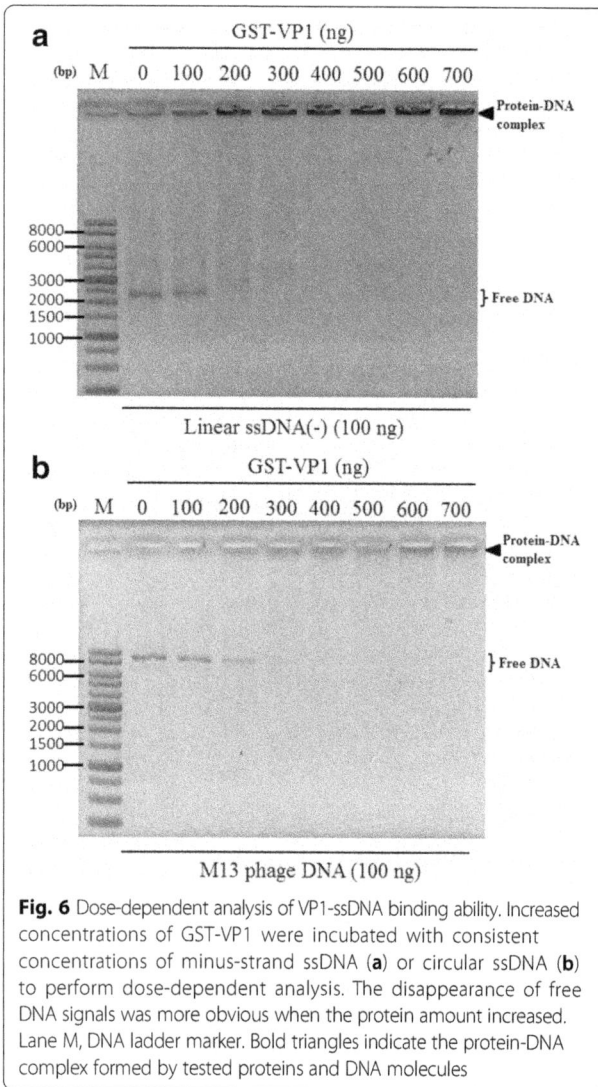

Fig. 6 Dose-dependent analysis of VP1-ssDNA binding ability. Increased concentrations of GST-VP1 were incubated with consistent concentrations of minus-strand ssDNA (**a**) or circular ssDNA (**b**) to perform dose-dependent analysis. The disappearance of free DNA signals was more obvious when the protein amount increased. Lane M, DNA ladder marker. Bold triangles indicate the protein-DNA complex formed by tested proteins and DNA molecules

Table 1 The ratio of unbound free DNA residue was determined by calculating the image intensity of free DNA fragments on an electrophoretic agarose gel after a VP1-DNA binding assay combining increasing amounts of recombinant GST-VP1 protein with certain DNA molecule conformations

	Concentrations of GST-VP1 protein (ng)				
	0	100	300	500	700
Open circular dsDNA (Form I)	100%	100%	100%	59.6%	11%
Supercoiled dsDNA (Form II)	100%	100%	82.6%	0%	0%
Linear dsDNA	100%	93.8%	84.7%	36.5%	0%
Linear ssDNA (−)	100%	92.9%	53.3%	0%	0%
M13 phage DNA	100%	90%	46.5%	0%	0%

The lower ratios indicate a higher preference of the VP1 protein for specific DNA conformations. The equation used in this study is presented under the table

after infecting the cells [16]. CAV is a non-enveloped, small DNA virus containing a circular ssDNA genome [2]. VP1, which is a major capsid protein of CAV, interacts with cells during virus infection, and the viral genome should theoretically be carried by VP1 to enter cells. The main question is what protein is conferred a function to direct a viral genome into the nucleus for sequential DNA replication? VP1 is thought to have a functional role to bind and direct the viral genome into the nucleus. However, there is still a lack of direct evidence to support this speculation. In our results from computational prediction, four putative DNA binding motifs were combined with the amino acid sequence of the CAV VP1 protein. Although the exact DNA binding motif was not determined in this study, it did not affect the characterization of VP1 DNA binding activity. After performing a DNA-binding assay to examine DNA migration, the VP1 protein of CAV was confirmed to have DNA-binding activity. Additional experiments are still needed for identifying major DNA binding motifs within VP1 protein. The confirmed DNA-binding activity in VP1 might be useful to further verify the underlying mechanisms of viral DNA replication. In addition, GFP-VP1 has demonstrated nucleo-cytoplasm shuttling activity (Fig. 3). The results imply VP1 is a nuclear protein to binds DNA molecules, such as those in the viral genome and travels into the nucleus during its early life cycle. In fact, we have not only predicted the presence of nuclear localization signals (NLSs) with PSORT II but also predicted nuclear exporting signals (NESs) with NetNES server (http://www.cbs.dtu.dk/services/NetNES/). These putative NLS motifs were within the amino acid sequence of VP1, which spanned from amino acid residues 3 to 19 (NLS1) and 24 to 47 (NLS2). For NESs, the putative motifs within the VP1 protein spanned from amino acid residues 76 to 84 (NES1), 109 to 119 (NES2) and 375 to 387 (NES3) (Fig. 1a). In terms of the observation

ssDNA (53.3% with respect to 300 μg of VP1), supercoiled circular dsDNA (82.6% with respect to 300 μg of VP1), linear dsDNA (84.7% with respect to 300 μg of VP1) and opened circular dsDNA (100% with respect to 300 μg of VP1). The comparison of binding preferences of VP1 protein to different conformations of DNA molecules are summarized in Table 1. More unbound DNA existed in the agarose gel, implying that this conformation of DNA exhibited a lower preference to interact with the VP1 protein. These results demonstrated that the interaction of VP1 with the DNA molecule exhibited various binding preferences that were dependent on the structural conformation of DNA.

Discussion

DNA replication in the DNA viruses usually occurs in the nuclei of infected cells. Thus, viral DNA needs to cross the nuclear envelope through protein-mediated nuclear transportation to establish a productive infection

of fluorescent of GFP-VP1, this result was confirmed according to computational predictions. In the early life cycle of the virus, viral DNA replication is an important stage for the establishment of productive infection [16]. Previous studies have reported circular, negative ssDNA of the CAV genome might be replicated through rolling-circle amplification [24]. During this stage, previous researchers isolated the viral replicative form (RF) DNA, which is an open circular dsDNA obtained from MDCC-MSB1 cells after being infected with the virus for 30 h [19]. In addition to the presence of closed and open circular dsDNA, circular ssDNA with genome-sized and small linear dsDNA of 800 bp were observed in the later stages of DNA replication [25]. In this study, all similar DNA molecules with different conformations, including linear ssDNA, which is derived from circular ssDNA, were used to examine the possible functional roles of the VP1 protein in DNA-binding activities. Additionally, other putative DNA binding motifs, especially for the initiation of rolling-circle amplification (RCR) within the VP1 protein, had also been reported and predicted in a previous study [26]. Three putative motifs were proposed that spanned from amino acid 313 to 320 (FATLTALG), 350 to 358 (GQRWHTLVP), and 399 to 408 (TATYALKEPV). These motifs might be interaction sites, such as the origin (Ori) site of the CAV genome, for interacting with VP1 for regulating DNA replication [26].

Generally, the DNA binding protein of most DNA viruses showed DNA binding activity with no sequence specificity. In this study, VP1 showed DNA binding characteristics with no sequence specificity similar to other DNA viruses, such as human papillomavirus and polyomavirus [27, 28]. Therefore, some other strategy of VP1 binding to the CAV viral genome should be adopted by the virus. With respect to the binding preferences of the VP1 protein to different DNA molecule conformations, VP1 was found to interact with circular ssDNA and exhibit a higher preference for this conformation, as shown in Table 1. This outcome might truly reflect the conditions of viral encapsidation for coating circular ssDNA of the CAV genome. Actually, circular ssDNA is prone to forming secondary structures with a high probability. This possibility was examined and confirmed with a computational prediction from the Mfold program (http://unafold.rna.albany .edu/?q=mfold/DNA-Folding-Form) (Additional file 1: Figure S1A, B). Similarly, linear ssDNA also has a high probability for forming secondary DNA structures (Additional file 1: Figure S2A, B). Thus, the binding preference of VP1 to linear ssDNA is surpassed only by the preference for circular ssDNA (Table 1). This difference truly meets our expectations. In fact, the sequence of linear ssDNA(−) was complemented with linear ssDNA(+). Then, linear ssDNA(+) was found to have significant VP1-DNA interaction in terms of the results

of the DNA migration assay. The binding preference of VP1 to the linear plus-strand of ssDNA(+) is slightly lower than the linear plus-strand of ssDNA(−) (53.1% for 300 μg of VP1) (Additional file 1: Figure S3A, B). Other DNA molecules, such as supercoiled dsDNA, linearized ds DNA and opened dsDNA, displayed lower binding preferences to VP1, which might be a mechanism for VP1 protein to competitively bind various DNA molecules during different stages of the life cycle.

Taken together, this report is the first to show that VP1 has no sequence specificity for its DNA-binding activity and that its particular binding preferences might play multiple roles in DNA replication or encapsidation during a viral life cycle.

Conclusion

In summary, the characterization of DNA binding activity of the VP1 protein was investigated in this study. VP1 was demonstrated to show DNA binding characteristics with no sequence specificity. In addition, the DNA binding activities of VP1 exhibited a differential preference to interact with various DNA molecules with different conformations. This information could be helpful for determining the biological roles of VP1 in the CAV viral life cycle.

Methods

Cell cultures, bacterial strains and plasmids

For Chinese Hamster Ovary (CHO-K1) cells, cells were purchased in 2014 from the Bioresource Collection and Research Center (BCRC 6006) in Taiwan. CHO-K1 cells were maintained in Ham's F12 medium (HyClone, USA) supplemented with 10% FBS (HyClone, USA), 1% P/S (Penicillin/Streptomycin solution) (Gibco, USA). Chicken lymphoblast MDCC-MSB1 cells were purchased from the CLS Cell Lines Service GmbH in Germany in 2015. MDCC-MSB1 cells were grown in RPMI 1640 medium (HyClone, USA) supplemented with 10% FBS (HyClone, USA) and 1% P/S (Penicillin/Streptomycin solution) (Gibco, USA). All cells were cultured in appropriate tissue culture flasks and maintained in a cell culture incubator with 5% CO_2 at 37 °C before experiments.

All expression constructs used in this study were maintained in the *E. coli* strain Top10F' (Invitrogen, USA). The *E. coli* strain BL21 (DE3)-*pLys*S was transformed with protein expression plasmids and followed by IPTG induction to produce recombinant proteins as described in a previous study [29].

The construction of an expressed plasmid used for detecting subcellular localization was described below. The full-length of CAV VP1 was amplified by PCR using the specific primer sets wt-VP1-f: 5'-CCCGAATTCATGG CAAGACGAGCTCGC-3', wt-VP1-r: 5'-CGCGTCGACT CAGGGCTGCGTCCCCCAGTA-3' from the CAV VP1 template that was kindly provided by Dr. Yi-Yang Lien. The PCR product was then cloned into expression vector pEGFP-C2 (#6083–1, Clontech, USA) between *EcoR* I and *Sal* I sites to generate a recombinant plasmid named pEGFP-VP1.

Expression and purification of the CAV VP1 and VP3 proteins

To purify the recombinant CAV VP1 and VP3 proteins, the previously created recombinant *E. coli* strains BL21 (DE3)-*pLys*S expressing VP1 and VP3 were used to express recombinant proteins [29]. The recombinant *E. coli* cells were cultured, and the harvested cells were disrupted and prepared following a previously described procedure [29]. Cells were spun down from 50 ml of culture supernatant and resuspended in GST resin binding buffer (140 mM NaCl, 2.7 mM KCl, 10 mM Na_2HPO_4, 1.8 mM KH_2PO_4, pH 7.3). After cell disruption, the resulting cell supernatant was loaded onto a GSTrap FF affinity column (GE healthcare, Piscataway, NJ) for protein purification following the operational conditions described in a previous study [29]. The total protein concentration of recombinant CAV VP1 and VP3 proteins was determined using a Micro BCA kit (Pierce, Rockford, IL) with bovine serum albumin as the reference protein. Purified VP1 and VP3 proteins were dialyzed against DNA-binding buffer (50 mM Tris-HCl, pH 7.5, 120 mM KCl, 1.0 mM EDTA, 0.5 mM DTT, and 30 mg/ml BSA) and analysed by sodium dodecyl sulfate-polyacrylamide gel electrophoresis (SDS-PAGE) and Western blotting. Purified proteins were stored at – 20 °C until required.

Generation of nucleic acids used for the DNA-binding assay

Different DNA species, including circular dsDNA, linear dsDNA and circular ssDNA were used for assessing DNA-binding activity. The circular dsDNA, pcDNA3.1 (#V80020, Invitrogen, USA), pGEM-T easy vector (#A1360, Promega, USA) and pCAV were used for the DNA binding assay. The pCAV plasmid DNA was composed of a full-length Australian CAV strain CAU269/7 (GenBank #AF227982). The linear dsDNA was prepared from a pcDNA3.1 and pGEM-T plasmid by the cutting restriction enzyme *EcoRI*. Pure M13mp18 single-stranded DNA along with circular ssDNA materials were purchased from New England BioLabs (#N4040S, NEB, USA). All DNA molecules were diluted to 50 ng/ml with DNA-binding buffer and store at – 20 °C until required.

Preparation of linear single-stranded DNA

The linear ssDNA was also used for the DNA binding assay. The preparation of linear ssDNA followed the protocol described in Marimuthu et al. [30] using the biotin-streptavidin separation method. First, a biotinylated DNA fragment containing the whole CAV genome was amplified

by PCR using the EmeraldAmp Max PCR Master kit (Takara, Japan) from pCAV with a designed primer set, including the reverse biotinylated primer Biotin-CAV-r: biotin-labelled-GATTGT GCGGTGAACGAA TTAG, and the forward regular primer CAV-f: GAAT TCCGAGTGGTTACTATTC. After PCR amplification, the biotinylated PCR product was then immobilized on 40 μl Dynabeads M-280 Streptavidin magnetic beads (Invitrogen, USA) and incubated at 4 °C overnight. After washing the DNA-bonded beads twice with B/W buffer (5 mM Tris-HCl, pH 7.5, 0.5 mM EDTA, 1 M NaCl), the washed beads were incubated in 150 μl elution buffer (0.1 M NaOH, 1 mM EDTA, pH 13.0) to perform alkaline denaturation. Under the high alkaline environment, the desired non-biotinylated strand can be separated from the biotinylated strand and suspended in the supernatant. After magnet adsorption, the supernatants were collected, and the linear ssDNA was further purified by a PCR clean-up kit (Geneaid, Taiwan). The linear ssDNA was diluted to 50 ng/ml with DNA-binding buffer and store at – 20 °C until required.

DNA binding assay

Purified GST, GST-VP1 and GST-VP3 proteins were diluted to 500 ng/μl with DNA-binding buffer and 500 ng of proteins were mixed with 100 ng of each DNA variant in a total of 20 μl of DNA-binding buffer. Then, each mixture was incubated for 30 min under 37 °C. The resulting sample was subjected to electrophoresis using a 0.8% agarose gel in a TAE buffer and, then the DNA was stained with ethidium bromide for the analysis of DNA migration.

Cell transfection of CHO-K1 and MDCC-MSB1 cells

CHO-K1 cells were transfected by X-tremeGene HP DNA transfection reagent (Sigma, USA) according the manual's protocol with a mixture containing 2 μg of plasmid pEGFP-VP1 and 4 μl of transfection reagent in 2.5 ml serum-free Opti-MEM medium (Gibco, USA). After incubating the mixture for 20 min at room temperature, the mixture was added drop by drop into cultured CHO-K1 cells in a 6-well plate. The 24 to 48 h posttransfection, the transfection effect was checked with a confocal fluorescent microscope.

For the transfection of MDCC-MSB1 cells, the 4×10^6 log-phase grown MDCC-MSB1 cells were gently pipetted with 15 μg of plasmid pEGFP-VP1 first in serum-free RPMI 1640 medium and then the mixture was transferred into a 0.4-cm gap electroporation cuvette and the cuvette was harvested on ice for 5 min. The electroporation of MDCC-MSB1 cells was performed with a Gene Pulser II (Bio-Rad, USA) with a Time Constant Protocol set at 34 ms and an operating voltage of 300 V. After electroporation, the transfected cells were then cultured into complete medium in a 6-well plate for 24 to 48 h. Post-transfection, the expression of recombinant EGFP-VP1 proteins were analysed by confocal fluorescence microscopy to make sure the transfection was effective.

Sample preparation for confocal microscopy observation

After transfecting CHO-K1 cells and MDCC-MSB1 cells with plasmid pEGFP-C2 or pEGFP-VP1, the fluorescent images were captured by a confocal fluorescence microscope in terms of the observation of protein fluorescent to verify the EGFP-expressed cells and EGFP-VP1-expressed cells. Transfected cells were collected and fixed with 4% formaldehyde in the dark. After washing the fixed cells twice to remove residual formaldehyde, the cells were stained in 0.1% PBS-T with 1 µg/ml DAPI for 5 min at 37 °C in the dark. Then, the stained cells were mounted with gelvatol medium (Sigma, USA) on a glass slide for confocal observation. Confocal laser scanning microscope (CLSM) images were captured from a Leica TCS SP8 confocal microscope and the images were integrated with LAS X Leica Confocal Software. EGFP fluorescence was observed through excitation at 488 nm and DAPI emitted blue fluorescence upon binding to DNA that was observed through excitation by UV light.

Comparing the DNA conformational preference of VP1 protein through DNA analysis

To obtain the ratio of unbound DNA residues, a dose-dependent DNA-binding experiment was performed by combining various concentrations (0 to 700 ng) of GST-VP1 proteins with consistent amounts of different DNA variants. Next, the signal intensities of free DNA from DNA migration images after DNA-binding experiments were obtained using ImageJ software observation. The equation used in this study is presented below.

$$\text{Unbound DNA residue (\%)}$$
$$= \frac{\text{Image intensity of free DNA (certain protein concentration)}}{\text{Reference image intensity (no protein)}} \times 100\%$$

After dividing the signal intensity at a certain concentration of GST-VP1 proteins by the signal intensity in the absence of proteins (blank), the calculated ratio of unbound DNA residue was obtained to determine the DNA conformational preference for the VP1 protein.

Acknowledgements
This work of the authors would like to thank the Ministry of Science and Technology, Taiwan (TW) for financially supporting. We also thank the Medical Research Core Facilities Center, office of Research & Development at China Medical University (Taichung, Taiwan, R.O.C.) for supporting experiments and data analysis on confocal microscopic observation.

Funding
This research was financially supported through grants from the Ministry of Science and Technology, Taiwan (TW) (grant number: NSC101–2321-B-039-007- and NSC102–2321-B-039-007) to Dr. Meng-Shiou Lee.

Authors' contributions
GHL and MKL contributed equally to first author in this work. MSL* participated in the study design, performed the experiments and drafted the manuscript. GHL and YYL performed the experiments and participated in the construction of the plasmids. MKL and JHC participated in the experiments on protein-DNA binding assay, and MSL[6] participated in the protein purification step. FCS participated in the data analysis, and the manuscript drafting. HJC, GHL and MKL coordinated the study and participated in drafting the manuscript. All authors read and approved the final manuscript.

Consent for publication
All data obtained from this work were approved by all authors for publication.

Competing interests
The authors declare that they have no competing interests. All information generated in this study there is no any commercial conflict of interests.

Author details
[1]Graduate Institute of Biotechnology, National Chung Hsing University, Taichung 40402, Taiwan. [2]Department of Chinese Pharmaceutical Science and Chinese Medicine Resources, China Medical University, 91, Hsueh-Shih Road, Taichung, Taiwan. [3]Department of Veterinary Medicine, National Pingtung University of Science and Technology, Pingtung 91201, Taiwan. [4]Center for Shockwave Medicine and Tissue Engineering, Department of Medical Research, Kaohsiung Chang Gung Memorial Hospital and Chang Gung University College of Medicine, Kaohsiung 83301, Taiwan. [5]Department of Bioresources, Da-Yeh University, Changhua 51591, Taiwan. [6]Research Assistance Center, Show Chwan Memorial Hospital, Changhua 500, Taiwan. [7]Department of Safety, Health and Environmental Engineering, Ming Chi University of Technology, New Taipei 24301, Taiwan.

References
1. Rosario K, Breitbart M, Harrach B, Segalés J, Delwart E, Biagini P, Varsani A. Revisiting the taxonomy of the family Circoviridae: establishment of the genus Cyclovirus and removal of the genus Gyrovirus. Arch Virol. 2017;162: 1447–63.
2. Noteborn MHM, De Boer GF, Van Roozelaar DJ, Karreman C, Kranenburg O, Vos JG, Jeurissen SHM, Hoeben RC, Zantema A, Koch G, Van Ormondt H, Van der Eb AJ. Characterization of cloned chicken anemia virus DNA that contains all elements for the infectious replication cycle. J Virol. 1991;65: 3131–9.
3. Adair BM. Immunopathogenesis of chicken anemia virus infection. Dev Comp Immunol. 2000;24:247–55.
4. Jeurissen SH, Wagenaar F, Pol JM, Van der Eb AJ, Noteborn MH. Chicken anemia virus causes apoptosis of thymocytes after in vivo infection and of cell lines after in vitro infection. J Virol. 1992;66:7383–8.
5. Noteborn MH, Todd D, Verschueren CA, De Gauw HW, Curran WL, Veldkamp S, Douglas AJ, McNulty MS, Ven der Eb AJ, Koch G. A single chicken anemia virus protein induces apoptosis. J Virol. 1994;68:346–51.
6. Hsu JP, Lee ML, Lu YP, Hung HT, Hung HH, Chein MS. Chicken infectious anemia in layer. J Chin Soc Vet Sci. 2002;28:153–60.
7. Todd D, Creelan JL, Mackie DP, Rixon F, McNulty MS. Purification and biochemical characterization of chicken anaemia agent. J Gen Virol. 1990;71: 819–23.
8. Peters MA, Jackson DC, Crabb BS, Browning GF. Chicken anemia virus VP2 is a novel dual specificity protein phosphatase. J Biol Chem. 2002;277:39566–73.
9. Peters MA, Jackson DC, Crabb BS, Browning GF. Mutation of chicken anemia virus VP2 differentially affects serine/threonine and tyrosine protein phosphatase activities. J Gen Virol. 2005;86:623–30.
10. Peters MA, Crabb BS, Washington EA, Browning GF. Site-directed mutagenesis of the VP2 gene of chicken anemia virus affects virus replication, cytopathology and host-cell MHC class I expression. J Gen Virol. 2006;87:823–31.
11. Zhuang SM, Landegent JE, Verschueren CAJ, Falkenburg JHF, van Ormondt H, Van der Eb AJ, Noteborn MHM. Apoptin, a protein encoded by chicken anemia virus, induces cell death in various human hematologic malignant cells in vitro. Leukemia. 1995;9:118–20.

12. Zhuang SM, Shvarts A, Van Ormondt H, Jochemsen AG, Van der Eb AJ, Noteborn MHM. Apoptin, a protein derived from chicken anemia virus, induces a p53-idependent apopotosis in human osteosarcoma cells. Cancer Res. 1995;55:486–9.

13. Danen-van Oorschot AA, van Der Eb AJ, Noteborn MH. The chicken anemia virus-derived protein apoptin requires activation of caspases for induction of apopotosis in human tumor cell. J Virol. 2000;74:75–7.

14. Danen-Van Oorschot AA, Zhang YH, Erkeland SJ, Fischer DF, Van der Eb AJ, Noteborn MHM. The effect of Bcl-2 on apoptin in normal cell versus transformed human cells. Leukemia. 1999;13:75–7.

15. Noteborn MH, Verschueren CAJ, Koch G, Van der Eb AJ. Simultaneous expression of recombinant baculovirus-encoded chicken anemia virus (CAV) proteins VP1 and VP2 is required for formation of the CAV-specific neutralizing epitope. J Gen Virol. 1998;79:3070–7.

16. Cohen S, Au S, Pante N. How viruses access the nucleus. Biochim Biophys Acta. 2011;1813:1634–45.

17. LaCasse EC, Lefevre YA. Nuclear localization signals overlap DNA- or RNA-binding domains in nucleic acid-binding proteins. Nucleic Acids Res. 1995;23:1647–56.

18. Cokol M, Nair R, Rost B. Finding nuclear localization signals. EMBO Rep. 2000;1:411–5.

19. Claessens JAJ, Schrier CC, Mockett APA, Jagt EHJM, Sondermeijer PJA. Molecular cloning and sequence analysis of the genome of chicken anaemia agent. J Gen Virol. 1991;72:2003–6.

20. Lacorte C, Lohuis H, Goldbach R, Prins M. Assessing the expression of chicken anemia virus proteins in plants. Virus Res. 2007;129:80–6.

21. Heath L, Williamson AL, Rybicki EP. The capsid protein of beak and feather disease virus binds to the viral DNA and is responsible for transporting the replication-associated protein into the nucleus. J Virol. 2006;80:7219–25.

22. Xiang QW, Zou JF, Wang X, Sun YN, Gao JM, Xie ZJ, Wang Y, Zhu YL, Jiang SJ. Identification of two functional nuclear localization signals in the capsid protein of duck circovirus. Virology. 2013;436:112–7.

23. Lelievld SR, Dame RT, Rohn JL, Noteborn MHM, Abrahams JP. Apoptin's functional N- and C-termini independently bind DNA. FEBS. 2004;557:155–8.

24. Noteborn MHM, Koch G. Chicken anaemia virus infection : molecular basis of pathogenicity. Avian Pathol. 1995;24:11–31.

25. Meehan BM, Creelan JL, Earle JAP, Hoey EM, McNulty MS. Characterization of viral DNAs from cells infected with chicken anaemia agent: sequence analysis of the cloned replicative form and transfection capabilities of cloned genome fragments. Arch Virol. 1992;124:301–19.

26. Ilyina TA, Koonin EV. Conserved sequence motifs in the initiator proteins for rolling circle DNA replication encoded by diverse replicons from eubacteria, eucaryotes and archaebacterial. Nucleic Acids Res. 1992;20:3279–85.

27. Zhou J, Sun X-Y, Louis K, Frazer H. Interaction of human papillomavirus (HPV) type 16 capsid proteins with HPV DNA requires an intact L2 N-terminal sequence. J Virol. 1994;68:619–25.

28. Chang D, Cai X, Consigli RA. Characterization of the DNA binding properties of polyomavirus capsid proteins. J Virol. 1993;67:6327–31.

29. Lai G-H, Lin M-K, Lien Y-Y, Fu J-H, Chen H-J, Huang C-H, Tzen J-T, Lee M-S. Expression and characterization of highly antigenic domains of chicken anemia virus viral VP2 and VP3 subunit proteins in a recombinant E. Coli for sero-diagnostic applications. BMC Vet Res. 2013;9:161.

30. Marimuthu C, Tang T-H, Tominaga J, Tan S-C, Gopinath SCB. Single-stranded DNA (ssDNA) production in DNA aptamer generation. Analyst. 2012;137:1307.

21

Immunohistochemical characterization of tuberculous lesions in sheep naturally infected with *Mycobacterium bovis*

Raquel Vallejo[1], Juan Francisco García Marín[1], Ramón Antonio Juste[2], Marta Muñoz-Mendoza[3], Francisco Javier Salguero[4] and Ana Balseiro[2]* (ORCID)

Abstract

Background: Sheep have been traditionally considered as less susceptible to *Mycobacterium bovis* (Mbovis) infection than other domestic ruminants such as cattle and goats. However, there is increasing evidence for the role of this species as a domestic Mbovis reservoir, mostly when sheep share grazing fields with infected cattle and goats. Nevertheless, there is a lack of information about the pathogenesis and the immune response of Mbovis infection in sheep. The goals of this study were to characterize the granuloma stages produced by the natural infection of Mbovis in sheep, to compare them with other species and to identify possible differences in the sheep immune response. Samples from bronchial lymph nodes from twelve Mbovis-naturally infected sheep were used. Four immunohistochemical protocols for the specific detection of T-lymphocytes, B-lymphocytes, plasma cells and macrophages were performed to study the local immune reaction within the granulomas.

Results: Differences were observed in the predominant cell type present in each type of granuloma, as well as differences and similarities with the development of tuberculous granulomas in other species. Very low numbers of T-lymphocytes were observed in all granuloma types indicating that specific cellular immune response mediated by T-cells might not be of much importance in sheep in the early stages of infection, when macrophages are the predominant cell type within lesions. Plasma cells and mainly B lymphocytes increased considerably as the granuloma developed being attracted to the lesions in a shift towards a Th2 response against the increasing amounts of mycobacteria. Therefore, we have proposed that the granulomas could be defined as initial, developed and terminal.

Conclusions: Results showed that the study of the lymphoid tissue granulomata reinforces the view that the three different types of granuloma represent stages of lesion progression and suggest an explanation to the higher resistance of sheep based on a higher effective innate immune response to control tuberculosis infection.

Keywords: Tuberculosis, Sheep, Immunohistochemistry, *Mycobacterium bovis*, Granuloma

Background

Recent studies have shown that sheep, traditionally considered a rare host for the *Mycobacterium tuberculosis* complex (MTBC), can be part of the multi-species system which can maintain tuberculosis (TB) in a region, at least in mixed farms where sheep share grazing fields with *Mycobacterium bovis* (Mbovis)-infected cattle and

goats [1]. Microscopic lesions in sheep infected with Mbovis are characterized by localized and well delimited granulomas mainly in lungs and bronchial lymph nodes [1], although limited information has been published about the pathogenesis and immune response in sheep against TB [2–4].

Granulomas are observed in active, latent and reactivation stages of TB. The TB lesion is highly dynamic and shaped both by the pathogen and the host defense elements [5]. In humans generally the TB granuloma successfully contains (but does not eliminate) the infectious

---

* Correspondence: abalseiro@serida.org
[2]SERIDA, Servicio Regional de Investigación y Desarrollo Agroalimentario, Centro de Biotecnología Animal, 33394 Gijón, Asturias, Spain

focus in more than 90% of cases, whereas 10% of individuals progresses towards clinical TB as a consequence of an unbalanced inflammatory reaction. The granuloma is capable of limiting growth of mycobacteria but also is a good environment from which the bacteria may disseminate [5]. Immunohistochemical techniques are valuable tools to advance the knowledge of the immunology of infection for a better understanding of the mechanisms of host-pathogen interactions and disease progression, as has been reported in Mbovis-infected cattle, badger, fallow deer and wild boar [6–10].

To gain further insight into the knowledge of the immunopathology of tuberculous lesions in sheep, the goals of this study were to characterize the granuloma stages produced by the natural infection of Mbovis in sheep, to compare them with other species and to identify possible differences in the sheep immune response.

Methods

Samples

A total of 12 sheep naturally infected with Mbovis (spolygotype SB0886) from the Galicia region (Northwestern Spain) were selected from a previous study [1]. In the former study at post mortem examination, small granulomatous nodules of less than 5 cm in diameter were histologically classified in stages I, II and III on the basis of cell inflammatory influx, inflammatory cell type, mineralization and calcification and degree of encapsulation. Briefly, type I were unencapsulated granulomas consisted mainly of epithelioid cells, macrophages, lymphocytes and few Langhan's multinucleated giant cells in which sometimes a minimum central focus of necrosis was observed. Type II were granulomas (often diffuse) composed of numerous inflammatory cells, mainly macrophages and Langhan's giant cells surrounded by a full thin fibrotic capsule with central necrotic areas which were caseous or partly calcified. Finally, type III were well encapsulated granulomas with areas of central caseous necrosis with mineralization occupying the majority of the lesion, surrounded by scarce inflammatory infiltrate consisted mainly of lymphocytes. Ziehl-Neelsen stain was used to identify and count acid fast bacilli (AFBs) within lesions in four random 400× magnification fields.

Twelve sheep showing granulomas only in bronchial lymph nodes were selected, four sheep with granuloma type I, four with granuloma type II and four with granuloma type III. We choose the bronchial lymph nodes because they were the most frequently affected tissues [1]. They were used for the immunohistochemical characterization of cellular types. For each tissue section and to equilibrate counts, the sum of four randomly selected different areas of 400× magnification were analyzed using a light microscope in order to count the total number of cells immunostained for each cell type. Based on the number of total cellular type counted, a semi-quantitative classification was elaborated as follows: (–) abscence of immunolabelled cells; (+) 1–10 immunolabelled cells; (++) 11–50 immunolabelled cells; (+++) 51–100 immunolabelled cells; (++++) > 100 immunolabelled cells.

Immunohistochemistry (IHC)

Four-μm sections were used for immunohistochemical detection of four different antigens (Table 1). The ABC Complex reagent-method (Vector Laboratories, California, USA) was used. Briefly, the sections were deparaffinised, rehydrated and rinsed with tap water. Afterwards, slides were treated to quench the endogenous peroxidase by incubation with methanol containing 3% H_2O_2 for 10 min at room temperature (RT) and washed with water for 10 min. Then, epitope demasking techniques (Table 1) were used to retrieve the antigens and samples were treated to prevent unspecific binding with a 20 min incubation at RT with 10% normal horse serum for CD3 protocol (DAKO, Glostrup, Denmark) or with 10% normal goat serum for CD20, Iba1 and Lambda protocols, and 3% bovine serum albumin (BSA) in tris buffered saline (TBS, 5 mM Tris/HCl pH 7.6, 136 mM NaCl). The tissue sections were incubated overnight at 4 °C with commercial monoclonal and polyclonal antibodies (Table 1) and then washed three times with TBS. Then, samples were incubated with horse anti-mouse serum or goat anti-rabbit serum (Table 1) (Vector Laboratories, California, USA) diluted 1:200 in TBS for 30 min at RT and washed three times with TBS followed by incubation with the ABC complex kit in TBS for 30 min at RT. Finally, the sections were incubated with the substrate 3,3′-diaminobenzidine tetrahydrochloride (DAB, Sigma, St. Louis, MO, USA) for 5 min and washed with TBS and water. After staining for 45 s with haematoxylin, slides were dehydrated and mounted with DPX (Fluka, Sigma, St. Louis, MO, USA). Stained slides were studied under light microscopy (Olympus BH-2) and photographed using a digital camera Olympus DP-12. Positive (where the target antigen was present in the control section and the specific antibody was used) and negative (additional slide with omission of the primary antibody) controls were used during each immunohistochemical run.

Statistical analysis

Results were submitted to analysis of variance to determine the statistical significance of differences between granuloma types for each cell type and then the means

Table 1 Immunohistochemical protocols used for cellular type characterization

Primary antibody (Ab)	Specifity	Dilution Ab	Epitope demasking	Secondary Ab
CD3 (Novocastra-CL-L-CD3–565), mouse monoclonal	Pan T cell marker	1:500 in TBS 1%	Microwave in citrate pH 6 20 min	Anti-mouse biotinylated (1:200)
CD20 (ThermoFisher.-PA516701), rabbit polyclonal	Pan B cell marker	1:200 in TBS + BSA 1%	Microwave in citrate pH 6 20 min	Anti-rabbit biotinylated (1:200)
Iba1 (WAKO 019_19741), rabbit polyclonal	Macrophages	1:1000 in TBS + BSA 1%	Los Angeles pH 9 40 min 95 °C	Anti-rabbit biotinylated (1:200)
Lambda (Dako A0193), rabbit polyclonal	Plasma cells	1:1000 in TBS + BSA 1%	Triton 1% 20 min room temperature	Anti-rabbit biotinylated (1:200)

TBS Tris-buffered saline, *BSA* Bovine serum albumin

were compared with the Tukey-Kramer test in the SAS statistical package.

Results

The number of AFBs identified in each type of granuloma was very low in three types of granulomas (< 10 AFBs in type I granuloma and 10–20 AFBs in types II and III granulomas) and no differences were observed between them.

Significant differences in the numbers of cell types were observed in each TB granuloma type (Table 2).

CD3 (T lymphocytes)

IHC analysis for CD3 showed very few scattered positively stained lymphocytes within type I and type II granulomas (Table 2, Figs. 1 and 2), randomly and peripherally distributed, respectively (Figs. 1 and 2), with no T cells observed within type III granuloma (Fig. 3).

CD20 (B lymphocytes)

B cells were found in all granulomas, mainly in type III granulomas, forming clusters surrounding the necrotic area (Fig. 3). In types I and II granulomas, these cells were sparsely distributed within the lesion.

Iba1 (macrophages)

Macrophages were the predominant cells observed within granulomas of the types I and II (Figs. 1 and 2). However, the number of macrophages decreased for the type III granuloma type and appeared peripherally distributed (Fig. 3).

Table 2 Cellular types found in tuberculous granulomas

	Initial granuloma	Developed granuloma	Terminal granuloma
T lymphocytes	+	+	–
B lymphocytes	++	+++	++++
Macrophages	+++	++++	++
Plasma cells	–	++	+++

-: abscence of immunolabelled cells; +: 1–10 immunolabelled cells; ++: 11–50 immunolabelled cells; +++: 51–100 immunolabelled cells; ++++: > 100 immunolabelled cells

Lambda chain (plasma cells)

No cells immunostained for lambda chains were observed in type I granuloma (Fig. 1). A few sparsely distributed cells at the periphery of the types II and III granulomas were observed (Figs. 2 and 3).

Statistical results

Significant differences in the mean number of cells were found for B lymphocytes and plasma cells ($p < 0.0001$), but not for T lymphocytes and macrophages (Fig. 4). The clearest differences were in the number of B lymphocytes that were observed at a different order of magnitude in each granuloma type. Plasma cells only showed differences between types I and III granulomas. In summary, the B lymphocyte was the cell that best discriminated between the three granuloma subtypes.

Therefore according to their cell composition we defined tuberculous granulomas as *initial* (type I), *developed* (type II) and *terminal* (type III). *Initial* granuloma would represent an initial or latent stage, *developed* granuloma a more mature stage and *terminal* granuloma a regressive stage, associated with mineralization and fibrosis indicative of resolution.

Discussion

This preliminary study describes, for the first time, the distribution of T cells, B cells, macrophages and plasma cells in tuberculous granulomas from sheep naturally infected with Mbovis. Differences were observed in the predominant cell type present in each type of granuloma, as well as differences and similarities with the development of tuberculous granulomas in other species [6–10].

It is important to note the scarce presence of T lymphocytes in all granuloma types (< 10 cells), which would suggest a clear difference with the spectrum of lesions of TB in other species, such as cattle, where T lymphocytes are essential in the initial cellular immune response against Mbovis [6, 7]. In the early or latent stages of the disease (namely initial granuloma in this study), a higher percentage of macrophages were observed, increasing considerably in developed granulomas. The composition

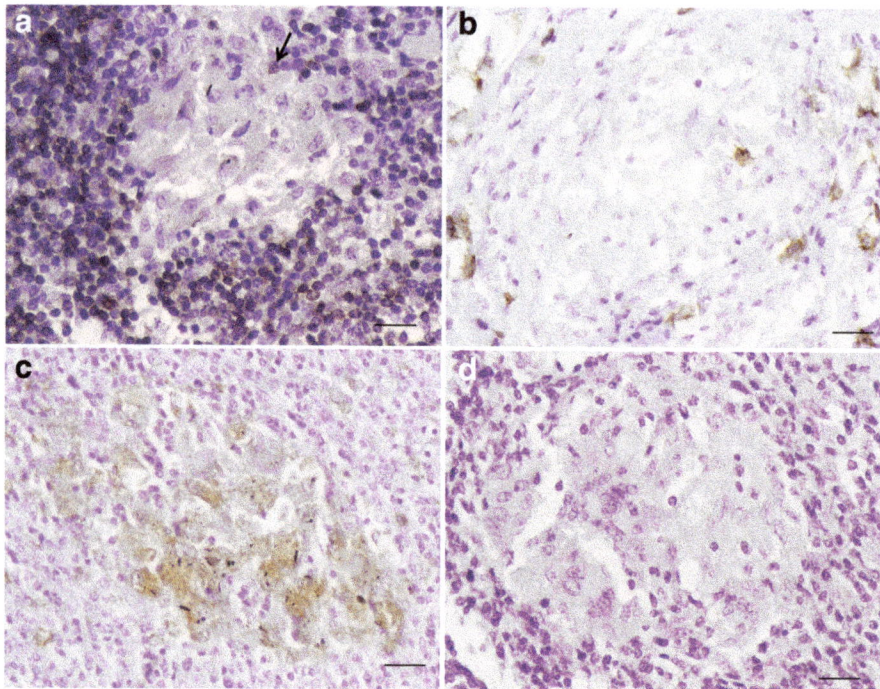

Fig. 1 Bronchial lymph node; immunohistochemical characterization of cellular populations in initial granuloma. (**a**) Isolated T lymphocytes (arrow) can be observed in the granuloma, bar = 20 μm. (**b**) A few scattered B lymphocytes are observed in this stage, bar = 20 μm. (**c**) Abundant immunolabelled macrophages are present, bar = 20 μm. (**d**) No plasma cells are observed, bar = 20 μm

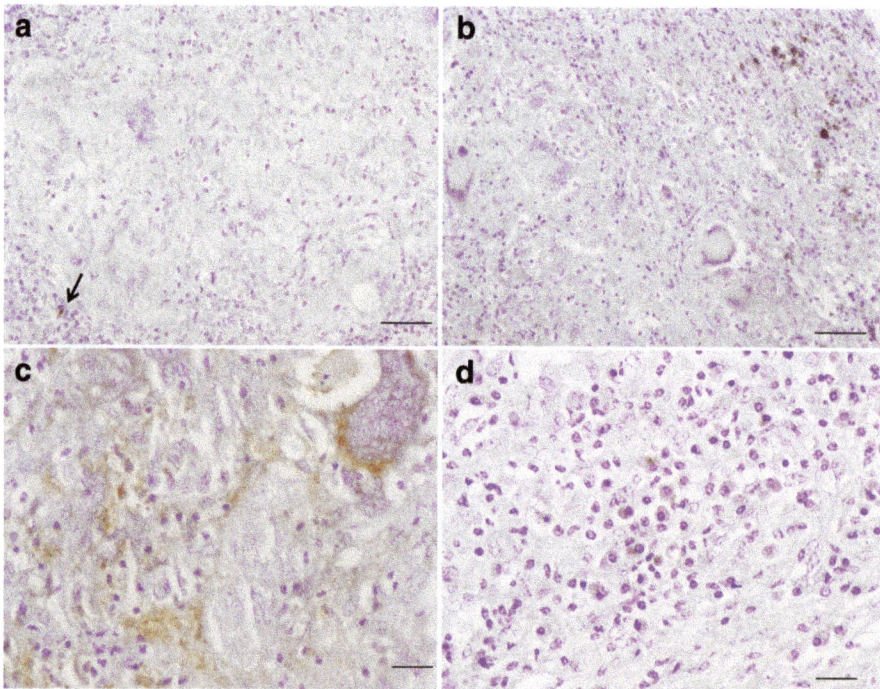

Fig. 2 Bronchial lymph node; immunohistochemical characterization of cellular populations in developed granuloma. Scattered T (arrow) and B lymphocytes are observed towards the periphery of the granuloma (**a** and **b**, bar = 50 μm). The presence of immunolabelled macrophages predominated over the whole granuloma (**c**, bar = 20 μm), while plasma cells were abundant and peripherally located (**d**, bar = 20 μm)

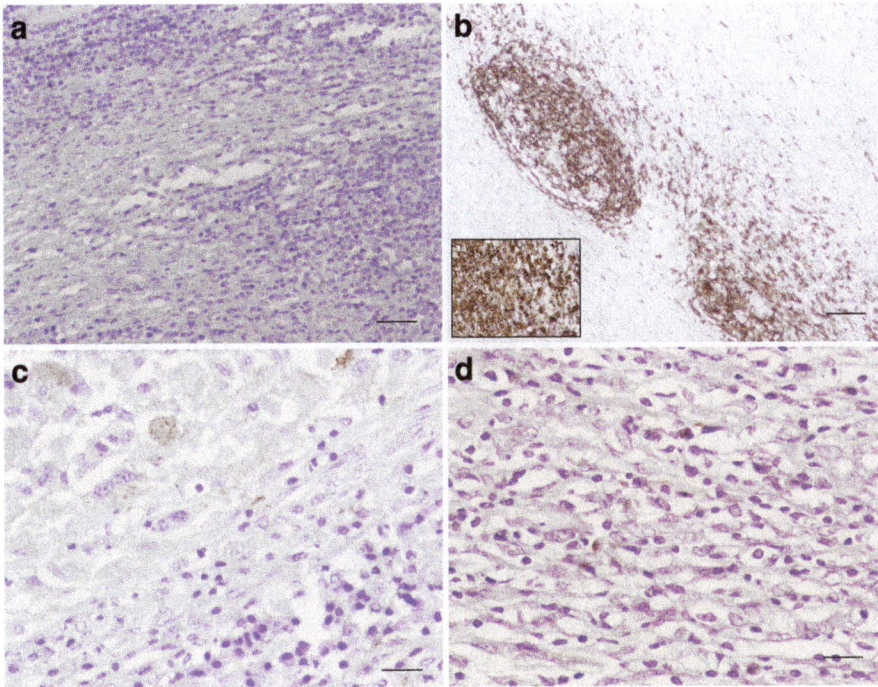

Fig. 3 Bronchial lymph node; immunohistochemical characterization of cellular populations in terminal granuloma. No T lymphocytes were observed within the granuloma (**a**, bar = 20 µm). The main cellular type were B lymphocytes forming clusters surrounding the necrotic area of the granuloma (**b**, bar = 200 µm; inset: bar = 20 µm). Positive immunolabelled macrophages and plasma cells were randomly and sparsely distributed (**c** and **d**, bar = 20 µm)

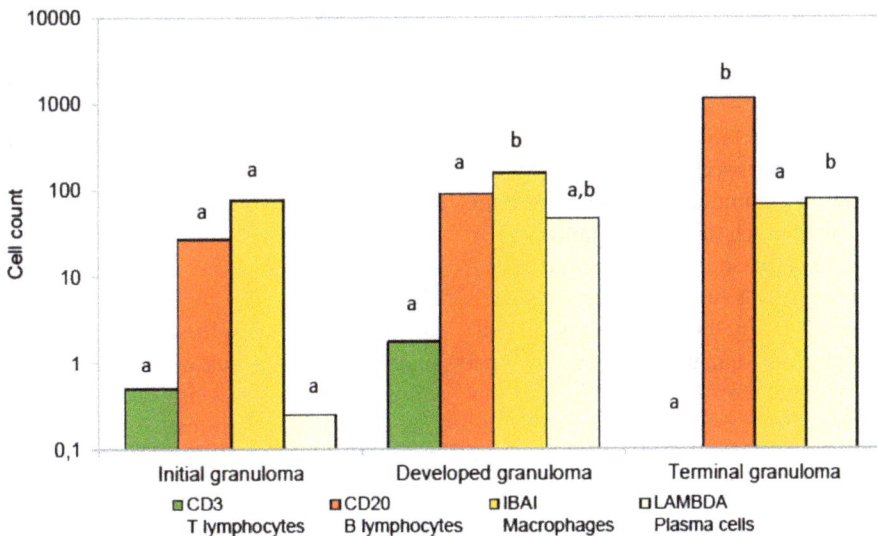

Fig. 4 Comparison between mean cell type counts in tuberculous granulomas. Cell counts are represented in a logarithmic scale. Cell type is identified both after the specific marker and the cell type name. Different letters indicate significant differences at $p < 0.05$ between granuloma type. No differences between granulomas for T lymphocyte marker; greater mean for B lymphocyte marker in terminal granulomas; greater mean for macrophages in developed granulomas than in initial granuloma, but no differences for the other comparisons; greater mean for plasma cells in terminal granuloma

of these initial granulomas is related to the initial host immune response [5, 11]. The high number of macrophages in the initial stages matches with that observed in early stages in other species such as wild boar [10] and fallow deer [9], although the former also presented a high percentage of T lymphocytes, not observed in sheep. This would suggest that the initial immune response in fallow deer is established by the combined and balanced action of T lymphocytes and macrophages, reducing their presence as the granuloma develops [9]. In sheep and wild boar the presence of macrophages was much greater than T lymphocytes, so the initial immune response in both species would be based on the phagocytosis and lysis of mycobacteria, which is the hallmark of an innate immune response. This, in turn, suggests that sheep response against Mbovis is biased towards the innate unspecific immune response represented by phagocytic cells instead of a well mounted adaptive specific immune response [5]. We have observed a shift towards an adaptive immune response in the more advanced phases of local response, where the larger presence of mycobacteria, showed using Ziehl-Neelsen stain, would drive a humoral rather than cellular immune response. This could lead to a stronger ability to clear initial colonization by small amounts of Mbovis and thus less chance of developing lesions and disease in the absence of an abundant source of Mbovis. The progression to active TB differs among individuals and even among granulomas in a single individual [11]. Before tubercle bacilli can be destroyed by macrophages, these cells must be activated by T lymphocytes and their cytokines, being the essence of cell-mediated immunity [11, 12]. Non-activated macrophages, however, would serve as sites where bacteria are protected within the nascent granuloma [11].

The number of B lymphocytes increased considerably as the granuloma developed. In terminal granulomas, the amount of macrophages decreased while B-lymphocytes increased, suggesting a humoral-mediated immune response. A calcified granuloma (terminal granuloma in this manuscript) generally represents a successful immune response and is associated with fewer inflammatory cells than other granulomas [11]. In the early stages, B cells appeared dispersed and in low quantity. However, in final stages they appeared in large numbers peripherally located, surrounding the caseous necrosis and mineralization, organized in clusters following the same pattern as in studies carried out in cattle [6, 13]. In cattle experimentally challenged with Mbovis, B cell accumulation has been observed together with increasing of chemokines which might suggest non-specific recruitment into lesions rather than a specific humoral response, although more studies are needed to confirm this hyphotesis [13]. In badgers, these final stages are characterized by

a large presence of T lymphocytes and scarce B lymphocytes [8], suggesting that the cellular immune response is more important in the elimination of the pathogen in this species than in the sheep. Plasma cells however were more abundant in developed granulomas, as occurs in other species such as badger [8]. These cells appear after activation of B lymphocytes by Th2 lymphocytes and release specific antibodies for antigen destruction [14]. This suggests that an ineffective humoral immune response is the host's answer to large amounts of antigen accumulated in the granulomata when the unspecific and cellular clearing of mycobacteria has failed. It would be of great interest to carry out subsequent studies in order to increase the sample size and quantify the expression of cytokines such as interferon-gamma (IFN-γ), iNOS or IL-17A in each type of granuloma in order to study the immune responses that accompany the progression of the infection, however, this was not possible in this study because of the excessive fixation time in the processing of the samples.

Lambs subjected to experimental challenge with *M. caprae* showed that the volume of gross pulmonary lesions, quantified by computed tomography and bacterial load in respiratory lymph nodes were similar to those observed in goats experimentally challenged with *M. caprae* at a similar dose suggesting that the susceptibility of sheep to TB infection is similar to that observed in goats [15]. In the present study the number of AFBs within developed and terminal granulomas was very low [1]. This finding would be common in cattle [16, 17] but not in goats with TB infections, where the presence of mycobacteria is higher in these types of lesions in natural cases [17]. Sheep might be considered as a host with a particular and unusual TB lesion development, with low presence of mycobacteria in lesions, indicating the presence of a negative environment for bacterial growth and less capacity of excretion of mycobacteria and, consequently, of spread and transmission [1]. More studies are needed to confirm this hypothesis.

Conclusions

In conclusion the three types of granulomas described could be the result of different combinations of the three immune response routes: innate, adaptive cellular and adaptive humoral [5]. Thus in the early or latent stages of tissue infection, the innate component would be the most prominent and efficient. If that fails to contain the Mbovis infection, the next step would be an increase in the number of macrophages that would attract more lymphocytes to the site of infection. Finally, a shift to a new phase in the immune response would occur where B lymphocytes and plasma cells would try to contain the infection with a humoral component that would not control the progression and may cause more local damage, as observed in sheep. All these observations could be very useful for identifying factors that may help us to understand the keys that lead to the reactivation of the disease or the persistence of latent infections [18].

Abbreviations
Ab: Antibody; AFBs: Acid fast bacilli; BSA: Bovine serum albumin; DAB: 3,3′-diaminobenzidine tetrahydrochloride; IHC: Immunohistochemistry; Mbovis: *Mycobacterium bovis*; MTBC: *Mycobacterium tuberculosis* complex; RT: Room temperature; TB: Tuberculosis; TBS: Tris buffered saline

Acknowledgements
Authors thank Alba García Delgado for her technical support and Kevin P. Dalton for critically reviewing the manuscript.

Funding
This paper was funded by a grant from INIA-RTA2014–00002-C02–01 (FEDER co-funded).

Authors' contributions
AB, RV, JFGM, RAJ, MMM, FJS: contributed to the conception, design, and data collection, laboratory work, drafting and writing of the manuscript. All authors have read and approved the final manuscript.

Competing interests
The authors declare that they have no competing interests.

Author details
[1]Universidad de León, Campus de Vegazana, León, Spain. [2]SERIDA, Servicio Regional de Investigación y Desarrollo Agroalimentario, Centro de Biotecnología Animal, 33394 Gijón, Asturias, Spain. [3]Xunta de Galicia, Santiago, A Coruña, Galicia, Spain. [4]School of Veterinary Medicine, University of Surrey, Guildford, UK.

References
1. Muñoz-Mendoza M, Romero B, del Cerro A, Gortázar C, García-Marín JF, Menéndez S, et al. Sheep as a potential source of bovine TB: epidemiology, pathology and evaluation of diagnostic techniques. Transbound Emerg Dis. 2016;63:635–46. https://doi.org/10.1111/tbed.12325.
2. Malone FE, Wilson EC, Pollock JM, Skuce RA. Investigations into an outbreak of tuberculosis in a flock of sheep in contact with tuberculosis cattle. J Veterinary Med Ser B. 2003;50:500–4.
3. Broughan JM, Downs SH, Crawshaw TR, Upton PA, Brewer J, Clifton-Hadley RS. Mycobacterium bovis infections in domesticated non-bovine mammalian species. Part 1: Review of epidemiology and laboratory submissions in Great Britain 2004-2010. Vet J. 2013;198:339–45. https://doi.org/10.1016/j.tvjl.2013.09.006.
4. Van der Burgt GM, Drummond F, Crawshaw T, Morris S. An outbreak of tuberculosis in Lleyn sheep in the UK associated with clinical signs. Vet Rec. 2013;172:69. https://doi.org/10.1136/vr.101048.
5. Ehlers S, Schaible UE. The granuloma in tuberculosis: dynamics of a host-pathogen collusion. Front Immunol. 2013;3:411. https://doi.org/10.3389/fimmu.2012.00411.
6. Salguero FJ, Gibson S, García Jiménez W, Gough J, Strickland TS, Vordermeier HM, et al. Differential cell composition and cytokine expression within lymph node granulomas from BCG-vaccinated and non-vaccinated cattle experimentally infected with *Mycobacterium bovis*. Transbound Emerg Dis. 2016. https://doi.org/10.1111/tbed.12561.
7. Palmer MV, Waters WR, Thacker TC. Lesion development and immunohistochemical changes in granulomas from cattle experimentally infected with *Mycobacterium bovis*. Vet Pathol. 2007;44:863–74.
8. Canfield PJ, Day MJ, Gavier-Widen D, Hewinson RG, Chambers MA. Immunohistochemical characterization of tuberculous and non-tuberculous lesions in naturally infected European badgers (*Meles meles*). J Comp Pathol. 2002;126:254–64.
9. García-Jiménez WL, Fernández-Llario P, Gómez L, Benítez-Medina JM, García-Sánchez A, Martínez R, et al. Histological and immunohistochemical characterisation of *Mycobacterium bovis* induced granulomas in naturally infected fallow deer (*Dama dama*). Vet Immunol Immunopathol. 2012;149:66–75. https://doi.org/10.1016/j.vetimm.2012.06.010.
10. García-Jiménez WL, Salguero FJ, Fernández-Llario P, Martínez R, Risco D, Gough J, et al. Immunopathology of granulomas produced by *Mycobacterium bovis* in naturally infected wild boar. Vet Immunol Immunopathol. 2013;156:54–63. https://doi.org/10.1016/j.vetimm.2013.09.008.
11. Flynn JL, Chan J, Lin PL. Macrophages and control of granulomatous inflammation in tuberculosis. Mucosal Immunol. 2011;4:271–8. https://doi.org/10.1038/mi.2011.14.
12. Dannenberg AM Jr, Rook GAW. Pathogenesis of pulmonary tuberculosis: an interplay of tissue-damaging and macrophages activating immune-response – dual mechanisms that control bacillary multiplication. In: Bloom BR, editor. Tuberculosis pathogenesis, protection, and control. Washington DC: American Society for Microbiology Press; 1994. p. 459–83.
13. Aranday-Cortes E, Bull NC, Villarreal-Ramos B, Gough J, Hicks D, Ortiz-Peláez A, et al. Upregulation of IL-17A, CXCL9 and CXCL10 in early-stage granulomas induced by *Mycobacterium bovis* in cattle. Transbound Emerg Dis. 2013;60:525–37. https://doi.org/10.1111/j.1865-1682.2012.01370.
14. Koolman J, Röhm KH. Bioquímica Humana texto y Atlas. 4th ed. Editorial panamericana; 2012. p. 294–6.
15. Balseiro A, Altuzarra R, Vidal E, Moll X, Espada Y, Sevilla IA, et al. Assessment of BCG and inactivated *Mycobacterium bovis* vaccines in an experimental tuberculosis infection model in sheep. PLoS One. 2017;12:e0180546. https://doi.org/10.1371/journal.pone.0180546.
16. Domingo M, Vidal E, Marco A. Pathology of bovine tuberculosis. Res Vet Sci. 2014;97:20–9. https://doi.org/10.1016/j.rvsc.2014.03.017.
17. Gutiérrez Cancela MM, García Marín JF. Comparison of Ziehl-Neelsen staining and immunohistochemistry for the detection of *Mycobacterium bovis* in bovine and caprine tuberculous lesions. J Comp Pathol. 1993;109:361–70.
18. Sakamoto K. The pathology of *Mycobacterium tuberculosis* infection. Vet Pathol. 2012;49:423–39. https://doi.org/10.1177/0300985811429313.

22

In vitro inhibition of porcine reproductive and respiratory syndrome virus replication by short antisense oligonucleotides with locked nucleic acid modification

Lingyun Zhu[1†], Junlong Bi[1,4†], Longlong Zheng[1,5], Qian Zhao[1], Xianghua Shu[1], Gang Guo[2], Jia Liu[1], Guishu Yang[1], Jianping Liu[3*] ⓘ and Gefen Yin[1*]

Abstract

Background: Porcine reproductive and respiratory syndrome virus (PRRSV) causes porcine reproductive and respiratory syndrome (PRRS), which is currently insufficiently controlled. From a previous small-scale screen we identified an effective DNA-based short antisense oligonucleotide (AS-ON) targeting viral NSP9, which could inhibit PRRSV replication in both Marc-145 cells and pulmonary alveolar macrophages (PAMs). The objective of this study was to explore the strategy of incorporating locked nucleic acids (LNAs) to achieve better inhibition of PRRSV replication in vitro.

Methods: The effective DNA-based AS-ON (YN8) was modified with LNAs at both ends as gap-mer (LNA-YN8-A) or as mix-mer (LNA-YN8-B). Marc-145 cells or PAMs were infected with PRRSV and subsequently transfected.

Results: Compared with the DNA-based YN8 control, the two AS-ONs modified with LNAs were found to be significantly more effective in decreasing the cytopathic effect (CPE) induced by PRRSV and thus in maintaining cell viability. LNA modifications conferred longer lifetimes to the AS-ON in the cell culture model. Viral ORF7 levels were more significantly reduced at both RNA and protein levels as shown by quantitative PCR, western blot and indirect immunofluorescence staining. Moreover, transfection with LNA modified AS-ON reduced the PRRSV titer by 10-fold compared with the YN8 control.

Conclusion: Taken together, incorporation of LNA into AS-ON technology holds higher therapeutic promise for PRRS control.

Keywords: Porcine reproductive and respiratory syndrome virus (PRRSV), Locked nucleic acids (LNAs), Virus replication, Antisense oligonucleotides, Marc-145, Pulmonary alveolar macrophages (PAMs)

Background

Porcine reproductive and respiratory syndrome (PRRS) is characterized by respiratory disorders in piglets and reproductive failure in sows [1]. The disease is one of the most economically significant problems in the swine industry. The responsible virus, porcine reproductive and respiratory syndrome virus (PRRSV) is a member of the family *Arteriviridae*, genus *Arterivirus* [2]. PRRSV is an enveloped, single-stranded positive-sense RNA virus. PRRSV's genome is about 15 kb long, consisting of nine open reading frames (ORFs) [3]. Among all the encoded viral proteins, NSP9 is a putative RNA-dependent RNA polymerase and plays central roles in viral replication [4]. In our previous small screen with nine candidate sequences [5], we identified YN8 as one of the most effective DNA-based short antisense oligonucleotides (AS-ONs) targeting NSP9, which could significantly

* Correspondence: jianping.liu@ki.se; yingefen@ynau.edu.cn;
13211658801@qq.com

†Equal contributors

3Department of Medical Biochemistry and Biophysics, Karolinska Institute,
-17177 Stockholm, SE, Sweden

1Department of Veterinary Medicine, College of Animal Science and
Technology, Yunnan Agricultural University, Yunnan province, Kunming
650201, China

inhibit PRRSV replication in both Marc-145 cells and pulmonary alveolar macrophages (PAMs).

Due to the disadvantages of DNA-based AS-ON, e.g. relatively low biostability, quick degradation by nucleases and low hybridization affinity with target sequences, the applications of antisense technologies in research and therapeutics are limited. The synthesis of Locked Nucleic Acid (LNA) [6] overcame these limitations as LNA modified nucleotides confer low cytotoxicity, high thermostability, resistance to nucleases and stable hybridization abilities with target sequences [7]. Enhanced nucleic acid recognition by LNA-containing oligonucleotides made them desirable for many applications in molecular biology, including genotyping [8] or single nucleotide polymorphism (SNP) analysis [9], hybridization [10, 11], decoy and fluorescence polarization [12], expression profiling or microarray [13], allele-specific PCR [14], fluorescent in situ hybridization (FISH) analysis [15], alteration of intron splicing and LNAzymes [16], 5′-nuclease assay [17], real-time PCR [18], siRNA [19], microRNA [20] and antisense [21].

In 2006, highly pathogenic PRRSV strains of the North American type were identified in more than 10 provinces in China, where they caused approximately four million fatal cases in 2006 [22]. At the beginning of 2007, the disease re-emerged and infected 310,000 pigs, of which more than 81,000 died in 26 provinces [23].The outbreak of PRRS in China has resulted in considerable economic losses and a rise in the price of pork. Therefore, it is urgent to develop more effective strategies to prevent and control PRRSV infection in the swine industry. In order to develop improved methods to manage PRRS, we selected the best antisense sequence YN8 from our previous small-scale screening [5] for LNA modifications and applied the two modified sequences to in vitro studies (lifetime of the antisense oligonucleotides, cytotoxicity, cytopathic effect observation, qPCR, virus titer assessment, western blot and indirect immunofluorescence) to evaluate the inhibitory effects on PRRSV replication in Marc-145 cells and in PAMs between the DNA- and LNA-based AS-ONs. Our data showed that incorporation of LNA into AS-ON technology holds higher therapeutic promise for PRRS control.

Methods

Ethics statement

In this study the pigs did not undergo any manipulations prior to standard industrial slaughter. Therefore, no specific ethical approval was required. All animal experiments were performed with the approval of the Animal Care Committee of Yunnan Agricultural University, China.

Virus and cells

From the lungs of an infected pig in Yunnan province (China) during a severe PRRSV outbreak in 2008, our research group isolated a highly pathogenic PRRSV field strain YN-1 (GenBank accession number: KJ747052), which belongs to the North American genotype. Both pulmonary alveolar macrophages (PAMs) and Marc-145 cells were applied in this study, as PRRSV can replicate in these two culture systems. The Marc-145 cells and PAMs were acquired and cultured as we previously described [5].

Locked nucleic acid modification in antisense oligonucleotide sequences

According to the statistic conclusion [24], we designed nine candidate AS-ONs with a length of 20 nt using RNA Structure 5.6 [25, 26] and YN8 was identified as the best antisense sequence inhibiting the replication of PRRSV in vitro [5]. In the present study, three types of antisense oligonucleotides based on YN-8 were investigated: (i) unmodified AS-ON (DNA, YN8, served as a control); (ii) LNA/DNA/LNA gap-mer with four LNAs at both ends (LNA-YN8-A) and (iii) LNA/DNA mix-mer (LNA-YN8-B). The three AS-ONs are listed in Table 1 and were synthesized by Shengong (Shanghai, China). The nucleotides in bold in the antisense oligonucleotides (LNA-YN8-A and LNA-YN8-B) were modified with locked nucleotide acids. The gene amplified using the primer pair ACTB-F and ACTB-R is beta-actin from *Macaca mulatta* (African green monkey). The NCBI Reference Sequence for mRNA of beta-actin is NM_001033084.1. The

Table 1 List of oligonucleotides used in this study

Name of oligonucleotides	Sequence (from 5′ to 3′)	Target gene	Position within the target gene	GC content (%)	Tm (°c)
YN8	TGCAGCATCCTCACAACCGT	Nsp9	704–723 bp	55	65.3
LNA-YN8-A	**TGCA**GCATCCTCACAA**CCGT**				71.8
LNA-YN8-B	**TG**CA**GC**ATC**CT**CAC**AA**CC**GT**				71.2
ACTB-F	agttgcgttacacccttcttga	beta-actin	1168–1190	43.5	58
ACTB-R	tgctgtcaccttcaccgttc		1297–1316	55	59.4
ORF7-F	aatggccagccagtcaatca	ORF7	14,844–14,863	50	58.5
ORF7-R	tcacgctgagggcgatgctg		15,154–15,173	65	65

gene amplified using the primer pair ORF7-F and ORF7-R is ORF7 from porcine reproductive and respiratory syndrome virus isolate YN-1 (NCBI Reference Sequence for the complete genome of YN-1 is GenBank: KJ747052.1).

Virus infection and transfection

Cell seeding, virus infection, transfection with AS-ONs and lifetime measurement of AS-ONs were performed [5]. In brief, the Marc-145 cells or PAMs were seeded in 96- or 6-well plates one day before PRRSV infection and AS-ON transfection. After inoculation with PRRSV YN1 strain (25 $TCID_{50}$/well) for one and half hours, the medium was removed and transfection with the desired concentrations of AS-ONs was performed. Four hours post transfection, the transfection medium was replaced with fresh full DMEM medium till further analysis. Each treatment was performed in triplicate.

Cy-3 labeled AS-ONs were used in the transfection for the measurement of the lifetime of AS-ONs in vitro, with cy-3 fluorescence signals recorded under the fluorescence microscopy (Olympus) at various time points (1, 2, 6, 8 and 10 h post transfection).

Cell viability analysis

The Marc-145 cells (10^4 cells/well in 100 µl) were seeded into 96-well plates and incubated for overnight. Transfection with the indicated concentrations of AS-ONs (32, 40, 48, 56, 64 and 72 µM) was performed as described above, with each treatment in triplicate. Seventy-two hours post transfection, the cell viability analysis using CCK-8 kit (Sigma-Aldrich, Cat. No. 96992) was performed according to the manufacturer's guide. In brief, 10 µl of CCK-8 solution was added to each well of the plate. Incubate the plate for 2 h in the incubator and measure the absorbance at 450 nm using a microplate reader. The wells without transfection were used as control for normalization.

Isolation of total RNA, reverse transcription and qPCR analysis

As we previously described [5], total RNA was isolated from Marc-145 cells or PAMs approximately 60 h post PRRSV infection and AS-ON transfection using the RNAiso Plus RNA isolation kit (Takara Dalian, China), and subjected to reverse transcription (Takara Primer-Script RT reagents kit) and qPCR analysis (SYBR Primer Ex Taq II kit). β-actin served as an internal control. The primer pairs used in this study are listed in Table 1. The ΔΔCt method [27] for relative quantification of gene expression was applied to determine viral RNA levels using SYBR Green real-time PCR. The relative amount of PRRSV RNA was normalized to β-actin mRNA. Amplification and detection of samples were performed with the CFX96 Touch Real-Time PCR Detection System (Bio-Rad, USA).

Western blot analysis

Total viral and cellular proteins were isolated from PAMs in 6-well plates approximately 60 h post PRRSV infection with or without AS-ON transfections. The cells were scraped after one time PBS wash and collected by centrifuge at 10,000 rpm for 5 min. Into the pellet 100 µl RIPA lysis and extraction buffer (ThermoFisher Scientific, Cat. no. 89900), 1 µl Halt Protease Inhibitor Cocktail (ThermoFisher Scientific, Cat. no. 78430) and 1 µl Halt Phosphatase Inhibitor Cocktail (ThermoFisher Scientific, Cat. no. 78420) was added. The pellet was lyzed by thorough pipetting followed by protein extraction at 4 °C for 15 min and then subjected to centrifuge at 10,000 rpm for 5 min. The protein-containing supernatants were collected, mixed with 20 µl loading buffer and 20 µl bromophenol blue (0.4%) and boiled for 10 min. The protein was stored at − 20 °C till further western blot analysis. Purified protein samples were resolved under reducing and denaturing conditions using sodium dodecyl sulfate-polyacrylamide gel electrophoresis (SDS-PAGE) and 8% Bis-Tris Novex NuPage gels in conjunction with running buffer. Resolved proteins were transferred to nitrocellulose membranes and blocked at room temperature for 1 h in PBST containing 5% (w/v) dehydrated milk and 0.05% Tween 20 with shaking. Viral N protein or cellular β-actin was probed by overnight incubation at 4 °C with rocking with the primary anti-N protein monoclonal antibody (VMRD, Cat. no. 080728–004, mouse origin) or with the primary anti- β-actin polyclonal antibody (Proteintech, Cat. No.20536–1-AP, rabbit origin). The antibodies were diluted in filtered 5% milk–PBST at a ratio of 1:500 (anti-N protein antibody) or 1:1,1000 (anti- β-actin antibody). The following incubation at room temperature with racking was with secondary goat anti-mouse conjugated horseradish peroxidase (HRP) (Proteintech) or goat anti-rabbit-conjugated horseradish peroxidase (Proteintech) antibody (1:2000 dilution in filtered 5% milk–PBST). Subsequently, western blots were treated with chemiluminescent ECL Plus substrate (Pierce, Rockford, IL) and imaged using chemiluminescent film (Kodak, Rochester, New York).

Indirect immunofluorescence staining

Sixty hours post PRRSV infection and transfection with AS-ONs, the Marc-145 cells were washed with PBS and fixed with 4% paraformaldehyde (PFA) at room temperature for 10–15 min. After three washes with PBS, the cells were permeabilized with PBS containing 0.3% Triton X-100 for 15 min and blocked with PBS containing 1% BSA for 2 h at 4 °C. The nuclei staining with 5 µg/ml of Hoechst 33,342 (Life Technology) was carried out for 20 min at room temperature. The cells were subsequently incubated at 4 °C overnight with

5 µg/ml of PRRSV antibody against N protein (encoded by ORF7) (VMRD, Cat. no. 080728–004, mouse origin), washed three times with PBS and incubated with Alexa Fluor 488 conjugated goat anti-mouse IgG (H + L) antibody (Proteintech, Cat. no. 861163) at 5 µg/ml for 1 h at 37 °C. After three times PBS wash, the cells were subjected to image analysis by fluorescence microscopy (Olympus). Images were processed to calculate the percentage of infected cells by ImageJ, which was downloaded from (https://imagej.nih.gov/ij/).

Virus titration

Marc-145 cells were seeded into 96-well plates (10^4 cells/well in 100 µl) one day before PRRSV infection and AS-ONs transfection. A 10× serial dilution of PRRSV YN-1 strain was prepared. Each dilution was added into six wells (100 µl/well). One and half hours post infection, the transfection was performed with 8 µM of AS-ONs. CPE was recorded using the inverted microscope over a period of 4 days post transfection. Cell number was counted and the 50% tissue culture infected dose ($TCID_{50}$) was determined by Reed–Muench method.

Statistical analysis

Statistical analysis was performed using GraphPad Prism 4.0 (GraphPad Software Inc., San Diego, CA, USA). Data were analyzed by using the t-test, with two-tailed distribution. $P < 0.05$ was considered statistically significant.

Results

LNA modification conferred longer lifetime to antisense oligonucleotides

To investigate how long AS-ONs with LNA modifications are present in the cell culture system, 32 µM cy-3 labelled LNA modified antisense oligonucleotides LNA-YN8-A and LNA-YN8-B were transfected into Marc-145 cells. Transfection with cy-3 labelled DNA antisense oligonucleotide YN8 at the same concentration served as the control for comparison. At different time points (1, 2, 6, 12 and 20 h) post transfection, the medium was aspirated and the cells were washed with PBS prior to imaging in fresh medium (Fig. 1). We found that the fluorescent signal from cy-3 labelled antisense oligonucleotides containing LNA modification (LNA-YN8-A and LNA-YN8-B) was still detectable 20 h post transfection while the signal from cy-3 labelled DNA antisense oligonucleotide YN8 was not visible 6 h post transfection, indicating the unmodified oligonucleotide was almost completely degraded after 6 h. The data presented in Fig. 1 demonstrated that LNA modification can add significant biostability to antisense oligonucleotides in cell cultures or reduce their sensitivity to nucleolytic degradation in biological media, and that protection with LNA significantly stabilized the antisense oligonucleotides against nucleolytic attack, which is consistent with the previously reported data [28, 29].

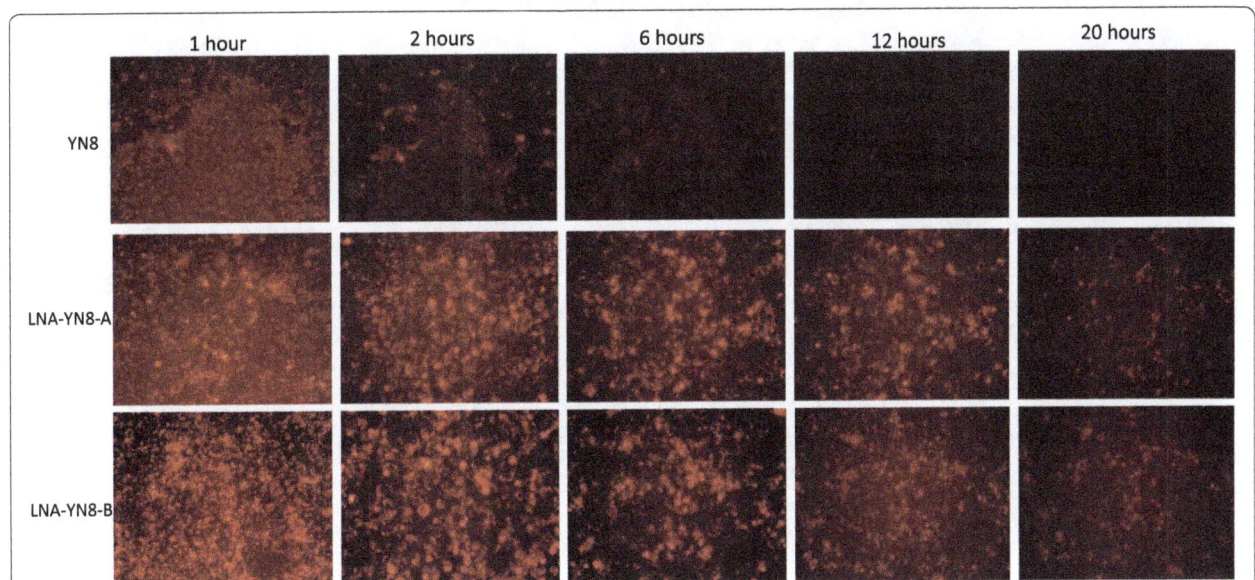

Fig. 1 AS-ONs with LNA modifications were present in cell culture twenty hours post transfection. Cy3-labelled antisense oligonucleotides (YN8, LNA-YN8-A and LNA-YN8-B, 32 µM) were used in the transfection in Marc-145 cells. The fluorescent signals were recorded with fluorescence microscopy (Olympus) at the indicated time points post transfection. The fluorescent signals decreased over the time, however, compared with the DNA AS-ON YN8, we could clearly observe the signals even 20 h post transfection, indicating that incorporating LNA into the antisense oligonucleotide could improve the stability in Marc-145 cells for sufficiently longer time periods to induce the degradation of viral RNA and thus to inhibit PRRSV replication. Shown here are the representative images from three independent experiments

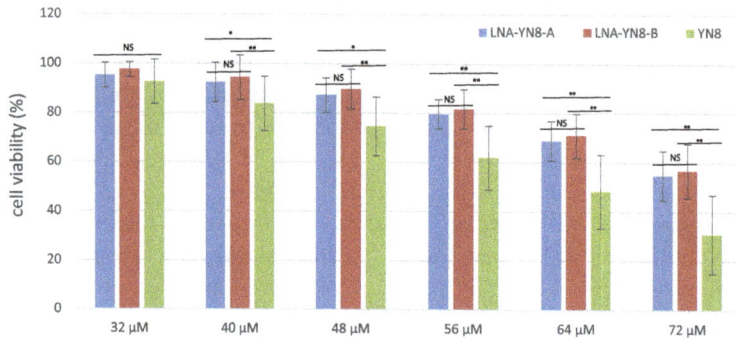

Fig. 2 Cell viability analysis of Marc-145 cells transfected with AS-ONs containing LNA modifications. The cell viability was assayed 72 h post transfection. The measurement of each concentration was normalized to the corresponding cell control (as 100%). Compared with the cell control, both DNA AS-ON (YN8, when concentration ≥ 48 μM) and the AS-ONs with LNA modification (LNA-YN8-A and LNA-YN8-B, when concentration ≥ 56 μM) showed some cytotoxicity. Higher concentration led to higher cytotoxicity. However, LNA modification conferred lower cytotoxicity when compared with the corresponding DNA AS-ON YN8. No difference in cell viability was observed between LNA-YN8-A and LNA-YN8-B. Shown here are the representative data from three independent experiments. NS: not significant; *: $P < 0.05$; **: $P < 0.01$

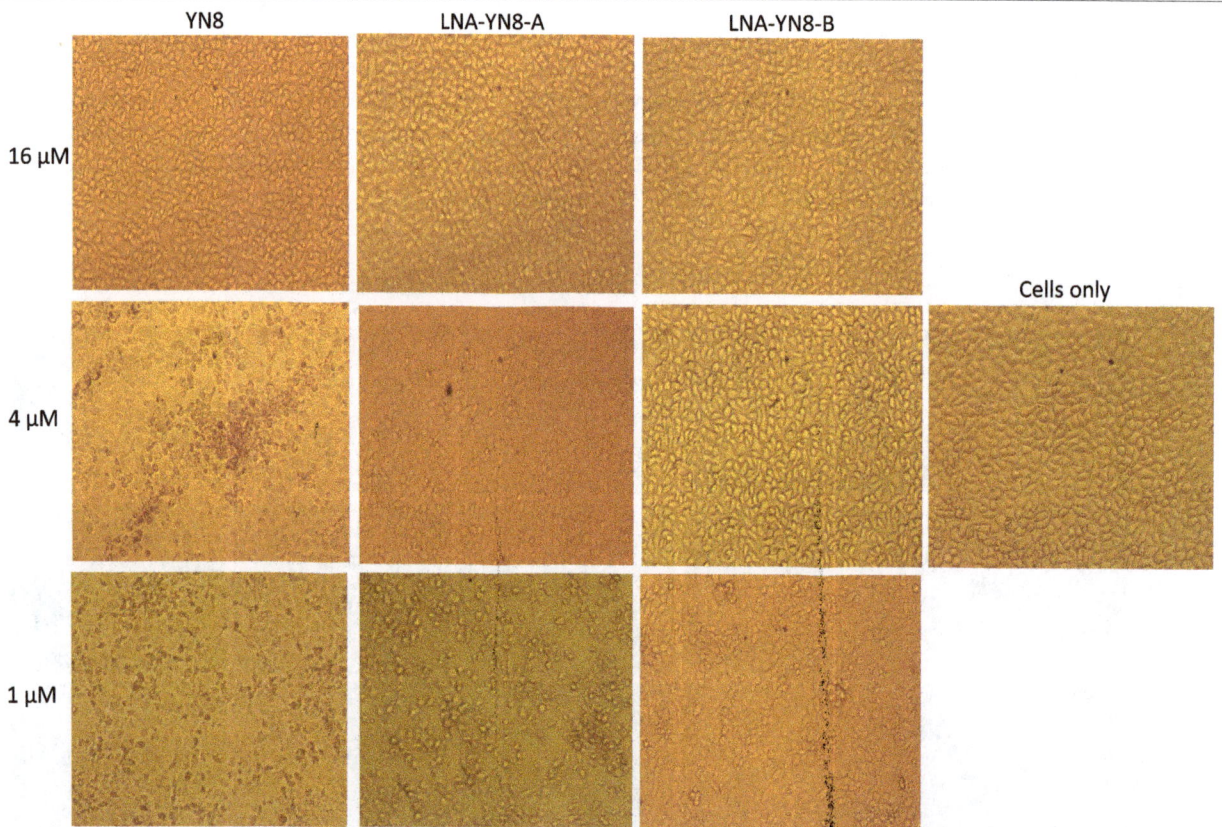

Fig. 3 Cytopathic effect (CPE) analysis of Marc-145 cells transfected with AS-ONs containing LNA modifications. Compared with the mock treatment (bottom panel), transfection with 16 μM DNA antisense oligonucleotide YN8 or 4 μM LNA-YN8-A and LNA-YN8-B showed significant protection of Marc-145 cells from PRRSV infection, while transfection with lower concentrations (4 or 1 μM of YN8, 1 μM of LNA-YN8-A or LNA-YN8-B) did not. Compared with the normal cells (cells only, neither PRRSV inoculation nor transfection was applied), the cells in the YN8 wells (4 or 1 μM) and 1 μM LNA-YN8-A (or LNA-YN8-B) wells aggregated, rounded up and detached from the monolayer, while the cells transfected with YN8 (16 μM) or LNA modified YN8 (16 or 4 μM) manifested overtly less CPE. Pictures were taken 72 h post infection with a Nikon E5400 camera mounted on an inverted microscope (Nikon TS100). Shown here are the representative images from three independent experiments

Antisense oligonucleotides containing LNAs showed lower cytotoxicity

The cell viability was assayed 72 h post transfection without infection using CCK-8 kit (Sigma-Aldrich, Cat. No. 96992) according to the manufacturer's instructions. The cell viability of cell control with mock transfection but without PRRSV challenge was set as 100% for normalization and comparison. Compared with the cell control, both DNA AS-ON (YN8) and the AS-ONs with LNA modification (LNA-YN8-A and LNA-YN8-B) showed dose-dependent cytotoxicity. The data shown in Fig. 2 demonstrated that LNA modification conferred lower cytotoxicity when compared with the corresponding DNA AS-ON YN8. The cell viability from transfection with approximate 64 μM of LNA-YN8-A or LNA-YN8-B was similar with that from 48 μM YN8. No difference in cell viability was observed between the two

LNA modified antisense sequences. The cell viability data implied that the antisense oligonucleotides with LNA modifications do not show obvious interference with cell viability with the concentration below 64 μM. Therefore the working concentrations of AS-ONs used in this study would not be toxic to the cells, excluding the possibility of inhibitory effects due to the toxicity of different AS-ONs.

LNA modification protected Marc-145 cells from CPE induced by PRRSV infection

To investigate whether AS-ONs with LNA modifications can further protect Marc-145 cells from the cytopathic effect (CPE) induced by PRRSV, one and half hours post challenge with 25 TCID$_{50}$ PRRSV, Marc-145 cells were transfected with LNA-YN8-A and LNA-YN8-B at different concentrations (16 μM, 4 μM and 1 μM). Transfection

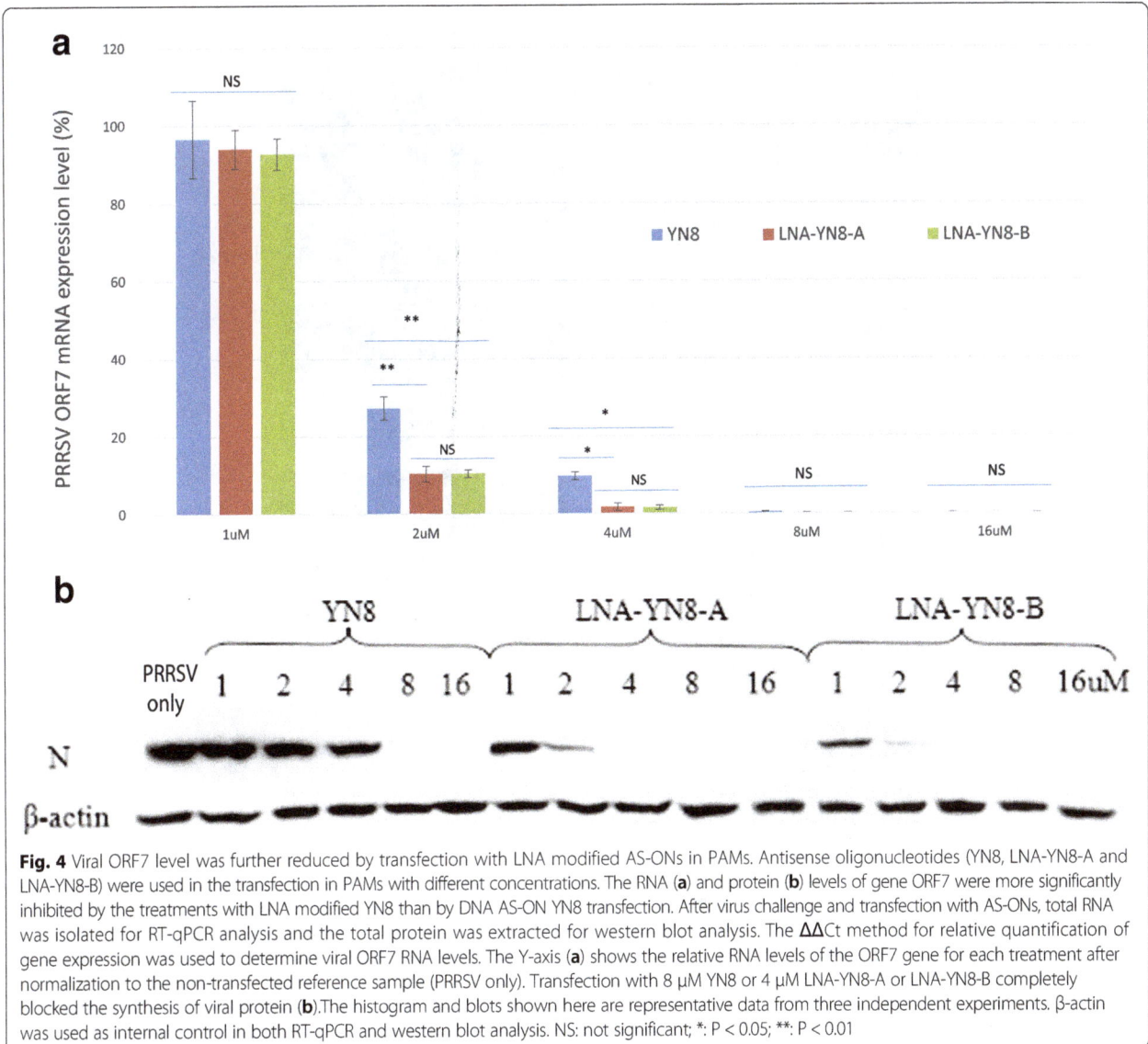

Fig. 4 Viral ORF7 level was further reduced by transfection with LNA modified AS-ONs in PAMs. Antisense oligonucleotides (YN8, LNA-YN8-A and LNA-YN8-B) were used in the transfection in PAMs with different concentrations. The RNA (a) and protein (b) levels of gene ORF7 were more significantly inhibited by the treatments with LNA modified YN8 than by DNA AS-ON YN8 transfection. After virus challenge and transfection with AS-ONs, total RNA was isolated for RT-qPCR analysis and the total protein was extracted for western blot analysis. The ΔΔCt method for relative quantification of gene expression was used to determine viral ORF7 RNA levels. The Y-axis (a) shows the relative RNA levels of the ORF7 gene for each treatment after normalization to the non-transfected reference sample (PRRSV only). Transfection with 8 μM YN8 or 4 μM LNA-YN8-A or LNA-YN8-B completely blocked the synthesis of viral protein (b).The histogram and blots shown here are representative data from three independent experiments. β-actin was used as internal control in both RT-qPCR and western blot analysis. NS: not significant; *: P < 0.05; **: P < 0.01

with YN8 and cells free of PRRSV infection and transfection (cells only) served as the controls for comparison. The CPE image for each treatment shown in Fig. 3 was acquired 72 h post transfection.

In Fig. 3, no significant cytotoxic effects were observed in the mock well or in the transfection well with YN8 (16 μM), LNA-YN8-A or LNA-YN8-B (16 or 4 μM for both LNA modified AS-ONs). However, the CPE was manifested in the transfection wells with YN8 (4 or 1 μM), LNA-YN8-A or LNA-YN8-B (1 μM). The data here indicates that 16 μM of DNA AS-ONs is protective for Marc-145 cells from PRRSV infection, which brought the working concentration (32 μM) [5] down to half (16 μM), while the inhibitory dose for AS-ON with LNA modification (LNA-YN8-A and LNA-YN8-B) is 4 μM. The CPE data suggested that the antisense oligonucleotides with LNA modifications were more potent in inhibiting PRRSV replication in Marc-145 cells.

Antisense oligonucleotides containing LNAs further inhibited PRRSV replication in PAMs

To investigate whether AS-ONs with LNA modification could further inhibit PRRSV replication in porcine alveolar macrophages (PAMs), which are the targets of PRRSV in the porcine lung, PAMs were infected with PRRSV in vitro and 90 min later transfected by the two AS-ONs containing LNA modifications at different doses (1 μM, 2 μM, 4 μM, 8 μM and 16 μM). PAMs transfected with DNA AS-ON YN8 at the same concentrations and challenged with PRRSV (25 $TCID_{50}$/well) were used as the control. Total RNA and protein was extracted from each treatment well 60 h post infection and subjected to RT-qPCR and western blot analysis, respectively. When normalized to β-actin for both mRNA and protein levels, the ORF7 RNA (Fig. 4a) and protein (Fig. 4b) levels in the PAMs transfected with any of the three AS-ONs were reduced in a dose dependent manner. Furthermore, Fig. 4

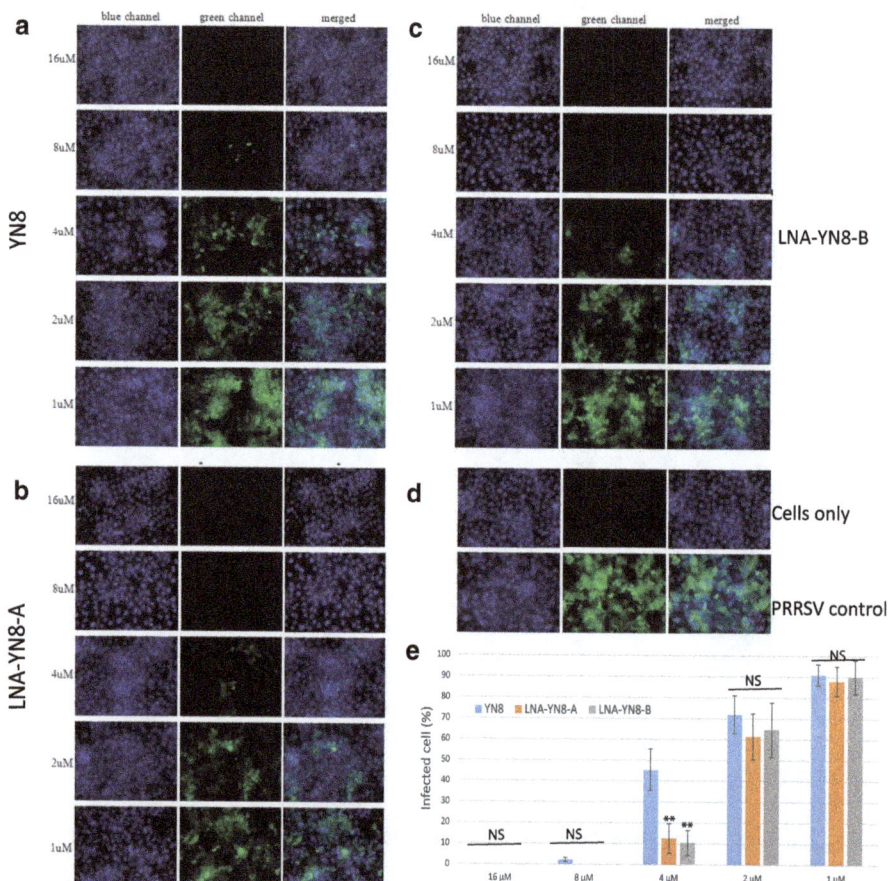

Fig. 5 Indirect immunofluorescence detection of PRRSV in Marc-145 cells transfected with antisense oligonucleotides YN8 (a), LNA-YN8-A (b) or LNA-YN8-B (c). Cells with neither infection nor transfection and cells with infection but without transfection (d) serve as controls. Transfection was performed 90 min post infection with PRRSV YN-1 strain (25 $TCID_{50}$). Cells were fixed 60 h post transfection. Anti-N monoclonal antibody (mAb) and Alexa Fluo488 conjugated secondary antibody were applied in the indirect immunofluorescence staining. Blue color stands for the nuclei and the green color indicates the expression of N protein in Marc-145 cells. The images of the same treatment from the two channels were merged. Treatment with LNA-YN8-A or LNA-YN8-B resulted in further reduction of virus replication, compared with the antisense oligonucleotide YN8 without LNA modifications. These images are representative for 3 independent experiments. Quantification data with statistical analysis of a, b and c was shown in e

showed that transfection with YN8 at 8 μM or LNA-YN8-A or LNA-YN8-B at 4 μM completely abrogated the PRRSV replications in PAMs and that transfection with only half of the amount (4 μM) of AS-ON YN8 containing LNA modification could achieve the same protective effects as with 8 μM DNA AS-ON YN8. In addition, LNA-YN8-A and LNA-YN8-B at 2 μM still showed moderate inhibition effects on PRRSV replication.

LNA modification conferred additional reduction of viral protein

To investigate the extra effect of AS-ONs containing LNA modification on inhibiting the expression of viral protein, indirect immunofluorescence assays were performed with anti-N protein mAb (encoded by gene ORF7) 60 h post transfection in Marc-145 cells with series dilution of AS-ONs (1, 2, 4, 8 and 16 μM). Indeed, as shown in Fig. 5, fewer fluorescing cells were seen in the monolayers treated with the two LNA modified AS-ONs than in the monolayers treated with the same amount of DNA AS-ON, and LNA-YN8-A or LNA-YN8-B at 4 μM displayed

similar inhibition on viral N protein synthesis as the counterpart YN8 did at 8 μM, which is concordance with the RT-qPCR and western blot data from PAMs (Fig. 4).

Reduction of viral titer by transfection with antisense oligonucleotides with LNA modifications

Change of viral titer is one of the most direct and convincing parameters in anti-virus research. In order to further determine the level of inhibition, PRRSV YN-1 strain was diluted by 1:10 and added into Marc-145 cells. Transfection with 8 μM of the three AS-ONs was performed 90 min post infection, respectively. CPE was monitored until 4 days post virus infection and transfection. Viral titers were measured by $TCID_{50}$ assay. We found that compared to the DNA AS-ON YN8 group, transfection with antisense oligonucleotide sequences containing LNA modifications (LNA-YN8-A or LNA-YN8-B) could more significantly protect Marc-145 cells from cytopathic effects (Fig. 6a) and reduced the viral titer by an extra 10-fold (Fig. 6b).

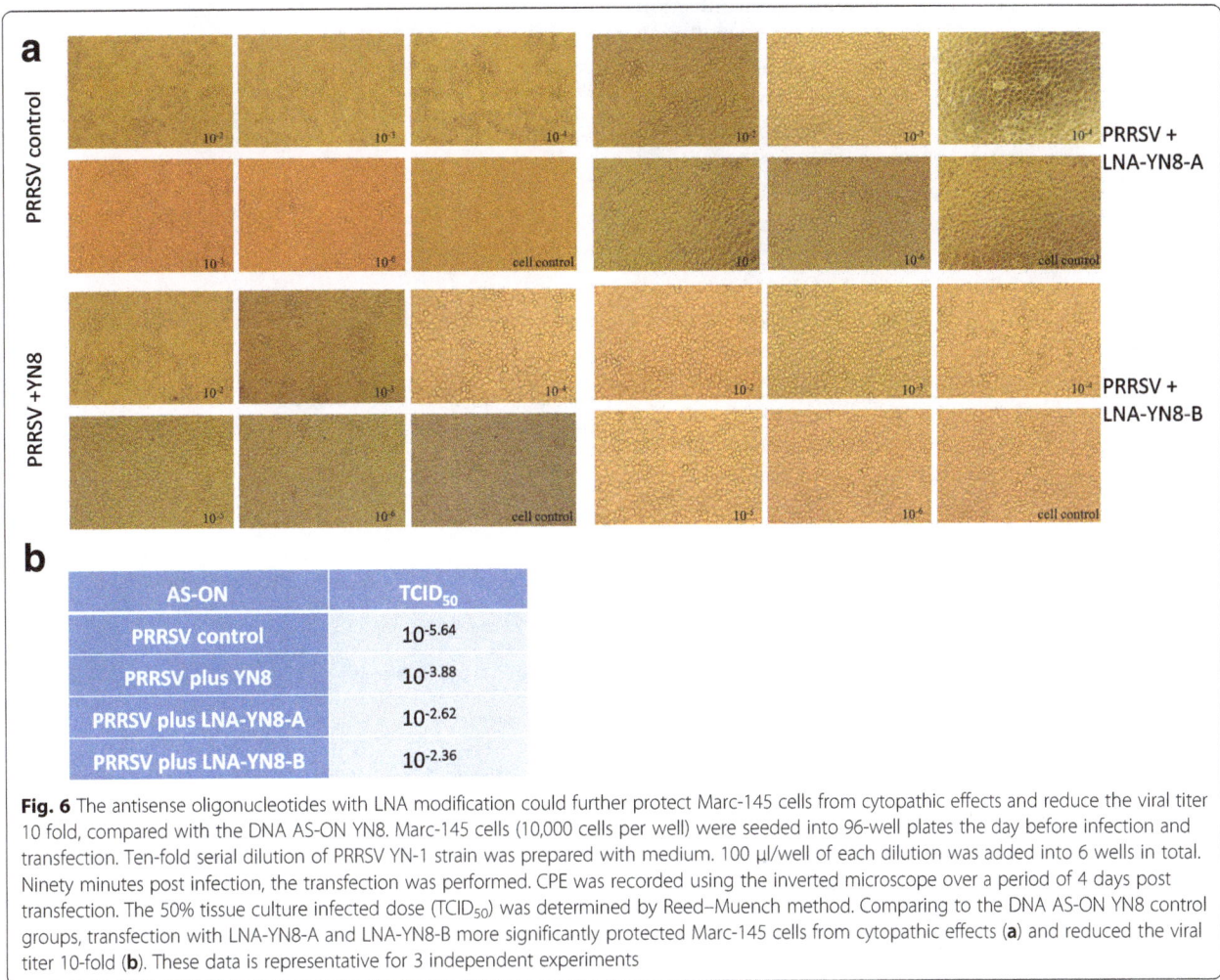

AS-ON	$TCID_{50}$
PRRSV control	$10^{-5.64}$
PRRSV plus YN8	$10^{-3.88}$
PRRSV plus LNA-YN8-A	$10^{-2.62}$
PRRSV plus LNA-YN8-B	$10^{-2.36}$

Fig. 6 The antisense oligonucleotides with LNA modification could further protect Marc-145 cells from cytopathic effects and reduce the viral titer 10 fold, compared with the DNA AS-ON YN8. Marc-145 cells (10,000 cells per well) were seeded into 96-well plates the day before infection and transfection. Ten-fold serial dilution of PRRSV YN-1 strain was prepared with medium. 100 μl/well of each dilution was added into 6 wells in total. Ninety minutes post infection, the transfection was performed. CPE was recorded using the inverted microscope over a period of 4 days post transfection. The 50% tissue culture infected dose ($TCID_{50}$) was determined by Reed–Muench method. Comparing to the DNA AS-ON YN8 control groups, transfection with LNA-YN8-A and LNA-YN8-B more significantly protected Marc-145 cells from cytopathic effects (a) and reduced the viral titer 10-fold (b). These data is representative for 3 independent experiments

Discussion and conclusion

Antisense therapy technology has been flourishing as a powerful therapeutic approach in the last 30 years with high expectations, exemplified by Merck's acquisition of Sirna Therapeutics in 2006. More than 130 clinical trials are listed on https://clinicaltrials.gov/ (access data January 25th, 2017), with three approved antisense drugs (Vitravene, Kynamro and Macugen) and several others under consideration for market approval [30]. Among the efforts to improve the efficacy of antisense oligonucleotides, LNA modification holds one of the most promising technologies in antisense oligonucleotide therapeutics, as LNAs show higher thermostability and hybridization abilities, DNA nuclease resistance, activation of RNase H activity, solubility and penetration into the cells, lower cytotoxicity and easy synthesis [6, 7, 31]. Among the different designs of LNA modifications in antisense oligonucleotides, a number of studies revealed that mix-mers and gap-mers showed the best target degrading effects [16, 32, 33].

PRRSV is recognized as one of the most important viruses for the swine industry, mainly due to its persistence in pigs for quite a long time after initial infection. In our previous study [5], we performed a small-scale screening with nine DNA based AS-ONs as candidates and found YN8 to be the best sequence. To explore the advantages of LNA, in this study we incorporated LNA modifications into the YN8 sequence as gap-mer (LNA-YN8-A) and mix-mer (LNA-YN8-B) and tested the inhibitory effects of these two AS-ONs in cell cultures compared with the corresponding DNA based AS-ON YN8. The Marc-145 cells or PAMs were challenged with PRRSV YN-1 strain and transfection with the three AS-ONs, respectively, followed by lifetime analysis, cytotoxicity measurement, CPE observation, RT-qPCR, western blot analysis and TCID$_{50}$ determination.

Taken together, under the experimental conditions in this study, we demonstrated that compared with the YN8 sequence, both gap-mer and mix-mer with LNA modifications showed longer lifetime (Fig. 1, 20 hs vs 6 hs), lower cytotoxicity (Fig. 2, 64 μM vs 48 μM) and higher protection from CPE induced by PRRSV infection (Fig. 3, 4 μM vs 16 μM), and more potent inhibitory effects on PRRSV replication in Marc-145 cells and PAMs at both mRNA and protein levels (2–4 μM vs 8 μM, Figs. 4, 5 and 6) with viral titer reduced by 10 fold. No significant difference was observed between the mix-mer and gap-mer. To our knowledge the only studies with modified antisense oligonucleotides in PRRSV research used phosphorodiamidate morpholono oligomers (PPMOs) [34–37]. The researchers tested the inhibition of some PPMOs on PRRSV replication in CRL11171 cells and found that the optimal working concentration was approximately 16 μM. In comparison, the LNA modified oligomers we designed and tested in both Marc-145 cells and PAMs inhibited the PRRSV replication at lower concentrations (2–4 μM), which demonstrated the advantages of incorporation of LNAs into AS-ONs to inhibit virus replication.

When it comes to the actual in vivo application of AS-ONs in the therapeutic control of PRRSV as shown in recent reports [37–39], we plan to integrate the methylation of cytosine, phosphorothioate modification in the backbone and the locked nucleic acid (LNA) modifications to redesign the YN8 sequence followed by testing the inhibition of PRRSV replication in piglets by gymnotic delivery of antisense oligonucleotides [40]. We believe this kind of systematical investigation will pave the way to further address the issue of PRRS control.

Abbreviations

AS-ONs: Antisense oligonucleotides; CPE: Cytopathic effect; FBS: Fetal bovine serum; h: Hour; hpi: Hours post infection; LNA: Locked nuclei acid; NSP: Non-structural protein; ORF: Open reading frame; PAMs: Pulmonary alveolar macrophages; PRRS: Porcine reproductive and respiratory syndrome; PRRSV: Porcine reproductive and respiratory syndrome virus; qPCR: Quantitative PCR; TCID$_{50}$: 50% Tissue Culture Infective Dose; UTR: Untranslated region

Acknowledgments

The work was jointly supported by the National Natural Science Foundation of China (grant no. 31160509 and 31560705) and Key Laboratory of Veterinary Public Health of Higher Education of Yunnan Province. The authors want to thank David Woods (The British Institute, Stockholm, Sweden), Drs. Bernhard Schmierer and Jenna Persson (Karolinska Insitute, Sweden) for proofreading this manuscript.

Authors' contributions

YGF and LJP conceived and designed the study. ZLY, BJL, LWG and YGS performed the experiments. BJL and GG analyzed the data. YGF and LJP wrote the paper with contributions from BJL and GG. All authors read and approved the manuscript.

Competing interests

The authors declare that they have no competing interests.

Author details

[1]Department of Veterinary Medicine, College of Animal Science and Technology, Yunnan Agricultural University, Yunnan province, Kunming 650201, China. [2]Haikou Experimental Station/Hainan Key Laboratory of Banana Genetic Improvement, Chinese Academy of Tropical Agricultural Sciences, Haikou 570102, Hainan, People's Republic of China. [3]Department of Medical Biochemistry and Biophysics, Karolinska Institute, -17177 Stockholm, SE, Sweden. [4]Present address: Center for Animal Disease Control and Prevention, City, 675000, Yunnan province, Chuxiong, China. [5]Present address: College of Animal Science and Technology, Shanxi Agricultural University, Shanxi province, Taigu 030801, China.

References

1. Pejsak Z, Markowska-Daniel I. Losses due to porcine reproductive and respiratory syndrome in a large swine farm. Comp Immunol Microbiol Infect Dis. 1997;20(4):345–52.
2. Meulenberg Janneke JM. PRRSV, the virus. Vet Res. 2000;31:11–21.
3. Dea S, Gagnon CA, Mardassi H, Pirzadeh B, Rogan D. Current knowledge on the structural proteins of porcine reproductive and respiratory syndrome

(PRRS) virus: comparison of the north American and European isolates. Arch Virol. 2000;145(4):659–88.

4. Wang FX, Wen YJ, Yang BC, Liu Z, Shi XC, Leng X, Song N, Wu H, Chen LZ, Cheng SP. Role of non-structural protein 2 in the regulation of the replication of the porcine reproductive and respiratory syndrome virus in MARC-145 cells: effect of gene silencing and over expression. Vet Microbiol. 2012;161:58–65.

5. Zheng L, Li X, Zhu L, Li W, Bi J, Yang G, Yin G, Liu J. Inhibition of porcine reproductive and respiratory syndrome virus replication in vitro using DNA-based short antisense oligonucleotides. BMC Vet Res. 2015;11:199.

6. Koshkin AA, Rajwanshi VK, Wengel J. Novel convenient syntheses of LNA [2.2.1]bicyclo nucleosides. Tetrahedron Lett. 1998;39(24):4381–4.

7. Wahlestedt C, Salmi P, Good L, Kela J, Johnsson T, Hökfelt T, Broberger C, Porreca F, Lai J, Ren K, Ossipov M, Koshkin A, Jakobsen N, Skouv J, Oerum H, Jacobsen MH, Wengel J. Potent and nontoxic antisense oligonucleotides containing locked nucleic acids. Proc Natl Acad Sci U S A. 2000;97(10):5633–8.

8. Fesenko EE, Heydarov RN, Stepanova EV, Abramov ME, Chudinov AV, Zasedatelev AS, Mikhailovich VM. Microarray with LNA-probes for genotyping of polymorphic variants of Gilbert's syndrome gene UGT1A1(TA)n. Clin Chem Lab Med. 2013;51(6):1177–84.

9. Karmakar S, Hrdlicka PJ. DNA strands with alternating incorporations of LNA and 2'-O-(pyren-1-yl)methyluridine: SNP-discriminating RNA detection probes. Chem Sci. 2013;4(9):3447–54.

10. Liu JP, Guerasimova A, Schwartz R, Lange M, Lehrach H, Nyársik L, Janitz M. LNA-modified Oligodeoxynucleotide hybridization with DNA microarrays printed on Nanoporous membrane slides. Comb Chem High Throughput Screen. 2006;9:591–7.

11. Liu JP, Drungowski M, Nyársik L, Schwartz R, Lehrach H, Herwig R, Janitz M. Oligonucleotide fingerprinting of arrayed genomic DNA sequences using LNA-modified hybridization probes. Comb Chem High Throughput Screen. 2007;10:269–76.

12. Karkare S, Bhatnagar D. Promising nucleic acid analogs and mimics: characteristic features and applications of PNA, LNA, and morpholino. Appl Microbiol Biotechnol. 2006;71(5):575–86.

13. Kakiuchi-Kiyota S, Whiteley LO, Ryan AM, Mathialagan N. Development of a method for profiling protein interactions with LNA-modified antisense oligonucleotides using protein microarrays. Nucleic Acid Ther. 2016;26(2):93–101.

14. Latorra D, Campbell K, Wolter A, Hurley JM. Enhanced allele-specific PCR discrimination in SNP genotyping using 3' locked nucleic acid (LNA) primers. Hum Mutat. 2003;22(1):79–85.

15. Vilas Boas D, Almeida C, Sillankorva S, Nicolau A, Azeredo J, Azevedo NF. Discrimination of bacteriophage infected cells using locked nucleic acid fluorescent in situ hybridization (LNA-FISH). Biofouling. 2016;32(2):179–90.

16. Jepsen JS, Sørensen MD, Wengel J. Locked nucleic acid: a potent nucleic acid analog in therapeutics and biotechnology. Oligonucleotides. 2004;14(2):130–46.

17. Letertre C, Perelle S, Dilasser F, Arar K, Fach P. Evaluation of the performance of LNA and MGB probes in 5'-nuclease PCR assays. Mol Cell Probes. 2003;17:307–11.

18. Palmano S, Mulholland V, Kenyon D, Saddler GS, Jeffries C. Diagnosis of Phytoplasmas by real-time PCR using locked nucleic acid (LNA) probes. Methods Mol Biol. 2015;1302:113–22.

19. Subramanian N, Kanwar JR, Kanwar RK, Krishnakumar S. Targeting Cancer cells using LNA-modified aptamer-siRNA chimeras. Nucleic Acid Ther. 2015;25(6):317–22.

20. Nielsen BS, Møller T, Holmstrøm K. Chromogen detection of microRNA in frozen clinical tissue samples using LNA™ probe technology. Methods Mol Biol. 2014;1211:77–84.

21. Guenther DC, Kumar P, Anderson BA, Hrdlicka PJ. C5-amino acid functionalized LNA: positively poised for antisense applications. Chem Commun (Camb). 2014;50(64):9007–9.

22. Tian K, Yu X, Zhao T, Feng Y, Cao Z, Wang C, Hu Y, Chen X, Hu D, Tian X, Liu D, Zhang S, Deng X, Ding Y, Yang L, Zhang Y, Xiao H, Qiao M, Wang B, Hou L, Wang X, Yang X, Kang L, Sun M, Jin P, Wang S, Kitamura Y, Yan J, Gao GF. Emergence of fatal PRRSV variants: unparalleled outbreaks of atypical PRRS in China and molecular dissection of the unique hallmark. PLoS One 2007;13;2(6):e526.

23. Zhou L, Yang HC. Porcine reproductive and respiratory syndrome in China. Virus Res. 2010;154(1–2):31–7.

24. Bo X, Lou S, Sun D, Shu W, Yang J, Wang S. Selection of antisense oligonucleotides based on multiple predicted target mRNA structures. BMC Bioinformatics. 2006;7:122.

25. Mathews DH. RNA secondary structure analysis using RNAstructure. Curr Protoc Bioinformatics. 2006;Chapter 12:Unit 12.6.

26. Reuter JS, Mathews DH. RNAstructure: software for RNA secondary structure prediction and analysis. BMC Bioinformatics. 2010;11:129.

27. Pfaffl MW. A new mathematical model for relative quantification in real time RT-PCR. Nucleic Acids Res. 2001;29:2002–7. ABI PRISM 7700 Sequence Detection System. 1999. User Bulletin # 2

28. Elayadi AN, Braasch DA, Corey DR. Implications of high-affinity hybridization by locked nucleic acid oligomers for inhibition of human telomerase. Biochemistry. 2002;41:9973–81.

29. Sazani P, Kole R. Therapeutic potential of antisense oligonucleotides as modulators of alternative splicing. J Clin Invest. 2003;112(4):481–6.

30. Aartsma-Rus A. New momentum for the field of oligonucleotide therapeutics. Mol Ther. 2016;24(2):193–4.

31. Kurreck J, Wyszko E, Gillen C, Erdmann VA. Design of antisense oligonucleotides stabilized by locked nucleic acids. Nucleic Acids Res. 2002;30(9):1911–8.

32. Fluiter K, ten Asbroek AL, de Wissel MB, Jakobs ME, Wissenbach M, Olsson H, Olsen O, Oerum H, Baas F. In vivo tumor growth inhibition and biodistribution studies of locked nucleic acid (LNA) antisense oligonucleotides. Nucleic Acids Res. 2003;31(3):953–62.

33. Jepsen JS, Pfundheller HM, Lykkesfeldt AE. Down-regulation of p21 (WAF1/CIP1) and estrogen receptor α in MCF-7 cells by antisense oligonucleotides containing locked nucleic acid (LNA). Oligonucleotides. 2004;14(2):147–56.

34. Zhang YJ, Stein DA, Fan SM, Wang KY, Kroeker AD, Xj M, Iversen PL, Matson DO. Suppression of porcine reproductive and respiratory syndrome virus replication by morpholino antisense oligomers. Vet Microbiol. 2006;117(2–4):117–29.

35. Patel D, Opriessnig T, Stein DA, Halbur PG, Meng XJ, Iversen PL, Zhang YJ. Peptide-conjugated morpholino oligomers inhibit porcine reproductive and respiratory syndrome virus replication. Antivir Res. 2008;77(2):95–107.

36. Han X, Fan S, Patel D, Zhang YJ. Enhanced inhibition of porcine reproductive and respiratory syndrome virus replication by combination of morpholino oligomers. Antivir Res. 2009;82(1):59–66.

37. Opriessnig T, Patel D, Wang R, Halbur PG, Meng XJ, Stein DA, Zhang YJ. Inhibition of porcine reproductive and respiratory syndrome virus infection in piglets by a peptide-conjugated morpholino oligomer. Antivir Res. 2011;91(1):36–42.

38. Stein CA, Hansen JB, Lai J, Wu SJ, Voskresenskiy A, Høg A, Worm J, Hedtja M, Souleimanian N, Miller P, Soifer HS, Castanotto D, Benimetskaya L, Ørum H, Koch T. Efficient gene silencing by delivery of locked nucleic acid antisense oligonucleotides, unassisted by transfection reagents. Nucleic Acids Res. 2010;38(1):e3.

39. Torres AG, Threlfall RN, Gait MJ. Potent and sustained cellular inhibition of miR-122 by lysine-derivatized peptide nucleic acids (PNA) and phosphorothioate locked nucleic acid (LNA)/2'-O-methyl (OMe) mixmer anti-miRs in the absence of transfection agents. Artif DNA PNA XNA. 2011;2(3):71–8.

40. Soifer HS, Koch T, Lai J, Hansen B, Hoeg A, Oerum H, Stein CA. Silencing of gene expression by Gymnotic delivery of antisense oligonucleotides. Methods Mol Biol. 2012;815:333–46.

23

Evaluation of the predictive value of tonsil examination by bacteriological culture for detecting positive lung colonization status of nursery pigs exposed to *Actinobacillus pleuropneumoniae* by experimental aerosol infection

Doris Hoeltig[1]* (iD), Florian Nietfeld[3], Katrin Strutzberg-Minder[4] and Judith Rohde[2]

Abstract

Background: *Actinobacillus (A.) pleuropneumoniae* is the causative agent of porcine pleuropneumonia. For control of the disease the detection of sub-clinically infected pigs is of major importance to avoid transmitting of subclinical infections. One method recommended is the testing of tonsillar samples for the presence of *A. pleuropneumoniae*. This is routinely done by PCR techniques. However, based upon PCR susceptibility testing and monitoring of resistance development is impossible. Therefore, in this study the informative values of bacteriological culture of tonsilar samples for the colonisation status of pigs were tested. In total, 163 German Landrace nursery pigs were experimentally exposed to *A. pleuropneumoniae* serotype 7 by aerosol and the rate of isolation from lung tissue and tonsils and the corresponding degree of lung lesions were investigated.

Results: Overall a significant correlation ($p < 0.001$) between degree of clinical disease, degree of lung alterations and degree of *A. pleuropneumoniae* isolation from tonsillar and lung tissue after exposure was detected. Of these animals tested, 74.8% were tested positive in tonsillar and lung samples, 7.4% remained completely negative and in 4.3% the tonsils were tested positive despite negative isolation results from lung tissue. In 13.5% of the pigs *A. pleuropneumoniae* could be isolated in lung tissue but not in tonsillar samples. In 36.4% of these animals a heavy colonization of the lungs and in 40.9% moderate to severe lung alterations were proven. Hence, the diagnostic sensitivity for the detection of a positive colonization status of the pigs by bacterial culture examination of tonsillar samples was 84.7%, the diagnostic specificity was 66.7% and the predictive values were 94.6% (positive) and 35.3% (negative). The overall sensitivity for *A. pleuropneumoniae* exposure was 78.2% (tonsils) and 88.0% (lung tissue).

Conclusions: In conclusion, tonsil examination alone for the detection of a positive colonization status of pigs performed might lead to false negative results as lungs might be heavily colonized despite negative tonsillar isolation results. Therefore culture of tonsillar samples should not be the sole test for the confirmation of a pigs' status but used in combination with methods also evaluating the colonization status of the lower respiratory tract.

Keywords: Detection, Sensitivity, Specificity, *A. pleuropneumoniae*, Culture method

* Correspondence: doris.hoeltig@tiho-hannover.de
[1]Clinic for Swine and Small Ruminants, Forensic Medicine and Ambulatory Service, University of Veterinary Medicine, Foundation, Bischofsholer Damm 15, D-30173 Hannover, Germany
Full list of author information is available at the end of the article

Background

Actinobacillus (*A.*) *pleuropneumoniae* is the causative agent of porcine pleuropneumia, a respiratory disease of pigs causing severe economic losses in pork production worldwide. The infection of pigs can lead to subclinical infection without any clinical signs of disease, peracute death, acute respiratory disease which is often accompanied by high mortality and chronic disease leading to recurrent coughing and wasting of the animals [1]. In cases of subclinical infection, often only the oropharyngeal cavity of the pigs is colonized [2]. Apart from tonsils, *A. pleuropneumoniae* is also able to survive in lungs and the affected pigs may appear clinically healthy. Additionally, even antibiotic treatment is not able to clear *A. pleuropneumoniae* completely from these localizations [3]. These subclinically infected pigs are the major sources for spreading of the disease, either by transmission of the pathogen from the sows to their offspring or by pig-to-pig contact during trading of living animals [4]. Studies show that more than 30% of piglets from infected sows are positive for *A. pleuropneumoniae* at the time of weaning, increasing to 50% or more at the age of 10 weeks [5, 6].

Several methods are described for the detection of these carrier pigs. For herd screening different serological tests are used routinely. However, serology has some disadvantages as seroconversion generally occurs 10 to 14 days after infection [1]. Additionally, *A. pleuropneumoniae* can colonize the upper respiratory tract without inducing a seroconversion [7, 8]. Moreover, it is noteworthy that the presence of specific antibodies is not able to totally clear *A. pleuropneumoniae* totally from the upper respiratory tract of the colonized pigs [1]. In cases with inconclusive serological results or cases where *A. pleuropneumoniae* is expected to be present in very low numbers attempts for the direct detection of the pathogen are being made [6]. For identifying sub-clinically infected herds the direct detection of *A. pleuropneumoniae* from tonsils is recommended. This direct bacterial detection can be based on bacteriological culture [8, 9] or molecular detection of the pathogen by several PCR techniques [10, 11]. PCR techniques might seem to be more appropriate as they can also detect low numbers of pathogen DNA from viable and non-viable pathogens, whereas the isolation by bacteriological culture depends on the viability of the bacteria [10]. Another advantage is that PCR techniques do not bear the risk of overgrowth of the specific pathogen by the abundant microflora as the upper respiratory tract is heavily colonized by different bacterial species [12]. However, a major disadvantage of PCR detection, especially in the area of antimicrobial stewardship is that susceptibility testing is only possible from cultured bacteria so far. As an alternative for classical bacterial isolation, the immunomagnetic

isolation of *A. pleuropneumoniae* [13, 14] is still very expensive and time-consuming. Thus the classical isolation of *A. pleuropneumoniae* by bacteriological culture from tonsillar samples retains importance.

As the transfer of the pathogen can either be from the sow to the offspring or by horizontal pig-to-pig contact [4] in this study tonsillar samples from experimentally infected nursery pigs were examined as nursery pigs are an age group often regrouped with a high risk of pathogen transfer. The tonsillar samples were investigated regarding applicability and the informative value of bacterial culture evaluating the hypothesis that the examination of tonsillar samples is an adequate tool to securely determine the colonization status of pigs in case of unclear serological results or recently suspected exposure.

Methods

For this investigation tonsils and samples from lung tissue were taken from 163 pigs experimentally infected with *A. pleuropneumoniae* serotype 7. All pigs belonged to a commercial German Landrace breeding line, vaccinated against *Mycoplasma hyopneumoniae* and PCV-2 and were delivered from the same piglet producing farm considered negative for *A. pleuropneumoniae*. Regarding sex 52 pigs were female and 111 pigs were male-castrated and all pigs were tail docked and considered clinically healthy on the day of delivery. The pigs were kept under standardized level 2 conditions in accordance with the Guidelines for Protection of Vertebrate Animals used for experimental and other Scientific Purposes, European Treaty Series, nos. 123 /170 (http://conventions.coe.int/treaty/EN/treaties/html/123.htm; http://conventions.coe.int/treaty/EN/treaties/html/170.htm). Study design and housing of the animals were approved by a local, independent committee on ethics (Commission for ethical estimation of animal research studies of the Lower Saxonian State Office for Consumer Protection and Food Safety; approval number: 33.12–42,502–04-15/1962). The pigs were kept on concrete floor with 8m^2 per 8 to 10 pigs. An area of 2.8m^2 of each stable was covered with a rubber mat to provide a more comfortable bedding area. This area was additionally heated by two infrared lamps. Allocation to the housing groups was performed by simple randomization [15]. Ambient temperature was 27.5 °C ± 1.9 °C (MEAN ± SD) and humidity was 33.2% ± 12.5% (MEAN ± SD). The pigs were fed a commercial standardized diet and each housing group was provided 2.5 kg of hay flakes (°AGROBS Pre Alpin Wiesenflakes, Co. AGROBS, Degerndorf, Germany) per day as material for rooting and manipulation. Potable water was constantly available for free choice and of drinking water quality. The pigs arrived three weeks prior to infection for an adequate adaptation to diet, new environment and clinical examination procedures. The pigs were seven weeks of age at the time of exposure (SD:

±1.5 days). They had an average body weight of 12.0 kg (± 1.9 kg).

Experimental exposure

Prior to experimental infection all pigs underwent a general clinical examination as well as a radiographic (80 kV no-scatter grid and automatic exposure; Precimat, Co. Picker International, Munich, Germany) and sonographic (8 MHz linear scanner, LOGIQ™ Book XP, Co. GE Medical Systems, Chalfont St. Giles, Great Britain) examination of the lungs. Radiography was performed in two views (latero-lateral and dorso-ventral). The film-focus distance for the x-ray examination was 1.5 m. Images were recorded with 240×300 mm image plate cassettes (Co. EURAS MedTech, PROVOTEC, X-RAY, Espelkamp, Hannover, Germany) and readout linearly with a digital non-contact image storage system (iCR 3600®, Co. EURAS MedTech, PROVOTEC, X-RAY, Espelkamp, Hannover, Germany). Bronchoalveolar lavage fluids (100 ml 0.9% NaCl-solution ad us. vet., Co. WDT, Garbsen, Germany) as well as serum samples were taken. The bronchoalveolar lavage fluid was examined by bacteriological culture and PCR [16] for the absence of A. pleuropneumoniae. Bronchoalveolar Lavage, radiographic and sonographic examination were performed under general anesthesia with 20 mg/kg Ketamine i.m. (Ketamin 100 mg/ml®, Co. CP-Pharma, Burgdorf, Germany) and 2 mg/kg Azaperon i.m. (Stresnil®, Co. Janssen-Cilag GmbH, Baar, Switzerland). Neuroleptanalgesia based upon ketamine and azaperon was chosen as this is the only official licensed anesthesia protocol for pigs in Germany. Serum samples were analyzed by ApxIV-ELISA (IDEXX APP-ApxIV Ab Test®, Co. IDEXX Laboratories, Maine, USA). All pigs were confirmed as clinically healthy and negative for A. pleuropneumoniae by direct and indirect screening procedures. This study was conducted as an uncontrolled exposure study; the experimental unit was the individual animal. The pigs were infected in groups of five or six pigs; each housing group was divided into two exposure groups. The assignment to the different infection groups was based upon simple randomization. In total the infection was independently repeated for 28 groups of pigs. The experimental procedure was identical for each exposure group. The experimental infection was conducted as an aerosol exposure with a total exposure time of 30 min [17]. Approximately 1×10^5 bacteria of A. pleuropneumoniae AP76 serotype 7 were nebulized resulting in an aerosol concentration of 1×10^2 colony forming units (cfu) per liter aerosol. After the aerosol exposure the pigs were clinically monitored every second hour for the first 48 h after exposure, and thereafter twice a day. Rescue criteria were defined to minimize the suffering of the infected pigs. These rescue criteria were a breathing frequency > 70/min and cyanosis and open-mouth breathing, apathy with no reactions to stimulation and standing with head down without lying down and vomiting as well as foam around nostrils or mouth, a rectal temperature of > 42.0°C or < 37.5 °C, a body temperature > 40.3 °C in combination with a breathing frequency > 46/min and dyspnea as well as decreased feed intake for more than two consecutive days, a body temperature > 40.3 °C and decreased general condition in combination with a breathing frequency > 35/min for more than two consecutive days from day three after exposure; any non-predictable event, reaction to treatment or disease leading to a moderate degree of general condition reduction or pain for more than 48 h; any non-predictable event, reaction to treatment or disease leading to high degree of general condition reduction or pain.

Necropsy and bacteriological examination

Seven days after infection or earlier in case of withdrawal due to rescue criteria the pigs were euthanized by intravenous injection of 80 mg/kg pentobarbital (Euthadorm®, Co. CP Pharma GmbH, Burgdorf, Germany). A necropsy was performed and the degree of developed lung lesions was assessed using a lung lesion score [18]. Each lung lobe could reach a maximum score of 5 resulting in an overall maximum score of 35. Additionally, samples for the isolation of A. pleuropneumoniae were taken from eight different localizations. In total, seven lung tissue samples (approximately 1 cm^2) of defined positions located in the outer third of each of the seven lung lobes as well as the palatine tonsils were collected.

Samples were plated on A. pleuropneumoniae-selective blood agar [19] using the quadrant streaking method. Abundance of growth was assessed semi-quantitatively. Bacterial isolates were identified as A. pleuropneumoniae by PCR amplification of the apxIV gene [16].

The level of isolation from the lungs was translated into a isolation score [20]. The amount of growth for each localization (0 = no growth; 1 = sparse growth; 2 = moderate growth; 3 = heavy growth) was added and divided by the total number of lung tissue samples, results indicating the level of the isolation from lung tissue (0–1: low-grade; > 1–2: moderate, > 2–3: high-grade).

Statistical methods

The collected data were transferred to a database based upon Excel ® (Co. Microsoft Cooperation, Dublin, Irland). Verification was assured by double data entry. Statistical analyses were carried out using Excel® and IBM SPSS Statistics® (Co. IBM Deutschland GmbH, Ehningen, Germany). For all continuous variables, sample size, mean (m), standard deviation (SD), median, quartiles, minimum and maximum were calculated. Categorical variables were displayed as absolute and relative frequencies. The applied level of significance was 5% ($p < 0.05$). For correlation analysis Spearman Rank Correlation was calculated. Primary criterion for the evaluation was the bacteriological re-isolation score of the whole lung. Secondary criteria for the analysis were the re-isolation score from different lung lobes, the lung lesion score and the severity of clinical symptoms.

Results

After infection 39 of the 163 pigs developed no or only slight signs of acute porcine pleuropneumonia, 89 pigs showed moderate symptoms of disease and 35 pigs had to be euthanized prior to day 7 after exposure due to the severity of clinical signs and the defined rescue criteria. No overgrowth by accompanying microflora appeared and all bacteriological cultures could be evaluated.

The lung lesion score (LLS) was 0 for 14 animals and 35 for 12 animals. Regarding the isolation of *A. pleuropneumoniae* it could be isolated from the tonsils of 129 pigs (ts+) and from the lungs, it was isolated of 144 pigs (lg+). In 7 cases of isolation from the tonsils the lungs of the pigs remained negative (ts+/lg-). In case of isolation from the lungs was isolated to the highest degree from the caudal lobes (Table 1). Isolation was significantly higher from these lung lobes ($p < 0.0001$) compared to all other lung lobes. There was also a difference ($p = 0.004$) between the right cranial lung lobe and the right middle lobe of the lung with significantly (p = 0.004) higher level of isolation from the middle lobe compared to the cranial lobe of the right lung. In 12 pigs, *A. pleuropneumoniae* could not be isolated neither from the lungs nor from the tonsils (ts−/lg-). 19 pigs (ts+/lg-;$n = 7$ and ts−/lg-;$n = 12$) were euthanized at the end of the trial, none of these pigs had to be euthanized due to rescue criteria .

Only 84.7% of the 144 pigs carrying *A. pleuropneumoniae* in the lungs were detected by bacteriological examination of the tonsils leaving 22 pigs positive for *A. pleuropneumoniae* (ts−/lg+) undetected. From the 15.3% of the pigs where *A. pleuropneumoniae* was not detected

Table 1 Distribution of *A. pleuropneumoniae* colonization status by bacteriological culture and corresponding clinical, pathomorphological and bacteriological examination results

| | Detected *Actinobacillus pleuropneumoniae* status by bacteriological culture (n = 163) | | | |
	Tonsils: negative Lungs: negative	Tonsils: negative Lungs: positive	Tonsils: positive Lungs: negative	Tonsils: positive Lungs: positive
Total number of animals	12	22	7	122
Degree of clinical disease[a] *(number of animals)*				
- No disease	8	0	4	4
- Mild disease	4	3	3	13
- Moderate disease	0	13	0	76
- Severe Disease	0	6	0	29
Euthanized due to Rescue criteria *(number of animals)*	0	6	0	29
Lung lesions score[a] *(LLS; number of animals)*				
- No lesions	5	0	4	4
- LLS: 0.1–12.5	6	13	3	49
- LLS: 12.5–25	1	7	0	41
- LLS: > 25	0	2	0	28
Degree of isolation from lung tissue[a] *(number of animals)*				
- No isolation	12	0	7	0
- Low-grade	0	7	0	16
- moderate	0	7	0	51
- high-grade	0	8	0	55
Positive test results for different lung lobes (%)				
- Cranial lobes	0	76,2	0	93,0
- Middle lobes	0	76,2	0	88,5
- Caudal lobes	0	83,3	0	93,2
- Lobus accessorius	0	85,7	0	91,8

[a]significant correlation between lung lesion score, overall isolation from lungs and overall isolation from tonsils

by tonsil examination one pig was tested positive in one lung lobe, two pigs in two lung lobes and one pig in three lung lobes. Three pigs had positive isolation results in the samples from five lung lobes, five pigs in samples from six lung lobes and in 10 pigs all lung lobes were tested positive for A. *pleuropneumoniae* by bacteriological examination. Of these ts–/lg + pigs 31.8% had an isolation score ≤ 1, 27.3% had a score between > 1 and ≤ 2 and 40.9% a score > 2 and ≤ 3. Six animals where A. *pleuropneumoniae* was detected in the lungs but not on the tonsils were euthanized because of the severity of symptoms prior to day 7 post infection and 16 animals were euthanized at the end of the trial .

Comparing the lung lesion score and the isolation results, A. *pleuropneumoniae* could be isolated from the lungs of all pigs showing LLS of 35 (*n* = 12), but only from the tonsils of 10 of these pigs. Within the group of pigs showing a LLS of 0 (*n* = 14), A. *pleuropneumoniae* could be isolated from the tonsils of 8 pigs but only from the lungs of 3 pigs.

A statistically significant correlation between lung lesion score, isolation from the lungs and isolation from the tonsils was detected (*p* < 0.001). The sensitivity for the detection of a colonization status of the pig by bacteriological examination of the tonsils was 84.7%, the specificity was 63.2%, the positive predictive value was 94.6%, and the negative predictive value was 35.3% (Table 2). The rate of detection of the previous A. *pleuropneumoniae* exposure of the animals was 79.1% for the tonsil examination and 88.3% for the examination of the lung tissue.

Discussion

As the exposed pigs showed the typical spectrum of clinical signs of porcine pleuropneumonia including death, as well as subclinical infection without any clinical signs of disease, the exposure protocol can be considered successful. As the evaluation of all cultures from tonsillar samples was possible and no sample had to be excluded due to overgrowth by accompanying microflora this study shows that in contrast to the results of other studies [11, 13] the bacteriological culture using selective media [18] can be utilized for the detection of A.

pleuropneumoniae in tonsillar samples. Nevertheless, resulting isolates should be confirmed by PCR [15] to avoid an incorrect classification of bacteria that are highly related to A. *pleuropneumoniae* [12].

Our isolation results confirm the findings of previous studies [2] that after an A. *pleuropneumoniae* infection the pathogen may still be present on the tonsils despite being cleared from the lung tissue by the host defense mechanisms. However, we also showed that the opposite may be possible, A. *pleuropneumoniae* may colonize the lungs without a detectable colonization of the tonsillar crypts. That A. *pleuropneumoniae* was not detected in some tonsillar samples may be due to the amount of living bacteria within the samples being too low to gain a positive result by bacteriological culture, and perhaps may have been detected using PCR protocols [10]. However, PCR results would also not have given insight to the viability of the A. *pleuropneumoniae* isolates as it also detects non-viable pathogen DNA and an antibiotic sensitivity testing would not be possible. As these ts-lg + results not only appeared in pigs evaluated at the end of the trial but also in animals that had been euthanized due to the severity of symptoms we would expect to be able to re-isolate from tonsils. That this was not possible from some pigs suggests they present a risk of transferring the bacteria to other pigs. These pigs would have been detected only if the tonsillar examination recommended for the detection of sub-clinically infected pigs [6, 8, 9, 13] had been combined with diagnostic techniques also examining the status of the lower respiratory tract.

Under field condition, especially in cases of the examination of trading pigs alive, a sampling of seven lung tissue samples, one from each lung lobe, in most cases is not practicable. A method developed for the examination of the bacterial colonization of the lungs from living animals is the bronchoalveolar lavage [21]. However it should be taken into account that information on detection rates of A. *pleuropneumoniae* from bronchoalveolar lavage (BAL) in comparison to lung tissue is sparse. The level of isolation of A. *pleuropneumoniae* was highest for the diaphragmatic lobes and these lobes are the

Table 2 2 × 2 contingency table for statistical calculation of SEN, SP, PPV and NPV for the detection of a positive A. *pleuropneumoniae* lung colonization status by bacteriological culture examination of tonsilar samples (*n* = 163)

	Samples positive by lung tissue examination	Samples negative by lung tissue examination
Samples positive by tonsil examination	122 (true positive)	7 (false positive)
Samples negative by tonsil examination	22 (false negative)	12 (true negative)
Sensitivity (SEN)	true positive / (true positive + false negative) =	84.72%
Specificity (SP)	true negative / (true negative + false positive) =	63.16%
Positive predictive value (PPV)	true positive /(true positive + false positive)=	94.57%
Negative predictive value (NPV)	True negative / (true negative + false negative) =	35.29%

main localizations reached by BAL without visual control [21, 22]. Therefore, it is reasonable to assume that colonization of the lung would have been detected by culturing BAL.

The observed correlation between isolation from the tonsils, isolation from the lung tissue and degree of lung lesions might lead to the assumption that the examination of the tonsils would be representative for the status of the whole respiratory tract regarding A. pleuropneumoniae colonization. Nevertheless, the negative predictive value was quite low. However, as our results are based upon an experimental exposure where all pigs presumably had contact with A. pleuropneumoniae, the prevalence of A. pleuropneumoniae colonization in this population might be higher than in a field population of swine. If so, with a lower prevalence of A. pleuropneumoniae infected animals in a field population, the negative predictive value of the culture of tonsillar samples would increase while the positive predictive value would decrease [23–25]. Another fact that might change the results under field conditions is the involved serotype of A. pleuropneumoniae. In this study only one strain of A. pleuropneumoniae serotype 7 was tested. Different factors are involved in the adhesion of A. pleuropneumoniae to the porcine respiratory tract including fimbrial structures, lipopolysaccharides, outer membrane proteins as well as genes involved in biofilm formation. Appearance and molecular structure of these virulence factors can be different between different serotypes and strains of the same serotypes of A. pleuropneumoniae. Additionally, their involvement in the adhesion process has mainly been shown for the lower respiratory tract and information on virulence factors responsible for the colonization of the upper respiratory tract is scarce [26]. Nevertheless such differences in virulence factors of the involved serotypes may change the test results under field conditions as often several strains of A. pleuropneumoniae are present in the same herd [27] and may also colonize the respiratory tract of the same pig.

Conclusions

In conclusion, bacteriological culture of tonsillar samples can be a useful tool for the detection of A. pleuropneumoniae if information on viability and antibiotic sensitivity of the isolates is needed. However, especially in herds with high traffic of animals or when trading of animals with a SPF status, culture of tonsils should not be performed as the only method for the detection of sub-clinically infected pigs. In cases of unclear serological results or if the possible date of infection is considered less than 10 days, the culturing of the tonsils should not be performed solely for the status evaluation of the pigs due to the low estimated negative predictive value of the tonsillar examination and the overall low sensitivity of the tonsillar examination regarding previous A. pleuropneumoniae contact.

Abbreviations
A. pleuropneumoniae: Actinobacillus pleuropneumoniae; BAL: Bronchoalveolar lavage; Co: Company; ELISA: Enzyme-Linked Immunosorbent Assay; i.m.: intramuscular; kg: kilogram; kV: kilo voltage; lg: lung; LLS: Lung Lesion Score; mg: milligram; min: minutes; ml: milliliter; NPV: Negative predictive value; p.inf.: post infectionem; PCR: Polymerase chain reaction; PPV: Positive predictive value; SEN: Sensitivity; SPF: Specific pathogen free; ts: Tonsillar

Acknowledgements
The authors would like to thank Karl-Heinz Waldmann (Clinic for Swine and Small Ruminants, University of Veterinary Medicine Hannover, Foundation, Hannover, Germany) for advice and support during the clinical studies.

Funding
Parts of this study were supported by the German Ministry of Agriculture (BLE) and the German Annuity Bank (Deutsche Rentenbank).

Authors' contributions
DH conceived, designed, coordinated and carried out the clinical studies, prepared and carried out the aerosol infection, performed the necropsies, participated in the re-isolation procedures, performed the statistical analysis and drafted the manuscript. FN participated in the clinical studies and necropsies, participated in the re-isolation work and helped to draft the manuscript. KSM carried out and interpreted the serological and PCR testing of all pigs entering the study and helped to draft the manuscript. JR participated in the design of the study, carried out the re-isolation of the bacterium from the post mortem samples of all animals and helped to draft the manuscript. All authors have read and approved the manuscript.

Consent for publication
Not applicable.

Author details
[1]Clinic for Swine and Small Ruminants, Forensic Medicine and Ambulatory Service, University of Veterinary Medicine, Foundation, Bischofsholer Damm 15, D-30173 Hannover, Germany. [2]Institute for Microbiology, University of Veterinary Medicine, Foundation, Bischofsholer Damm 15, D-30173 Hannover, Germany. [3]Clinic for Swine, Department of Veterinary Medicine, Justus-Liebig-University Giessen, Frankfurter Str. 112, D- 35392 Giessen, Germany. [4]Innovative Veterinary Diagnostics (IVD-GmbH), Albert-Einstein-Str. 5, 30926 Seelze, Germany.

References
1. Gottschalk M: Actinobacillosis. Disease of Swine, 10th ed(Straw, BE, Zimmerman, JJ, D'Allaire, S and Taylor, eds), Wiley-Blackwell, Oxford, UK 2012:653–669.
2. Tobias T, Klinkenberg D, Bouma A, Van Den Broek J, Daemen A, Wagenaar J, Stegeman J. A cohort study on Actinobacillus pleuropneumoniae colonisation in suckling piglets. Preventive veterinary medicine. 2014;114(3):223–30.
3. Angen Ø, Andreasen M, Nielsen E, Stockmarr A, Baekbo P. Effect of tulathromycin on the carrier status of Actinobacillus pleuropneumoniae serotype 2 in the tonsils of pigs. The Veterinary record. 2008;163(15):445-7.
4. Tobias T. Actinobacillus pleuropneumoniae transmission and clinical outbreaks: Utrecht University; 2014.

5.   Klinkenberg D, Tobias T, Bouma A, Van Leengoed L, Stegeman J. Simulation study of the mechanisms underlying outbreaks of clinical disease caused by Actinobacillus pleuropneumoniae in finishing pigs. Vet J. 2014;202(1):99–105.

6.   Gottschalk M. The challenge of detecting herds sub-clinically infected with Actinobacillus pleuropneumoniae. Vet J. 2015;206(1):30–8.

7.   Kume K, Nakai T, Sawata A. Isolation of Haemophilus pleuropneumoniae from the nasal cavities of healthy pigs. Nihon Juigaku Zasshi. 1984;46(5): 641–7.

8.   Sidibe M, Messier S, Lariviere S, Gottschalk M, Mittal KR. Detection of Actinobacillus pleuropneumoniae in the porcine upper respiratory tract as a complement to serological tests. Can J Vet Res. 1993;57(3):204.

9.   Chiers K, Donné E, Van Overbeke I, Ducatelle R, Haesebrouck F. Evaluation of serology, bacteriological isolation and polymerase chain reaction for the detection of pigs carrying Actinobacillus pleuropneumoniae in the upper respiratory tract after experimental infection. Vet Microbiol. 2002;88(4):385–92.

10. Gram T, Ahrens P, Nielsen J. Evaluation of a PCR for detection of Actinobacillus pleuropneumoniae in mixed bacterial cultures from tonsils. Vet Microbiol. 1996;51(1–2):95–104.

11. Chiers K, Van Overbeke I, Donné E, Baele M, Ducatelle R, De Baere T, Haesebrouck F. Detection of Actinobacillus pleuropneumoniae in cultures from nasal and tonsillar swabs of pigs by a PCR assay based on the nucleotide sequence of a dsbE-like gene. Vet Microbiol. 2001;83(2):147–59.

12. Gottschalk M, Broes A, Mittal K, Kobisch M, Kuhnert P, Lebrun A, Frey J. Non-pathogenic Actinobacillus isolates antigenically and biochemically similar to Actinobacillus pleuropneumoniae: a novel species? Vet Microbiol. 2003; 92(1):87–101.

13. Gagné A, Lacouture S, Broes A, D'Allaire S, Gottschalk M. Development of an immunomagnetic method for selective isolation of Actinobacillus pleuropneumoniae serotype 1 from tonsils. J Clin Microbiol. 1998;36(1):251–4.

14. Angen Ø, Heegaard PM, Lavritsen DT, Sørensen V. Isolation of Actinobacillus pleuropneumoniae serotype 2 by immunomagnetic separation. Vet Microbiol. 2001;79(1):19–29.

15. Suresh KP. An overview of randomization techniques: an unbiased assessment of outcome in clinical research. J Hum Reprod Sci. 2011;4(1):8–11.

16. Frey J. Detection, identification, and subtyping of Actinobacillus pleuropneumoniae. PCR Detection of Microbial Pathogens. 2003:87–95.

17. Jacobsen M, Nielsen J, Nielsen R. Comparison of virulence of different Actinobacillus pleuropneumoniae serotypes and biotypes using an aerosol infection model. Vet Microbiol. 1996;49:159–68.

18. Hannan P, Bhogal B, Fish J. Tylosin tartrate and tiamutilin effects on experimental piglet pneumonia induced with pneumonic pig lung homogenate containing mycoplasmas, bacteria and viruses. Res Vet Sci. 1982;33:76–88.

19. Jacobsen MJ, Nielsen JP. Development and evaluation of a selective and indicative medium for isolation of Actinobacillus pleuropneumoniae from tonsils. Vet Microbiol. 1995;47(1–2):191–7.

20. Maas A, Jacobsen ID, Meens J, Gerlach G-F. Use of an Actinobacillus pleuropneumoniae multiple mutant as a vaccine that allows differentiation of vaccinated and infected animals. Infect Immun. 2006;74(7):4124–32.

21. van Leengoed LA, Kamp EM. A method for bronchoalveolar lavage in live pigs. Vet Q. 1989;11(2):65–72.

22. Moorkamp L, Nathues H, Spergser J, Tegeler R, grosse Beilage E. Detection of respiratory pathogens in porcine lung tissue and lavage fluid. Vet J. 2008; 175(2):273–5.

23. Altman DG, Bland JM. Statistics notes: diagnostic tests 2: predictive values. Bmj. 1994;309(6947):102.

24. Usher-Smith JA, Sharp SJ, Griffin SJ. The spectrum effect in tests for risk prediction, screening, and diagnosis. BMJ. 2016;353.

25. Brenner H, Gefeller O. Variation of sensitivity, specificity, likelihood ratios and predictive values with disease prevalence. Stat Med. 1997;16(9):981–91.

26. Chiers K, De Waele T, Pasmans F, Ducatelle R, Haesebrouck F. Virulence factors of Actinobacillus pleuropneumoniae involved in colonization, persistence and induction of lesions in its porcine host. Vet Res. 2010;41(5):65.

27. Jens Brackmann KB, Lücken C, Baier S. Zur Verbreitung und Diagnostik von Actinobacillus pleuropneumoniae. Prakt Tierarzt. 2015;96:372–81.

Permissions

All chapters in this book were first published in VR, by BioMed Central; hereby published with permission under the Creative Commons Attribution License or equivalent. Every chapter published in this book has been scrutinized by our experts. Their significance has been extensively debated. The topics covered herein carry significant findings which will fuel the growth of the discipline. They may even be implemented as practical applications or may be referred to as a beginning point for another development.

The contributors of this book come from diverse backgrounds, making this book a truly international effort. This book will bring forth new frontiers with its revolutionizing research information and detailed analysis of the nascent developments around the world.

We would like to thank all the contributing authors for lending their expertise to make the book truly unique. They have played a crucial role in the development of this book. Without their invaluable contributions this book wouldn't have been possible. They have made vital efforts to compile up to date information on the varied aspects of this subject to make this book a valuable addition to the collection of many professionals and students.

This book was conceptualized with the vision of imparting up-to-date information and advanced data in this field. To ensure the same, a matchless editorial board was set up. Every individual on the board went through rigorous rounds of assessment to prove their worth. After which they invested a large part of their time researching and compiling the most relevant data for our readers.

The editorial board has been involved in producing this book since its inception. They have spent rigorous hours researching and exploring the diverse topics which have resulted in the successful publishing of this book. They have passed on their knowledge of decades through this book. To expedite this challenging task, the publisher supported the team at every step. A small team of assistant editors was also appointed to further simplify the editing procedure and attain best results for the readers.

Apart from the editorial board, the designing team has also invested a significant amount of their time in understanding the subject and creating the most relevant covers. They scrutinized every image to scout for the most suitable representation of the subject and create an appropriate cover for the book.

The publishing team has been an ardent support to the editorial, designing and production team. Their endless efforts to recruit the best for this project, has resulted in the accomplishment of this book. They are a veteran in the field of academics and their pool of knowledge is as vast as their experience in printing. Their expertise and guidance has proved useful at every step. Their uncompromising quality standards have made this book an exceptional effort. Their encouragement from time to time has been an inspiration for everyone.

The publisher and the editorial board hope that this book will prove to be a valuable piece of knowledge for researchers, students, practitioners and scholars across the globe.

List of Contributors

Patric Maurer and Ernst Lücker
Institute of Food Hygiene, Centre of Veterinary Public Health, Faculty of Veterinary Medicine, University of Leipzig, An den Tierkliniken 1, 04103 Leipzig, Germany

Katharina Riehn
Faculty of Life Sciences, Hamburg University of Applied Sciences, Ulmenliet 20, 21033 Hamburg, Germany
Zhiyong XuCollege of Animal Science and Technology, Shanxi Agricultural University, Taigu 030801, China College of Animal Science and Technology, Henan Institute of Science and Technology, Xinxiang 453003, China

Mingxue Zheng, Li Zhang, Xuesong Zhang, Yan Zhang, Xiaozhen Cui, Xin Gong, Rou Xi and Rui Bai
College of Animal Science and Technology, Shanxi Agricultural University, Taigu 030801, China

Ting Chen, Qian-Yun Xi, Jia-Jie Sun, Rui-Song Ye, Xiao Cheng, Song-Bo Wang, Gang Shu, Li-Na Wang, Xiao-Tong Zhu, Qing-Yan Jiang and Yong-Liang Zhang
National Engineering Research Center For Breeding Swine Industry, Guandong Provincial Key Laboratory of Agro-Animal Genomics and Molecular Breeding, Guandong Province Research Center of Woody Forage Engineering and Technology, South China Agricultural University, 483 Wushan Road, Guangzhou 510642, China

Rui-Ping Sun
National Engineering Research Center For Breeding Swine Industry, Guandong Provincial Key Laboratory of Agro-Animal Genomics and Molecular Breeding, Guandong Province Research Center of Woody Forage Engineering and Technology, South China Agricultural University, 483 Wushan Road, Guangzhou 510642, China

Anwar Nuru
Aklilu Lemma Institute of Pathobiology, Addis Ababa University, Addis Ababa, Ethiopia
College of Veterinary Medicine and Animal Sciences, University of Gondar, Gondar, Ethiopia

Aboma Zewude, Temesgen Mohammed, Biniam Wondale, Girmay Medhin and Gobena Ameni
Aklilu Lemma Institute of Pathobiology, Addis Ababa University, Addis Ababa, Ethiopia

Laikemariam Teshome
Animal Diseases Investigation and Diagnostic Laboratory, Amhara Region Bureau of Agriculture, Bahir Dar, Ethiopia

Muluwork Getahun
Ethiopian Public Health Institute, Addis Ababa, Ethiopia

Gezahegne Mamo
College of Veterinary Medicine and Agriculture, Addis Ababa University, Debre Zeit, Ethiopia

Rembert Pieper
J. Craig Venter Institute, 9704 Medical Center Drive, Rockville, MD, USA

Kristiane Barington, Kristine Dich-Jørgensen and Henrik Elvang Jensen
Department of Veterinary Disease Biology, Faculty of Health and Medical Sciences, University of Copenhagen, Ridebanevej 3, DK-1870 Frederiksberg C, Denmark

Jens Frederik Gramstrup Agger and Søren Saxmose Nielsen
Department of Large Animal Sciences, Faculty of Health and Medical Sciences, University of Copenhagen, Frederiksberg C, Denmark

Yuchi Chen
School of Animal and Veterinary Sciences, Charles Sturt University, Wagga Wagga, NSW 2650, Australia

Jane C. Quinn, Muhammad Shoaib Tufail and Panayiotis Loukopoulos
School of Animal and Veterinary Sciences, Charles Sturt University, Wagga Wagga, NSW 2650, Australia
Graham Centre for Agricultural Innovation; Charles Sturt University and NSW Department of Primary Industries, Wagga Wagga, NSW 2650, Australia

Belinda Hackney and Leslie A. Weston
Graham Centre for Agricultural Innovation; Charles Sturt University and NSW Department of Primary Industries, Wagga Wagga, NSW 2650, Australia

Mohamed Tharwat and Fahd Al-Sobayil
Department of Veterinary Medicine, College of Agriculture and Veterinary Medicine, Qassim University, Buraydah, Saudi Arabia

Sergio Ghidini, Emanuela Zanardi, Pierluigi Aldo Di Ciccio and Adriana Ianieri
Department of Food and Drug, Parma University, Via Del Taglio, 10, 43126 Parma, Italy

Silvio Borrello and Sarah Guizzardi
Italian Ministry of Health, Via Giorgio Ribotta, 5, 00144 Rome, Italy

Giancarlo Belluzi
Italian Ministry of Health, Viale Tanara 31/A, 43100 Parma (PR), Italy

Tenzin Tenzin
Disease Prevention and Control Unit, National Centre for Animal Health, Department of Livestock, Thimphu, Bhutan

Chador Wangdi
Bhutan Agriculture and Food Regulatory Authority, Ministry of Agriculture & Forests, Thimphu, Bhutan

Purna Bdr Rai
Laboratory Service Unit, National Centre for Animal Health, Department of Livestock, Thimphu, Bhutan

Pornpiroon Chinson
Department of Livestock Development (DLD), Bangkok 10400, Thailand

Weerapong Thanapongtharm
Department of Livestock Development (DLD), Bangkok 10400, Thailand
Lutte biologique et Ecologie spatiale (LUBIES), Université Libre de Bruxelles, Brussels 1050, Belgium

Marius Gilbert and Catherine Linard
Lutte biologique et Ecologie spatiale (LUBIES), Université Libre de Bruxelles, Brussels 1050, Belgium
Fonds National de la Recherche Scientifique (FNRS), Brussels 1050, Belgium

Suwicha Kasemsuwan
Faculty of Veterinary Medicine, Kasetsart University, Kampangsaen Campus, Nakornpatom 73140, Thailand

Marjolein Visser
Research Unit of Landscape Ecology AND Plant Production Systems (EPSPV), University of Brussels, 1050 Brussels, Belgium

Andrea E. Gaughan
Department of Geography and Geosciences, University of Louisville, Louisville 40292, USA

Michael Epprech
Centre for Development and Environment (CDE), Country office in the Lao PDR, Vientiane 6101, Lao PDR

Timothy P. Robinson
Livestock Systems and Environment (LSE), International Livestock Research Institute (ILRI), Nairobi 30709, Kenya

N. Moyen and G. Fournie
Department of Pathobiology and Population Sciences, Royal Veterinary College, University of London, Hatfield, Hertfordshire AL9 7TA, UK

D. U. Pfeiffer
Department of Pathobiology and Population Sciences, Royal Veterinary College, University of London, Hatfield, Hertfordshire AL9 7TA, UK
College of Veterinary Medicine and Life Sciences, City University of Hong Kong, Tat Chee Avenue, Kowloon, Hong Kong

S. Gupta
School of Veterinary Science, The University of Queensland, Gatton 4343, Qld, Australia.
Emergency Centre for Transboundary Animal Diseases, Food and Agriculture Organisation of the United Nations, Dhaka, Bangladesh

R. Khan, T. Khan, N. Debnath, T. Tenzin, M. Yamage and G. Ahmed
Emergency Centre for Transboundary Animal Diseases, Food and Agriculture Organisation of the United Nations, Dhaka, Bangladesh
National Centre for Animal Health, Thimphu, Bhutan

Dietmar Hamel, Michael Kellermann, Katrin Kley, Sandra Mayr, Renate Rauh, Martin Visser, Thea Wiefel and Steffen Rehbein
Merial GmbH, Kathrinenhof Research Center, Walchenseestr. 8-12, 83101 Rohrdorf, Germany

Antonio Bosco, Laura Rinaldi and Giuseppe Cringoli
Department of Veterinary Medicine and Animal Production, University of Naples Federico II, Via della Veterinaria, 1, 80137 Naples, Italy

Karl-Heinz Kaulfuß
Tierarztpraxis Hoffmann, Untere Schulstraße 8, 38875 Elbingerode, Germany

James Fischer
Merial, Inc., North Brunswick Research Center, 631 Route 1 South, North Brunswick, NJ 08902, USA

Hailun Wang and Becky Fankhauser
Merial, Inc., 3239 Satellite Blvd., Duluth, GA 30096-4640, USA

Wan-Chen Lee
Department of Veterinary Medicine, School of Veterinary Medicine, College of Bio-Resources and Agriculture, National Taiwan University, Taipei 106, Taiwan

Kuang-Sheng Yeh
Department of Veterinary Medicine, School of Veterinary Medicine, College of Bio-Resources and Agriculture, National Taiwan University, Taipei 106, Taiwan
National Taiwan University Veterinary Hospital, Taipei 106, Taiwan

Carla Surlis, Keelan McNamara, Eoin O'Hara, Sinead Waters and David Kenny
Animal and Grassland Research and Innovation Centre, Teagasc, Grange, Dunsany, Co. Meath, Ireland

Marijke Beltman and Joseph Cassidy
School of Veterinary Medicine, University College Dublin, Belfield, Dublin 4, Ireland

Myriam Esteban-Ballesteros, Francisco A. Rojo-Vázquez, Camino González-Lanza and María Martínez-Valladares
Departamento de Sanidad Animal, Facultad de Veterinaria, Universidad de León, Campus de Vegazana s/n, 24071 León, Spain

Instituto de Ganadería de Montaña (CSIC-Universidad de León), Finca Marzanas, Grulleros, 24346 León, Spain

Philip J. Skuce and Lynsey Melville
Moredun Research Institute, Pentland Science Park, Bush Loan, Edinburgh, UK

Alexander Grahofer and Heiko Nathues
Clinic for Swine, Department of Clinical Veterinary Medicine, Vetsuisse Faculty, University of Bern, Bremgartenstrasse 109a, 3012 Bern, CH, Switzerland

Jeanette Bannoehr
Dermatology Department, Animal Health Trust, Lanwades Park, Kentford, Newmarket, Suffolk, Cardiff CB8 7UU, UK

Petra Roosje
Division of Clinical Dermatology, Department of Clinical Veterinary Medicine, Vetsuisse Faculty, University of Bern, Bern, Länggassstrasse 128, 3012 Berne, CH, Switzerland

Lijiao Zhang, Zhanhong Li, Huan Jin and Jingliang Su
Key Laboratory of Animal Epidemiology of the Ministry of Agriculture, College of Veterinary Medicine, China Agricultural University, Beijing 100193, China

Xueying Hu
College of Veterinary Medicine, Huazhong Agricultural University, Wuhan 430070, China

Rakesh Kumar, Rinku Sharma and Gorakh Mal
Disease Investigation Laboratory, ICAR-Indian Veterinary Research Institute, Regional Station, Palampur, Himachal Pradesh, India

Rajendra D. Patil and Adarsh Kumar
DGCN COVAS, CSK HPKV, Palampur, Himachal Pradesh, India

Vikram Patial, Pawan Kumar and Bikram Singh
CSIR-IHBT, Palampur, Himachal Pradesh, India

Guan-Hua Lai
Graduate Institute of Biotechnology, National Chung Hsing University, Taichung 40402, Taiwan

Meng-Shiou Lee and Ming-Kuem Lin
Department of Chinese Pharmaceutical Science and Chinese Medicine Resources, China Medical University, 91, Hsueh-Shih Road, Taichung, Taiwan

Yi-Yang Lien
Department of Veterinary Medicine, National Pingtung University of Science and Technology, Pingtung 91201, Taiwan

Jai-Hong Cheng
Center for Shockwave Medicine and Tissue Engineering, Department of Medical Research, Kaohsiung Chang Gung Memorial Hospital and Chang Gung University College of Medicine, Kaohsiung 83301, Taiwan

Fang-Chun Sun
Department of Bioresources, Da-Yeh University, Changhua 51591, Taiwan

Meng-Shiunn Lee
Research Assistance Center, Show Chwan Memorial Hospital, Changhua 500, Taiwan

Hsi-Jien Chen
Department of Safety, Health and Environmental Engineering, Ming Chi University of Technology, New Taipei 24301, Taiwan

Raquel Vallejo and Juan Francisco García Marín
Universidad de León, Campus de Vegazana, León, Spain

Ramón Antonio and Juste Ana Balseiro
SERIDA, Servicio Regional de Investigación y Desarrollo Agroalimentario, Centro de Biotecnología Animal, 33394 Gijón, Asturias, Spain

Marta Muñoz-Mendoza
Xunta de Galicia, Santiago, A Coruña, Galicia, Spain

Francisco Javier Salguero
School of Veterinary Medicine, University of Surrey, Guildford, UK

Lingyun Zh, Qian Zhao, Xianghua Shu, Jia Liu, Guishu Yang and Gefen Yin
Department of Veterinary Medicine, College of Animal Science and Technology, Yunnan Agricultural University, Yunnan province, Kunming 650201, China

Junlong Bi
Department of Veterinary Medicine, College of Animal Science and Technology, Yunnan Agricultural University, Yunnan province, Kunming 650201, China
Center for Animal Disease Control and Prevention, City, 675000, Yunnan province, Chuxiong, China

Longlong Zheng
Department of Veterinary Medicine, College of Animal Science and Technology, Yunnan Agricultural University, Yunnan province, Kunming 650201, China
College of Animal Science and Technology, Shanxi Agricultural University, Shanxi province, Taigu 030801, China

Gang Guo
Haikou Experimental Station/Hainan Key Laboratory of Banana Genetic Improvement, Chinese Academy of Tropical Agricultural Sciences, Haikou 570102, Hainan, People's Republic of China

Jianping Liu
Department of Medical Biochemistry and Biophysics, Karolinska Institute, -17177 Stockholm, SE, Sweden

Doris Hoeltig
Clinic for Swine and Small Ruminants, Forensic Medicine and Ambulatory Service, University of Veterinary Medicine, Foundation, Bischofsholer Damm 15, D-30173 Hannover, Germany

Judith Rohde
Institute for Microbiology, University of Veterinary Medicine, Foundation, Bischofsholer Damm 15, D-30173 Hannover, Germany

Florian Nietfeld
Clinic for Swine, Department of Veterinary Medicine, Justus-Liebig-University Giessen, Frankfurter Str. 112, D-35392 Giessen, Germany

Katrin Strutzberg-Minder
Innovative Veterinary Diagnostics (IVD-GmbH), Albert-Einstein-Str. 5, 30926 Seelze, Germany

Index

Porcine Reproductive And Respiratory Syndrome Virus, 202, 210-211

Poultry Birds, 82-88

Poultry Farming Systems, 82

Primary Photosensitization, 62

Pulmonary Alveolar Macrophages, 202

Pyrosequencing, 150, 152, 154, 156-157

R

Real-time Quantitative Pcr, 142

Reduced Lantadene A, 171, 183

Reduced Lantadene B, 171

S

Sarcoptes Infestation, 158, 160

Sarcoptes Mites, 158-161

Sarcoptes Scabiei, 158, 160-162

Sarcoptic Mange, 158-162

Scabies, 158-162

Severe Subcutaneous Facial Oedema, 55, 60

Short Antisense Oligonucleotides, 202, 211

Swine Influenza, 92, 104

T

Teladorsagia Circumcincta, 118, 123-124, 150-151, 156-157

Tembusu Virus, 163, 166, 169-170

Transmissible Gastroenteritis, 93, 103, 131

Trichostrongylus Colubriformis, 123-124, 150-152, 156-157

V

Veterinarians, 6, 50, 69, 72, 74, 78-81, 160

Veterinary Clinics, 33

Veterinary Medicine, 1, 3-4, 8, 31, 42-43, 47, 64, 68, 104, 116, 127, 129-130, 148, 158, 161, 163, 169, 193, 201-202, 210, 212, 217

Veterinary Public Health, 1, 8

Veterinary Service, 33, 74

Z

Zoonotic Diseases, 92

www.ingramcontent.com/pod-product-compliance
Lightning Source LLC
Chambersburg PA
CBHW082100190326
41458CB00010B/3533